Advanced Fluid Mechanics

Advanced Fluid Mechanics

W. P. Graebel

Professor Emeritus, The University of Michigan

AMSTERDAM • BOSTON • HEIDELBERG • LONDON
NEW YORK • OXFORD • PARIS • SAN DIEGO
SAN FRANCISCO • SINGAPORE • SYDNEY • TOKYO

Academic Press is an imprint of Elsevier

Academic Press is an imprint of Elsevier
30 Corporate Drive, Suite 400, Burlington, MA 01803, USA
525 B Street, Suite 1900, San Diego, California 92101-4495, USA
84 Theobald's Road, London WC1X 8RR, UK

This book is printed on acid-free paper. ∞

Library of Congress Cataloging-in-Publication Data
Application submitted

British Library Cataloguing-in-Publication Data
A catalogue record for this book is available from the British Library.

ISBN: 978-0-12-370885-4

For information on all Academic Press publications, visit
our Web site at www.books.elsevier.com

I maintained my edge by always being a student.
You will always have ideas, have something new to learn.
Jackie Joyner-Kersee

Education is not the filling of a pail, but the lighting of the fire.
William Butler Yeats

I have always believed that 98% of a student's progress is due to his own efforts,
and 2% to his teacher.
John Philip Sousa

The one thing that matters is the effort.
Antoine de Saint-Exupery

Contents

Chapter 1

Fundamentals

Chapter 2

Inviscid Irrotational Flows

Chapter 3

Irrotational Two-Dimensional Flows

Chapter 4

Surface and Interfacial Waves

Chapter 5

Exact Solutions of the Navier-Stokes Equations

Chapter 6

The Boundary Layer Approximation

Chapter 7

Thermal Effects

Chapter 8

Low Reynolds Number Flows

Chapter 9

Flow Stability

Chapter 10

Turbulent Flows

Chapter 11

Computational Methods—Ordinary Differential Equations

Chapter 12

Multidimensional Computational Methods

Appendix

Preface

This book covers material for second fluid dynamics courses at the senior/graduate level. Students are introduced to three-dimensional fluid mechanics and classical theory, with an introduction to modern computational methods. Problems discussed in the text are accompanied by examples and computer programs illustrating how classical theory can be applied to solve practical problems with techniques that are well within the capabilities of present-day personal computers.

Modern fluid dynamics covers a wide range of subject areas and facets—far too many to include in a single book. Therefore, this book concentrates on incompressible fluid dynamics. Because it is an introduction to basic computational fluid dynamics, it does not go into great depth on the various methods that exist today. Rather, it focuses on how theory and computation can be combined and applied to problems to demonstrate and give insight into how various describing parameters affect the behavior of the flow. Many large and expensive computer programs are used in industry today that serve as major tools in industrial design. In many cases the user does not have any information about the program developers' assumptions. This book shows students how to test various methods and ask the right questions when evaluating such programs.

The references in this book are quite extensive—for three reasons. First, the originator of the work deserves due credit. Many of the originators' names have become associated with their work, so referring to an equation as the Orr-Sommerfeld equation is common shorthand.

A more subversive reason for the number of references is to entice students to explore the history of the subject and how the world has been affected by the growth of science. Isaac Newton (1643–1747) is credited with providing the first solid footings of fluid dynamics. Newton, who applied algebra to geometry and established the fields of analytical geometry and the calculus, combined mathematical proof with physical observation. His treatise *Philosophiae Naturalis Principia Mathematica* not only firmly established the concept of the scientific method, but it led to what is called the Age of Enlightenment, which became the intellectual framework for the American and French Revolutions and led to the birth of the Industrial Revolution.

The Industrial Revolution, which started in Great Britain, produced a revolution in science (in those days called "natural philosophy" in reference to Newton's treatise) of gigantic magnitude. In just a few decades, theories of dynamics, solid mechanics, fluid dynamics, thermodynamics, electricity, magnetism, mathematics, medical science, and many other sciences were born, grew, and thrived with an intellectual verve never before found in the history of mankind. As a result, the world saw the invention of steam engines and locomotives, electric motors and light, automobiles, the telephone, manned flight, and other advances that had only existed in dreams before then. A chronologic and geographic study of the references would show how ideas jumped from country to country and how the time interval between the advances shortened dramatically in time. Truly, Newton's work was directly responsible for bringing civilization from the dark ages to the founding of democracy and the downfall of tyranny.

This book is the product of material covered in many classes over a period of five decades, mostly at The University of Michigan. I arrived there as a student at the same time as Professor Chia-Shun Yih, who over the years I was fortunate to have as a teacher, colleague, and good friend. His lively presentations lured many of us to the excitement of fluid dynamics. I can only hope that this book has a similar effect on its readers.

I give much credit for this book to my wife, June, who encouraged me greatly during this work—in fact, during all of our 50+ years of marriage! Her proofreading removed some of the most egregious errors. I take full credit for any that remain.

Fundamentals

1.1   Introduction

A few basic laws are fundamental to the subject of fluid mechanics: the law of conservation of mass, Newton's laws, and the laws of thermodynamics. These laws bear a remarkable similarity to one another in their structure. They all state that if a given volume of the fluid is investigated, quantities such as mass, momentum, and energy will change due to internal causes, net change in that quantity entering and leaving the volume, and action on the surface of the volume due to external agents. In fluid mechanics, these laws are best expressed in rate form.

Since these laws have to do with some quantities entering and leaving the volume and other quantities changing inside the volume, in applying these fundamental laws to a finite-size volume it can be expected that both terms involving surface and volume integrals would result. In some cases a global description is satisfactory for carrying out further analysis, but often a local statement of the laws in the form of differential equations is preferred to obtain more detailed information on the behavior of the quantity under investigation.

There is a way to convert certain types of surface integrals to volume integrals that is extremely useful for developing the derivations. This is the ***divergence theorem***, expressed in the form

$$\iint_S \mathbf{n} \cdot \mathbf{U} dS = \iiint_V \nabla \cdot \mathbf{U} dV. \tag{1.1.1}$$

In this expression $\mathbf{U}$[1] is an arbitrary vector and $\mathbf{n}$ is a unit normal to the surface S. The closed surface completely surrounding the volume V is S. The unit normal is positive if drawn outward from the volume.

The theorem assumes that the scalar and vector quantities are finite and continuous within V and on S. It is sometimes useful to add appropriate singularities in these functions, generating additional terms that can be simply evaluated. This will be discussed in later chapters in connection with inviscid flows.

In studying fluid mechanics three of the preceding laws lead to differential equations—namely, the law of conservation of mass, Newton's law of momentum, and the first law of thermodynamics. They can be expressed in the following descriptive form:

$$\text{Rate at which} \begin{bmatrix} \text{mass} \\ \text{momentum} \\ \text{energy} \end{bmatrix} \text{accumulates within the volume}$$

$$+ \text{ rate at which} \begin{bmatrix} \text{mass} \\ \text{momentum} \\ \text{energy} \end{bmatrix} \text{enters the volume}$$

$$- \text{ rate at which} \begin{bmatrix} \text{mass} \\ \text{momentum} \\ \text{energy} \end{bmatrix} \text{leaves the volume}$$

$$= \begin{bmatrix} 0 \\ \text{net force acting on volume} \\ \text{rate of heat addition} + \text{rate of work done on volume} \end{bmatrix}.$$

These can be looked at as a balance sheet for mass, momentum, and energy, accounting for rate changes on the left side of the equation by describing on the right side the external events that cause the changes. As simple as these laws may appear to us today, it took many centuries to arrive at these fundamental results.

Some of the quantities we will see in the following pages, like pressure and temperature, have magnitude but zero directional property. Others have various degrees of directional properties. Quantities like velocity have magnitude and one direction associated with them, whereas others, like stress, have magnitude and two directions associated with them: the direction of the force and the direction of the area on which it acts.

The general term used to classify these quantities is ***tensor***. The ***order*** of a tensor refers to the number of directions associated with them. Thus, pressure and temperature are tensors of order zero (also referred to as scalars), velocity is a tensor of order one (also referred to as a vector), and stress is a tensor of order two. Tensors of order higher

---

[1] Vectors are denoted by boldface.

than two usually are derivatives or products of lower-order tensors. A famous one is the fourth-order Einstein curvature tensor of relativity theory.

To qualify as a tensor, a quantity must have more than just magnitude and directionality. When the components of the tensor are compared in two coordinate systems that have origins at the same point, the components must relate to one another in a specific manner. In the case of a tensor of order zero, the transformation law is simply that the magnitudes are the same in both coordinate systems. Components of tensors of order one must transform according to the parallelogram law, which is another way of stating that the components in one coordinate system are the sum of products of direction cosines of the angles between the two sets of axes and the components in the second system. For components of second-order tensors, the transformation law involves the sum of products of two of the direction cosines between the axes and the components in the second system. (You may already be familiar with Mohr's circle, which is a graphical representation of this law in two dimensions.) In general, the transformation law for a tensor of order N then will involve the sum of N products of direction cosines and the components in the second system. More detail on this is given in the Appendix.

One example of a quantity that has both directionality and magnitude but is not a tensor is finite angle rotations. A branch of mathematics called *quaternions* was invented by the Irish mathematician Sir William Rowan Hamilton in 1843 to deal with these and other problems in spherical trigonometry and body rotations. Information about quaternions can be found on the Internet.

In dealing with the general equations of fluid mechanics, the equations are easiest to understand when written in their most compact form—that is, in vector form. This makes it easy to see the grouping of terms, the physical interpretation of them, and subsequent manipulation of the equations to obtain other interpretations. This general form, however, is usually not the form best suited to solving particular problems. For such applications the component form is better. This, of course, involves the selection of an appropriate coordinate system, which is dictated by the geometry of the problem.

When dealing with flows that involve flat surfaces, the proper choice of a coordinate system is *Cartesian coordinates*. Boundary conditions are most easily satisfied, manipulations are easiest, and equations generally have the fewest number of terms when expressed in these coordinates. Trigonometric, exponential, and logarithmic functions are often encountered. The conventions used to represent the components of a vector, for example, are typically (v_x, v_y, v_z), (v_1, v_2, v_3), and (u, v, w). The first of these conventions use (x, y, z) to refer to the coordinate system, while the second convention uses (x_1, x_2, x_3). This is referred to either as *index notation* or as *indicial notation*, and it is used extensively in tensor analysis, matrix theory, and computer programming. It frequently is more compact than the x, y, z notation.

For geometries that involve either circular cylinders, ellipses, spheres, or ellipsoids, cylindrical polar, spherical polar, or ellipsoidal coordinates are the appropriate choice, since they make satisfaction of boundary conditions easiest. The mathematical functions and the length and complexity of equations become more complicated than in Cartesian coordinates.

Beyond these systems, general tensor analysis must be used to obtain governing equations, particularly if nonorthogonal coordinates are used. While it is easy to write the general equations in tensor form, breaking down these equations into component form in a specific non-Cartesian coordinate frame frequently involves a fair amount of work. This is discussed in more detail in the Appendix.

1.2 Velocity, Acceleration, and the Material Derivative

A fluid is defined as a material that will undergo sustained motion when shearing forces are applied, the motion continuing as long as the shearing forces are maintained. The general study of fluid mechanics considers a fluid to be a **continuum**. That is, the fact that the fluid is made up of molecules is ignored but rather the fluid is taken to be a continuous media.

In solid and rigid body mechanics, it is convenient to start the geometric discussion of motion and deformation by considering the continuum to be made up of a collection of particles and consider their subsequent displacement. This is called a **Lagrangian**, or **material, description**, named after Joseph Louis Lagrange (1736–1836). To illustrate its usage, let $(X(X_0, Y_0, Z_0, t), Y(X_0, Y_0, Z_0, t), Z(X_0, Y_0, Z_0, t))$ be the position at time t of a particle initially at the point (X_0, Y_0, Z_0). Then the velocity and acceleration of that particle is given by

$$v_x(X_0, Y_0, Z_0, t) = \frac{\partial X(X_0, Y_0, Z_0, t)}{\partial t}$$

$$v_y(X_0, Y_0, Z_0, t) = \frac{\partial X(X_0, Y_0, Z_0, t)}{\partial t}$$

$$v_z(X_0, Y_0, Z_0, t) = \frac{\partial X(X_0, Y_0, Z_0, t)}{\partial t}$$

and (1.2.1)

$$a_x(X_0, Y_0, Z_0, t) = \frac{\partial v_x(X_0, Y_0, Z_0, t)}{\partial t} = \frac{\partial^2 X(X_0, Y_0, Z_0, t)}{\partial t^2},$$

$$a_y(X_0, Y_0, Z_0, t) = \frac{\partial v_y(X_0, Y_0, Z_0, t)}{\partial t} = \frac{\partial^2 Y(X_0, Y_0, Z_0, t)}{\partial t^2},$$

$$a_z(X_0, Y_0, Z_0, t) = \frac{\partial v_z(X_0, Y_0, Z_0, t)}{\partial t} = \frac{\partial^2 Z(X_0, Y_0, Z_0, t)}{\partial t^2}.$$

The partial derivatives signify that differentiation is performed holding X_0, Y_0 and Z_0 fixed.

This description works well for particle dynamics, but since fluids consist of an infinite number of flowing particles in the continuum hypothesis, it is not convenient to label the various fluid particles and then follow each particle as it moves. Experimental techniques certainly would be hard pressed to perform measurements that are suited to such a description. Also, since displacement itself does not enter into stress-geometric relations for fluids, there is seldom a need to consider using this descriptive method.

Instead, an **Eularian**, or **spatial, description**, named after Leonard Euler (1707–1783), is used. This description starts with velocity, written as $\mathbf{v} = \mathbf{v}(\mathbf{x}, t)$, where $\mathbf{x}$ refers to the position of a fixed point in space, as the basic descriptor rather than displacement. To find acceleration, recognize that *acceleration* means the rate of change of the velocity of a particular fluid particle at a position while noting that the particle is in the process of moving from that position at the time it is being studied. Thus, for instance, the acceleration component in the x direction is defined as

$$a_x = \lim_{\Delta t \to 0} \frac{v_x(x + v_x \Delta t, y + v_y \Delta t, z + v_z \Delta t, t + \Delta t) - v_x(x, y, z, t)}{\Delta t}$$

$$= \frac{\partial v_x}{\partial t} + v_x \frac{\partial v_x}{\partial x} + v_y \frac{\partial v_x}{\partial y} + v_z \frac{\partial v_x}{\partial z},$$

since the rate at which the particle leaves this position is $\mathbf{v}dt$. Similar results can be obtained in the y and z direction, leading to the general vector form of the acceleration as

$$\mathbf{a} = \frac{\partial \mathbf{v}}{\partial t} + (\mathbf{v} \cdot \nabla)\,\mathbf{v}. \tag{1.2.2}$$

The first term in equation (1.2.2) is referred to as the **temporal acceleration**, and the second as the **convective**, or occasionally **advective, acceleration**.

Note that the convective acceleration terms are quadratic in the velocity components and hence mathematically nonlinear. This introduces a major difficulty in the solution of the governing equations of fluid flow. At this point it might be thought that since the Lagrangian approach has no nonlinearities in the acceleration expression, it could be more convenient. Such, however, is not the case, as the various force terms introduced by Newton's laws all become nonlinear in the Lagrangian approach. In fact, these nonlinearities are even worse than those found using the Eularian approach.

The convective acceleration term $(\mathbf{v} \cdot \nabla)\,\mathbf{v}$ can also be written as

$$(\mathbf{v} \cdot \nabla)\,\mathbf{v} = \frac{1}{2}\nabla\,(\mathbf{v} \cdot \mathbf{v}) + \mathbf{v} \times (\nabla \times \mathbf{v}). \tag{1.2.3}$$

This can be shown to be true by writing out the left- and right-hand sides.

The operator $\frac{\partial}{\partial t} + (\mathbf{v} \cdot \nabla)$, which appears in equation (1.2.2), is often seen in fluid mechanics. It has been variously called the **material**, or **substantial, derivative**, and represents differentiation as a fluid particle is followed. It is often written as

$$\frac{D}{Dt} = \frac{\partial}{\partial t} + (\mathbf{v} \cdot \nabla). \tag{1.2.4}$$

Note that the operator $\mathbf{v} \cdot \nabla$ is not a strictly correct vector operator, as it does not obey the commutative rule. That is, $\mathbf{v} \cdot \nabla \neq \nabla \cdot \mathbf{v}$. This operator is sometimes referred to as a **pseudo-vector**. Nevertheless, when it is used to operate on a scalar like mass density or a vector such as velocity, the result is a proper vector as long as no attempt is made to commute it.

1.3 The Local Continuity Equation

To derive local equations that hold true at any point in our fluid, a volume of arbitrary shape is constructed and referred to as a **control volume**. A control volume is a device used in analyzing fluid flows to account for mass, momentum, and energy balances. It is usually a volume of fixed size, attached to a specified coordinate system. A **control surface** is the bounding surface of the control volume. Fluid enters and leaves the control volume through the control surface. The density and velocity inside and on the surface of the control volume are represented by ρ and $\mathbf{v}$. These quantities may vary throughout the control volume and so are generally functions of the spatial coordinates as well as time.

The mass of the fluid inside our control volume is $\iiint_V \rho\,dV$. For a control volume fixed in space, the rate of change of mass inside of our control volume is

$$\frac{d}{dt}\iiint_V \rho\,dV = \iiint_V \frac{\partial \rho}{\partial t}\,dV. \tag{1.3.1}$$

The rate at which mass enters the control volume through its surface is

$$\iint_S \rho\mathbf{v} \cdot \mathbf{n}\,dS, \tag{1.3.2}$$

where $\rho \mathbf{v} \cdot \mathbf{n}\, dS$ is the mass rate of flow out of the small area dS. The quantity $\mathbf{v} \cdot \mathbf{n}$ is the normal component of the velocity to the surface. Therefore, a positive value of $\mathbf{v} \cdot \mathbf{n}$ means the $\mathbf{v} \cdot \mathbf{n}$ flow locally is out of the volume, whereas a negative value means that it is into the volume.

The net rate of change of mass inside and entering the control volume is then found by adding together equations (1.3.1) and (1.3.2) and setting the sum to zero. This gives

$$\iiint_V \frac{\partial \rho}{\partial t}\, dV + \iint_S \rho \mathbf{v} \cdot \mathbf{n}\, dS = 0. \tag{1.3.3}$$

To obtain the local continuity equation in differential equation form, use (1.1.1) to transform the surface integral to a volume integral. This gives

$$\iint_S \rho \mathbf{v} \cdot \mathbf{n}\, dS = \iiint_V \nabla \cdot (\rho \mathbf{v})\, dV.$$

Making this replacement in (1.3.3) gives

$$\iiint_V \frac{\partial \rho}{\partial t}\, dV + \iiint_V \nabla \cdot (\rho \mathbf{v})\, dV = 0.$$

Rearranging this to put everything under the same integral sign gives

$$\iiint_V \left[\frac{\partial \rho}{\partial t} + \nabla \cdot (\rho \mathbf{v}) \right] dV = 0. \tag{1.3.4}$$

Since the choice of the control volume was arbitrary and since the integral must vanish no matter what choice of control volume was made, the only way this integral can vanish is for the integrand to vanish. Thus,

$$\frac{\partial \rho}{\partial t} + \nabla \cdot (\rho \mathbf{v}) = 0. \tag{1.3.5}$$

Equation (1.3.5) is the local form of the **continuity equation**. An alternate expression of it can be obtained by expanding the divergence term to obtain

$$\frac{\partial \rho}{\partial t} + \mathbf{v} \cdot \nabla \rho + \rho \nabla \cdot \mathbf{v} = 0.$$

The first two terms represent the material derivative of ρ. It is the change of mass density as we follow an individual fluid particle for an infinitesimal time. Writing $\frac{D}{Dt} \equiv \frac{\partial}{\partial t} + \mathbf{v} \cdot \nabla$, equation (1.3.5) becomes

$$\frac{D\rho}{Dt} + \rho \nabla \cdot \mathbf{v} = 0. \tag{1.3.6}$$

The first term in equation (1.3.6) represents the change in density following a particle, and the second represents (as shall be shown later) mass density times the change in volume per unit volume as a mass particle is followed.

An **incompressible flow** is defined as one where the mass density of a fluid particle does not change as the particle is followed. This can be expressed as

$$\frac{D\rho}{Dt} = 0. \tag{1.3.7}$$

Thus, the continuity equation for an incompressible flow is

$$\nabla \cdot \mathbf{v} = 0. \tag{1.3.8}$$

1.4   Path Lines, Streamlines, and Stream Functions

A *path line* is a line along which a fluid particle actually travels. Since it is the time history of the position of a fluid particle, it is best described using the Lagrangian description. Since the particle incrementally moves in the direction of the velocity vector, the equation of a path line is given by

$$dt = \frac{dX}{V_x} = \frac{dY}{V_y} = \frac{dZ}{V_z},$$ (1.4.1)

the integration being performed with X_0, Y_0, and Z_0 held fixed.

A *streamline* is defined as a line drawn in the flow at a given instant of time such that the fluid velocity vector at any point on the streamline is tangent to the line at that point. The requirement of tangency means that the streamlines are given by the equation

$$\frac{dx}{v_x} = \frac{dy}{v_y} = \frac{dz}{v_z}.$$ (1.4.2)

While in principle the streamlines can be found from equation (1.4.2), it is usually easier to pursue a method utilizing the continuity equation and stream functions described in the following.

A *stream surface* (or *stream sheet*) is a collection of adjacent streamlines, providing a surface through which there is no flow. A *stream tube* is a tube made up of adjoining streamlines.

For *steady flows* (time-independent), path lines and streamlines coincide. For *unsteady flows* (time-dependent), path lines and streamlines may differ. Generally path lines are more difficult to find analytically than are streamlines, and they are of less use in practical applications.

The continuity equation imposes a restriction on the velocity components. It is a relation between the various velocity components and mass density. It is not possible to directly integrate the continuity equation for one of the velocity components in terms of the others. Rather, this is handled through the use of an intermediary scalar function called a *stream function*.

Stream functions are used principally in connection with incompressible flows— that is, flows where the density of individual fluid particles does not change as the particle moves in the flow. In such a flow, equation (1.3.8) showed that the continuity equation reduces to

$$\frac{\partial v_x}{\partial x} + \frac{\partial v_y}{\partial y} + \frac{\partial v_z}{\partial z} = 0.$$ (1.4.3)

Because the equations that relate stream functions to velocity differ in two and three dimensions, the two cases will be considered separately.

1.4.1   Lagrange's Stream Function for Two-Dimensional Flows

For two-dimensional flows, equation (1.4.3) reduces to

$$\frac{\partial v_x}{\partial x} + \frac{\partial v_y}{\partial y} = 0,$$ (1.4.4)

indicating that one of the velocity components can be expressed in terms of the other. To attempt to do this directly by integration, the result would be

$$v_y = - \int \frac{\partial v_x}{\partial x} dy.$$

Unfortunately, this cannot be integrated in any straightforward manner. Instead, it is more convenient to introduce an intermediary, a scalar function ψ called ***Lagrange's stream function***, that allows the integration to be carried out explicitly.

Let

$$v_x = \frac{\partial \psi}{\partial y}. \tag{1.4.5}$$

Then equation (1.4.4) becomes

$$\frac{\partial v_y}{\partial y} + \frac{\partial^2 \psi}{\partial x \partial y} = 0,$$

which can be integrated with respect to y to give

$$v_y = - \frac{\partial \psi}{\partial x}. \tag{1.4.6}$$

Therefore, expressing the two velocity components in terms of ψ in the manner of equations (1.4.5) and (1.4.6) guarantees that continuity is satisfied for an incompressible flow.

In two dimensions, the tangency requirement equation (1.4.2) reduces to $dx/v_x = dy/v_y$ on the streamline. Using equations (1.4.5) and (1.4.6) reduces this to $\frac{dx}{\frac{\partial \psi}{\partial y}} = \frac{dy}{-\frac{\partial \psi}{\partial x}}$, or, upon rearrangement,

$$0 = \frac{\partial \psi}{\partial x} dx + \frac{\partial \psi}{\partial y} dy = d\psi. \tag{1.4.7}$$

Equation (1.4.7) states that along a streamline $d\psi$ vanishes. In other words, on a streamline, ψ is constant. This is the motivation for the name stream function for ψ.

Sometimes it is more convenient in a given problem to use cylindrical polar coordinates rather than Cartesian coordinates. The relation of cylindrical polar coordinates to a Cartesian frame is shown in Figure 1.4.1, with the appropriate tangent unit vectors shown at point P. The suitable expressions for the velocity components in cylindrical polar coordinates using Lagrange's stream function are

$$v_r = \frac{\partial \psi}{r \partial \theta} \quad \text{and} \quad v_\theta = - \frac{\partial \psi}{\partial r}. \tag{1.4.8}$$

Note from either equation (1.4.5), (1.4.6), or (1.4.8) that the dimension of ψ is length squared per unit time.

A physical understanding of ψ can also be found by looking at the discharge through a curve C as seen in Figure 1.4.2. The discharge through the curve C per unit distance into the paper is

$$Q = \int_A^B (v_x \, dy - v_y \, dx). \tag{1.4.9}$$

When the velocity components are replaced by their expressions in terms of ψ, equation (1.4.9) becomes

$$Q = \int_A^B \left[\left(\frac{\partial \psi}{\partial y} \right) dy - \left(- \frac{\partial \psi}{\partial x} \right) dx \right] = \int_A^B d\psi = \psi_B - \psi_A. \tag{1.4.10}$$

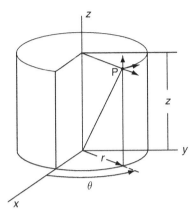

Figure 1.4.1   Cylindrical coordinate conventions

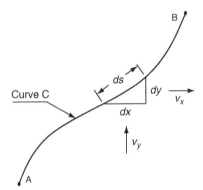

Figure 1.4.2   Discharge through a curve

Thus, the discharge through the curve C is equal to the difference of the value of ψ at the endpoints of C.

Example 1.4.1   Lagrange stream function

Find the stream function associated with the two-dimensional incompressible flow:

$$v_r = U\left(1 - \frac{a^2}{r^2}\right)\cos\theta,$$

$$v_\theta = -U\left(1 + \frac{a^2}{r^2}\right)\sin\theta.$$

Solution. Since $v_r = \partial\psi/r\partial\theta$, by integration of v_r with respect to θ, find

$$\psi = \int U\left(1 - \frac{a^2}{r^2}\right)\cos\theta\,dr = U\left(r - \frac{a^2}{r}\right)\sin\theta + f(r).$$

The "constant of integration" f here possibly depends on r, since we have integrated a partial derivative that had been taken with respect to θ, the derivative being taken with r held constant in the process.

Differentiating this ψ with respect to r gives

$$\frac{\partial \psi}{\partial r} = -v_\theta = U\left(1 + \frac{a^2}{r^2}\right)\sin\theta + \frac{df(r)}{dr}.$$

Comparing the two expressions for v_θ, we see that df/dr must vanish, so f must be a constant. We can set this constant to any convenient value without affecting the velocity components or the discharge. Here, for simplicity, we set it to zero. This gives

$$\psi = U\left(r - \frac{a^2}{r}\right)\sin\theta.$$

Example 1.4.2 Path lines

Find the path lines for the flow of Example 1.4.1. Find also equations for the position of a fluid particle along the path line as a function of time.

Solution. For steady flows, path lines and streamlines coincide. Therefore, on a path line, ψ is constant. Using the stream function from the previous example, we have

$$\psi = U\left(r - \frac{a^2}{r}\right)\sin\theta = \text{constant} = B \text{ (say)}.$$

The equations for a path line in cylindrical polar coordinates are

$$dt = \frac{dr}{v_r} = r\frac{d\theta}{v_\theta}.$$

Since on the path line $\sin\theta = \frac{\phi_0}{r - \frac{a^2}{r}}$, then

$$v_r = U\left(1 - \frac{a^2}{r^2}\right)\sqrt{1 - \sin^2\theta} = \frac{U}{r^2}\sqrt{(r^2 - a^2)^2 - (\phi_0 r)^2}.$$

If for convenience we introduce new constants

$$F^2 = a^2 + \frac{\phi_0^2}{2} + \phi_0\sqrt{a^2 + \frac{\phi_0^2}{4}}, \quad G^2 = a^2 + \frac{\phi_0^2}{2} - \phi_0\sqrt{a^2 + \frac{\phi_0^2}{4}}, \quad k = \frac{F}{G},$$

and a new variable $s = r/F$. Then, since $a^4 = (FG)^2$ and $2a^2 + \psi_0^2 = F^2 + G^2$,

$$v_r = \frac{U}{F^2 s^2}\sqrt{F^4 s^4 - (F^2 + G^2)F^2 s^2 + F^2 G^2} = \frac{UG}{F^2 s^2}\sqrt{(1 - s)(1 - k^2 s^2)}.$$

Consequently, we have

$$dt = \frac{F^2 s^2}{UG(1 - s)(1 - k^2 s^2)}\,ds.$$

Upon integration this gives

$$t - t_0 = \frac{F^2}{UG}\int_{r_0/F}^{r/F}\frac{s^2}{(1 - s)(1 - k^2 s^2)}\,ds, \quad \text{where } r_0 \text{ is the value of } r \text{ at } t_0.$$

The integral in the previous expression is related to what are called elliptic integrals. Its values can be found tabulated in many handbooks or by numerical integration. Once r is found as a function of t on a path line (albeit in an inverse manner, since we have t as a function of r), the angle is found from $\theta = \sin^{-1} \frac{\psi_0}{r - \frac{a^2}{r}}$. From this we see that the constant ψ_0 can also be interpreted as $\psi_0 = \left(r_0 - \frac{a^2}{r_0} \right) \sin \theta_0$. Note that calculation of path lines usually is much more difficult than are calculations for streamlines.

1.4.2   Stream Functions for Three-Dimensional Flows, Including Stokes Stream Function

For three dimensions, equation (1.4.3) states that there is one relationship between the three velocity components, so it is expected that the velocity can be expressed in terms of two scalar functions. There are at least two ways to do this. The one that retains the interpretation of stream function as introduced in two dimensions is

$$\mathbf{v} = \nabla\psi \times \nabla\xi \tag{1.4.11}$$

where ψ and ξ are each constant on stream surfaces. The intersection of these stream surfaces is a streamline. Equations (1.4.4) and (1.4.5) are in fact a special case of this result with $\xi = z$.

Although equation (1.4.11) preserves the advantage of having the stream functions constant on a streamline, it has the disadvantage that $\mathbf{v}$ is a nonlinear function of ψ and ξ. Introducing further nonlinearities into an already highly nonlinear problem is not usually helpful!

Another possibility for solving the continuity equation is to write

$$\mathbf{v} = \nabla \times \mathbf{A}, \tag{1.4.12}$$

which guarantees $\nabla \cdot \mathbf{v} = 0$ for any vector $\mathbf{A}$. For the two-dimensional case, $\mathbf{A} = (0, 0, \psi)$ corresponds to our Lagrange stream function. Since only two scalars are needed, in three dimensions one component of $\mathbf{A}$ can be arbitrarily set to zero. (Some thought must be used in doing this. Obviously, in the two-dimensional case, difficulty would be encountered if one of the components of $\mathbf{A}$ that we set to zero was the z component.) The form of equation (1.4.12), while guaranteeing satisfaction of continuity, has not been much used, since the appropriate boundary conditions to be imposed on $\mathbf{A}$ can be awkward.

A particular three-dimensional case in which a stream function is useful is that of axisymmetric flow. Taking the z-axis as the axis of symmetry, either spherical polar coordinates,

$$R = \sqrt{x^2 + y^2 + z^2}, \quad \beta = \cos^{-1} \frac{z}{R}, \quad \theta = \tan^{-1} \frac{y}{z}, \tag{1.4.13}$$

(see Figure 1.4.3, with the appropriate tangent unit vectors shown at point P), or cylindrical polar coordinates with

$$r = \sqrt{x^2 + y^2}, \quad \theta = \tan^{-1} \frac{y}{x}, \quad z = z \tag{1.4.14}$$

(see Figure 1.4.1) can be used. The term *axisymmetry* means that the flow appears the same in any $\theta = $ constant plane, and the velocity component normal to that plane is zero. (There could in fact be a swirl velocity component v_θ without changing anything

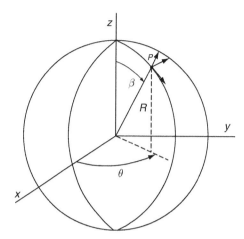

Figure 1.4.3 Spherical coordinates conventions

we have said. It would not be related to or determined by ψ.) Since any plane given by θ equal to a constant is therefore a stream surface, we can use equation (1.4.11) with $\xi = \theta$, giving

$$v_R = \frac{1}{R^2 \sin \beta} \frac{\partial \psi}{\partial \beta}, \quad v_\beta = \frac{-1}{R \sin \beta} \frac{\partial \psi}{\partial R} \tag{1.4.15}$$

in spherical coordinates with $\psi = \psi(R, \beta)$, or

$$v_r = -\frac{1}{r} \frac{\partial \psi}{\partial z}, \quad v_z = \frac{1}{r} \frac{\partial \psi}{\partial r} \tag{1.4.16}$$

in cylindrical coordinates, with $\psi = \psi(r, z)$. This stream function in either a cylindrical or spherical coordinate system is called the ***Stokes stream function***. Note that the dimensions of the Stokes stream function is length cubed per unit time, thus differing from Lagrange's stream function by a length dimension.

The volumetric discharge through an annular region is given in terms of the Stokes stream function by

$$Q = \int_A^B 2\pi r(v_z \, dr - v_r \, dz) = \int_A^B 2\pi r \left(\frac{1}{r} \frac{\partial \psi}{\partial r} dr - \frac{-1}{r} \frac{\partial \psi}{\partial z} dz \right)$$

$$= 2\pi \int_A^B \left(\frac{\partial \psi}{\partial r} dr + \frac{\partial \psi}{\partial z} dz \right) = 2\pi \int_A^B d\psi = 2\pi(\psi_B - \psi_A). \tag{1.4.17}$$

Example 1.4.3   Stokes stream function
A flow field in cylindrical polar coordinates is given by

$$v_r = -\frac{1.5 U a^3 r z}{(r^2 + z^2)^{5/2}}, \quad v_z = \frac{U a^3 (r^2 - 2z^2)}{2(r^2 + z^2)^{5/2}}.$$

Find the Stokes stream function for this flow.

Solution. Since $v_r = \frac{1}{r} \frac{\partial \psi}{\partial z}$, integration of v_r with respect to z gives

$$\psi = -1.5 U a^3 \int \frac{rz}{(r^2 + z^2)^{5/2}} r \, dz = -U a^3 \frac{r^2}{2(r^2 + z^2)^{3/2}} + f(r).$$

Differentiating this with respect to r, we have

$$\frac{\partial \psi}{\partial r} = rv_z = 0.5Ua^3\frac{r(r^2 - 2z^2)}{(r^2 + z^2)^{5/2}} + \frac{df(r)}{dr}.$$

Comparing this with the preceding expression for v_z, we see that $df(r)/dr = rU$, and so $df/dr = 0.5Ur^2$. Thus, finally

$$\psi = -\frac{Ua^3r^2}{2(r^2 + z^2)^{3/2}} + 0.5Ur^2 = \frac{Ur^2}{2}\left(1 - \frac{a^3}{(r^2 + z^2)^{3/2}}\right).$$

The preceding derivations were all for incompressible flow, whether the flow is steady or unsteady. The derivations can be extended to **steady** compressible flow by recognizing that since for these flows the continuity equation can be written as $\nabla \cdot (\rho\mathbf{v}) = 0$, the previous stream functions can be used simply by replacing $\mathbf{v}$ by $\rho\mathbf{v}$.

A further extension, to **unsteady** compressible flows, is possible by regarding time as a fourth dimension and using the extended four-dimensional vector $\mathbf{V} = (\rho v_x, \rho v_y, \rho v_z, \rho)$ together with the augmented del operator $\nabla_a = (\frac{\partial}{\partial x}, \frac{\partial}{\partial y}, \frac{\partial}{\partial z}, \frac{\partial}{\partial t})$. The continuity equation (1.3.5) can then be written as

$$\nabla_a \cdot \mathbf{V} = 0 \tag{1.4.18}$$

The previous results can be extended to this general case in a straightforward manner.

1.5   Newton's Momentum Equation

Next apply Newton's momentum equation to the control volume. The momentum in the interior of the control volume is $\iiint_V \rho\mathbf{v}\,dV$. The rate at which momentum enters the control volume through its surface is $\iint_S \rho\mathbf{v}\,\mathbf{v} \cdot \mathbf{n}\,dS$. The net rate of change of momentum is then

$$\frac{d}{dt}\iiint_V \rho\mathbf{v}\,dV + \iint_S \rho\mathbf{v}\,\mathbf{v} \cdot \mathbf{n}\,dS = \iiint_V \frac{\partial(\rho\mathbf{v})}{\partial t}dV + \iiint_V \nabla \cdot (\rho\mathbf{v}\mathbf{v})dV$$

$$= \iiint_V \left[\rho\frac{\partial\mathbf{v}}{\partial t} + \mathbf{v}\nabla \cdot (\rho\mathbf{v}) + \rho(\mathbf{v} \cdot \nabla)\mathbf{v}\right]dV \tag{1.5.1}$$

$$= \iiint_V \left\{\mathbf{v}\left[\frac{\partial\rho}{\partial t} + \nabla \cdot (\rho\mathbf{v})\right] + \rho\left[\frac{\partial\mathbf{v}}{\partial t} + \mathbf{v} \cdot \nabla\mathbf{v}\right]\right\}dV = \iiint_V \rho\frac{D\mathbf{v}}{Dt}dV.$$

The first bracket in the third line of equation (1.5.1) vanishes by virtue of equation (1.3.5), the continuity equation. The second bracket represents the material derivative of the velocity, which is the acceleration.

Note that in converting the surface integral to a volume integral using equation (1.1.1) the "vector" was $\mathbf{vv}$, which is a product of two vectors but is neither the dot nor cross product. It is sometimes referred to as the ***indefinite product***. This usage is in fact a slight generalization of equation (1.1.1), which can be verified by writing out the term in component form.

The forces applied to the surface of the control volume are due to pressure and viscous forces on the surface and gravitational force distributed throughout the volume. The pressure force is normal to the surface and points toward the volume. On an infinitesimal element its value is thus $-p\mathbf{n}\,dS$—the minus sign because the pressure force acts toward the area dS and thus is opposite to the unit normal.

The viscous force in general will have both normal and tangential components. For now, simply write it as $\boldsymbol{\tau}^{(n)}dS$ acting on a small portion of the surface, where $\boldsymbol{\tau}^{(n)}$ is called the **stress vector**. It is the force per unit area acting on the surface dS. The n superscript reminds us that the stress vector is applied to a surface with normal pointing in the $\mathbf{n}$ direction.

The gravity force per unit volume is written as $\rho\mathbf{g}\,dV$, where the magnitude of $\mathbf{g}$ is 9.80 or 32.17, depending on whether the units used are SI or British. The net force is then

$$\iint_S (-p\mathbf{n} + \boldsymbol{\tau}^{(n)})dS + \iiint_V \rho\mathbf{g}\,dV.$$

Equating this to the net change in momentum gives

$$\iiint_V \rho\frac{D\mathbf{v}}{Dt}dV = \iint_S (-p\mathbf{n} + \boldsymbol{\tau}^{(n)})dS + \iiint_V \rho\mathbf{g}\,dV. \qquad (1.5.2)$$

Use of the divergence theorem allows us to write

$$\iint_S -p\mathbf{n}\,dS = \iiint_V -\nabla p\,dV. \qquad (1.5.3)$$

This reduces equation (1.5.2) to

$$\iiint_V \rho\frac{D\mathbf{v}}{Dt}dV = \iiint_V (-\nabla p + \rho\mathbf{g})dV + \iint_S \boldsymbol{\tau}^{(n)}\,dS. \qquad (1.5.4)$$

At this point a similar simplification for the viscous term is not possible. This term is investigated further in the next section.

1.6 Stress

Stress is defined as a force applied to an area divided by that area. Thus, two directions are associated with stress: the direction of the force and the direction (orientation) of the area. Therefore, stress has a more complicated mathematical structure than does either a scalar or a vector. To put this into its simplest form, three special stress vectors will be introduced that act on mutually orthogonal surfaces whose faces are orientated with normals along our coordinate axes.

When a material is treated as a continuum, a force must be applied as a quantity distributed over an area. (In analysis, a concentrated force or load can sometimes be a convenient idealization. In a real material, any concentrated force would provide very large changes—in fact, infinite changes—both in deformation and in the material.) The previously introduced stress vector $\boldsymbol{\tau}^{(n)}$, for example, is defined as

$$\boldsymbol{\tau}^{(n)} = \lim_{\Delta S \to 0} \frac{\Delta\mathbf{F}}{\Delta S}, \qquad (1.6.1)$$

where ΔS is the magnitude of the infinitesimal area. In the limit as ΔS approaches zero the direction of the normal to ΔS is held fixed.

It appears that at a given point in the fluid there can be an infinity of different stress vectors, corresponding to the infinitely many orientations of $\mathbf{n}$ that are possible. To bring order out of such confusion, we consider three very special orientations of $\mathbf{n}$ and then show that all other orientations of $\mathbf{n}$ produce stress vectors that are simply related to the first three.

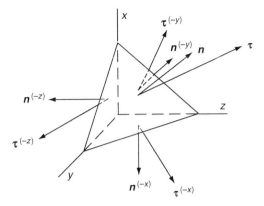

Figure 1.6.1 Stress vector

First consider the three special stress vectors $\boldsymbol{\tau}^{(x)}$, $\boldsymbol{\tau}^{(y)}$, $\boldsymbol{\tau}^{(z)}$, corresponding to forces acting on areas with unit normals pointing in the x, y, and z directions, respectively. These act on the small tetrahedron shown in Figure 1.6.1. For a force $\Delta\mathbf{F}$ acting on a surface ΔS_x with unit normal pointing in the x direction (thus, $\mathbf{n} = \mathbf{i}$), write the stress on this face of our tetrahedron as

$$\boldsymbol{\tau}^{(x)} = \lim_{\Delta A_x \to 0} \frac{\Delta\mathbf{F}}{\Delta A_x} = \tau_{xx}\,\mathbf{i} + \tau_{xy}\,\mathbf{j} + \tau_{xz}\,\mathbf{k}, \qquad (1.6.2)$$

where τ_{xx} is the limit of the x component of the force acting on this face, τ_{xy} is the limit of the y component of the force acting on this face, and τ_{xz} is the limit of the z component of the force acting on this face.

Similarly, for normals pointing in the y ($\mathbf{n} = \mathbf{j}$) and z ($\mathbf{n} = \mathbf{k}$) directions we have

$$\boldsymbol{\tau}^{(y)} = \lim_{\Delta A_y \to 0} \frac{\Delta\mathbf{F}}{\Delta A_y} = \tau_{yx}\,\mathbf{i} + \tau_{yy}\,\mathbf{j} + \tau_{yz}\,\mathbf{k}, \qquad (1.6.3)$$

and

$$\boldsymbol{\tau}^{(z)} = \lim_{\Delta A_z \to 0} \frac{\Delta\mathbf{F}}{\Delta A_z} = \tau_{zx}\,\mathbf{i} + \tau_{zy}\,\mathbf{j} + \tau_{zz}\,\mathbf{k}. \qquad (1.6.4)$$

As in equation (1.6.2), the first subscript on the components tells the direction that the area faces, and the second subscript gives the direction of the force component on that face.

Upon examination of the three stress vectors shown in equations (1.6.2), (1.6.3), and (1.6.4), we see that they can be summarized in the matrix form

$$\boldsymbol{\tau} = \begin{pmatrix} \boldsymbol{\tau}^{(x)} \\ \boldsymbol{\tau}^{(y)} \\ \boldsymbol{\tau}^{(z)} \end{pmatrix} = \begin{pmatrix} \tau_{xx}\,\tau_{xy}\,\tau_{xz} \\ \tau_{yx}\,\tau_{yy}\,\tau_{yz} \\ \tau_{zx}\,\tau_{zy}\,\tau_{zz} \end{pmatrix} \begin{pmatrix} \mathbf{i} \\ \mathbf{j} \\ \mathbf{k} \end{pmatrix}.$$

The nine quantities in this 3 by 3 matrix are the components of the second-order stress tensor.

In the limit, as the areas are taken smaller and smaller, the forces acting on the four faces of the tetrahedron are $-\boldsymbol{\tau}^{(x)}\,dS_x$, $-\boldsymbol{\tau}^{(y)}\,dS_y$, $-\boldsymbol{\tau}^{(z)}\,dS_z$, and $\boldsymbol{\tau}^{(n)}\,dS$. The first three of these forces act on faces whose normals are in the $-x$, $-y$, $-z$ directions. In writing them we have used Newton's third law, which tells us that $\boldsymbol{\tau}^{(-x)} = -\boldsymbol{\tau}^{(x)}$, $\boldsymbol{\tau}^{(-y)} = -\boldsymbol{\tau}^{(y)}$, and $\boldsymbol{\tau}^{(-z)} = -\boldsymbol{\tau}^{(z)}$. The fourth of these forces acts on the slant face with area dS and normal $\mathbf{n}$.

Summing the surface forces and using $dS_x = n_x \, dS, \; dS_y = n_y \, dS, \; dS_z = n_z \, dS$, the result is

$$
\begin{aligned}
\Sigma \mathbf{F}_S &= -\boldsymbol{\tau}^{(x)} dS_x - \boldsymbol{\tau}^{(y)} dS_y - \boldsymbol{\tau}^{(z)} dS_z + \boldsymbol{\tau}^{(n)} \, dS \\
&= [-\boldsymbol{\tau}^{(x)} n_x - \boldsymbol{\tau}^{(y)} n_y - \boldsymbol{\tau}^{(z)} n_z + \boldsymbol{\tau}^{(n)}] dS.
\end{aligned}
\tag{1.6.5}
$$

Considering the summation of forces as the tetrahedron shrinks to zero, we have

$$
\mathbf{F}_S + \mathbf{F}_V = \rho \frac{D\mathbf{v}}{Dt} dV.
$$

Our surface forces depend on dS, and thus are second order in the tetrahedron dimensions. The body forces and the mass times acceleration terms depend on dV, which is of the third order in the tetrahedron dimensions. If in our analysis the tetrahedron is decreased in size, the dV terms approach zero faster than do the dS terms. Thus, the body force and acceleration terms are of higher (third) order in the dimensions of the tetrahedron than the surface force terms (second). In the limit as the tetrahedron goes to zero, we are thus left with only the second-order terms, resulting in

$$
\boldsymbol{\tau}^{(n)} = \boldsymbol{\tau}^{(x)} n_x + \boldsymbol{\tau}^{(y)} n_y + \boldsymbol{\tau}^{(z)} n_z.
\tag{1.6.6}
$$

Therefore, from knowledge of the three special stress vectors $\boldsymbol{\tau}^{(x)}, \; \boldsymbol{\tau}^{(y)}, \; \boldsymbol{\tau}^{(z)}$, we can find the stress vector in the direction of any n at the same point.

Next, combine equations (1.6.2), (1.6.3), and (1.6.4) with (1.6.6). This gives

$$
\begin{aligned}
\boldsymbol{\tau}^{(n)} &= n_x(\tau_{xx}\mathbf{i} + \tau_{xy}\mathbf{j} + \tau_{xz}\mathbf{k}) + n_y(\tau_{yx}\mathbf{i} + \tau_{yy}\mathbf{j} + \tau_{yz}\mathbf{k}) \\
&\quad + n_z(\tau_{zx}\mathbf{i} + \tau_{zy}\mathbf{j} + \tau_{zz}\mathbf{k}) \\
&= \mathbf{i}(n_x\tau_{xx} + n_y\tau_{yx} + n_z\tau_{zx}) + \mathbf{j}(n_x\tau_{xy} + n_y\tau_{yy} + n_z\tau_{zy}) \\
&\quad + \mathbf{k}(n_x\tau_{xz} + n_y\tau_{yz} + n_z\tau_{zz}),
\end{aligned}
\tag{1.6.7}
$$

the expression after the second equals sign being a rearrangement of the preceding.

Taking the dot product of the three unit vectors $\mathbf{i}, \mathbf{j}, \mathbf{k}$ with (1.6.7) gives

$$
\begin{aligned}
\tau_{xx} &= \mathbf{i} \cdot \boldsymbol{\tau}^{(x)}, & \tau_{xy} &= \mathbf{j} \cdot \boldsymbol{\tau}^{(x)}, & \tau_{xz} &= \mathbf{k} \cdot \boldsymbol{\tau}^{(x)}, \\
\tau_{yx} &= \mathbf{i} \cdot \boldsymbol{\tau}^{(y)}, & \tau_{yy} &= \mathbf{j} \cdot \boldsymbol{\tau}^{(y)}, & \tau_{yz} &= \mathbf{k} \cdot \boldsymbol{\tau}^{(y)}, \\
\tau_{zx} &= \mathbf{i} \cdot \boldsymbol{\tau}^{(z)}, & \tau_{zy} &= \mathbf{j} \cdot \boldsymbol{\tau}^{(z)}, & \tau_{zz} &= \mathbf{k} \cdot \boldsymbol{\tau}^{(z)}.
\end{aligned}
\tag{1.6.8}
$$

This is in agreement with our definition of τ_{xx} (and so on) being the components of the various stress vectors.

Stress components associated with force components that act in the same direction as their normal (i.e., $\tau_{xx}, \; \tau_{yy}, \; \tau_{zz}$) are referred to as **normal stresses**. Stress components associated with force components that act perpendicular to their normal (i.e., $\tau_{yx}, \; \tau_{yz}, \; \tau_{zx}, \; \tau_{xy}, \; \tau_{zx}, \; \tau_{xz}$) are referred to as **shear stresses**. Note that the first subscript on the components tells us the direction in which the area faces, and the second subscript gives us the direction of the force component on that face. Positive sign conventions for viscous stress components are given in Figure 1.6.2. The nine components $(\tau_{xx}, \; \tau_{xy}, \; \tau_{xz}, \; \tau_{yx}, \; \tau_{yy}, \; \tau_{yz}, \; \tau_{zx}, \; \tau_{zy}, \; \tau_{zzj})$ of our three special stress vectors $(\boldsymbol{\tau}^{(x)}, \; \boldsymbol{\tau}^{(y)}, \; \boldsymbol{\tau}^{\tau(z)})$ can be shown to be components of a second-order tensor called the

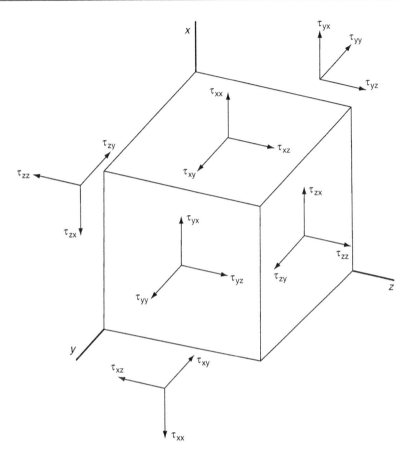

Figure 1.6.2 Stress tensor sign convention

stress tensor. The stress vectors $\boldsymbol{\tau}^{(x)}$, $\boldsymbol{\tau}^{(y)}$, and $\boldsymbol{\tau}^{(z)}$ are thus expressible in terms of these nine components, as shown above. As seen from the manner in which the components of the three stress vectors were introduced while carrying out the development, the nine-component stress tensor is the collection of the three x, y, z components of each of these three stress vectors. Thus, the stress tensor is the collection of nine components of the three special stress vectors $\boldsymbol{\tau}^{(x)}$, $\boldsymbol{\tau}^{(y)}$, $\boldsymbol{\tau}^{(z)}$.

Besides its use in formulating the basic equations of fluid dynamics, the stress vector is also used to apply conditions at the boundary of the fluid, as will be seen when we consider boundary conditions. The stress tensor is used to describe the state of stress in the interior of the fluid.

Return now to equation (1.5.4), and change the surface integral to a volume integral. From equation (1.6.7) we have

$$
\begin{aligned}
\iint_{S} \boldsymbol{\tau}^{(n)}\,dS &= \iint_{S}[n_{x}(\tau_{xx}\mathbf{i}+\tau_{xy}\mathbf{j}+\tau_{xz}\mathbf{k})+n_{y}(\tau_{yx}\mathbf{i}+\tau_{yy}\mathbf{j}+\tau_{yz}\mathbf{k})+n_{z}(\tau_{zx}\mathbf{i}+\tau_{zy}\mathbf{j}+\tau_{zz}\mathbf{k})]\,dS \\
&= \iiint_{V}\left[\frac{\partial}{\partial x}\left(\tau_{xx}\mathbf{i}+\tau_{xy}\mathbf{j}+\tau_{xz}\mathbf{k}\right)+\frac{\partial}{\partial y}\left(\tau_{yx}\mathbf{i}+\tau_{yy}\mathbf{j}+\tau_{yz}\mathbf{k}\right)\right. \\
&\qquad\left.+\frac{\partial}{\partial z}(\tau_{zx}\mathbf{i}+\tau_{zy}\mathbf{j}+\tau_{zz}\mathbf{k})\right]dV
\end{aligned}
\tag{1.6.9}
$$

$$= \iiint_V \left[\mathbf{i} \left(\frac{\partial \tau_{xx}}{\partial x} + \frac{\partial \tau_{yx}}{\partial y} + \frac{\partial \tau_{zx}}{\partial z} \right) + \mathbf{j} \left(\frac{\partial \tau_{xy}}{\partial x} + \frac{\partial \tau_{yy}}{\partial y} + \frac{\partial \tau_{zy}}{\partial z} \right) \right.$$
$$\left. + \mathbf{k} \left(\frac{\partial \tau_{xz}}{\partial x} + \frac{\partial \tau_{yz}}{\partial y} + \frac{\partial \tau_{zz}}{\partial z} \right) \right] dV.$$

Inserting this into equation (1.5.4) gives

$$\iiint_V \rho \frac{D\mathbf{v}}{Dt} dV = \iiint_V (-\nabla p + \rho \mathbf{g}) dV + \iiint_V \left[\mathbf{i} \left(\frac{\partial \tau_{xx}}{\partial x} + \frac{\partial \tau_{yx}}{\partial y} + \frac{\partial \tau_{zx}}{\partial z} \right) \right.$$
$$\left. + \mathbf{j} \left(\frac{\partial \tau_{xy}}{\partial x} + \frac{\partial \tau_{yy}}{\partial y} + \frac{\partial \tau_{zy}}{\partial z} \right) + \mathbf{k} \left(\frac{\partial \tau_{xz}}{\partial x} + \frac{\partial \tau_{yz}}{\partial y} + \frac{\partial \tau_{zz}}{\partial z} \right) \right] dV. \tag{1.6.10}$$

As was the case for the continuity equation, equation (1.6.10) must be valid no matter what volume we choose. Therefore, it must be that

$$\rho \frac{D\mathbf{v}}{Dt} = -\nabla p + \rho \mathbf{g} + \mathbf{i} \left(\frac{\partial \tau_{xx}}{\partial x} + \frac{\partial \tau_{yx}}{\partial y} + \frac{\partial \tau_{zx}}{\partial z} \right)$$
$$+ \mathbf{j} \left(\frac{\partial \tau_{xy}}{\partial x} + \frac{\partial \tau_{yy}}{\partial y} + \frac{\partial \tau_{zy}}{\partial z} \right) + \mathbf{k} \left(\frac{\partial \tau_{xz}}{\partial x} + \frac{\partial \tau_{yz}}{\partial y} + \frac{\partial \tau_{zz}}{\partial z} \right). \tag{1.6.11}$$

Before leaving these equations, note that the continuity equation (1.3.5) and the momentum equation (1.6.11) can be written in a combined matrix form as

$$\frac{\partial}{\partial t} \begin{pmatrix} \rho \\ \rho v_x \\ \rho v_y \\ \rho v_z \end{pmatrix} + \frac{\partial}{\partial x} \begin{pmatrix} \rho v_x \\ p - \rho \Omega_x - \tau_{xx} \\ -\tau_{xy} \\ -\tau_{xz} \end{pmatrix} + \frac{\partial}{\partial y} \begin{pmatrix} \rho v_y \\ -\tau_{yx} \\ p - \rho \Omega_y - \tau_{yy} \\ -\tau_{yz} \end{pmatrix}$$
$$+ \frac{\partial}{\partial z} \begin{pmatrix} \rho v_z \\ -\tau_{zx} \\ -\tau_{zy} \\ p - \rho \Omega_z - \tau_{zx} \end{pmatrix} = 0. \tag{1.6.12}$$

This form of the equations is referred to as the **conservative form** and is frequently used in computational fluid dynamics.

Moments can be balanced in the same manner as forces. Using a finite control mass and taking $\mathbf{R}$ as a position vector drawn from the point about which moments are being taken to a general fluid particle, equating moments to the time rate of change of moment of momentum gives

$$\frac{d}{dt} \iiint_V \rho \mathbf{R} \times \mathbf{v} \, dV + \iint_S \rho \mathbf{R} \times \mathbf{v}(\mathbf{v} \cdot \mathbf{n}) dS$$
$$= \iint_S \mathbf{R} \times (-p\mathbf{n}) dS + \iiint_V \mathbf{R} \times \rho \mathbf{g} \, dV + \iint_S \mathbf{R} \times \boldsymbol{\tau}^{(n)} dS \tag{1.6.13}$$

But

$$\frac{d}{dt} \iiint_V \rho \mathbf{R} \times \mathbf{v} dV = \iiint_V \left[\frac{D\rho}{Dt} \mathbf{R} \times \mathbf{v} + \rho \frac{D(\mathbf{R} \times \mathbf{v})}{Dt} \right] dV \tag{1.6.14}$$

and using the product rule and the fact that a vector crossed with itself is zero,

$$\frac{D(\mathbf{R} \times \mathbf{v})}{Dt} = \frac{D\mathbf{R}}{Dt} \times \mathbf{v} + \mathbf{R} \times \frac{D\mathbf{v}}{Dt} = \mathbf{v} \times \mathbf{v} + \mathbf{R} \times \frac{D\mathbf{v}}{Dt} = \mathbf{R} \times \frac{D\mathbf{v}}{Dt}. \qquad (1.6.15)$$

Also, using equation (1.6.7), $\mathbf{R} \times \boldsymbol{\tau}^{(n)}$ can be written as

$$
\begin{aligned}
\mathbf{R} \times \boldsymbol{\tau}^{(n)} &= \mathbf{R} \times [\mathbf{i}(n_x\tau_{xx} + n_y\tau_{yx} + n_z\tau_{zx}) + \mathbf{j}(n_x\tau_{xy} + n_y\tau_{yy} + n_z\tau_{zy}) \\
&\quad + \mathbf{k}(n_x\tau_{xz} + n_y\tau_{yz} + n_z\tau_{zz})] \\
&= n_x[\mathbf{i}(y\tau_{xz} - z\tau_{xy}) + \mathbf{j}(z\tau_{xx} - x\tau_{xz}) + \mathbf{k}(x\tau_{xy} - y\tau_{xx})] \qquad (1.6.16) \\
&\quad + n_y[\mathbf{i}(y\tau_{yz} - z\tau_{yy}) + \mathbf{j}(z\tau_{yx} - x\tau_{yz}) + \mathbf{k}(x\tau_{yy} - y\tau_{yx})] \\
&\quad + n_z[\mathbf{i}(y\tau_{zz} - z\tau_{zy}) + \mathbf{j}(z\tau_{zx} - x\tau_{zz}) + \mathbf{k}(x\tau_{zy} - y\tau_{zx})].
\end{aligned}
$$

The surface integral involving $\boldsymbol{\tau}^{(n)}$ then becomes

$$
\begin{aligned}
\iint_S \mathbf{R} \times \boldsymbol{\tau}^{(n)} \, dS &= \iint_S \{n_x[\mathbf{i}(y\tau_{xz} - z\tau_{xy}) + \mathbf{j}(z\tau_{xx} - x\tau_{xz}) + \mathbf{k}(x\tau_{xy} - y\tau_{xx})] \\
&\quad + n_y[\mathbf{i}(y\tau_{yz} - z\tau_{yy}) + \mathbf{j}(z\tau_{yx} - x\tau_{yz}) + \mathbf{k}(x\tau_{yy} - y\tau_{yx})] \\
&\quad + n_z[\mathbf{i}(y\tau_{zz} - z\tau_{zy}) + \mathbf{j}(z\tau_{zx} - x\tau_{zz}) + \mathbf{k}(x\tau_{zy} - y\tau_{zx})]\} dS \\
&= \iiint_V \left\{ \frac{\partial}{\partial x}[\mathbf{i}(y\tau_{xz} - z\tau_{xy}) + \mathbf{j}(z\tau_{xx} - x\tau_{xz}) + \mathbf{k}(x\tau_{xy} - y\tau_{xx})] \right. \\
&\quad + \frac{\partial}{\partial y}[\mathbf{i}(y\tau_{yz} - z\tau_{yy}) + \mathbf{j}(z\tau_{yx} - x\tau_{yz}) + \mathbf{k}(x\tau_{yy} - y\tau_{yx})] \\
&\quad \left. + \frac{\partial}{\partial z}[\mathbf{i}(y\tau_{zz} - z\tau_{zy}) + \mathbf{j}(z\tau_{zx} - x\tau_{zz}) + \mathbf{k}(x\tau_{zy} - y\tau_{zx})] \right\} dV \quad (1.6.17) \\
&= \iiint_V \left\{ \mathbf{i}(\tau_{yz} - \tau_{zy}) + \mathbf{j}(\tau_{zx} - \tau_{xz}) + \mathbf{k}(\tau_{xy} - \tau_{yx}) \right. \\
&\quad + \mathbf{i}\left[y\left(\frac{\partial \tau_{xz}}{\partial x} + \frac{\partial \tau_{yz}}{\partial y} + \frac{\partial \tau_{zz}}{\partial z}\right) - z\left(\frac{\partial \tau_{xy}}{\partial x} + \frac{\partial \tau_{yy}}{\partial y} + \frac{\partial \tau_{zy}}{\partial z}\right) \right] \\
&\quad + \mathbf{j}\left[z\left(\frac{\partial \tau_{xx}}{\partial x} + \frac{\partial \tau_{yx}}{\partial y} + \frac{\partial \tau_{zx}}{\partial z}\right) - x\left(\frac{\partial \tau_{xz}}{\partial x} + \frac{\partial \tau_{yz}}{\partial y} + \frac{\partial \tau_{zz}}{\partial z}\right) \right] \\
&\quad \left. + \mathbf{k}\left[x\left(\frac{\partial \tau_{xy}}{\partial x} + \frac{\partial \tau_{yy}}{\partial y} + \frac{\partial \tau_{zy}}{\partial z}\right) - y\left(\frac{\partial \tau_{xx}}{\partial x} + \frac{\partial \tau_{yx}}{\partial y} + \frac{\partial \tau_{zx}}{\partial z}\right) \right] \right\} dV \\
&= \iiint_V \left\{ \mathbf{i}(\tau_{yz} - \tau_{zy}) + \mathbf{j}(\tau_{zx} - \tau_{xz}) + \mathbf{k}(\tau_{xy} - \tau_{yx}) \right. \\
&\quad + \mathbf{R} \times \left[\mathbf{i}\left(\frac{\partial \tau_{xx}}{\partial x} + \frac{\partial \tau_{yx}}{\partial y} + \frac{\partial \tau_{zx}}{\partial z}\right) + \mathbf{j}\left(\frac{\partial \tau_{xy}}{\partial x} + \frac{\partial \tau_{yy}}{\partial y} + \frac{\partial \tau_{zy}}{\partial z}\right) \right. \\
&\quad \left. \left. + \mathbf{k}\left(\frac{\partial \tau_{xz}}{\partial x} + \frac{\partial \tau_{yz}}{\partial y} + \frac{\partial \tau_{zz}}{\partial z}\right) \right] \right\} dV.
\end{aligned}
$$

Thus, using equations (1.6.14), (1.6.15), (1.6.16), (1.6.17), equation (1.6.13) becomes

$$\iiint_V \left[\frac{\partial \rho}{\partial t} \mathbf{R} \times \mathbf{v} + \rho \mathbf{R} \times \frac{D\mathbf{v}}{Dt} \right] dV + \iiint_V \nabla \cdot (\rho \mathbf{v}) \mathbf{R} \times \mathbf{v} dV = \iiint_V \mathbf{R} \times (-\nabla p) dV$$

$$+ \iiint_V \mathbf{R} \times \rho \mathbf{g}\, dV + \iiint_V \left\{ \mathbf{i}(\tau_{yz} - \tau_{zy}) + \mathbf{j}(\tau_{zx} - \tau_{xz}) + \mathbf{k}(\tau_{xy} - \tau_{yx}) \right.$$

$$+ \mathbf{R} \times \left[\mathbf{i} \left(\frac{\partial \tau_{xx}}{\partial x} + \frac{\partial \tau_{yx}}{\partial y} + \frac{\partial \tau_{zx}}{\partial z} \right) + \mathbf{j} \left(\frac{\partial \tau_{xy}}{\partial x} + \frac{\partial \tau_{yy}}{\partial y} + \frac{\partial \tau_{zy}}{\partial z} \right) \right.$$

$$\left. \left. + \mathbf{k} \left(\frac{\partial \tau_{xz}}{\partial x} + \frac{\partial \tau_{yz}}{\partial y} + \frac{\partial \tau_{zz}}{\partial z} \right) \right] \right\} dV.$$

Collecting terms and rearranging, the result is

$$\iiint_V R \times \mathbf{v} \left[\frac{\partial \rho}{\partial t} + \nabla \cdot (\rho\, \mathbf{v}) \right] dV + \iiint_V \rho \mathbf{R} \times \frac{D\mathbf{v}}{Dt} dV$$

$$= \iiint_V \mathbf{R} \times \left[- \nabla p + \rho\, \mathbf{g} + \mathbf{i} \left(\frac{\partial \tau_{xx}}{\partial x} + \frac{\partial \tau_{yx}}{\partial y} + \frac{\partial \tau_{zx}}{\partial z} \right) \right. \qquad (1.6.18)$$

$$+ \mathbf{j} \left(\frac{\partial \tau_{xy}}{\partial x} + \frac{\partial \tau_{yy}}{\partial y} + \frac{\partial \tau_{zy}}{\partial z} \right) + \mathbf{k} \left(\frac{\partial \tau_{xz}}{\partial x} + \frac{\partial \tau_{yz}}{\partial y} + \frac{\partial \tau_{zz}}{\partial z} \right) \bigg] dV$$

$$+ \iiint_V [i(\tau_y z - \tau_{zy}) + \mathbf{j}(\tau_{zx} - \tau_{xz}) + \mathbf{k}(\tau_{xy} - \tau_{yx})] dV.$$

Examining the various terms, it is seen that the integrand in the first integral is zero by virtue of the continuity equation. The integrand in the last integral on the left side of the equals sign is **R** crossed with the mass density times the acceleration, which is the left side of equation (1.6.11). The integrand in the first integral on the right of the integral sign is **R** crossed with the right side of equation (1.6.11). Thus, the two integrals cancel and we are left with

$$0 = \iiint_V [\mathbf{i}(\tau_{yz} - \tau_{zy}) + \mathbf{j}(\tau_{zx} - \tau_{xz}) + \mathbf{k}(\tau_{xy} - \tau_{yx})] dV.$$

Again, for this to be true for any volume it is the integrand that must vanish. This yields

$$0 = \mathbf{i}(\tau_{yz} - \tau_{zy}) + \mathbf{j}(\tau_{zx} - \tau_{xz}) + \mathbf{k}(\tau_{xy} - \tau_{yx}),$$

or

$$\tau_{yz} = \tau_{zy}, \quad \tau_{zx} = \tau_{xz}, \quad \tau_{xy} = \tau_{yx}. \qquad (1.6.19)$$

Thus, equation (1.6.19), the result of summing moments, tells us that the order of writing the subscripts is immaterial, since the stress due to the y force component acting on the face with normal in the x direction is equal to the stress due to the x force component acting on the face with normal in the y direction, and so on for the other two faces. This interchangeability of the indices tells us that the stress tensor is a *symmetric tensor*, and there are only six rather than nine numerically unique values for its components at a point. (It has been proposed that there is the possibility of a magnetic material to have an antisymmetric stress tensor. Such a material would indeed have a complex mathematical description.)

1.7 Rates of Deformation

In the equations developed so far, we have not identified the nature of the material we are studying. In fact, our equations are so general at this stage that they apply to solid materials as well as to fluids. To narrow the subject to fluids, it is necessary to show how the fluid behaves under applied stresses. The important geometric quantity that describes the fluids' behavior under stress is the *rate of deformation*.

To define the deformation of a fluid under acting stresses, first consider motion in the two-dimensional *xy* plane. Choose three neighboring points ABC (Figure 1.7.1) selected to make up a right angle at an initial time *t*. A is a distance Δx along the *x*-axis from B, and C is a distance Δy above B along the *y*-axis.

As the flow evolves through a short time interval Δt, the angle ABC will change, as will the distance between the three points. At this later time $t + \Delta t$, these points will have moved to A', B', and C', as shown. Using Taylor series expansions to the first order, the fluid that initially was at point A will have *x* and *y* velocity components given by $v_x + \frac{\partial v_x}{\partial x}\Delta x$ and $v_y + \frac{\partial v_y}{\partial x}\Delta x$. The fluid initially at point A will move a distance $\left(v_x + \frac{\partial v_x}{\partial x}\Delta x\right)\Delta t$ to the right of A, and a distance $\left(v_y + \frac{\partial v_y}{\partial x}\Delta x\right)\Delta t$ above A. Similarly, the fluid initially occupying point B has moved to point B'—that is, $v_x\Delta t$ to the right of B and $v_y\Delta t$ above B, and the fluid initially at point C has moved to C', which is $\left(v_y + \frac{\partial v_y}{\partial y}\Delta y\right)\Delta t$ above C and $\left(v_x + \frac{\partial v_x}{\partial y}\Delta y\right)\Delta t$ to the right of C.

Looking first at the time rate of change of lengths, it is seen that after a time interval Δt the rate of change of length along the *x*-axis per unit length, which will be denoted by d_{xx}, is the final length minus the original length divided by the original length, all divided by Δt. In the limit this is

$$d_{xx} = \frac{\left[\Delta x + \left(v_x + \frac{\partial v_x}{\partial x}\Delta x\right)\Delta t\right] - [\Delta x + v_x\Delta t]}{\Delta x \Delta t} = \frac{\partial v_x}{\partial x}. \tag{1.7.1}$$

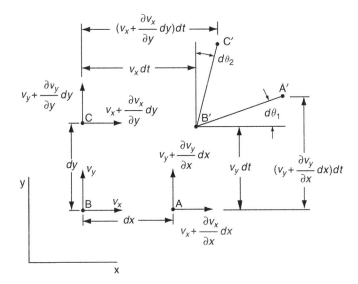

Figure 1.7.1   Rates of deformation—two dimensions

A similar analysis along the y-axis would give the rate of change of length per unit length as measured along the y-axis, d_{yy}, as

$$d_{yy} = \lim_{\Delta t \to 0} \frac{\left[\Delta y + \left(v_y + \frac{\partial v_y}{\partial y}\Delta y\right)\Delta t\right] - [\Delta y + v_y\Delta t]}{\Delta y \Delta t} = \frac{\partial v_y}{\partial y} \qquad (1.7.2)$$

and similarly in the z direction

$$d_{zz} = \lim_{\Delta t \to 0} \frac{\left[\Delta z + \left(v_z + \frac{\partial v_z}{\partial x}\Delta z\right)\Delta t\right] - [\Delta z + v_z\Delta t]}{\Delta z \Delta t} = \frac{\partial v_z}{\partial z}. \qquad (1.7.3)$$

The d_{xx}, d_{yy}, and d_{zz} are the ***normal rates of deformation*** and can loosely be thought of as rates of normal, or extensional, strain. The term "loosely" is used, since the definitions of strain you might be familiar with from the study of solid mechanics are for infinitesimal strains. For finite strain, many different definitions are used in solid mechanics for rates of strain. Since in fluid mechanics strains are always finite—and also usually very large—using the term "rates of deformation" avoids confusion.

Note that the rate of change of volume per unit volume is the volume at $t + \Delta t$ minus the original volume at t divided by the original volume times Δt, or

$$\lim_{\Delta x, \Delta y, \Delta z, \Delta t \to 0} \frac{(\Delta x + d_{xx}\Delta x\Delta t)(\Delta + d_{yy}\Delta y\Delta t)(\Delta z + d_{zz}\Delta z\Delta t) - \Delta x\Delta y\Delta z}{\Delta x\Delta y\Delta z\Delta t} \qquad (1.7.4)$$

$$= d_{xx} + d_{yy} + d_{zz} = \nabla \cdot \mathbf{v}.$$

This is the ***dilatational strain rate***.

Besides changes of length, changes of angles are involved in the deformation. Looking at the angles that AB and BC have rotated through, it is seen that

$$\dot{\theta}_1 \approx \frac{\tan^{-1}\Delta\theta_1}{\Delta t} \approx \frac{\frac{\partial v_y}{\partial x}\Delta x\Delta t}{\Delta x\Delta t(1 + \frac{\partial v_x}{\partial x}\Delta t)} \approx \frac{\partial v_y}{\partial x} \qquad (1.7.5)$$

and similarly

$$\dot{\theta}_2 \approx \frac{\tan^{-1}\Delta\theta_2}{\Delta t} \approx \frac{\frac{\partial v_x}{\partial y}\Delta y\Delta t}{\Delta y\Delta t(1 + \frac{\partial v_y}{\partial y}\Delta t)} \approx \frac{\partial v_x}{\partial y}. \qquad (1.7.6)$$

The sum of the two angular rates, $\dot{\theta}_1 + \dot{\theta}_2$, represents the rate of change of the angle ABC. Let

$$d_{xy} = d_{yx} = 0.5(\dot{\theta}_1 + \dot{\theta}_2) = \frac{1}{2}\left(\frac{\partial v_y}{\partial x} + \frac{\partial v_x}{\partial y}\right) \qquad (1.7.7)$$

be defined as the ***rate of shear deformation*** as measured in the xy plane. Similarly define

$$d_{yz} = d_{zy} = \frac{1}{2}\left(\frac{\partial v_y}{\partial z} + \frac{\partial v_z}{\partial y}\right) \qquad (1.7.8)$$

and

$$d_{zx} = d_{xz} = \frac{1}{2}\left(\frac{\partial v_z}{\partial x} + \frac{\partial v_x}{\partial z}\right) \qquad (1.7.9)$$

as the rates of shear deformation as measured in the yx and xz planes, respectively.

The one-half factor in the definition of the rate of deformation components is introduced so that the components transform independent of our selection of axes. Since the definition of angular rate of deformation is to some degree our option, any choice that gives a measure of the deformation is suitable. In this case we wish to relate the deformation rate to stress, and we would like to do it in such a manner that if we change to another coordinate system, all quantities change correctly.

The rate of deformation components that have been arrived at can be shown to transform as a second-order tensor. Note that if we interchange the order in which the subscripts are written in the definitions, there is no change in the various components. That is,

$$d_{xy} = d_{yx}, \quad d_{xz} = d_{zx}, \quad \text{and} \quad d_{yz} = d_{zy}. \tag{1.7.10}$$

As in the case of the stress tensor, such a tensor is said to be a ***symmetric tensor***.

It may be helpful to your physical understanding of rate of deformation to look at what is happening from a slightly different viewpoint. Consider any two neighboring fluid particles a distance $d\mathbf{r}$ apart, where the distance $d\mathbf{r}$ changes with time but must remain small because the particles were initially close together. To find the rate at which the particles separate, take the time derivative of $d\mathbf{r}$, obtaining

$$\frac{D(d\mathbf{r})}{Dt} = d\frac{D\mathbf{r}}{Dt} = d\mathbf{v}, \tag{1.7.11}$$

where $d\mathbf{v}$ is the difference in velocity between the two points, as shown in Figure 1.7.2. Since the magnitude of the distance between the two points, or more conveniently its square, is $|\mathbf{dr}|^2 = \mathbf{dr} \cdot \mathbf{dr}$, we have

$$
\begin{aligned}
\frac{D(\mathbf{dr}^2)}{Dt} &= 2\mathbf{dr} \cdot \frac{D(\mathbf{dr})}{Dt} = 2\mathbf{dr} \cdot \mathbf{dv} \\
&= 2dx\left(\frac{\partial v_x}{\partial x}dx + \frac{\partial v_x}{\partial y}dy + \frac{\partial v_x}{\partial z}dz\right) + 2dy\left(\frac{\partial v_y}{\partial x}dx + \frac{\partial v_y}{\partial y}dy + \frac{\partial v_y}{\partial z}dz\right) \\
&\quad + 2dz\left(\frac{\partial v_z}{\partial x}dx + \frac{\partial v_z}{\partial y}dy + \frac{\partial v_z}{\partial z}dz\right) \\
&= 2\left[\frac{\partial v_x}{\partial x}dx^2 + \frac{\partial v_y}{\partial y}dy^2 + \frac{\partial v_z}{\partial z}dz^2 + \left(\frac{\partial v_x}{\partial y} + \frac{\partial v_y}{\partial x}\right)dxdy + \left(\frac{\partial v_x}{\partial z} + \frac{\partial v_z}{\partial x}\right)dxdz\right. \\
&\quad \left. + \left(\frac{\partial v_z}{\partial y} + \frac{\partial v_y}{\partial z}\right)dzdy\right] \\
&= 2\left(d_{xx}dx^2 + d_{yy}dy^2 + d_{zz}dz^2 + 2d_{xy}dxdy + 2d_{xz}dxdz + 2d_{yz}dydz\right).
\end{aligned}
\tag{1.7.12}
$$

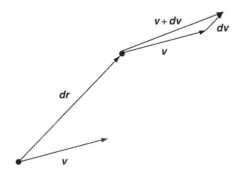

Figure 1.7.2 Velocity changes

Thus, after choosing the two neighboring points that we wish to describe (i.e., selecting $d\mathbf{r}$) to find the rate at which the distance between the points change, we need to know the local values of the six components of the rate of deformation—that is, d_{xx}, d_{yy}, d_{zz}, d_{xy}, d_{yz}, d_{zx}.

Example 1.7.1   Rigid-body rotation
Find the rate of deformation for rigid-body rotation as given by the velocity field $(-y\Omega,\ x\Omega,\ 0)$.

   Solution. From the definition of rate of deformation, $d_{xx} = d_{yy} = d_{zz} = 0$, $d_{xy} = d_{yz} = d_{xz} = 0$. The absence of rates of deformation confirms that the fluid is behaving as a rotating rigid body.

———————

Example 1.7.2   Vortex motion
Find the rate of deformation for a line vortex with velocity $\mathbf{v} = \left(\dfrac{-yG}{x^2 + y^2}, \dfrac{xG}{x^2 + y^2}, 0 \right)$.

   Solution. Again from the definition,

$$d_{xx} = -d_{yy} = \frac{2xyG}{(x^2 + y^2)^2}, \quad d_{zz} = 0, \quad d_{xy} = \frac{xyG}{(x^2 + y^2)^2}, \quad d_{xz} = d_{yz} = 0.$$

———————

1.8   Constitutive Relations

Although the physical laws presented so far are of general validity, very little can be said concerning the behavior of any substance, as can be ascertained by noting that at this point there are many more unknown quantities than there are equations. The "missing" relations are those that describe the connection between how a material is made up, or constituted, and the relation between stress and the geometric and thermodynamic variables. Hooke's law and the state equations of an ideal gas are two familiar examples of *constitutive equations*. Whereas in a few cases constitutive relations can be derived from statistical mechanics considerations using special mathematical models for the molecular structure, the usual procedure is to decide, based on experiments, which quantities must go into the constitutive equation and then formulate from these a set of equations that agree with fundamental ideas, such as invariance with respect to the observer and the like.

   There is much inventiveness in proceeding along the path of determining constitutive equations, and much attention has been given to the subject in past decades. The mental process of generating a description of a particular fluid involves a continuous interchange between theory and practice. Once a constitutive model is proposed, mathematical predictions can be made that then, it is hoped, can be compared with the experiment. Such a procedure can show a model to be wrong, but it cannot guarantee that it will always be correct, since many models can predict the same velocity field for the very simple flows used in *viscometry* (tests used in measuring the coefficient of viscosity) and *rheogoniometry* (measuring the properties of complicated molecular structures such as polymers). As an example, if a fluid is contained between two large plane sheets placed parallel to each other, and the sheets are allowed to move parallel to one another at different velocities, many constitutive models will predict a fluid velocity that varies linearly with the distance from one plate, with all models giving the same shear stress. The various models, however, usually predict quite different normal stresses.

The familiar model presented by Newton and elaborated on by Navier and Stokes has withstood many of these tests for fluids of relatively simple molecular structure and holds for many of the fluids that one normally encounters. This model is not valid for fluids such as polymers, suspensions, or many of the fluids encountered in the kitchen, such as cake batter, catsup, and the like. A good rule of thumb is that if the molecular weight of the fluid is less than a half million or so, and if the distance between molecules is not too great (as in rarefied gases), a Newtonian model is very likely to be valid.

We will not delve deeper into a rigorous justification of a particular constitutive equation here. We simply put down the minimum requirements that we expect of our constitutive law and give a partial justification of the results.

Considering a fluid of simple molecular structure, such as water or air, experience and many experiments suggest the following:

1. Stress will depend explicitly only on pressure and the rate of deformation. Temperature can enter only implicitly through coefficients such as viscosity.

2. When the rate of deformation is identically zero, all shear stresses vanish, and the normal stresses are each equal to the negative of the pressure.

3. The fluid is *isotropic*. That is, the material properties of a fluid at any given point are the same in all directions.

4. The stress must depend on rate of deformation in a linear manner, according to the original concepts of Newton.

The most general constitutive relation satisfying all of the above requirements is

$$\tau_{xx} = -p + \mu' \nabla \cdot \mathbf{v} + 2\mu d_{xx}, \quad \tau_{yy} = -p + \mu' \nabla \cdot \mathbf{v} + 2\mu d_{yy}, \quad \tau_{zz} = -p + \mu' \nabla \cdot \mathbf{v} + 2\mu d_{zz},$$

$$\tau_{xy} = \tau_{yx} = 2\mu d_{xy}, \quad \tau_{xy} = \tau_{yx} = 2\mu d_{xz}, \quad \tau_{yz} = \tau_{zy} = 2\mu d_{yz}, \tag{1.8.1}$$

where we have used the abbreviation

$$d_{xx} + d_{yy} + d_{zz} = \frac{\partial v_x}{\partial x} + \frac{\partial v_y}{\partial y} + \frac{\partial v_z}{\partial z} = \nabla \cdot \mathbf{v}.$$

Here, μ is the *viscosity* and μ' is the *second viscosity coefficient*. Both of these viscosities can depend on temperature and even pressure. The fluid described by equation (1.8.1) is called a *Newtonian fluid*, although the term *Navier-Stokes fluid* is also used.

Up until now, pressure has deliberately been left undefined. The definition of pressure varies in different instances. For instance, in elementary thermodynamics texts, the term *pressure* is commonly used for the negative of mean normal stress. Summing our constitutive equations gives

$$\text{Mean normal stress} - \frac{\text{sum of the total normal stress components}}{3} = -p + \frac{\tau_{xx} + \tau_{yy} + \tau_{zz}}{3}.$$

From equation (1.8.1), however, we see that

$$\frac{\tau_{xx} + \tau_{yy} + \tau_{zz}}{3} = \left(\mu' + \frac{2\mu}{3} \right) \nabla \cdot \mathbf{v}. \tag{1.8.2}$$

The coefficient $\mu' + \frac{2\mu}{3}$ is called the *bulk viscosity*, or *volume viscosity*, since it represents the amount of normal stress change needed to get a unit specific volume rate change. The second law of thermodynamics can be used to show that $\mu' + 2\mu/3 \geq 0$. If we are to have the mean normal stress equal to the negative of pressure, it follows that the bulk viscosity must be zero. At one time Stokes suggested that this might in general be true but later wrote that he never put much faith in this relationship. Since for most flows the term is numerically much smaller than p, Stokes's assumption is still widely used.

Usually one thinks of p as being the thermodynamic pressure, given by an equation of state (e.g., $p = \rho RT$ for an ideal gas). In such a case, the bulk viscosity is not necessarily zero, and so the thermodynamic pressure generally must differ from the mean normal stress. Values of μ' for various fluids have been determined experimentally in flows involving very high-speed sound waves,[2] but the data is quite sparse. Statistical mechanics tells us that for a monatomic gas, the bulk viscosity is zero. In any case, in flows where both the dilatation and the bulk viscosity tend to be very small compared to pressure, the effects of the bulk viscosity can usually be neglected.

Some elaboration of the four postulates for determining our constitutive equation are in order. We have said that the only kinematic quantity appearing in the stress is the rate of deformation. What about strain or rate of rotation of fluid elements?

The type of fluid we are considering is completely without a sense of history. (One class of fluids with a sense of history of their straining is the *viscoelastic fluids*. For these fluids strain and strain rates appear explicitly in their constitutive equations.) A Newtonian fluid is only aware of the present. It cannot remember the past—even the immediate past—so strain cannot enter the model. Although such a model predicts most of the basic features of flows, it does have its disturbing aspects, such as infinite speed of propagation of information if compressibility is ignored. For most flows, however, it seems a reasonable assumption.

Our assumption that when the rate of deformation is zero the stresses reduce to the pressure is simply a reaffirmation of the principles of hydrostatics and is a basic law used for practically all materials.

The isotropy of a fluid is a realistic assumption for a fluid of simple molecular structure. If we had in mind materials made up of small rods, ellipsoids, or complicated molecular chains, all of which have directional properties, other models would be called for, and this constraint would have to be relaxed.

The linearity assumption can be justified only by experiment. The remarkable thing is that it quite often works! If we relax this point but retain all the other assumptions, the effect is to add only one term to the right side of equation (1.8.1), making stress quadratic in the rate of deformation. Also, the various viscosities could also depend on invariant combinations of the rate of strain, as well as on the thermodynamic variables. While this adds to the mathematical generality, no fluids are presently known to behave according to this second-order law.

We have already remarked that a state equation is a necessary addition to the constitutive description of our fluids. Examples frequently used are incompressibility ($D\rho/Dt = 0$) and the ideal gas law ($p = \rho RT$). Additionally, information on the heat flux and internal energy must be added to the list. Familiar laws are Fourier's law of heat conduction, where the heat flux is proportional to the negative of the temperature gradient, or

$$\mathbf{q} = -k\nabla T, \tag{1.8.3}$$

---

[2] See, for example, Lieberman (1955).

and the ideal gas relation for internal energy,

$$U = U(T). \tag{1.8.4}$$

The latter is frequently simplified further by the assumption of the internal energy depending linearly on temperature so that

$$U = U_0 + c_p(T - T_0), \tag{1.8.5}$$

c_p being the specific heat at constant density or volume.

Knowledge of the constitutive behavior of non-Newtonian fluids is unfortunately much sparser than our knowledge of Newtonian fluids. Particularly with the ever-increasing use of plastics in our modern society, the ability to predict the behavior of such fluids is of great economic importance in manufacturing processes. Although many theoretical models have been put forth over the last century, the situation in general is far from satisfactory. In principle, from a few simple experiments the parameters in a given constitutive model can be found. Then predictions of other flow geometries can be made from this model and compared with further experiments. The result more often than not is that the predictions may be valid only for a very few simple flows whose nature is closely related to the flows from which the parameters in the constitutive model were determined. There are many gaps in our fundamental understanding of these fluids.

1.9   Equations for Newtonian Fluids

Substitution of equation (1.8.1) into equation (1.6.5) gives the result

$$\rho \frac{Dv_x}{Dt} = -\frac{\partial(p - \mu' \nabla \cdot \mathbf{v})}{\partial x} + \rho g_x + \frac{\partial}{\partial x}\left[2\mu\left(\frac{\partial v_x}{\partial x}\right)\right] + \frac{\partial}{\partial y}\left[\mu\left(\frac{\partial v_x}{\partial y} + \frac{\partial v_y}{\partial x}\right)\right]$$
$$+ \frac{\partial}{\partial z}\left[\mu\left(\frac{\partial v_x}{\partial z} + \frac{\partial v_z}{\partial x}\right)\right],$$

$$\rho \frac{Dv_y}{Dt} = -\frac{\partial(p - \mu' \nabla \cdot \mathbf{v})}{\partial y} + \rho g_y + \frac{\partial}{\partial x}\left[\mu\left(\frac{\partial v_y}{\partial x} + \frac{\partial v_x}{\partial y}\right)\right] + \frac{\partial}{\partial y}\left[2\mu\left(\frac{\partial v_y}{\partial y}\right)\right]$$
$$+ \frac{\partial}{\partial z}\left[\mu\left(\frac{\partial v_y}{\partial z} + \frac{\partial v_z}{\partial y}\right)\right],$$

$$\rho \frac{Dv_z}{Dt} = -\frac{\partial(p - \mu' \nabla \cdot \mathbf{v})}{\partial z} + \rho g_z + \frac{\partial}{\partial x}\left[\mu\left(\frac{\partial v_z}{\partial x} + \frac{\partial v_x}{\partial z}\right)\right] + \frac{\partial}{\partial y}\left[\mu\left(\frac{\partial v_z}{\partial y} + \frac{\partial v_y}{\partial z}\right)\right]$$
$$+ \frac{\partial}{\partial z}\left[2\mu\left(\frac{\partial v_z}{\partial z}\right)\right]. \tag{1.9.1}$$

When ρ and μ are constant, and for incompressible flows, this simplifies greatly with the help of the continuity condition $\nabla \cdot \mathbf{v} = 0$ to the vector form

$$\rho \frac{D\mathbf{v}}{Dt} = -\nabla p + \rho \mathbf{g} + \mu \nabla^2 \mathbf{v}, \tag{1.9.2}$$

where

$$\nabla^2 \equiv \frac{\partial^2}{\partial x^2} + \frac{\partial^2}{\partial y^2} + \frac{\partial^2}{\partial z^2}$$

is called the Laplace operator.

Equation (1.9.2) can be written in component notation as

$$\rho\left(\frac{\partial v_x}{\partial t}+v_x\frac{\partial v_x}{\partial x}+v_y\frac{\partial v_x}{\partial y}+v_z\frac{\partial v_x}{\partial z}\right)=-\frac{\partial p}{\partial x}+\rho g_x+\mu\nabla^2 v_x,$$

$$\rho\left(\frac{\partial v_y}{\partial t}+v_x\frac{\partial v_y}{\partial x}+v_y\frac{\partial v_y}{\partial y}+v_z\frac{\partial v_y}{\partial z}\right)=-\frac{\partial p}{\partial y}+\rho g_y+\mu\nabla^2 v_y, \qquad (1.9.3)$$

$$\rho\left(\frac{\partial v_z}{\partial t}+v_x\frac{\partial v_z}{\partial x}+v_y\frac{\partial v_z}{\partial y}+v_z\frac{\partial v_z}{\partial z}\right)=-\frac{\partial p}{\partial z}+\rho g_z+\mu\nabla^2 v_z.$$

Either form (1.9.2) or (1.9.3) is referred to as the **Navier-Stokes equation**.

The last form (1.9.3) is the form most useful for problem solving. Form (1.9.2) is useful for physical understanding and other manipulations of our equations. The Navier-Stokes equations in non-Cartesian coordinate systems are given in the Appendix.

1.10 Boundary Conditions

To obtain a solution of the Navier-Stokes equations that suits a particular problem, it is necessary to add conditions that need to be satisfied on the boundaries of the region of interest. The conditions that are most commonly encountered are the following:

1. The fluid velocity component normal to an impenetrable boundary is always equal to the normal velocity of the boundary. If **n** is the unit normal to the boundary, then

$$\mathbf{n}\cdot(\mathbf{v}_{\text{fluid}}-\mathbf{v}_{\text{boundary}})=0 \qquad (1.10.1)$$

on the boundary. If this condition were not true, fluid would pass through the boundary. This condition must hold true even in the case of vanishing viscosity ("inviscid flows").

If the boundary is moving, as in the case of a flow with a free surface or moving body, then, with $F(\mathbf{x},\ t)=0$ as the equation of the bounding surface, (1.10.1) is satisfied if

$$\frac{DF}{Dt}=0 \text{ on the surface } F=0. \qquad (1.10.2)$$

This condition is necessary to establish that $F=0$ is a **material surface**—that is, a surface moving with the fluid that always contains the same fluid particles. An important special case of the material surface is the **free surface**, a constant pressure surface, typically the interface between a liquid and a gas.

2. Stress must be continuous everywhere within the fluid. If stress were not continuous, an infinitesimal layer of fluid with an infinitesimal mass would be acted upon by a finite force, giving rise to infinite acceleration of that layer.

At interfaces where fluid properties such as density are discontinuous, however, a discontinuity in stress can exist. This stress difference is related to the surface tension of the interface. Write the stress in the direction normal to the interface as $\boldsymbol{\tau}^{(\mathbf{n})}$ and denote the **surface tension** by σ (a force per unit length). By summing forces on an area of the interface, if the surface tensile force acts outwardly along the edge of S in a direction locally tangent to both S and C, the result is

$$\iint_S\left[\tau^{(n)}_{\text{lower fluid}}-\tau^{(n)}_{\text{upper fluid}}\right]dS=\oint\sigma\mathbf{t}\times\mathbf{n}\,ds,$$

where **t** and **n** are the unit tangent and normal vectors, respectively.

A variation of the curl theorem allows us to write the line integral as a surface integral. Letting

$$\Delta \tau^{(n)} - \tau^{(n)}_{\text{lower fluid}} - \tau^{(n)}_{\text{upper fluid}},$$

we have

$$\iint_S \Delta \boldsymbol{\tau}^{(n)} dS = \iint_S [-\nabla \sigma + \mathbf{n}(\mathbf{n} \cdot \nabla \sigma + \sigma \nabla \cdot \mathbf{n})] dS = 0. \qquad (1.10.3)$$

But $\mathbf{n} \cdot \nabla \sigma$ is zero, since σ is defined only on the surface S and $\mathbf{n} \cdot \nabla$ is perpendicular to S. Also, $\nabla \cdot \mathbf{n} = -\left(\frac{1}{R_1} + \frac{1}{R_2}\right)$, where R_1 and R_2 are the principal radii of curvature of the surfaces. (For information on this, see McConnell (1957), pp. 202–204.) By taking components of this equation in directions locally normal and tangential to the surface, equation (1.10.3) can be conveniently split into

$$\mathbf{n} \cdot \nabla \boldsymbol{\tau}^{(n)} = \sigma \left(\frac{1}{R_1} + \frac{1}{R_2}\right),$$

$$\mathbf{t} \cdot \nabla \boldsymbol{\tau}^{(n)} = -\mathbf{t} \cdot \nabla \sigma, \qquad (1.10.4)$$

where $\mathbf{n}$ is the unit normal and $\mathbf{t}$ is a unit tangent to the surface. In words, if surface tension is present, the difference in normal stress is proportional to the local surface curvature. If gradients in the surface tension can exist, shear stress discontinuities can also be present across an interface.

3. Velocity must be continuous everywhere. That is, in the interior of a fluid, there can be no discrete changes in $\mathbf{v}$. If there were such changes, it would give rise to discontinuous deformation gradients and, from the constitutive equations, result in discontinuous stresses.

The velocity of most fluids at a solid boundary must have the same velocity tangential to the boundary as the boundary itself. This is the *"no-slip" condition* that has been observed over and over experimentally. The molecular forces required to peel away fluid from a boundary are quite large, due to molecular attraction of dissimilar molecules. The only exceptions observed to this are in extreme cases of rarified gas flow, when the continuum concept is no longer completely valid.[3]

1.11 Vorticity and Circulation

Any motion of a small region of a fluid can be thought of as a combination of translation, rotation, and deformation. *Translation* is described by velocity of a point. *Deformation* is described by rates of deformation, as in Section 1.7. Here, we consider the rotation of a fluid element.

In rigid-body mechanics, the concept of angular rotation is an extremely important one—and rather intuitive. Along with translational velocity, it is one of the basic descriptors of the motion. In fluid mechanics we can introduce a similar concept in the following manner.

---

[3] Those interested in the history of this once-controversial condition should read the note at the end of Goldstein's *Modern Developments in Fluid Dynamics* (1965) for an interesting account.

Consider again the two-dimensional picture shown in Figure 1.7.1. In Section 1.7, we saw that the instantaneous rates of rotation of lines AB and BC were $\dot{\theta}_1$ and $\dot{\theta}_2$, given by equations (1.7.5) and (1.7.6).

Since equations (1.7.5) and (1.7.6) generally will differ, it is seen that the angular velocity of a line depends on the initial orientation of that line. To develop our analogy of angular velocity, we want a definition that is independent of orientation and direction at a point, and depends only on local conditions at the point itself. In considering the transformation of ABC into A′B′C′, it is seen that two things have happened: The angle has changed, or deformed, by an amount $d\theta_1 + d\theta_2$, and the bisector of the angle ABC has rotated an amount $0.5(d\theta_1 - d\theta_2)$. Considering only this rotation, the rate of rotation of the bisector is seen to be

$$\frac{1}{2}\left(\frac{\partial v_y}{\partial x} - \frac{\partial v_x}{\partial y}\right) \tag{1.11.1}$$

Writing the curl of **v** in Cartesian coordinates, we have

$$\text{curl } \mathbf{v} = \nabla \times \mathbf{v} = \mathbf{i}\left(\frac{\partial v_z}{\partial y} - \frac{\partial v_y}{\partial z}\right) + \mathbf{j}\left(\frac{\partial v_x}{\partial z} - \frac{\partial v_z}{\partial x}\right) + \mathbf{k}\left(\frac{\partial v_y}{\partial x} - \frac{\partial v_x}{\partial y}\right).$$

We see by comparison that equation (1.11.1) is one-half the z component of the curl of **v**.

If we were to consider similar neighboring points in the yz and xz planes, we would find that similar arguments would yield the x and y components of one-half the curl of the velocity. We therefore define the ***vorticity vector*** as being the curl of the velocity—that is,

$$\omega = \text{curl } \mathbf{v} = \nabla \times \mathbf{v}, \tag{1.11.2}$$

and note that the vorticity is twice the local angular rotation of the fluid. (Omitting the one-half from the definition saves us some unimportant arithmetic and does not obscure the physical interpretation of the concept.) Since vorticity is a vector, it is independent of the coordinate frame used. Our definition agrees with the usual "right-hand-rule" sign convention of angular rotation.

Vorticity also can be represented as a second-order tensor. Writing

$$\omega_{xx} = \omega_{yy} = \omega_{zz} = 0, \qquad \omega_{xy} = \frac{1}{2}\left(\frac{\partial v_x}{\partial y} - \frac{\partial v_y}{\partial x}\right),$$

$$\omega_{xz} = \frac{1}{2}\left(\frac{\partial v_x}{\partial z} - \frac{\partial v_z}{\partial x}\right), \qquad \omega_{yz} = \frac{1}{2}\left(\frac{\partial v_y}{\partial z} - \frac{\partial v_z}{\partial y}\right), \tag{1.11.3}$$

note that $\nabla \times \mathbf{v} = 2\left(\omega_{zy}\mathbf{i} + \omega_{xz}\mathbf{j} + \omega_{yx}\mathbf{k}\right)$. Thus, the vector and the second-order tensor contain the same information.

Since interchanging the indices in the definitions in equation (1.11.3) only changes the sign, the vorticity (second-order) tensor is said to be ***skew-symmetric***.

If we represent the rate of deformation and vorticity in index notation, we can summarize some of our findings as follows:

$$d_{ij} = \frac{1}{2}\left(\frac{\partial v_i}{\partial x_j} + \frac{\partial v_j}{\partial x_i}\right) = d_{ji}, \qquad \omega_{ij} = \frac{1}{2}\left(\frac{\partial v_i}{\partial x_j} - \frac{\partial v_j}{\partial x_i}\right) = -\omega_{ji},$$

$$d_{ij} + \omega_{ij} = \frac{\partial v_i}{\partial x_j}. \tag{1.11.4}$$

Flows with vorticity are said to be **rotational flows**; flows without vorticity are said to be **irrotational flows**. Note for later use that, from a well-known vector identity, it follows that

$$\mathrm{div}\boldsymbol{\omega} = \nabla \cdot (\nabla \times \mathbf{v}) = 0. \qquad (1.11.5)$$

You might wonder whether vorticity should enter the constitutive equation along with the rates of deformation. What has been called the "principle of material objectivity" or "principle of isotropy of space" or "material frame indifference," among other things, states that all observers, regardless of their frame of reference (inertial or otherwise), must observe the same material behavior. Therefore, an observer stationed on a rotating platform, for example, sees the same fluid behavior as an observer standing on the floor of the laboratory. As we have seen in the last section, vorticity is not satisfactory in this regard in that it is sensitive to rigid rotations. (If you find the idea of material frame indifference unsettling, see Truesdell (1966), page 6. Truesdell was one of the first expounders of this concept, but he apparently had many doubts on the same question initially.) Many constitutive equations that have been proposed violate this principle (both intentionally and unintentionally). Present work in constitutive equations tends to obey the principle religiously, although doubts are still sometimes expressed.

Example 1.11.1   Rigid body rotation.
Find the vorticity associated with the velocity field $\mathbf{v} = (-y\Omega, x\Omega, 0)$.

Solution. This is the case of rigid body rotation, with the speed being given by Ω times the distance of the point from the origin and the stream lines being concentric circles. Taking the curl of the velocity we have

$$\boldsymbol{\omega} = \nabla \times (-y\Omega\, \mathbf{i} + x\Omega\, \mathbf{j}) = 2\Omega\, \mathbf{k}.$$

Thus, the flow everywhere has the same vorticity.

———————

Example 1.11.2   Vortex motion
Find the vorticity associated with the velocity field $\mathbf{v} = \left(\dfrac{-yG}{x^2+y^2}, \dfrac{xG}{x^2+y^2}, 0 \right)$.

Solution. This is a velocity field again with streamlines that are concentric circles but with the speed now being proportional to the reciprocal distance from the origin. This flow is called a **line vortex**. Taking the curl of this velocity, we have

$$\boldsymbol{\omega} = \nabla \times \left(\frac{-yG}{x^2+y^2}\mathbf{i} + \frac{xG}{x^2+y^2}\mathbf{j} \right) = 0.$$

We see that the vorticity is zero, except perhaps at the origin, where the derivatives become infinite, and a more careful examination must be made.

———————

The preceding two examples point out what vorticity is and what it is not. In both flows, the streamlines are concentric circles, and a fluid particle travels around the origin of the coordinate system. In the rigid-body rotation case, particles on two neighboring streamlines travel at slightly different velocities, the particle on the streamline with the greater radius having the greater velocity. An arrow connecting the two particles on different streamlines will travel around the origin, as shown in Figure 1.11.1.

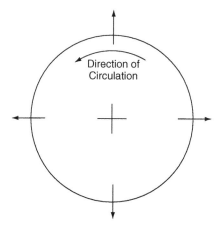

Figure 1.11.1 Rigid body rotation

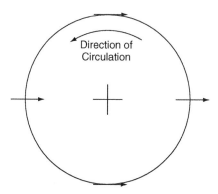

Figure 1.11.2 Line vortex

Considering a similar pair of neighboring streamlines in the line vortex case, the outer streamline has a slower velocity and the outer particle will lag behind the inner one (Figure 1.11.2). An arrow connecting the two particles in the limit of zero distance will always point in the same direction, much as the needle of a compass would. It is this rotation that vorticity deals with and not the rotation of a point about some arbitrary reference point, such as the origin.

Since vorticity is a vector, many of the concepts encountered with velocity and stream functions can be carried over. Thus, *vortex lines* can be defined as being lines instantaneously tangent to the vorticity vector, satisfying the equations

$$\frac{dx}{\omega_x} = \frac{dy}{\omega_y} = \frac{dz}{\omega_z}. \tag{1.11.6}$$

Vortex sheets are surfaces of vortex lines lying side by side. *Vortex tubes* are closed vortex sheets with vorticity entering and leaving through the ends of the tube.

Analogous to the concept of volume flow through an area, $\iint_S \mathbf{v} \cdot d\mathbf{A}$, the vorticity flow through an area, termed *circulation*, is defined as

$$\text{circulation} = \Gamma = \oint_C \mathbf{v} \cdot d\mathbf{s} = \iint_S \boldsymbol{\omega} \cdot d\mathbf{S}. \tag{1.11.7}$$

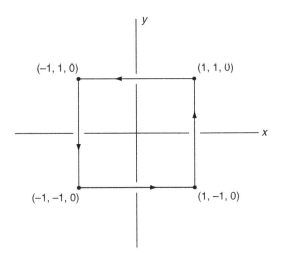

Figure 1.11.3 Calculation of circulation about a square

The relation between the line and surface integral forms follow from Stokes theorem, where C is a closed path bounding the area S.

Example 1.11.3   Circulation for a rigid rotation
Find the circulation through the square with corners at $(+1, +1, 0), (-1, +1, 0),$ $(-1, -1, 0), (+1, -1, 0)$ for rigid-body rotation flow shown in Figure 1.11.3.

Solution. Starting first with the line integral form of the definition in equation (1.11.7), we see that

$$\Gamma = \oint_C v \cdot ds = \int_{+1}^{-1} v_x|_{y=1}\, dx + \int_{+1}^{-1} v_y|_{x=-1}\, dx + \int_{+1}^{+1} v_x|_{y=-1}\, dx + \int_{+1}^{+1} v_y|_{x=1}\, dy$$

$$= \int_{+1}^{-1} -\Omega dx + \int_{+1}^{-1} -\Omega dx + \int_{+1}^{+1} \Omega dx + \int_{+1}^{+1} \Omega dy$$

$$= -\Omega(-2) - \Omega(-2) + \Omega(2) + \Omega(2) = 8\Omega.$$

Of course, we could have obtained this result much faster by using the area integral form of the definition (1.11.7),

$$\Gamma = \iint_S \boldsymbol{\omega} \cdot d\mathbf{A}.$$

Since the integrand is constant (2Ω) and the area is 4, the result for the circulation follows from a simple arithmetic multiplication.

Example 1.11.4   Circulation for a vortex motion
Using the square given in Example 1.11.3, find the circulation for the vortex flow given in Example 1.5.2—that is, $\mathbf{v} = \left(\frac{-yG}{x^2+y^2}, \frac{xG}{x^2+y^2}, 0 \right).$

Solution. In this case, we cannot easily use the area form of the definition, since the vorticity is not defined at the origin. The line integral form of equation (1.11.7) gives us

$$\Gamma = \oint_C \mathbf{v} \cdot d\mathbf{s} = \int_1^{-1} \frac{-G}{x^2+1} \, dx + \int_1^{-1} \frac{-G}{1+y^2} \, dy + \int_{-1}^{1} \frac{G}{x^2+1} \, dx + \int_{-1}^{1} \frac{G}{1+y^2} dy$$

$$= -G\left(\frac{-1}{4} - \frac{1}{4}\right) - G\left(\frac{-1}{4} - \frac{1}{4}\right) + G\left(\frac{1}{4} + \frac{1}{4}\right) + G\left(\frac{1}{4} + \frac{1}{4}\right) = 2G.$$

If we take any path that does not include the origin, we could in fact use the area form of the definition, and we would conclude that the circulation about that path was zero. Any path containing the origin would have circulation $2\pi G$. Therefore, we say that the vorticity at the origin is infinite for the line vortex, being infinite in such a manner that the infinite vorticity times the zero area gives a finite, nonzero, value for the circulation.

1.12 The Vorticity Equation

Differential equations governing the change of vorticity can be formed from the Navier-Stokes equations. Dividing equation (1.9.2) by the mass density and then taking the curl of the equation the result after some manipulation and use of the continuity equation is

$$\frac{D\boldsymbol{\omega}}{Dt} = (\boldsymbol{\omega} \cdot \nabla)\mathbf{v} - \nabla\left(\frac{1}{\rho}\right) \times \nabla p + \frac{\mu}{\rho}\nabla^2\boldsymbol{\omega}. \tag{1.12.1}$$

The right-hand side of equation (1.12.1) tells us that as we follow a fluid particle, there are three mechanisms by which its vorticity can change. The first term, $(\boldsymbol{\omega} \cdot \nabla)\mathbf{v}$, is vorticity change due to *vortex line stretching*. The operator $\boldsymbol{\omega} \cdot \nabla$ is the magnitude of the vorticity times the derivative in the direction of the vortex line. Consequently, if the velocity vector changes along the vortex line (thus "stretching" the vortex line), there will be a contribution to the change of vorticity. The second term says that unless the pressure gradient and the density gradient are aligned so that they are parallel to one another, the local vorticity will be changed by the density gradient. The third term says that vorticity will be diffused by viscosity.

Some further insights into the nature of vorticity and circulation can be gained by considering several theorems introduced first by Helmholtz. The first states the following:

The circulation taken over any cross-sectional area of a vortex tube is a constant.

The proof is simple. Use a vortex tube segment with ends S_1 and S_2, and side surface S_0 as shown in Figure 1.11.4. By one of Green's theorems,

$$\iint_{S_0} \boldsymbol{\omega} \cdot d\mathbf{A} + \iint_{S_1} \boldsymbol{\omega} \cdot d\mathbf{A} + \iint_{S_2} \boldsymbol{\omega} \cdot d\mathbf{A} = \iiint_V \nabla \cdot \boldsymbol{\omega} dV = 0$$

by virtue of equation (1.11.2). On S_0, vorticity is normal to the surface, so the integral is zero. Then

$$\iint_{S_1} \boldsymbol{\omega} \cdot d\mathbf{A} + \iint_{S_2} \boldsymbol{\omega} \cdot d\mathbf{A} = 0,$$

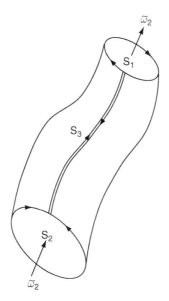

Figure 1.11.4   Vortex tube

from which it follows that

$$\iint_{S_2} \boldsymbol{\omega} \cdot d\mathbf{A} = -\iint_{S_1} \boldsymbol{\omega} \cdot d\mathbf{A}.$$

With the proper interpretation of signs of the outward normal, this proves the theorem.

A very important corollary of this theorem follows:

Vortex lines can neither originate nor terminate in the interior of the flow. Either they are closed curves (e.g., smoke rings) or they originate at the boundary.

Notice in this proof we did not use any dynamical information. Only kinematics and definitions were used.

Another useful theorem has to do with the rate of change of circulations. Starting with the line integral definition of circulation and noting that $D d\mathbf{s}/Dt = d\mathbf{v}$, then

$$\frac{D\Gamma}{Dt} = \frac{D}{Dt} \oint_C \mathbf{v} \cdot d\mathbf{s} = \oint_C \left(\left(\frac{D\mathbf{v}}{Dt}\right) \cdot d\mathbf{s} + \mathbf{v} \cdot d\mathbf{v} \right) = \oint_C \left(\frac{D\mathbf{v}}{Dt} + \frac{\nabla(\mathbf{v} \cdot \mathbf{v})}{2} \right) \cdot d\mathbf{s}.$$

The second term in the last integral is an exact differential, so integrating it around a closed path gives zero. Thus, the rate of change of circulation is given by

$$\frac{D\Gamma}{Dt} = \oint_C \frac{D\mathbf{v}}{Dt} \cdot d\mathbf{s}, \qquad (1.12.2a)$$

or, with the help of the Navier-Stokes equations,

$$\frac{D\Gamma}{Dt} = \oint_C \left(-\frac{1}{\rho}\nabla p + \frac{\mu}{\rho}\nabla^2 \mathbf{v} \right) \cdot d\mathbf{s}. \qquad (1.12.2b)$$

Here, the body force terms vanish because they are an exact differential, so an integration around a closed path gives zero. Therefore, as we follow a curve C drawn in the flow as it is carried along with a flow, the pressure gradient can change the circulation associated with the curve only if $\frac{1}{\rho}\nabla p \cdot d\mathbf{s}$ is not an exact differential. Thus, pressure gradients must be accompanied by density variations to affect the circulation.

The understanding of the nature of vorticity is extremely important to the study of fluid mechanics. Vorticity plays a central role in the transition from laminar to turbulent flow. It can affect the performance of pumps and the lift of airfoils. Being able to control vorticity, either by causing it or by eliminating it, is at the core of many engineering problems.

1.13   The Work-Energy Equation

Another useful equation derived from the Navier-Stokes equations is a work-energy statement. If we take the dot product of equation (1.9.1) with the velocity, with the help of the product rule of calculus we have

$$
\begin{aligned}
\rho\frac{D}{Dt}\left(\frac{\mathbf{v}\cdot\mathbf{v}}{2}\right) = {}& \rho\mathbf{v}\cdot\mathbf{g} + \mathbf{v}\cdot\left[\mathbf{i}\left(\frac{\partial\tau_{xx}}{\partial x}+\frac{\partial\tau_{yx}}{\partial y}+\frac{\partial\tau_{zx}}{\partial z}\right)\right.\\
& \left. +\,\mathbf{j}\left(\frac{\partial\tau_{xy}}{\partial x}+\frac{\partial\tau_{yy}}{\partial y}+\frac{\partial\tau_{zy}}{\partial z}\right)+\mathbf{k}\left(\frac{\partial\tau_{xz}}{\partial x}+\frac{\partial\tau_{yz}}{\partial y}+\frac{\partial\tau_{zz}}{\partial z}\right)\right]\\
= {}& \rho\mathbf{v}\cdot\mathbf{g} + \left[\frac{\partial(v_x\tau_{xx})}{\partial x}+\frac{\partial(v_x\tau_{yx})}{\partial y}+\frac{\partial(v_x\tau_{zx})}{\partial z}\right]\\
& +\left[\frac{\partial(v_y\tau_{xy})}{\partial x}+\frac{\partial(v_y\tau_{yy})}{\partial y}+\frac{\partial(v_y\tau_{zy})}{\partial z}\right]+\left[\frac{\partial(v_z\tau_{xz})}{\partial x}+\frac{\partial(v_z\tau_{yz})}{\partial y}+\frac{\partial(v_z\tau_{zz})}{\partial z}\right]\\
& -\left(\tau_{xx}\frac{\partial v_x}{\partial x}+\tau_{yx}\frac{\partial v_x}{\partial y}+\tau_{zx}\frac{\partial v_x}{\partial z}\right)-\left(\tau_{xy}\frac{\partial v_y}{\partial x}+\tau_{yy}\frac{\partial v_y}{\partial y}+\tau_{zy}\frac{\partial v_y}{\partial z}\right)\\
& -\left(\tau_{xz}\frac{\partial v_z}{\partial x}+\tau_{yz}\frac{\partial v_z}{\partial y}+\tau_{zz}\frac{\partial v_z}{\partial z}\right)\\
= {}& \rho\mathbf{v}\cdot\mathbf{g} + \frac{\partial}{\partial x}\left(v_x\tau_{xx}+v_y\tau_{xy}+v_z\tau_{xz}\right)\\
& +\frac{\partial}{\partial y}\left(v_x\tau_{yx}+v_y\tau_{yy}+v_z\tau_{yz}\right)+\frac{\partial}{\partial z}\left(v_x\tau_{zx}+v_y\tau_{zy}+v_z\tau_{zz}\right)\\
& -\left[\tau_{xx}\frac{\partial v_x}{\partial x}+\tau_{yx}\left(\frac{\partial v_x}{\partial y}+\frac{\partial v_y}{\partial x}\right)+\tau_{zx}\left(\frac{\partial v_x}{\partial z}+\frac{\partial v_z}{\partial x}\right)\right.\\
& \left. +\,\tau_{yy}\frac{\partial v_y}{\partial y}+\tau_{zy}\left(\frac{\partial v_y}{\partial z}+\frac{\partial v_z}{\partial y}\right)+\tau_{zz}\frac{\partial v_z}{\partial z}\right]\\
= {}& (p-\mu'\nabla\cdot\mathbf{v})\nabla\mathbf{v} + \rho\mathbf{v}\cdot\mathbf{g}+\frac{\partial}{\partial x}\left(v_x\tau_{xx}+v_y\tau_{xy}+v_z\tau_{xz}\right)\\
& +\frac{\partial}{\partial y}\left(v_x\tau_{yx}+v_y\tau_{yy}+v_z\tau_{yz}\right)+\frac{\partial}{\partial z}\left(v_x\tau_{zx}+v_y\tau_{zy}+v_z\tau_{zz}\right)-\Phi,
\end{aligned}
$$

$$\text{(1.13.1)}$$

where

$$\Phi = \tau_{xx}\frac{\partial v_x}{\partial x} + \tau_{yx}\left(\frac{\partial v_x}{\partial y} + \frac{\partial v_y}{\partial x}\right) + \tau_{zx}\left(\frac{\partial v_x}{\partial z} + \frac{\partial v_z}{\partial x}\right) + \tau_{yy}\frac{\partial v_y}{\partial y}$$

$$+ \tau_{zy}\left(\frac{\partial v_y}{\partial z} + \frac{\partial v_z}{\partial y}\right) + \tau_{zz}\frac{\partial v_z}{\partial z} + (p - \mu'\nabla \cdot \mathbf{v})\,\nabla \cdot \mathbf{v} \qquad (1.13.2)$$

$$= \tau_{xx}d_{xx} + \tau_{yy}d_{yy} + \tau_{zz}d_{zz} + 2\left(\tau_{yx}d_{yx} + \tau_{zx}d_{zx} + \tau_{zy}d_{zy}\right) + (p - \mu'\nabla \cdot \mathbf{v})\,\nabla \cdot \mathbf{v}$$

$$= 2\mu\left[d_{xx} + d_{yy} + d_{zz} + 2\left(d_{xy}^2 + d_{yz}^2 + d_{zx}^2\right)\right].$$

The function Φ represents the rate of dissipation of energy by viscosity and is called the ***dissipation function***. Since the viscosity is always positive, the quantity Φ is positive definite. That is, no matter what the velocity field is, Φ is greater or equal to zero.

Note that in deriving (1.13.1) we used only the Navier-Stokes equations—that is, Newton's law. We next combine (1.13.1) with the ***first law of thermodynamics***.[4]

1.14   The First Law of Thermodynamics

The conservation of energy principal (first law of thermodynamics) in its rate form states that the rate of change of energy of the system is equal to the rate of heat addition to the system due to conduction from the surroundings, radiation, and internal reactions plus the rate at which work is done on the system. In equation form, this is

$$\frac{dE}{dt} = \frac{dQ}{dt} + \frac{dW}{dt}. \qquad (1.14.1)$$

For the rate of energy change we have

$$\frac{dE}{dt} = \iiint_V \frac{\partial(\rho e)}{\partial t}\,dV + \iint_S \rho e\mathbf{v} \cdot d\mathbf{S}, \qquad (1.14.2)$$

where e is the ***specific energy*** (energy per unit mass), given by

$$e = \frac{\mathbf{v} \cdot \mathbf{v}}{2} + u,$$

with u being the specific internal energy. Also, write

$$\frac{dQ}{dt} = -\iint_S \mathbf{q} \cdot d\mathbf{S} + \iiint_V \frac{dr}{dt}\,dV, \qquad (1.14.3)$$

where $\mathbf{q}$ is the heat flux vector represent heat transfer from the surroundings. The body term dr/dt represents heat generated either internally or transferred by radiation. Often Fourier's law of conductivity is used to relate the heat flux vector to the temperature. This law states that $\mathbf{q} = -k\nabla T$, T being the temperature and k the coefficient of thermal conduction.

---

[4] The idea of conservation of energy was first published by Émilie du Châtelet (1706–1749), a physicist and mathematician who made the first translation into French (along with her commentary) of Newton's *Principia Mathematica*. Her book *Lessons in Physics* was written for her 13-year-old son. It elaborated on and advanced the leading scientific ideas of the time.

The rate at which work is being done by the various forces can be written as

$$\frac{dW}{dt} = \iiint_V \rho \mathbf{g} \cdot \mathbf{v} \, dV + \iint_S \mathbf{v} \cdot \boldsymbol{\tau}^{(n)} dS. \tag{1.14.4}$$

Putting these expressions into equation (1.14.1), we have

$$\iiint_V \frac{\partial(\rho e)}{\partial t} dV + \iint_S \rho e \mathbf{v} \cdot d\mathbf{S}$$

$$= -\iint_S \mathbf{q} \cdot d\mathbf{S} + \iiint_V \frac{dr}{dt} dV + \iiint_V \rho \mathbf{g} \cdot \mathbf{v} dV + \iint_S \mathbf{v} \cdot \boldsymbol{\tau}^{(n)} dS.$$

With the help of the divergence theorem, this becomes

$$\iiint_V \left[\frac{\partial(\rho e)}{\partial t} + \nabla \cdot (\rho e \mathbf{v}) \right] dV$$

$$= \iiint_V \left[-\nabla \cdot \mathbf{q} + \frac{dr}{dt} + \rho \mathbf{g} \cdot \mathbf{v} + \frac{\partial}{\partial x}(v_x \tau_{xx} + v_y \tau_{yx} + v_z \tau_{zx}) \right.$$

$$\left. + \frac{\partial}{\partial y}(v_x \tau_{xy} + v_y \tau_{yy} + v_z \tau_{zy}) + \frac{\partial}{\partial z}(v_x \tau_{xz} + v_y \tau_{yz} + v_z \tau_{zz}) \right] dV.$$

Again, the choice of the control volume is arbitrary, so the integrand on the right-hand side of the equation must equal the integrand on the left. The result is

$$\frac{\partial(\rho e)}{\partial t} + \nabla \cdot (\rho e \mathbf{v}) = \rho \frac{De}{Dt} + e \frac{D\rho}{Dt}$$

$$= -\nabla \cdot \mathbf{q} + \frac{dr}{dt} + \rho \mathbf{g} \cdot \mathbf{v} + \frac{\partial}{\partial x}(v_x \tau_{xx} + v_y \tau_{yx} + v_z \tau_{zx})$$

$$+ \frac{\partial}{\partial y}(v_x \tau_{xy} + v_y \tau_{yy} + v_z \tau_{zy}) + \frac{\partial}{\partial z}(v_x \tau_{xz} + v_y \tau_{yz} + v_z \tau_{zz}). \tag{1.14.5}$$

Since many of the viscous terms on the right-hand side of equation (1.14.5) are also in equation (1.13.1), that equation can be used to eliminate terms here. The result after subtracting equation (1.13.1) from equation (1.14.5) is

$$\rho \frac{De}{Dt} + e \frac{D\rho}{Dt} - \rho \frac{D(\frac{\mathbf{v} \cdot \mathbf{v}}{2})}{Dt} = -\nabla \cdot \mathbf{q} + \frac{dr}{dt} - (p - \mu' \nabla \cdot \mathbf{v}) \nabla \cdot \mathbf{v} + \Phi.$$

The kinetic energies cancel and we are left with

$$\rho \frac{Du}{Dt} = -\nabla \cdot \mathbf{q} + \frac{dr}{dt} + \Phi \, (-p + \rho e + \mu' \nabla \cdot \mathbf{v}) \nabla \cdot \mathbf{v}. \tag{1.14.6}$$

In words, this equation states that the internal energy of the fluid will be changed by the addition of heat transfer from the surroundings (first term), heat generated internally (second term), viscosity (third term), and compressibility effects (fourth term). For incompressible flow, the fourth term on the right side is zero by virtue of the continuity equation. Equation (1.14.6) can then be rewritten as

$$\rho \frac{Du}{Dt} = -\nabla \cdot \mathbf{q} + \frac{dr}{dt} + \Phi. \tag{1.14.7}$$

These represent the thermodynamic portion of the first law. The thermodynamic field quantities are thus seen to be coupled to the mechanical portion through the convective change of the internal energy, the viscous dissipation, and the pressure.

1.15   Dimensionless Parameters

The development of fluid mechanics over the years has relied on both experimentation and analysis, with the former leading the way in many cases. To express data in the most useful form, dimensional analysis was used extensively. It is also extremely useful in analysis. The number of such parameters introduced from time to time in the literature is certainly in the hundreds, if not thousands. There are a few, however, that predominate in general usage. Most of them come from the Navier-Stokes equations and their boundary conditions.

The *Reynolds number*, named after Osborne Reynolds (1842–1912), a British scientist/mathematician, represents the relative importance of the convective acceleration terms in the Navier-Stokes equations to the viscous terms. Reynolds used the term in 1883 in a paper presenting his results on the transition from laminar to turbulent flow of liquids in round pipes. It is typically found in the form $\rho VD/\mu$.

The *Froude number* was named after William Froude (1810–1879), a British mathematics professor who became interested in ship construction. He started his studies of ship resistance by building scale models and then towing them in long, narrow basins he had constructed himself. Towing basins are now used extensively in the design of ships to determine the proportion of ship resistance to waves. They also are used to study wave forces on offshore drilling cables and pipelines. The Froude number (actually never used by him) represents the ratio of the convective acceleration terms in the Navier-Stokes equations to the wave forces as represented by the gravity terms. It is typically used in the form $V/\sqrt{gh}$, although the square of this is also used. Froude's ideas on model studies were initially ridiculed by his peers, but his perseverance led to their acceptance.

The *Richardson number* $V/\sqrt{\Delta\rho gh/\rho}$, named after colonel A. R. Richardson, a faculty member of the University of London, is a variant of the Froude number. It is used in studying waves in flows with density stratification. He introduced it in 1920.

The *Strouhal number* is named after C. Strouhal (1850–1922), a German physicist who studied the aeolian sounds generated by wind blowing through trees. It is an important parameter in studying the shedding of vortices and is written as $\omega D/V$ or fD/V, where f is the frequency and ω the circular frequency.

The *pressure coefficient* $\Delta p/\frac{1}{2}\rho V^2$, sometimes called the *Euler coefficient* (Leonard Euler, 1707–1783, a Swiss mathematician) is a form suited for the presentation of pressure data. It is the ratio of pressure forces to the convective acceleration.

The *drag coefficient* $F_D/\frac{1}{2}\rho V^2 A$ is used for presenting the drag force (the force in the direction of motion), where A is the projected area.

The *lift coefficient* $F_L/\frac{1}{2}\rho V^2 A$ is similar to the drag force but perpendicular to the direction of motion.

The *moment coefficient* $M/\frac{1}{2}\rho V^2 AL$ is convenient for measuring the moment on a wing or rudder.

The *Weber number* $\rho V^2 D/\sigma$ was named after Moritz Weber (1871–1951), a professor of naval mechanics at the Polytechnic Institute of Berlin. He introduced the name *similitude* to describe model studies that were scaled both geometrically and using dimensionless parameter for forces, and introduced a capillary parameter, including surface tension.

1.16   Non-Newtonian Fluids

Fluids such as large molecular weight polymers that do not obey the Newtonian constitutive equation are encountered frequently in the chemical and plastics industry. Paints, slurries, toothpastes, blood, drilling mud, lubricants, nylon, and colloids all exhibit non-Newton behavior. Many such fluids also can be found in the kitchen: catsup, suspensions of corn or rice starch in water, melted chocolate, eggwhites, mayonnaise, milk, and gelatine all are examples! They exhibit such effects as die swell when exiting a tube, climbing of a rotating rod in an otherwise still container of fluid, self-siphoning, drag reduction, and transformation into a semisolid when an electric or magnetic field is applied.

As will be seen, finding solutions to the Navier-Stokes equations is more than a sufficient challenge. For these fluids of greater constitutive complexity, we can hope to find solutions for only the very simplest flows. Such flows are called *viscometric flows* and are the flows found in simple viscometers. They are the flows involving pipe flow, flow between rotating cylinders, and flows between a rotating cone and a plate.

Section 1.8 listed the four considerations that a constitutive equation for a fluid must satisfy. The last one, the requirement that the stress-rate of deformation equation be linear in the rate of deformation, can be broadened to include quadratic terms in the rate of deformation. (There is no need to go to higher powers: The Cayley-Hamilton theorem states that powers higher than the second can be expressed in lower powers.) Thus, the most general form of constitutive equation we could propose for flows that are described by only stress and rate of deformation is given by

$$\tau_{ij} = (-p + \mu' \nabla \cdot \mathbf{v})\delta_{ij} + \mu d_{ij} + \mu'' d_{ik} d_{kj}, \qquad (1.16.1)$$

where μ'' is an additional viscosity coefficient. Note that its dimensions differ from the standard viscosity by an additional unit of time. This constitutive law describes what are called *Stokesian fluids*, after George Stokes, or sometimes *Reiner-Rivlin fluids*, after Markus Reiner of Israel and Ronald Rivlin of the United States. Because of the added difficulty that this second power term introduces into problem solution, it is fortunate that no fluids have so far been found that obey this law better than they do the first power law!

Suspensions such as paint, clays, and wood pulp solutions[5] appear to behave as if they must have a certain level of stress applied before the fluid deforms. Such fluids have been referred to as *Bingham fluids*, or sometimes *visco-plastic fluids*. The constitutive equation that has been proposed for them is

$$\tau_{ij} = \mathrm{T}_{ij} \quad \text{if } \frac{1}{2}T_{i_j}T_{jk} \le T^2, \text{ or}$$
$$\tau_{ij} = \mathrm{T}_{ij} + (-p + \nabla \cdot \mathbf{v})\delta_{ij} + \mu d_{ij} \text{ if } \quad \frac{1}{2}T_{mn}T_{mn} > T^2. \qquad (1.16.2)$$

In the latter case, $\mathrm{T}_{ij} = \frac{2T}{\sqrt{2T_{mn}T_{mn}}} d_{ij}$. Here, T_{ij} are components of the yield stress tensor and T is the yield stress according to the von Mises yield criterion.

---

[5] There is some doubt as to whether such fluids are indeed Bingham fluids. It has been claimed that the only true Bingham fluid that has been discovered is an aluminum soap dispersed in a petroleum fraction (McMillan, 1948). Controversy is not unheard of in the discussion of non-Newtonian fluids.

The non-Newtonian fluids that are more interesting in practical use are polymers. Their long chain molecules act like springs and have often been modeled as springs or dumbbells where the weights are connected by springs. Such fluids can also have memory effects, where they do not respond immediately to changes in the applied force. In viscometric flows they typically exhibit stress effects normal to the direction of flow, and precise measurement of such effects is difficult.

A substantial number of constitutive models have been introduced for such fluids (Macosko, 1994). They involve either time derivatives or integrals of the stress and/or rate of deformation. The time derivative used is a convective one, a derivative that is concerned with the relative rate of motion between two molecules, and thus is more complicated than the time derivative involved with computing acceleration from velocity. Since a number of different convected time derivatives have been introduced over the years, deciding on the "correct" one is not a simple task.

Much of the terminology used in the field is associated with simple experiments, where the substance is placed in either simple extension or simple shear. Based on such experiments, classification can be carried out as shown in Table 1.16.1.

Table 1.16.1 shows that between the states of elastic (Hookean) solids and viscous (Newtonian) fluids, there is a continuum of effects exhibited that take on many weird and wonderful forms. Applications including drug delivery, drag reduction, body armor, and automobile brakes and suspensions have been proposed and investigated. Heat and stress over time tend to degrade long chain polymers, which have made their practical application a nontrivial task.

It is fair to say that the problems found in fitting constitutive equations to such fluids is an extremely difficult one, and a one-size-fits-all solution is unlikely to be found. The best hope appears to be to fit a given constitutive relationship to a specialized set of circumstances and applications.

1.17 Moving Coordinate Systems

Occasionally, it is necessary to use moving coordinate systems to understand a given flow. For instance, the flow past an aircraft appears different to a person on the ground than it does to a person in the aircraft. The changes that are introduced when moving coordinates are used is primarily in the velocity and acceleration.

Consider first a coordinate system rotating with an angular velocity Ω_0 and whose origin translates with a velocity $\mathbf{v}_0$. Then the velocity is given by

$$\mathbf{v}_{abs} = \mathbf{v}_{rel} + \mathbf{v}_0 + \Omega_0 \times \mathbf{r}. \tag{1.17.1}$$

Here $\mathbf{v}_{abs}$ is the velocity with respect to a nonmoving axes system, $\mathbf{v}_{rel}$ is the velocity measured in the moving system, and $\mathbf{r}$ is the position in the moving system. The acceleration is given by

$$\frac{D\mathbf{v}_{abs}}{Dt} = \frac{D\mathbf{v}_{rel}}{Dt} + \frac{d\mathbf{v}_0}{dt} + \Omega_0 \times \mathbf{v}_0 + \frac{d\Omega_0}{dt} \times \mathbf{r} + \Omega_0 \times (\Omega_0 \times \mathbf{r}) + 2\Omega_0 \times \mathbf{v}_{rel}. \tag{1.17.2}$$

These follow from well-known results in dynamics. The last two terms on the right-hand side represent the centripetal and Coriolis accelerations, while the second, third and fourth represent the acceleration due to the coordinate system.

TABLE 1.16.1   Classification of non-Newtonian fluids

Type of fluid	Classification	Stress/rate of deformation behavior	Examples
Elastic solids	Hookean	Linear stress-strain relation	Most solids below the yield stress
Plastic solids	Perfectly plastic	Strain continues with no additional stress	Ductile metals stressed above the yield point
	Bingham plastic	Behaves Newtonian when threshold shear is exceeded	Iron oxide suspensions
	Visco-plastic	Yield like the Bingham plastic, but the relation between stress and rate of deformation is not linear	Drilling mud, nuclear fuel slurries, mayonnaise, toothpaste, blood
	Yield dilatant*	Dilatant when threshold shear is exceeded	
	Visco-elastic	Exhibits both viscous and elastic effects	Eggwhite, polymer melts, and solutions
Power-law fluids	Shear thinning	Apparent viscosity reduces as shear rate increases	Some colloids, clay, milk, gelatine, blood, liquid cement, molten polystyrene, polyethylene oxide in water
	Dilatant or shear thickening	Apparent viscosity increases as shear rate increases	Concentrated solutions of sugar in water, suspensions of rice or cornstarch, solutions of certain surfactants
Time-dependent viscosity	Rheopectic	Apparent viscosity increases the longer stress is applied	Some lubricants
	Thixotropic	Apparent viscosity decreases the longer stress is applied	Nondrip paints, tomato catsup
Electromagnetic	Electrorheologic	Becomes dilatant when an electric field is applied	Melted chocolate bars, single- or polycrystalline suspensions in insulating fluids
	Magnetorheologic	Becomes dilatant when a magnetic field is applied	Colloids with nanosize silica particles suspended in polyethylene glycol
Newtonian fluids		Linear stress-rate of deformation relationship	Water, air

* Dilatant here refers to shear thickening as stress increases.

This result for the acceleration can be put in a more useful form by a bit of rearranging. Consider the following:

$$\frac{D\mathbf{v}_{abs}}{Dt} = \frac{\partial \mathbf{v}_{rel}}{\partial t} + (\mathbf{v}_{rel} \cdot \nabla)\,\mathbf{v}_{rel} + \frac{d\mathbf{v}_0}{dt} + \mathbf{\Omega}_0 \times \mathbf{v}_0 + \frac{d\mathbf{\Omega}_0}{dt} \times \mathbf{r} + \mathbf{\Omega}_0 \times (\mathbf{\Omega}_0 \times \mathbf{r}) + 2\mathbf{\Omega}_0 \times \mathbf{v}_{rel}$$

$$= \frac{\partial}{\partial t}[\mathbf{v}_{rel} + \mathbf{v}_0 + \mathbf{\Omega}_0 \times \mathbf{r}] + \mathbf{\Omega}_0 \times [\mathbf{v}_{rel} + \mathbf{v}_0 + \mathbf{\Omega}_0 \times \mathbf{r}] + \mathbf{v}_{rel} \cdot \nabla[\mathbf{v}_{rel} + \mathbf{v}_0 + \mathbf{\Omega}_0 \times \mathbf{r}]$$

$$= \left(\frac{\partial}{\partial t} + \mathbf{\Omega}_0 \times + \mathbf{v}_{rel} \cdot \nabla\right)\mathbf{v}_{abs}, \tag{1.17.3}$$

since

$$(\mathbf{v}_{rel} \cdot \nabla)\,\mathbf{v}_0 = 0 \text{ and } (\mathbf{v}_{rel} \cdot \nabla)(\mathbf{\Omega}_0 \times \mathbf{r}) = \mathbf{\Omega}_0 \times \mathbf{v}_{rel}.$$

This can also be written as

$$\frac{D\mathbf{v}_{abs}}{Dt} = \frac{\partial \mathbf{v}_{abs}}{\partial t} + \mathbf{\Omega}_0 \times \mathbf{v}_{abs} + (\mathbf{v}_{abs} - \mathbf{v}_{rel} - \mathbf{\Omega}_0 \times \mathbf{r}) \cdot \nabla\mathbf{v}_{abs}. \tag{1.17.4}$$

Problems—Chapter 1

1.1 For the two-dimensional flow field defined by the velocity components $v_x = \frac{1}{1+t}$, $v_y = 1$, $v_z = 0$, find the Lagrangian representation of the paths taken by the fluid particles.

1.2 Find the acceleration at point $(1, 1, 1)$ for the velocity $\mathbf{v} = (yz + t,\ xz - t,\ xy)$.

1.3 a. Find the relationship between velocity components in cylindrical polar coordinates in terms of components in Cartesian coordinates, as well as the inverse relations. Use Figure 1.4.1.

   b. Find the relationships between velocity components in spherical polar coordinates in terms of components in Cartesian coordinates, as well as the inverse relations. Use Figure 1.4.3.

1.4 Derive the continuity equation in cylindrical coordinates by examining a control volume bounded by the following: two cylinders perpendicular to the x-y plane, the first of radius r, the second of radius $r + dr$; two planes perpendicular to the x-y plane, the first making an angle θ with the x-axis, the second an angle $\theta + d\theta$; two planes parallel to the x-y plane, the first above it an amount z, the second an amount $z + dz$.

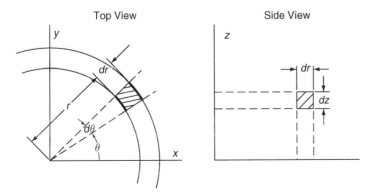

Figure P1.4 Problem 1.4—Cylindrical coordinates

1.5 a. Find the stream function for the two-dimensional incompressible flow with velocity components $\mathbf{v} = (x^2 - 2xy \cos y^2, \ -2xy + \sin y^2, \ 0)$.

b. Find the discharge per unit between the points $(1, \ \pi)$ and $(0, 0)$.

1.6 For the following flows, find the missing velocity component needed for the flow to satisfy the incompressible continuity equation.

a. $v_x = x^2 + y^2 + a^2, \quad v_y = -xy - yz - xz, \quad v_z = ?$

b. $v_x = \ln\left(y^2 + z^2\right), \quad v_y = \sin\left(x^2 + z^2\right), \quad v_z = ?$

c. $v_x = ?, \qquad v_y = \dfrac{y}{\left(x^2 + y^2 + z^2\right)^{3/2}}, \quad v_z = \dfrac{z}{\left(x^2 + y^2 + z^2\right)^{3/2}}.$

1.7 For the flow field given by $\psi = A \ln(x^2 + y^2) + yS$, find the discharge per unit width in the z direction between the points $(1, 1, 0)$ and $(-1, -1, 0)$.

1.8 Find the stream function for the two-dimensional incompressible flow with a radial velocity (cylindrical polar coordinates) given by $v_r = \frac{A}{\sqrt{r}} \cos \theta$. Also find the missing velocity component.

1.9 Find the stream function for the two-dimensional incompressible flow with velocity components given by $v_r = U\left(1 - \frac{a^2}{r^2}\right) \cos \theta, \ v_\theta = -U\left(1 + \frac{a^2}{r^2}\right) \sin \theta$.

1.10 Find the stream function for the two-dimensional incompressible flow with radial velocity component $v_r = \frac{3}{2} A r^{3/2} \cos \frac{3\theta}{2}$. Also find the missing velocity component.

1.11 Find the stream function for the velocity field $\mathbf{v} = (x^2 - 2x + 1, \ -2xy + 2y - x, 0)$.

1.12 For steady, incompressible inviscid flows with body forces neglected, it was shown by Yih (1958) that the flow can be reduced to that of a constant-density flow by the transformation

$$v_x^* = \sqrt{\frac{\rho}{\rho_0}} v_x, \ v_y^* = \sqrt{\frac{\rho}{\rho_0}} v_y, \ v_z^* = \sqrt{\frac{\rho}{\rho_0}} v_z,$$

where ρ_0 is a constant reference density. Verify this result.

1.13 Find the rates of deformation and vorticity in Cartesian coordinates for the velocity field

$$\mathbf{v} = \left(\frac{Bx}{\left(x^2 + y^2 + z^2\right)^{3/2}}, \ \frac{By}{\left(x^2 + y^2 + z^2\right)^{3/2}}, \ \frac{Bz}{\left(x^2 + y^2 + z^2\right)^{3/2}}\right).$$

1.14 a. Compute the unit normal for the inclined surface shown.

b. For the stress vector with components $\boldsymbol{\tau}_b = (10, 3, 7)$, calculate the normal and tangential components of the stress vector on the inclined surface.

1.15 Given the velocity field $\mathbf{v} = 5y\mathbf{i}$, find the circulation about a rectangle 6 units long and 4 units high, centered at the origin. Compute it two ways, once by the line integral and again using the area formula.

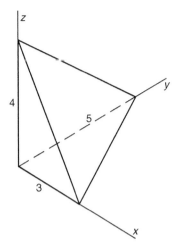

Figure P1.14   Problem1.14—Stress tetrahedron

1.16 For two dimensional incompressible flow, insert the stream function

$$v_x = \frac{\partial \psi}{\partial y}, \quad v_y = -\frac{\partial \psi}{\partial x}$$

into the Navier-Stokes equation, and eliminate the pressure to find the governing equation for the stream function.

Chapter 2

Inviscid Irrotational Flows

2.1 Inviscid Flows

Finding solutions with the Navier-Stokes equations that were introduced in Chapter 1 is a formidable challenge, particularly for flows where convective acceleration is present. When, however, the Reynolds number is sufficiently high, of the order of 10^5 or more, viscosity effects usually are of importance to the flow only in the boundary layer near a body or a wall or possibly in confined regions in the wake of a body. In many problems, such as the case of waves on a free surface, viscosity effects many be of secondary importance in most of the flow field.

In solving such flows, it is convenient—and useful—to first omit viscosity terms completely. Since this reduces the order of the differential equations, this means that fewer boundary conditions can be applied. The zero normal velocity condition generally is the most important condition and so is retained, whereas the no-slip velocity condition is ignored. For many flows, viscous effects can be included later by considering the boundary layer flow using the slip velocity found from the inviscid flow at the outer edge of the boundary layer.

Most 19th-century fluid mechanics was concerned with the study of inviscid flows.[1] There was no clear understanding of the effects of the Reynolds number on the flow, and the study of turbulence was left largely untouched. This was, however, a time of great ferment in the fields of fluid mechanics, electricity and magnetism, and thermodynamics. Particularly in the first two areas, scientists discovered a great similarity in their fields, and it was not unusual for a researcher to make contributions in both. As a result, today both fields share many terms such as *source*, *sink*, *potential*, and *current*, among others. Workers in one field often use analogies in the other, perhaps feeling that the other field can give a clearer understanding. In this chapter, the electrical and magnetic analogies are set aside, but keep in mind that despite this, a minor change in terminology is all that is needed to change the topic to electrostatics, electrodynamics, and magnetostatics.

If viscosity terms are omitted in the Navier-Stokes equations, they reduce to the form

$$\rho\frac{D\mathbf{v}}{Dt} = -\nabla p + \rho\mathbf{g}. \qquad (2.1.1)$$

This is called the **Euler equation**.

2.2   Irrotational Flows and the Velocity Potential

From the circulation theorem (equation 1.12.2b), if mass density is constant and viscous effects can be neglected, $D\Gamma/Dt = \oint_C -\left(\frac{1}{\rho}\nabla p\right)\cdot d\mathbf{s}$. The integrand on the right-hand side of the equation is an exact differential, so the integral around the path vanishes. Consequently, for a flow with no upstream circulation, as the flow moves downstream, it must continue to be vorticity-free, or **irrotational**. This is called the **persistence of irrotationality**. (In a real fluid, viscosity effects will introduce vorticity at a boundary, but at high Reynolds numbers this vorticity will be convected downstream and chiefly confined to the vicinity of the boundary and the wake.)

By the definition of irrotational flows,

$$\boldsymbol{\omega} = \nabla \times \mathbf{v} = 0. \qquad (2.2.1)$$

This suggests that two of the velocity components can be solved for in terms of the third component or, alternatively, that, as in the case of the continuity equation, scalar functions can be introduced that have the effect of accomplishing this.

An easier approach is to realize that since for irrotational flows $\Gamma = \oint_C \mathbf{v} \cdot d\mathbf{s} = 0$ for any C, it follows that the integrand $\mathbf{v} \cdot d\mathbf{s}$ must be an exact differential, and therefore for an irrotational flow field with velocity $\mathbf{v}$, it must be expressible as the gradient of a scalar. This allows us to write

$$\mathbf{v} = \nabla\phi, \qquad (2.2.2)$$

---

[1] *Inviscid fluids* do not exist, but fluids can flow in such a manner that viscosity effects are negligible in most of the region of flow.

where ϕ is called the **velocity potential**. For any velocity field written as the gradient of a scalar as in equation (2.2.2), it is guaranteed that for any scalar function ϕ, $\mathbf{v}$ will automatically be an irrotational velocity field. Because of equation (2.2.2), irrotational flows are often also called **potential flows**.

From equation (2.2.2) it follows that surfaces of constant ϕ are locally orthogonal to the velocity vector. Thus, surfaces of constant ϕ must be orthogonal to stream surfaces.

The introduction of a velocity potential guarantees irrotationality, but it is still required that the flow field satisfy the basic dynamical equations. To simplify the discussion, we will consider only incompressible flows. Then continuity requires that the divergence of the velocity field vanish. Therefore, the continuity equation for an irrotational incompressible flow is

$$0 = \nabla \cdot \mathbf{v} = \nabla^2 \phi. \tag{2.2.3}$$

This equation, called the **Laplace equation**, is used to determine ϕ for a given flow situation.

What then of the dynamics of the flow? Our flow field at this point seems to be completely determined from irrotationality and continuity, yet we have not considered Euler's equation. From equations (1.2.3) and (2.1.1), since $\nabla \times \mathbf{v} = 0$, Euler's equation (2.1.1) can be written in the form

$$\rho \left(\frac{\partial \mathbf{v}}{\partial t} + \frac{\nabla |\mathbf{v}|^2}{2} \right) = -\nabla p + \rho \mathbf{g}.$$

For irrotational flows $\mathbf{v} = \nabla \phi$, and $\mathbf{g}$ can be written as $\mathbf{g} = -g \nabla h$, where h is the elevation of the point in the direction in which gravity acts. Euler's equation can then be rearranged and, after dividing by ρ, becomes the form $\nabla \left(\frac{\partial \phi}{\partial t} + \frac{1}{2} |\mathbf{v}|^2 + gh + \frac{p}{\rho} \right) = 0$. Upon integration this gives

$$\frac{\partial \phi}{\partial t} + \frac{1}{2} |\mathbf{v}|^2 + gh + \frac{p}{\rho} = f(t), \tag{2.2.4}$$

where $f(t)$ is a constant or at most a function of time and is determined from either conditions at a reference point or a point far upstream. Equation (2.2.4) is known as the **Bernoulli equation** for irrotational flows.

The Bernoulli equation could also have been derived for the case where mass density depends only on pressure. In that case, the p/ρ term in equation (2.2.4) would be replaced by $\int dp/\rho$.

For most incompressible flows, then, the velocity field is found using only the conditions of irrotationality (usually by the introduction of the velocity potential ϕ) and the continuity equation in the form of equation (2.2.3), along with the imposition of conditions on the normal velocity at boundaries. Once ϕ is known, pressure can be found from the Bernoulli equation (2.2.4). Note that all of the mathematical nonlinearities appear only in the Bernoulli equation. For interface problems, however, further nonlinearities can be introduced by boundary conditions.

The linearity of equation (2.2.3) allows for the superposition of velocity fields. The nonlinearity of equation (2.2.4) means that pressure fields cannot be superposed in a linear manner.

Note that for irrotational flows equation (2.2.1) can be written as

$$\Gamma = \oint_C \nabla\phi \cdot ds = \phi_2 - \phi_1 = \Delta\phi, \qquad (2.2.5)$$

where ϕ_1 and ϕ_2 are the values of the velocity potential at the start and end points of the traverse of C. Thus, if ϕ is a single-valued function (i.e., if we go around a closed loop and the value of ϕ has not changed), since the curve C is closed, Γ will be zero.

It is, however, possible that ϕ can be multivalued if there exist points or isolated regions where ϕ is either singular or not uniquely defined. A line vortex, described following, has one such nonuniquely defined potential function. For such functions the circulation will usually be different from zero. Multivalued velocity potentials are associated with the presence of vorticity, corresponding to multivalued stream functions that are associated with discharge.

Since we will be looking at some methods for solving Laplace's equation, you may wonder whether any solution you might have obtained for a given flow is unique. That is, if you and your neighbor both solve the same problem but use different methods, will you end up with the same velocity field? The answer is yes, provided that you both stay with the same set of rules (and, of course, both do your work accurately). The irrotational flow field around a body is unique for a given set of boundary conditions, provided we also specify the circulation and do not allow the development of cavities that do not contain fluid. Always remember that the primary interest is in finding a flow field that models a real physical phenomenon to some degree of accuracy. Since vorticity, and hence circulation, is present in the boundary layer, it may need to be included in a model to give a realistic model of the flow field. Cavities might possibly be a reasonable model for wake flows. As long as the circulation is prescribed, together with rules concerning whether or not cavities are present, the flow field will be unique.

It is possible that the methods used to solve Laplace's equation will introduce a "mathematical" flow inside a body as well. In that region the flow is not unique. Any flows that methods generate inside bodies lie outside our domain of interest and are artifices of the mathematics with little if any physical meaning.

2.2.1   Intersection of Velocity Potential Lines and Streamlines in Two Dimensions

Surfaces of constant velocity potential and of constant stream function intersect one another throughout the flow. Since by definition the velocity is always normal to a constant potential surface, it follows that the constant stream surfaces are generally perpendicular to the constant potential surfaces. There are some exceptions to this, such as near stagnation points. We next investigate this for two-dimensional flows to see at what angles these intersections can take place.

To see the relationship between ϕ and ψ lines, start with

$$d\phi = \frac{\partial\phi}{\partial x}dx + \frac{\partial\phi}{\partial y}dy = v_x dx + v_y dy$$

and

$$d\psi = \frac{\partial\psi}{\partial x}dx + \frac{\partial\psi}{\partial y}dy = -v_y dx + v_x dy.$$

From these it follows that the slope of a $\phi = $ constant line $(d\phi = 0)$ is given by

$$\left(\frac{dy}{dx}\right)_\phi = -\frac{v_x}{v_y},$$

and the slope of a $\psi = $ constant line $(d\psi = 0)$ is given by

$$\left(\frac{dy}{dx}\right)_\psi = +\frac{v_y}{v_x}.$$

Multiplying the two slopes gives

$$\left(\frac{dy}{dx}\right)_\psi \left(\frac{dy}{dx}\right)_\phi = -1,$$

which leads to the conclusion that constant ϕ and ψ lines are orthogonal to one another except possibly at places where the velocity is either zero (stagnation points) or infinite (singularities).

In either of these two special cases we can investigate the situation further by considering the second-order terms to see what occurs. For example, at a stagnation point a Taylor series expansion gives to second-order terms in dx and dy

$$d\phi = \frac{\partial\phi}{\partial x}dx + \frac{\partial\phi}{\partial y}dy + \frac{\partial^2\phi}{\partial x^2}\frac{dx^2}{2} + \frac{\partial^2\phi}{\partial x\partial y}dxdy + \frac{\partial^2\phi}{\partial y^2}\frac{dy^2}{2}.$$

Since the first derivatives vanish at the stagnation point, on a line of constant ϕ passing through a stagnation point this expression becomes

$$d\phi = \frac{\partial^2\phi}{\partial x^2}\frac{dx^2}{2} + \frac{\partial^2\phi}{\partial x\partial y}dxdy + \frac{\partial^2\phi}{\partial y^2}\frac{dy^2}{2}.$$

First-order terms vanish at the stagnation point, so this becomes

$$0 = \left[\frac{\partial^2\phi}{\partial x^2} + 2\frac{\partial^2\phi}{\partial x\partial y}\frac{dy}{dx} + \frac{\partial^2\phi}{\partial y^2}\left(\frac{dy}{dx}\right)^2\right]\frac{dx^2}{2}.$$

Solving this quadratic equation for the slope at the stagnation point, we have

$$\left(\frac{dy}{dx}\right)_\phi = \frac{-\dfrac{\partial^2\phi}{\partial x\partial y} \pm \sqrt{\left(\dfrac{\partial^2\phi}{\partial x\partial y}\right)^2 - \dfrac{\partial^2\phi}{\partial x^2}\dfrac{\partial^2\phi}{\partial y^2}}}{\dfrac{\partial^2\phi}{\partial y^2}}.$$

A similar expression can be found involving ψ by the same process.

Note that the term underneath the square root sign must always be positive, since by Laplace's equation

$$\frac{\partial^2\phi}{\partial x^2} = -\frac{\partial^2\phi}{\partial y^2},$$

and so this term is the sum of two squares.

The conclusion then is that there will be two values for the slope at the stagnation point, hence the ϕ line divides, or bifurcates, at that point. For details of the angle between ϕ and ψ, individual examples must be considered. Since the Laplace equation is a great averager of things (for instance, it can be shown from the Gauss-Green theorems that the value of ϕ at the center of a circle is the average of all the values it takes on the circle), the constant ϕ lines can be expected to fall midway between the constant ψ lines.

2.2.2   Basic Two-Dimensional Irrotational Flows

Next, consider several basic simple flows that are the building blocks of potential flow theory from which all other potential flows can be constructed. These basic flows have their counterparts in electrostatics and electromagnetics (point charges, dipoles), beam deflection theory (concentrated loads), and many other branches of engineering physics, and they are special cases of what are termed ***Green's functions***.

In two-dimensional flows the basic solutions must satisfy the equations

$$v_x = \frac{\partial \phi}{\partial x} = \frac{\partial \psi}{\partial y} \tag{2.2.6a}$$

and

$$v_y = \frac{\partial \phi}{\partial y} = -\frac{\partial \psi}{\partial x}. \tag{2.2.6b}$$

Also, the equation

$$\nabla^2 \phi = 0 \tag{2.2.7}$$

is the incompressible continuity equation for irrotational flows, corresponding to $\nabla \cdot \mathbf{v} = 0$, while

$$\nabla^2 \psi = 0 \tag{2.2.8}$$

is the irrotationality condition for two-dimensional incompressible flows satisfying continuity, corresponding to $\omega_z = 0$. These must be satisfied as well.

Uniform stream

A uniform stream is a flow whose velocity is the same at every point in space. Therefore, the velocity components are

$$v_x = U_x = \frac{\partial \phi}{\partial x} \tag{2.2.9a}$$

and

$$v_y = U_y = \frac{\partial \phi}{\partial y}. \tag{2.2.9b}$$

Integrating equations (2.2.9a) and (2.2.9b), we find that

$$\phi_{\text{uniform stream}} = xU_x + yU_y, \tag{2.2.10}$$

where the constant of integration has arbitrarily been set to zero, since it does not contribute to the velocity field in any way. Lines of constant ϕ and ψ (both are straight lines and mutually orthogonal) are shown in Figure 2.2.1. The stream function is found from use of equations (2.2.6a) and (2.2.6b) to be

$$\psi_{\text{uniform stream}} = yU_x - xU_y. \tag{2.2.11}$$

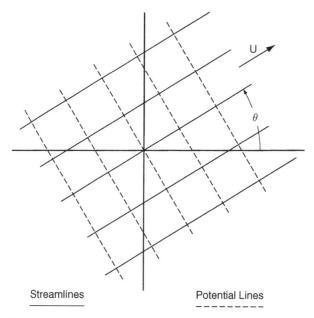

Figure 2.2.1   Uniform stream streamlines and iso-potential lines

Line source or sink (monopole)
The velocity potential

$$\phi = \frac{m}{2\pi} \ln |\mathbf{r} - \mathbf{r}_0| = \frac{m}{2\pi} \ln \sqrt{(x - x_0)^2 + (y - y_0)^2} \qquad (2.2.12)$$

is called a **line source** (if m is positive) or **line sink** (if m is negative) of strength m, located at the point (x_0, y_0). It extends from $-\infty$ to $+\infty$ in the z direction. The velocity components are given by

$$v_x = \frac{\partial \phi}{\partial x} = \frac{m(x - x_0)}{2\pi |\mathbf{r} - \mathbf{r}_0|^2} \qquad (2.2.13a)$$

and

$$v_y = \frac{\partial \phi}{\partial y} = \frac{m(y - y_0)}{2\pi |\mathbf{r} - \mathbf{r}_0|^2}. \qquad (2.2.13b)$$

Differentiating the velocity components gives

$$\frac{\partial^2 \phi}{\partial x^2} = \frac{m}{2\pi |\mathbf{r} - \mathbf{r}_0|^4} \left(|\mathbf{r} - \mathbf{r}_0|^2 - 2(x - x_0)^2 \right)$$

and

$$\frac{\partial^2 \phi}{\partial y^2} = \frac{m}{2\pi |\mathbf{r} - \mathbf{r}_0|^4} \left(|\mathbf{r} - \mathbf{r}_0|^2 - 2(y - y_0)^2 \right) = -\frac{\partial^2 \phi}{\partial x^2}.$$

Therefore, continuity is satisfied everywhere except possibly at the point (x_0, y_0).

To investigate what is happening at the location of the source itself, integrate the flow normal velocity about a 1 by 1 square centered at (x_0, y_0), as is seen in Figure 2.2.2. (The size and shape of the square is actually arbitrary, since the same result would be

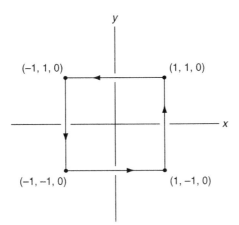

Figure 2.2.2   Source/sink discharge calculation

obtained for a contour of any size or shape that encloses the source.) Starting from the definition of discharge, we have

$$
Q = \int_{y_0-1}^{y_0+1} [v_x(1, y) - v_x(-1, y)]dy + \int_{x_0-1}^{x_0+1} [v_y(x, 1) - v_y(x, -1)]dx
$$
$$
= \int_{y_0-1}^{y_0+1} \frac{m}{2\pi} \left(\frac{1}{1 + (y - y_0)^2} - \frac{-1}{1 + (y - y_0)^2} \right)dy
$$
$$
+ \int_{x_0-1}^{x_0+1} \frac{m}{2\pi} \left(\frac{1}{1 + (x - x_0)^2} - \frac{-1}{1 + (x - x_0)^2} \right)dx = m.
$$

Thus, m represents the flow rate per unit length in the z direction being emitted from the source at (x_0, y_0), and is called the **strength** of the source. Lines of constant ϕ (concentric circles) and ψ (radial lines) are shown in Figure 2.2.3.

The issue of rotationality at (x_0, y_0) should also be considered. If the circulation around a square similar to the one we just used is computed, the result is zero. Therefore, there is a concentrated source of mass at (x_0, y_0) but no concentrated source of vorticity there.

From equations (2.2.6a) and (2.2.6b) the stream function for a source is seen to be

$$
\psi = \frac{m}{2\pi} \tan^{-1} \frac{y - y_0}{x - x_0}. \tag{2.2.14}
$$

The arctangent is a multivalued function, changing by 2π as we go around a contour enclosing from equations (2.2.6a) and (2.2.6b). Therefore, the change in ψ as we go around the contour is m, consistent with its representation of rate of flow.

From equation (2.2.12) we see that constant ϕ lines are concentric circles centered at (x_0, y_0). From equation (2.2.14), we see that the constant ψ lines are radial lines emanating from (x_0, y_0). Thus, the flow will be along the radial streamlines emanating from (x_0, y_0). To conserve mass, the velocity must decrease inversely with the distance from (x_0, y_0).

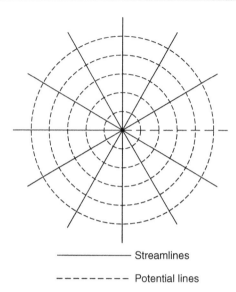

Streamlines

Potential lines

Figure 2.2.3   Source/sink streamlines and iso-potential lines

Line doublet (dipole)

Consider a source and sink pair of equal strengths, and let them approach one another along a connecting line in such a manner that their strengths increase inversely as the distance between them increases. The pair then has the combined potential

$$\frac{\phi_{\text{source}}(\mathbf{r}+\mathbf{a})+\phi_{\text{sink}}(\mathbf{r}-\mathbf{a})}{2\,|\mathbf{a}|},$$

and since the sink is negative with respect to the source, in the limit as $|\mathbf{a}|$ goes to zero this becomes a derivative. Therefore, define the potential of a **line doublet** as

$$\phi_{\text{doublet}} = \mathbf{B}\cdot\nabla\phi_{\text{source of strength }2\pi} = \frac{B_x(x-x_0)+B_y(y-y_0)}{|\mathbf{r}-\mathbf{r}_0|^2}, \qquad (2.2.15)$$

where $\mathbf{B}$ gives the strength and direction of the doublet. Often a doublet is denoted by a half-filled circle, the filled part representing the "source end" and the unfilled part the "sink end," to show the source-sink nature and the directionality. From its relation to the source, we expect that there is no net discharge or vorticity at $(x_0,\ y_0)$. This can be verified by taking appropriate integrations around that point. To find what lines of constant ϕ are like, rearrange equation (2.2.15) as

$$(x-x_0)^2+(y-y_0)^2 = \frac{B_x(x-x_0)+B_y(y-y_0)}{\phi},$$

or equivalently

$$\left(x-x_0-\frac{B_x\,(x-x_0)}{\phi}\right)^2+\left(y-y_0-\frac{B_y\,(y-y_0)}{\phi}\right)^2=\left(\frac{B_x\,(x-x_0)}{\phi}\right)^2+\left(\frac{B_y\,(y-y_0)}{\phi}\right)^2.$$

For constant values of ϕ this is the equation of a circle whose radius is $|\mathbf{B}|/2\phi$ centered at $(x_0+B_x/2\phi,\ y_0+B_y/2\phi)$. Therefore, it is on a line through $(x_0,\ y_0)$ in the direction of $\mathbf{B}$. Constant ϕ and ψ lines are shown in Figure 2.2.4. The constant ϕ lines are nested circles centered at $(B_x/2\phi,\ B_y/2\phi)$ and with radius $[(B_x/2\phi)^2+(B_y/2\phi)^2]^{1/2}$. Constant

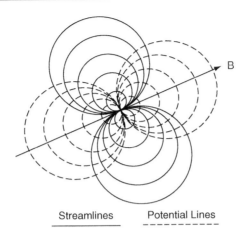

Figure 2.2.4   Doublet streamlines and iso-potential lines

ψ lines are a similar family of circles, but rotated 90 degrees with respect to the ϕ circles. They are centered at $(B_y/2\psi, -B_x/2\psi)$ and have radius $[(B_x/2\psi)^2 + (B_y/2\psi)^2]^{1/2}$.

The velocity field associated with the doublet is

$$v_x = \frac{\partial \phi}{\partial x} = \frac{B_x[(y-y_0)^2 - (x-x_0)^2] - 2B_y(x-x_0)(y-y_0)}{(x-x_0)^2 + (y-y_0)^2},$$

$$v_y = \frac{\partial \phi}{\partial y} = \frac{B_y[(x-x_0)^2 - (y-y_0)^2] - 2B_x(x-x_0)(y-y_0)}{(x-x_0)^2 + (y-y_0)^2}. \tag{2.2.16}$$

Along a line in the direction of **B**, we see from equation (2.2.16) that the velocity dies out as the square of the distance. Since ϕ_{source} satisfies Laplace's equation, and since $\nabla(\nabla^2 \phi_{\text{source}}) = \nabla(0) = 0$, therefore

$$0 = \mathbf{B} \cdot \nabla(\nabla^2 \phi_{\text{source}}) = \mathbf{B} \cdot \nabla^2(\nabla \phi_{\text{source}}) = \nabla^2(\mathbf{B} \cdot \nabla \phi_{\text{source}}) = \nabla^2 \phi_{\text{doublet}}. \tag{2.2.17}$$

The velocity potential for a line doublet therefore satisfies Laplace's equation.

From equation (2.2.6) the stream function for a line doublet is

$$\psi = \frac{B_y(x-x_0) - B_x(y-y_0)}{|\mathbf{r} - \mathbf{r}_0|^2}. \tag{2.2.18}$$

Since we have also labeled the source and doublet as monopole and dipole, you might wonder whether taking even higher derivatives would be useful. The derivative of the dipole is the **quadrapole**, which is mainly of interest in acoustic problems. The dipole is usually sufficient for fluid mechanics use.

Line vortex

The vortex is a "reverse analog" of the source, in that it has concentrated vorticity rather than concentrated discharge and also because the constant ϕ lines are radial lines while the lines of constant ψ are concentric circles. Its velocity potential is

$$\phi_{\text{vortex}} = \frac{\Gamma}{2\pi} \tan^{-1} \frac{y - y_0}{x - x_0} \tag{2.2.19}$$

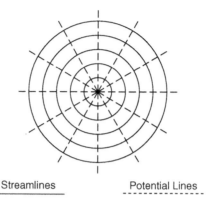

Figure 2.2.5 Vortex streamlines and iso-potential lines

with velocity components

$$v_x = \frac{\partial \phi}{\partial x} = \frac{-\Gamma(y - y_0)}{2\pi \, |\mathbf{r} - \mathbf{r}_0|^2},$$

$$v_y = \frac{\partial \phi}{\partial y} = \frac{\Gamma(x - x_0)}{2\pi \, |\mathbf{r} - \mathbf{r}_0|^2}.$$

(2.2.20)

Thus, the velocity decreases inversely with the distance from (x_0, y_0). Constant ϕ (radial lines) and ψ (concentric circles) lines are shown in Figure 2.2.5.

Since

$$\frac{\partial^2 \phi}{\partial x^2} = \frac{\Gamma(y - y_0)(x - x_0)}{\pi [(x - x_0)^2 + (y - y_0)^2]^2}$$

and

$$\frac{\partial^2 \phi}{\partial y^2} = \frac{\Gamma(y - y_0)(x - x_0)}{\pi [(x - x_0)^2 + (y - y_0)^2]^2} = -\frac{\partial^2 \phi}{\partial x^2},$$

Laplace's equation is seen to be satisfied. The vortex is counterclockwise if Γ is positive, clockwise if Γ is negative.

Since ϕ_{vortex} is multivalued (in traversing a path around the point $\mathbf{r}_0$ ϕ_{vortex} changes by Γ), we anticipate that there is circulation associated with vortex flows. Checking this by calculating the circulation around a 2 by 2 square centered at $(x_0, \, y_0)$, we have

$$\text{Circulation} = \int_{y_0-1}^{y_0+1} v_y(x_0 - 1, y) dy + \int_{x_0+1}^{x_0-1} v_x(x, y_0 + 1) dx + \int_{y_0+1}^{y_0-1} v_y(x_0 - 1, y) dy$$

$$+ \int_{x_0-1}^{x_0+1} v_x(x, y_0 - 1) dx$$

$$= \frac{\Gamma}{2\pi} \left[\int_{y_0-1}^{y_0+1} \frac{dy}{1 + (y - y_0)^2} + \int_{x_0+1}^{x_0-1} \frac{-dx}{1 + (x - x_0)^2} + \int_{y_0+1}^{y_0-1} \frac{-dy}{1 + (y - y_0)^2} \right.$$

$$\left. + \int_{x_0-1}^{x_0+1} \frac{dx}{1 + (x - x_0)^2} \right] = \Gamma,$$

a result that depends only on the fact that the path encircles (x_0, y_0), not on any other details of the path such as shape or size. Any path not enclosing the vortex has zero

TABLE 2.2.1   Velocity potentials and stream functions for irrotational flows

Flow Element	Two-dimensional	
	ϕ	ψ
Uniform stream	$xU_x + yU_{y'}$	$yU_x - xU_{y'}$
Source or sink	$\dfrac{m}{2\pi} \ln \sqrt{(x-x_0)^2 + (y-y_0)^2}$	$\dfrac{m}{2\pi} \tan^{-1} \dfrac{y-y_0}{x-x_0}$
Doublet	$\dfrac{B_x(x-x_0) + B_y(y-y_0)}{|\mathbf{r} - \mathbf{r}_0|^2}$	$\dfrac{B_y(x-x_0) - B_x(y-y_0)}{|\mathbf{r} - \mathbf{r}_0|^2}$
Line vortex	$\dfrac{\Gamma}{2\pi} \tan^{-1} \dfrac{y-y_0}{x-x_0}$	$-\dfrac{\Gamma}{2\pi} \tan^{-1} \dfrac{y-y_0}{x-x_0}$

Flow Element	Three-dimensional	
	ϕ	ψ(axisymmetric)
Uniform stream	$xU_x + yU_y + zU_z$	$0.5U_z r^2$
Source or sink	$\dfrac{-m}{4\pi |\mathbf{r} - \mathbf{r}_0|}$	$\dfrac{-m(z-z_0)}{4\pi\sqrt{r^2 + (z-z_0)^2}}$
Doublet	$\dfrac{\mathbf{B} \cdot (\mathbf{r} - \mathbf{r}_0)}{|\mathbf{r} - \mathbf{r}_0|^3}$	$\dfrac{-B_z r^2}{[r^2 + (z-z_0)^2]}$
Line vortex	None	None

circulation, as can be verified by Stokes's theorem. Therefore, a vortex has concentrated vorticity but no concentrated mass discharge.

From equation (2.2.6) the stream function for a vortex is

$$\psi_{\text{vortex}} = -\frac{\Gamma}{2\pi} \ln |\mathbf{r} - \mathbf{r}_0| = -\frac{\Gamma}{2\pi} \ln \sqrt{(x-x_0)^2 + (y-y_0)^2}. \qquad (2.2.21)$$

The velocity potentials and stream functions for all of these flows plus their three-dimensional counterparts are summarized in Table 2.2.1.

2.2.3   Hele-Shaw Flows

The four solutions we have just seen—uniform stream, source/sink, doublet, and vortex—are the fundamental solutions for two-dimensional potential flow. Before considering combinations of these to obtain physically interesting flows, it is useful to see how these basic flows can be produced in a laboratory.

One way of producing them is by means of a Hele-Shaw table, which consists of a flat horizontal floor with a trough at one end for introducing a uniform stream and another trough at the other end for removing it. Holes in the bottom of the floor are connected to bottles that may be raised or lowered. These provide the sources and sinks, with the elevation of the bottle controlling their strength. Doublets can be produced by putting a source and sink pair almost together, having the source height above the table being equal to the sink height below. Sometimes a top transparent plate is added to ensure that the flow is of the same depth everywhere. The vorticity component perpendicular to the floor will be essentially zero; components in the plane of the floor will be nonzero.

Vortices are more difficult to produce in Hele-Shaw flows than are sources or sinks. One way of producing them is to use a vertical circular rod driven by an electric motor.

Hele-Shaw flows are slow flows, and because the velocity will vary roughly parabolically with the coordinate in the direction normal to the floor inherent to these flows, there is a great deal of vorticity. However, this vorticity is largely parallel to the bed, the vorticity component normal to the floor being virtually zero. Thus, Hele-Shaw flows viewed perpendicular to the floor are good models of two-dimensional irrotational flows.

Streamlines can be traced by inserting dye into the flow. A permanent record of these streamlines can be made by photography. Alternatively, one method that has been used is to cast the bed of the table out of a hard plaster, with inlets for the source and sink permanently cast into the plaster. If the bed is painted with a white latex paint, streamlines can be recorded by carefully placing potassium permanganate crystals on the bed. A record of the streamlines remains as dark stains on the paint.

2.2.4 Basic Three-Dimensional Irrotational Flows

Except for the vortex, all our two-dimensional irrotational flows have three-dimensional counterparts that qualitatively are much like their two-dimensional counterparts. (An example of a three-dimensional vortex is a smoke ring. Mathematical representation of such a three-dimensional phenomenon is more complicated than in the case of two-dimensional flows.) Here, we simply list these counterparts. The analysis proceeds as in the two-dimensional case.

Uniform stream
The velocity potential for a uniform stream is

$$\phi_{\text{uniform stream}} = xU_x + yU_y + zU_z = \mathbf{r} \cdot \mathbf{U} \tag{2.2.22}$$

with a velocity field

$$\mathbf{v} = \nabla\phi = \mathbf{U}.$$

Surfaces of constant ϕ are planes perpendicular to $\mathbf{U}$.

When $\mathbf{U}$ has only a component in the z direction, a Stokes stream function can be found in the form

$$\psi_{\text{uniform stream}} = 0.5U_z R^2 \sin^2\beta = 0.5U_z r^2. \tag{2.2.23}$$

Point source or sink (point monopole)
The velocity potential for a point source of strength m at $\mathbf{r}_0$ is

$$\phi_{\text{source}} = \frac{-m}{4\pi|\mathbf{r}-\mathbf{r}_0|} = \frac{-m}{4\pi\sqrt{(x-x_0)^2+(y-y_0)^2+(z-z_0)^2}} \tag{2.2.24}$$

Here, m is the volume discharge from the source, with continuity satisfied everywhere except at $\mathbf{r}_0$. If m is positive, ϕ represents a source. If negative, it represents a sink. Irrotationality is satisfied everywhere.

Surfaces of constant ϕ are concentric spheres centered at $\mathbf{r}_0$. The velocity is directed along the radius of these spheres and dies out like the reciprocal of the distance squared to satisfy continuity. The velocity is given by

$$\mathbf{v} = \nabla\phi = \frac{m(\mathbf{r}-\mathbf{r}_0)}{4\pi|\mathbf{r}-\mathbf{r}_0|^3}. \tag{2.2.25}$$

When the source lies on the z-axis, a Stokes stream function can be found in the form

$$\psi_{\text{source}} = \frac{-m(z - z_0)}{4\pi\sqrt{r^2 + (z - z_0)^2}}.$$ (2.2.26)

Point doublet (point dipole)

The velocity potential for a doublet can again be found by differentiating the potential for a source, giving

$$\phi_{\text{doublet}} = \mathbf{B} \cdot \nabla\phi_{\text{source of strength } 4\pi} = \frac{\mathbf{B} \cdot (\mathbf{r} - \mathbf{r}_0)}{|\mathbf{r} - \mathbf{r}_0|^3}$$ (2.2.27)

with velocity components

$$\mathbf{v} = \frac{\mathbf{B}}{|\mathbf{r} - \mathbf{r}_0|^3} - \frac{3(\mathbf{r} - \mathbf{r}_0)\mathbf{B} \cdot (\mathbf{r} - \mathbf{r}_0)}{|\mathbf{r} - \mathbf{r}_0|^5}.$$ (2.2.28)

The constant surfaces are no longer simple geometries like circles and spheres.

When the doublet lies on the z-axis and additionally $\mathbf{B}$ points parallel to the z-axis, a Stokes stream function for a doublet can be found in the form

$$\psi_{\text{doublet}} = \frac{-B_z r^2}{[r^2 + (z - z_0)^2]^{3/2}}.$$ (2.2.29)

These results are summarized in Table 2.2.1.

2.2.5   Superposition and the Method of Images

In a number of simple cases, the solution for flow past a given body shape can be obtained by recognizing an analogy between potential flow and geometrical optics, since the Laplace equation also governs the passage of light waves. Boundaries can be thought of as mirrors, with images of the fundamental solutions appearing at appropriate points to generate the "mirror" boundary.

Line source near a plane wall

Suppose we have a source of strength m a distance b from a plane wall. According to the method of images, the plane wall can be regarded as a mirror. As the source "looks" into the mirror, it sees an ***image source*** of the same strength at distance b behind the mirror (see Figure 2.2.6). A mirror interchanges right and left. This does not affect the source, but will affect signs of vortices and doublets. The potential for the two-dimensional case is, with the x-axis acting as the wall,

$$\phi = \frac{m}{2\pi}\left(\ln\sqrt{x^2 + (y - b)^2} + \ln\sqrt{x^2 + (y + b)^2}\right)$$ (2.2.30)

with velocity components

$$v_x = \frac{m}{2\pi}\left(\frac{x}{x^2 + (y - b)^2} + \frac{x}{x^2 + (y + b)^2}\right),$$

$$v_y = \frac{m}{2\pi}\left(\frac{y - b}{x^2 + (y - b)^2} + \frac{y + b}{x^2 + (y + b)^2}\right).$$ (2.2.31)

Note that v_y vanishes on $y = 0$, satisfying the necessary boundary condition that the velocity normal to the stationary wall be zero.

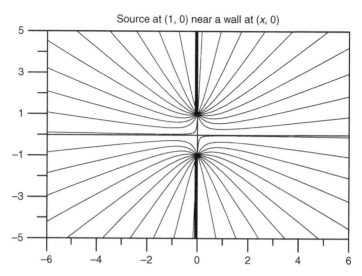

Figure 2.2.6 Source and its image by a wall—streamlines

This solution can be used as a simple model of physical problems such as water intakes or pollution sources near a straight coastline. Additional sources for more complicated flows can be easily added.

Note that if the source is in a corner, both walls act as mirrors, and there will be three image sources plus the original source. The third image comes about because the mirrors extend to plus and minus infinity, so the two images that you might think would be sufficient in turn have their own image. The preceding procedure can be easily extended to vortices and doublets near a wall and to three-dimensional examples. It can also be used for acoustic sources. Some loudspeaker enclosures are designed to be placed in corners or near walls to increase their apparent power output by the image speakers.

Example 2.2.1 Point source near walls

A point source of strength m is located at the point $(1, 2, 3)$. There is a wall at $x = 0$ and a second wall at $y = 0$. What is the velocity potential for this flow?

Solution. From equation (2.2.24), the given source has a velocity potential

$$\phi_0 = -\frac{m}{4\,|\mathbf{r} - \mathbf{r}_0|}, \quad \text{where } \mathbf{r}_0 = \mathbf{i} + 2\mathbf{j} + 3\mathbf{k}.$$

To generate the first wall, an image source of strength m at $(-1, 2, 3)$ is needed. This will have a velocity potential $\phi_1 = -\frac{m}{4|\mathbf{r}-\mathbf{r}_1|}$, where $\mathbf{r}_1 = -\mathbf{i} + 2\mathbf{j} + 3\mathbf{k}$. When the second wall is added, both the original source and the image at $(-1, 2, 3)$ will have images, with a combined velocity potential

$$\phi_2 + \phi_3 = -\frac{m}{4\,|\mathbf{r} - \mathbf{r}_2|} - \frac{m}{4\,|\mathbf{r} - \mathbf{r}_3|}, \quad \text{where } \mathbf{r}_2 = \mathbf{i} - 2\mathbf{j} + 3\mathbf{k} \text{ and } \mathbf{r}_3 = -\mathbf{i} - 2\mathbf{j} + 3\mathbf{k}.$$

Thus, our velocity potential is the sum of the potentials of four sources, each of strength m, and located at the points $(1, 2, 3)$, $(-1, 2, 3)$, $(1, -2, 3)$, and $(-1, -2, 3)$. The total potential is

$$\phi_{\text{total}} = \phi_0 + \phi_1 + \phi_2 + \phi_3 = -\frac{m}{4\,|r - r_0|} - \frac{m}{4\,|r - r_1|} - \frac{m}{4\,|r - r_2|} - \frac{m}{4\,|r - r_3|}.$$

Notice that if there were a third wall at $z = 0$, eight sources would be needed.

2.2.6 Vortices Near Walls

An interesting variation of the source near a wall is that of a vortex near a wall. The image will have a circulation in the reverse direction of the original vortex because left and right are interchanged in a reflection (see Figure 2.2.7). For a wall at $x = 0$ the velocity potential for the original vortex plus its image is

$$\phi = \frac{\Gamma}{2\pi} \left(\tan^{-1} \frac{y - b}{x} - \tan^{-1} \frac{y + b}{x} \right) \tag{2.2.32}$$

with velocity components

$$v_x = \frac{\partial \phi}{\partial x} = \frac{\Gamma}{2\pi} \left(\frac{-(y - b)}{x^2 + (y - b)^2} + \frac{y + b}{x^2 + (y + b)^2} \right),$$

$$v_y = \frac{\partial \phi}{\partial y} = \frac{\Gamma}{2\pi} \left(\frac{x}{x^2 + (y - b)^2} + \frac{x}{x^2 + (y + b)^2} \right). \tag{2.2.33}$$

The velocity at $(0, b)$ due to the image vortex is $(\Gamma/4\pi b, 0)$. This is called the ***induced velocity***. The vortex at $(0, b)$ will tend to travel with the induced velocity, carrying its image with it, since the induced velocity at the image will be the same value.

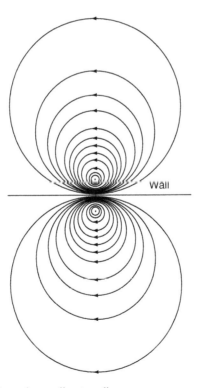

Figure 2.2.7   Vortex and its image by a wall—streamlines

The stream function to accompany equation (2.2.31) is

$$\psi = \frac{\Gamma}{2\pi}\left[\ln \sqrt{x^2+(y-b)^2}+\ln \sqrt{x^2+(y+b)^2}\right]$$
$$= \frac{\Gamma}{4\pi}\ln \frac{x^2+(y-b)^2}{x^2+(y+b)^2}.$$

(2.2.34)

Therefore, lines of constant ψ are given by

$$x^2+(y-b)^2 = C[x^2+(y+b)^2],$$

(2.2.35)

where C is a constant related to the value of the stream function by $\psi = (\Gamma/4\pi)\ell n\ C$. Expanding and rearranging, equation (4.3.33) becomes

$$x^2+\left(y-\frac{b(1+C)}{1-C}\right)^2 = \frac{4Cb^2}{(1-C)^2}.$$

Thus, the streamlines are circles of radius $r = 2b\sqrt{C/(1-C)}$ centered at $(0, d)$, where $d = b\frac{1+C}{1-C}$.

Since the streamlines in this example are circles, a physical realization of this flow is a single vortex in a cup of radius r, the vortex being a distance b from the center of the cup. The vortex will travel around the cup on a circular path of radius r at a speed of $\Gamma/4\pi b$.

Still another realization of a vortex flow is a vortex pair near a wall, the vortices being of equal but opposite circulation. Take them to be a distance a from the wall and separated by a distance $2b$ (Figure 2.2.8). To generate the wall, it is necessary to add a vortex image pair as well. The total velocity potential is then

$$\phi = \frac{\Gamma}{2\pi}\left(\tan^{-1}\frac{y-a}{x-b}-\tan^{-1}\frac{y-a}{x+b}-\tan^{-1}\frac{y+a}{x-b}+\tan^{-1}\frac{y+a}{x+b}\right)$$

(2.2.36)

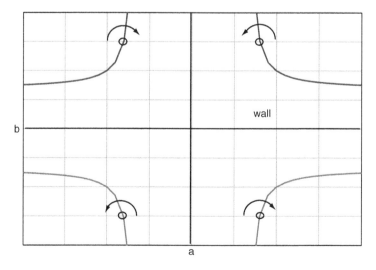

Figure 2.2.8 Trajectories of a vortex pair and its images by a wall

with stream function

$$\psi = \frac{\Gamma}{4\pi}\{\ln[(x-b)^2+(y-a)^2]-\ln[(x+b)^2+(y-a)^2]$$
$$-\ln[(x-b)^2+(y+a)^2]+\ln[(x+b)^2+(y+a)^2]\} \qquad (2.2.37)$$

and with velocity components

$$v_x = \frac{\partial\phi}{\partial x} = \frac{\Gamma}{2\pi}\left[\frac{-(y-a)}{(x-b)^2+(y-a)^2}+\frac{y-a}{(x+b)^2+(y-a)^2}\right.$$
$$\left.+\frac{y+a}{(x-b)^2+(y+a)^2}-\frac{y+a}{(x+b)^2+(y+a)^2}\right],$$

$$v_y = \frac{\partial\phi}{\partial y} = \frac{\Gamma}{2\pi}\left[\frac{x-b}{(x-b)^2+(y-a)^2}-\frac{x+b}{(x+b)^2+(y-a)^2}\right. \qquad (2.2.38)$$
$$\left.-\frac{x-b}{(x-b)^2+(y+a)^2}+\frac{x+b}{(x+b)^2+(y+a)^2}\right].$$

The induced velocity at $(b,\ a)$ is

$$\mathbf{V}_{\text{induced}} = \frac{\Gamma}{4\pi}\left[\left(\frac{1}{a}-\frac{a}{a^2+b^2}\right)\mathbf{i}+\left(\frac{-1}{b}+\frac{b}{a^2+b^2}\right)\mathbf{j}\right].$$

Therefore, the equations of motion for this vortex are

$$\frac{db}{dt} = \frac{\Gamma}{4\pi}\left(\frac{1}{a}-\frac{a}{a^2+b^2}\right),$$
$$\frac{da}{dt} = \frac{\Gamma}{4\pi}\left(\frac{-1}{b}+\frac{b}{a^2+b^2}\right). \qquad (2.2.39)$$

With appropriate sign changes, similar equations hold for the other three vortices.

The path the vortices travel can be found by rearranging (2.2.39) in the form

$$\frac{\Gamma}{4\pi}\frac{dt}{a^2+b^2} = \frac{a}{b^2}db = -\frac{b}{a^2}da.$$

The variables a and b can be separated and the resulting equation integrated, giving as the path the vortex travels

$$\frac{1}{b^2} = \frac{1}{b_0^2}+\frac{1}{a_0^2}-\frac{1}{a^2}, \qquad (2.2.40)$$

$(a_0,\ b_0)$ being the initial position of the vortex.

This model of a traveling vortex pair is useful in describing the spreading of the vortex pair left by the wing tips of an aircraft on takeoff. In this case, the wall represents the ground. It is also responsible for the ***ground effect***—the increased lift an airplane experiences on takeoff due to the proximity of the ground.

Example 2.2.2   A vortex near a plane wall
A wall is located at $x=0$. A vortex with circulation 20π meter^2/second is placed 1 meter above the wall. What is the velocity potential, and at what speed does the vortex move?

Solution. For convenience, take the instantaneous position of the vortex to be $(0, 1, 0)$. Then the velocity potential for the original vortex is $\phi_{\text{orig}} = 10\tan^{-1}\frac{y-1}{x}$.

Remembering that the circulation of the image vortex is reversed, from equation (2.2.19) the velocity potential for the original vortex plus its image is

$$\phi_{\text{total}} = \phi_{\text{orig}} + \phi_{\text{image}} = 10\left(\tan^{-1}\frac{y-1}{x} - \tan^{-1}\frac{y+1}{x}\right).$$

Taking the gradient of the velocity potential, the velocity components are found to be

$$v_x = 10\left(\frac{-(y-1)}{x^2+(y-1)^2} + \frac{(y+1)}{x^2+(y+1)^2}\right),$$

$$v_y = 10\left(\frac{x}{x^2+(y-1)^2} - \frac{x}{x^2+(y+1)^2}\right).$$

The induced velocity at $(0, 1, 0)$ is the velocity at that point due to the image vortex. This gives a velocity at $(0, 1, 0)$ of $v_x = 10 \cdot 2/2^2 = 5\,\text{m/s}$, $v_y = 0$. The vortex thus moves parallel to the wall at a speed of 5 m/s.

Example 2.2.3 A vortex pair in a cup

A vortex pair is generated in a cup of coffee of radius c by brushing the tip of a spoon lightly across the surface of the coffee. The pair so generated will have opposite circulations (try it!). If the vortex with positive circulation is at (a, b), and the vortex with negative circulation at $(a, -b)$, verify that the flow with the cup is generated by an image vortex with positive circulation at $(ac^2/(a^2+b^2), -bc^2/(a^2+b^2))$, plus an image vortex with negative circulation at $(ac^2/(a^2+b^2), bc^2/(a^2+b^2))$. These image points are located at what are called the inverse points of our cup.

Solution. What is needed is a pair of opposite-rotating vortices plus the images needed to generate the cup. Each vortex moves because of the induced velocity generated by the images and the other vortex. The given vortex pair has a stream function

$$\phi = -\frac{\Gamma}{4\pi}\left\{\ln[(x-a)^2+(y-b)^2] - \ln[(x-a)^2+(y+b)^2]\right\}$$

$$= -\frac{\Gamma}{4\pi}\ln\frac{(x-a)^2+(y-b)^2}{(x-a)^2+(y+b)^2}.$$

The proposed stream function consists of the original stream function plus the stream function due to a pair of vortices at (ak, bk) and $(ak, -bk)$, where $k = c^2/(a^2+b^2)$. The combined stream function is then

$$\phi = -\frac{\Gamma}{4\pi}\left\{\ln[(x-a)^2+(y-b)^2] - \ln[(x-a)^2+(y+b)^2]\right.$$

$$\left. + \ln[(x-ak)^2+(y-bk)^2] - \ln[(x-ak)^2+(y+bk)^2]\right\}.$$

As a check of the result, on the circle of radius c

$$x = c\cos\theta \quad \text{and} \quad y = c\sin\theta, \text{ and so}$$

$$\phi = -\frac{\Gamma}{4\pi}\left\{\ln[(c\cos\theta - a)^2+(c\sin\theta - b)^2] + \ln[(c\cos\theta - ak)^2+(c\sin\theta + bk)^2]\right.$$

$$\left. - \ln[(c\cos\theta - a)^2+(c\sin\theta + b)^2] - \ln[(c\cos\theta - ak)^2+(c\sin\theta - bk)^2]\right\}.$$

But since $\cos^2\theta + \sin^2\theta = 1$ and using the definition of k

$$
\begin{aligned}
(c\cos\theta - ak)^2 + (c\sin\theta - bk)^2 &= c^2 - 2k(a\cos\theta + b\sin\theta) + (a^2 + b^2)k^2 \\
&= k[c^2/k - 2(a\cos\theta + b\sin\theta) + (a^2 + b^2)k] \\
&= k[(a^2 + b^2 - 2(a\cos\theta + b\sin\theta) + c^2] \\
&= k[(c\cos\theta - a)^2 + (c\sin\theta - b)^2]
\end{aligned}
$$

and

$$
\begin{aligned}
(c\cos\theta - ak)^2 + (c\sin\theta + bk)^2 &= c^2 - 2k(a\cos\theta - b\sin\theta) + (a^2 + b^2)k^2 \\
&= k[c^2/k - 2(a\cos\theta - b\sin\theta) + (a^2 + b^2)k] \\
&= k[(a^2 + b^2 - 2(a\cos\theta - b\sin\theta) + c^2] \\
&= k[(c\cos\theta - a)^2 + (c\sin\theta + b)^2].
\end{aligned}
$$

Substituting these into the expression of the stream function, we have

$$
\begin{aligned}
\psi = -\frac{\Gamma}{4\pi}\Big\{ &\ln[(c\cos\theta - a)^2 + (c\sin\theta - b)^2] + \ln k[(c\cos\theta - a)^2 + (c\sin\theta + b)^2] \\
&- \ln[(c\cos\theta - a)^2 + (c\sin\theta + b)^2] - \ln k[(c\cos\theta - a)^2 + (c\sin\theta - b)^2]\Big\} \\
&= 0.
\end{aligned}
$$

Thus, the cup of radius c is a streamline with the stream function equal to zero.

To find the equations that govern how the vortex at $(a,\ b)$ moves, the induced velocity components are computed by taking the derivatives of ψ, omitting the term from the vortex at (a, b), and then letting $x = a$, $y = b$. The result is

$$
a = \frac{\Gamma b}{2}\left(\frac{1}{2b^2} - \frac{(a^2 + b^2)(a^2 + b^2 + c^2)}{a^2(a^2 + b^2 - c^2)^2 + b^2(a^2 + b^2 + c^2)^2} + \frac{1}{a^2 + b^2 - c^2}\right),
$$

$$
b = \frac{\Gamma a}{2}\left(\frac{(a^2 + b^2)(a^2 + b^2 + c^2)}{a^2(a^2 + b^2 - c^2)^2 + b^2(a^2 + b^2 + c^2)^2} - \frac{1}{a^2 + b^2 - c^2}\right).
$$

2.2.7   Rankine Half-Body

A source located at the origin in a uniform stream (Figure 2.2.9) has the velocity potential and a stream function

$$
\phi = xU + \frac{m}{2\pi}\ln\sqrt{x^2 + y^2},
$$
$$
\psi = yU + \frac{m}{2\pi}\tan^{-1}\frac{y}{x}
$$

(2.2.41)

in a two-dimensional flow, and

$$
\phi = zU - \frac{m}{4\pi\sqrt{r^2 + z^2}},
$$
$$
\psi = \frac{1}{2}r^2U - \frac{m}{4\pi\sqrt{r^2 + z^2}}
$$

(2.2.42)

in a three-dimensional flow.

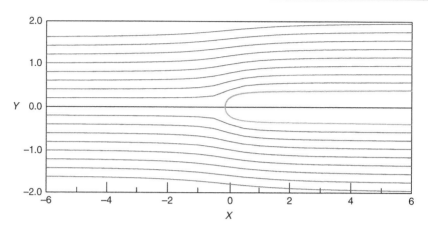

Figure 2.2.9   Rankine half-body—two dimensions

Examining the three-dimensional case in further detail, we find the velocity components to be

$$v_r = \frac{\partial \phi}{\partial r} = \frac{mr}{4\pi(r^2+z^2)^{3/2}}, \, v_z = \frac{\partial \phi}{\partial x} = U + \frac{mz}{4\pi(r^2+z^2)^{3/2}}. \qquad (2.2.43)$$

It is seen from equation (2.2.43) that there is a stagnation point ($\mathbf{v}_{\text{stagnation point}} = 0$) at the point $r = 0, z = -\sqrt{m/4\pi U}$. On $r = 0$, the stream function takes on values $\psi = -m/4\pi$ for z positive and $\psi = m/4\pi$ for z negative. The streamline $\psi = m/4\pi$ goes from the source to minus infinity. At the stagnation point, however, it bifurcates and goes along the curve $z^2 = (b^2-2r^2)^2/4(b^2-r^2)z^2 = (b^2-2r^2)^2/4(b^2-r^2)$ with $b^2 = m/\pi U$. This follows by putting $\psi = m/4\pi$ into equation (2.2.42) and solving for z. The radius of the body goes from 0 at the stagnation point to $r_\infty = b$ far downstream from the source. This last result can be obtained either from searching for the value of r needed to make z become infinite in equation (2.2.43) or by realizing that far downstream from the source the velocity must be U and all the discharge from the source must be contained within the body. In either case, the result is that far downstream the body radius is $b = \sqrt{m/\pi U}$.

This flow could be considered as a model for flow past a ***pitot tube*** of a slightly unusual shape. A pitot tube determines velocity by measuring the pressure at the stagnation point and another point far enough down the body so that the speed is essentially U. The difference in pressure between these two points is proportional to U^2. This pressure difference can be found from our analytical results by writing the Bernoulli equation between the stagnation point and infinity.

In the two-dimensional counterpart of this, the velocity is

$$v_x = \frac{\partial \phi}{\partial x} = U + \frac{mx}{2\pi(x^2+y^2)}, \quad v_y = \frac{\partial \phi}{\partial y} = \frac{my}{2\pi(x^2+y^2)},$$

with the stagnation point located at $(-m/2\pi U, \, 0)$. On $y = 0$, $\psi = 0$ for x positive and $m/2$ for x negative. The streamline $\psi = m/2$ starts at $(-\infty, 0)$, going along the x-axis to the source. At the stagnation point the streamline bifurcates, having the shape given by

$$2\pi Uy/m = \pi - \tan^{-1} y/x. \qquad (2.2.44)$$

The asymptotic half-width of the body is $y_\infty = m/2U$.

It is instructive to examine the behavior of the ϕ and ψ lines at the stagnation point. For the two-dimensional case, first expand ψ about the stagnation point in a Taylor series about $(-m/2\pi U, \ 0)$, giving to second order in $x + m/2\pi U$ and y

$$\psi = \frac{m}{2} - \frac{4\pi U^2}{m} \left(x + \frac{m}{2\pi U} \right) y + \cdots .$$

This tells us that at the bifurcation point, the $\psi = m/2$ line is either along $y = 0$ or perpendicular to it (i.e., locally either on $y = 0$ or on $x = -m/2\pi U$).

Expanding ϕ about $(-m/2\pi U, \ 0)$ to second order in $x + m/2\pi U$ and y, we have

$$\phi = \frac{m}{2\pi} \left(-1 + \ln \frac{m}{2\pi U} \right) + \frac{2\pi U^2}{m} \left[-\left(x + \frac{m}{2\pi U} \right)^2 + y^2 \right] + \cdots .$$

This tells us that $\phi = \frac{m}{2\pi} \left(-1 + \ln \frac{m}{2\pi U} \right)$ along lines with slope $\pm 45°$ $[y = \pm(x + m/2\pi U)]$, thereby bisecting the ψ lines as we had earlier expected would happen.

Notice that if we wish to model this flow on a Hele-Shaw table, one way of accomplishing this without the necessity of drilling any holes in our table would be to cut out a solid obstacle of the shape given by setting $\psi = m/2$ and then placing it on the table aligned with the flow. The flow exterior to the obstacle is the same as if we had drilled a hole and inserted the source.

2.2.8   Rankine Oval

The previous Rankine half-body was not closed because there was a net unbalance in mass discharge. By putting an aligned source and sink pair in a uniform stream with the source upstream of the sink, a closed oval shape is obtained. The velocity potential and stream function then become

$$\phi = xU + \frac{m}{2\pi} \left[\ln \sqrt{(x+a)^2 + y^2} - \ln \sqrt{(x-a)^2 + y^2} \right],$$

$$\psi = yU + \frac{m}{2\pi} \left(\tan^{-1} \frac{y}{x+a} - \tan^{-1} \frac{y}{x-a} \right) \tag{2.2.45}$$

for a two-dimensional body, and

$$\phi - zU + \frac{m}{4\pi} \left(\frac{1}{\sqrt{r^2 + (z+a)^2}} - \frac{1}{\sqrt{r^2 + (z-a)^2}} \right),$$

$$\psi = \frac{1}{2} r^2 U + \frac{m}{4\pi} \left(\frac{z+a}{\sqrt{r^2 + (z+a)^2}} - \frac{z-a}{\sqrt{r^2 + (z-a)^2}} \right), \tag{2.2.46}$$

for a three-dimensional body.

The velocity for the two-dimensional body is

$$v_x = \frac{\partial \phi}{\partial x} = U + \frac{m}{2\pi} \left(\frac{x+a}{(x+a)^2 + y^2} - \frac{x-a}{(x-a)^2 + y^2} \right),$$

$$v_y = \frac{\partial \phi}{\partial y} = \frac{m}{2\pi} \left(\frac{y}{(x+a)^2 + y^2} - \frac{y}{(x-a)^2 + y^2} \right).$$

The stagnation points are therefore at $(\pm \sqrt{a^2 + ma/\pi U}, 0)$.

For the two-dimensional case, according to equation (2.2.45) the streamline that makes up the body is given by $\psi = 0$. (Notice that along $y = 0$, $\psi = 0$, except in the range $-a < x < a$, where $\psi = -m/2$.) Therefore, the equation giving the body shape is

$$0 = yU + \frac{m}{2\pi}\left(\tan^{-1}\frac{y}{x+a} - \tan^{-1}\frac{y}{x-a}\right). \tag{2.2.47}$$

From symmetry, the maximum height of the body will be at $x = 0$. This height is given from equation (2.2.47) as a solution of the equation

$$y_{max} = \frac{m}{2\pi U}\left(\pi - 2\tan^{-1}\frac{y_{max}}{a}\right). \tag{2.2.48}$$

Similar results hold for the three-dimensional case. The parameter governing shape is m/Ua in the two-dimensional case and m/Ua^2 in the three-dimensional case. If m/Ua is large, the body is long and slender. If m/Ua is small, the body is short and rounded.

More complicated Rankine ovals can be formed by putting more sources and sinks in a uniform stream. For the body to close, it is necessary that the sum of the source and sink strengths be zero. This, however, is not a sufficient condition. Notice that in our simple example that if the source and sink were interchanged (just change the sign of m), there will be no closed streamlines about the source and sink pair.

2.2.9 Circular Cylinder or Sphere in a Uniform Stream

The Rankine oval is an unfamiliar geometrical shape, but if you plot its shape for various values of the separation a and the shape parameter m/Ua, you can see that as the source-sink pair get closer together while the shape parameter is held constant, the oval shape becomes more and more circular. This suggests that in the limit as the source-sink pair becomes a doublet, a circular shape would be achieved. The source portion of the doublet should be facing upstream and the sink portion facing downstream in order to generate a closed stream surface.

The velocity potential and stream function for a uniform stream plus a doublet is

$$\phi = xU + \frac{xB_x}{x^2+y^2}, \quad \psi = yU - \frac{yB_y}{x^2+y^2}, \tag{2.2.49}$$

in a two-dimensional flow, and

$$\phi = zU + \frac{zB_z}{(r^2+z^2)^{3/2}}, \quad \psi = \frac{1}{2}r^2U - \frac{r^2B_z}{(r^2+z^2)^{3/2}} \tag{2.2.50}$$

in a three-dimensional flow.

If we let $B_x = Ub^2$ in the two-dimensional case, equation (2.2.50) shows that $\psi = 0$ both on $y = 0$ and on $x^2 + y^2 = b^2$. Therefore, we have flow past a circular cylinder of radius b. Letting $B_z = Ub^3/2$ in three dimensions gives $\psi = 0$ on a sphere of radius b.

How does this solution relate to the method of images? The interpretation is complicated by having to consider a curved mirror, but the flow can be thought of as the body that focuses the uniform stream upstream (a very large distributed source) into the source part of the doublet. The downstream part of the uniform stream (a very large distributed sink) is focused into the sink part of the doublet.

More orderly ways of distributing sources to generate flows about given body shapes are known. For thin bodies such as wings or airplane fuselages, sources are distributed on the centerline, the strength of the source distribution per unit length being proportional to the rate that the cross-sectional area changes. For more complicated shapes, sources are distributed on the surface of the body. Vortices are included if lift forces are needed, as indicated in the following section.

2.3   Singularity Distribution Methods

When generating flows past bodies by placing sources, sinks, and other basic flows inside bodies, as was just done for Rankine bodies, we are solving inverse problems. That is, we first place the basic flow elements, and *then* find the location of the closed streamline corresponding to this distribution of mathematical singularities. We would prefer to solve the more difficult direct problem, where we first specify a body shape and then find where we must put the singularities. Since generally we do not want the flow to be singular within the flow domain (i.e., to have infinite velocities outside of the body), that leaves two possibilities: singularities distributed either within the interior of the body or on its surface. We shall discuss both possibilities.

2.3.1   Two- and Three-Dimensional Slender Body Theory

First consider the possibility of putting singularities within the body. For simplicity, restrict attention to two-dimensional, symmetric, slender bodies. (The symmetry restriction can easily be removed but with some complication in the mathematics.) By *slender*, we mean that the thickness of the body is small compared to its length and that the rate at which the thickness changes is also small. To emphasize this, write the body thickness as

$$y = \pm \varepsilon T(x) \quad \text{for} \quad 0 \le x \le L. \tag{2.3.1}$$

where ε, the **slenderness ratio**, is defined as the maximum thickness divided by the length L. For slender body theory to be valid, the slenderness ratio must be a small parameter. Saying that the rate of thickness change is small implies that dy/dx is everywhere small or that dT/dx is everywhere of order one or less.

To generate the flow past this body, assume that this can be accomplished by using a uniform stream together with a continuous distribution of sources and sinks along the center line of the body. Thus, write

$$\phi = xU + \frac{1}{2\pi} \int_0^L m(\xi) \ln \sqrt{(x - \xi)^2 + y^2} d\xi, \tag{2.3.2}$$

where m is the source strength per unit length along the body. We know that the velocity derived from equation (2.3.2) automatically satisfies Laplace's equation (continuity), and, since it is a velocity potential, it will also be an irrotational flow. If we also satisfy the boundary condition of zero normal velocity on the boundary, or

$$\frac{dx}{\frac{\partial \phi}{\partial x}} = \frac{dy}{\frac{\partial \phi}{\partial y}} \quad \text{on} \quad y = \pm \varepsilon T(x), 0 \le x \le L, \tag{2.3.3}$$

we have solved the problem.

Rearranging equation (2.3.3), and recognizing that on the body $dy/dx = \pm \varepsilon T'$, where the prime denotes differentiation with respect to x, equation (2.3.3) becomes

$$\frac{\partial \phi}{\partial y} = \pm \varepsilon \frac{\partial \phi}{\partial x} T' \quad \text{on} \quad y = \pm \varepsilon T. \tag{2.3.4}$$

Using equation (2.3.2), this results in

$$\frac{\varepsilon T(x)}{2\pi} = \int_0^L \frac{m(\xi)}{(x-\xi)^2 + (\varepsilon T)^2} d\xi = \varepsilon T'(x) \left[U + \frac{1}{2\pi} \int_0^L \frac{(x-\xi)m(\xi)}{(x-\xi)^2 + (\varepsilon T)^2} d\xi \right] \quad (2.3.5)$$

for $0 \leq x \leq L$.

The task now is to solve equation (2.3.5) for m, given a thickness distribution T. This is by no means a trivial job. This type of equation is called a Fredholm integral equation of the first kind, and its solution by either analytic or numerical means requires care and expertise.

To simplify the mathematics and obtain an approximate solution valid when ε is small, notice that the effect of the source distribution is to give a y velocity component (the left-hand side of equation (2.3.5)) that is small compared with U and also an accompanying small x velocity component (the second term in the right-hand side of equation (2.3.5)). The small y component cannot be neglected, since this term controls the slope of the streamline in the boundary, but it might be reasonable to neglect the x component compared to U. (This may sound a little ad hoc, but bear with me). The validity of the assumptions can be checked once the solution has been found.

Making this approximation, and again with a little rearranging, equation (2.3.5) becomes

$$\frac{1}{2\pi} \int_{-x/\varepsilon T}^{(L-x)/\varepsilon T} \frac{m(x + \varepsilon \eta T)}{1 + \eta^2} d\eta = \varepsilon U T'(x), \quad (2.3.6)$$

where the substitution $\eta = (\xi - x)/\varepsilon T$ has been made. Taking the limit as ε becomes very small, equation (2.3.6) reduces to

$$\frac{m(x)}{2\pi} \int_{-\infty}^{\infty} \frac{d\eta}{1 + \eta^2} \simeq \varepsilon U T'(x), \quad (2.3.7)$$

or

$$m(x) \simeq \varepsilon U \frac{d[2T(x)]}{dx}. \quad (2.3.8)$$

Thus, the local source strength is the stream velocity times the slenderness ratio times the local rate of change of total thickness of the body ($2T$). The appearance of ε in equation (2.3.8) justifies the neglect of the x velocity component due to the source distribution.

It might be argued that we were a little cavalier in bringing m outside the integral sign in equation (2.3.6) because even though ε is small, η becomes large in part of the range of integration, and, in any case, $\varepsilon\eta$ is finite. Recall, however, that m is multiplied by $1/1 + \eta^2$ and that in these regions this factor guarantees that the integrand is small.

Another question that arises is whether the body shape that we have generated closes. To check this, in order for the body to close, the total source/sink strength must be zero. Thus, we must have

$$\int_0^L m(\xi) d\xi = 0. \quad (2.3.9)$$

From equation (2.3.8), we see that this condition is met, providing $T(0) = T(L)$. In that part of the body where the thickness increases, from equation (2.3.8) it is seen that m is positive, so it is a source. Where the body thickness decreases, m is negative, so it is a sink. Where the body thickness does not change, we have neither a source nor a sink.

This analysis can be easily extended to curved two-dimensional slender bodies and to three-dimensional bodies. For the three-dimensional case, a similar analysis shows that the local source strength is again proportional to the rate of change of the body's area. For a uniform stream parallel to the z-axis and the body along the z-axis between a and b, the result is

$$\phi(r, z) = W\left[z - \frac{1}{4\pi}\int_a^b \frac{1}{[r^2 + (z-\xi)^2]^{3/2}}\frac{dA(\xi)}{d\xi}d\xi\right]. \qquad (2.3.10)$$

Cross-flow can be included in the three-dimensional case with a little more effort, the result being (since close to the slender body the flow appears to see what looks like a long cylinder of circular cross-section) a doublet facing the cross-flow and with strength equal to the cross-flow speed times the radius squared. For the cross-flow in the x direction, the result is

$$\phi(r, \theta, z) = Ur\cos\theta\left[1 + \frac{1}{2\pi}\int_a^b \frac{A(\xi)}{[r^2 + (z-\xi)^2]^{3/2}}d\xi\right].$$

Slender body theory has also been extended to curved, nonsymmetric bodies with circulation, and the procedure for obtaining accuracy to any order of ε has been explained. See van Dyke (1964), Cole (1968), and Moran (1984).

2.3.2   Panel Methods

The previous development is unfortunately only valid for slender bodies and, it turns out, bodies of relatively simple shape. Even in such cases, it is likely that you will not be able to carry out the integration of the source distribution potential by analytic means and must resort to numerical integration to be able to proceed with applying the results.

For more general shapes, surface distributions of sources and doublets allow us to find solutions for any body shape—in three as well as two dimensions. Computer solution of the resulting algebraic equations is a necessity, and the availability of high-speed computers has revolutionized our approach to these problems. All of the mathematical theory needed to carry out this approach was well known in the middle of the nineteenth century. However, use of this mathematics was impractical before computers.

To see the theory behind this, consider Green's theorem from Appendix A. If $\mathbf{x}$ is a point within the flow field, and $\mathbf{x}_s$ is a generic point on the surrounding surface S, then the potential corresponding to the distribution is

$$\phi(\mathbf{r}) = \iint_S [g(\mathbf{r}, \mathbf{r}_S)\mathbf{n}\cdot\nabla\phi(\mathbf{r}_S) - \phi(\mathbf{r}_S)\mathbf{n}\cdot\nabla g(\mathbf{r}, \mathbf{r}_S)]dS, \qquad (2.3.11)$$

where $g(\mathbf{r}, \mathbf{r}_S)$ is the potential for a source of unit strength located at $\mathbf{x}_s$ on the surface and $\mathbf{n}$ is a unit normal pointing *into* the flow. Thus, the Green's function in the two cases is given by

$$g(\mathbf{r}, \mathbf{r}_S) = \frac{-1}{4\pi\,|\mathbf{r} - \mathbf{r}_s|}\text{for three-dimensional flows,}$$

$$= \frac{1}{2\pi}\ln|\mathbf{r} - \mathbf{r}_s|\text{ for two-dimensional flows.} \qquad (2.3.12)$$

The operation $\mathbf{n} \cdot \nabla$ is equivalent to $\partial/\partial n$, the derivative in the direction locally normal to the surface dS. If $\mathbf{r}$ is the position vector on the interior of S, the left side of equation (2.3.11) is replaced by zero. Since g is the potential for a source, the first term in the integral can be interpreted as a source distribution of strength $\partial\phi/\partial n$ and the second as an outward-facing doublet distribution (recall that the derivative of the source potential is a doublet potential) of strength ϕ.

We could now proceed to set up the boundary conditions as was done for the slender body case, knowing that our solution already satisfies irrotationality and continuity exactly, since we are working with velocity potentials that have already been shown to satisfy the Laplace equation. Our effort, then, need only be directed toward satisfying the boundary conditions.

If we were to proceed directly, however, we would find ourselves with an embarrassment of riches. We would have at any point on the surface two unknowns: the source strength and the doublet strength. Roughly speaking, we would end up with twice as many unknowns as we have conditions. Presumably we could discard half of our unknowns—but which half is "best" discarded? And what are the consequences of our actions and their physical interpretation?

To clarify this, realize that as long as the body surface is a stream surface, we really don't care about any "flow" that our velocity potential may give inside the body. Thus, think of S as being a double surface—that is, a surface made up of our original surface plus a very slightly smaller surface of the same shape inside it. On this inner surface, we will construct a second source distribution that will have a velocity potential ϕ' inside of S and zero outside of S. Then, with $\mathbf{r}_s$ referring to a point outside of S, we have

$$0 = \iint_S [g(\mathbf{r}, \mathbf{r}_s)\mathbf{n} \cdot \nabla\phi'(\mathbf{r}_s) - \phi'(\mathbf{r}_s)\mathbf{n} \cdot \nabla g(\mathbf{r}, \mathbf{r}_s)]dS.$$

Adding this to equation (2.3.12), and recognizing that the preceding normal vector is the negative of the one in equation (2.3.11), the result is

$$\phi(\mathbf{r}_S) = \iint_S \{g(\mathbf{r}, \mathbf{r}_s)\mathbf{n} \cdot \nabla[\Delta\phi'(\mathbf{r}_s)] - [\Delta\phi'(\mathbf{r}_s)]\mathbf{n} \cdot \nabla g(\mathbf{r}, \mathbf{r}_s)\}dS, \qquad (2.3.13)$$

where $\Delta\phi$ represents the discontinuity in the velocity potential across the surface S, and $\mathbf{r}$ lies outside of S.

There are then two distinct choices that can be made concerning the flow inside the body:

1. $\Delta\phi = 0$ on S. In this case, the velocity potential is continuous across S. Then the second term in equation (2.3.13) vanishes, and only a source distribution remains.

2. $\mathbf{n} \cdot \Delta\phi = 0$ on S. In this case, the normal velocity is continuous across S. Then the first term in equation (2.3.13) vanishes, and only a doublet distribution remains.

If we are interested in flows with no lift forces, in principle either choice is correct, although the numerical implementation in one case may be easier and/or have greater accuracy than the other. If, however, we are interested in lifting flows, doublets must be included.

You may wonder why the only doublets that appear in equation (2.3.13) are those oriented perpendicular to the surface S. Why are there no doublets tangent to the surface? The reason for this is that if we were to include a tangent doublet distribution of the form

$$\phi_{td} = \iint_S \mathbf{B}_t \cdot \nabla_s g \, dS, \qquad (2.3.14)$$

where $\mathbf{B}_t$ is tangent to S and ∇_s involves derivatives with respect to $\mathbf{x}_s$, then with ϕ_{td} as the potential due to these tangent doublets, since

$$\mathbf{B}_t \cdot \nabla_s g = \nabla_s \cdot (\mathbf{B}_t g) - g \nabla_s \cdot \mathbf{B}_t,$$

$$\phi_{td} = \iint_S [\nabla_s \cdot (\mathbf{B}_t g) - g \nabla_s \cdot \mathbf{B}_t] dS \qquad (2.3.15)$$

$$= -\oint_C g \mathbf{B}_t \cdot (\mathbf{n} \times d\mathbf{s}) - \iint_S g \nabla_s \cdot \mathbf{B}_t dS.$$

The first integral on the right-hand side of equation (2.3.15) represents a ring of sources bounding the surface S. If the surface is closed, the integral disappears. The second integral is a surface source distribution of strength $\nabla_s \cdot \mathbf{B}_t$. Thus, the tangent doublet distribution is already realized in equation (2.3.13) through the source distribution.

The normal doublet distribution is equivalent to a ring vortex bounding the surface S plus a surface vortex distribution on S. This should not be surprising, as the velocity potential jump across S is typical of a vortex sheet, which itself is nothing more than a velocity potential discontinuity.

The proof of this equivalence is quite involved in three dimensions but elementary in two dimensions. Suppose we have a normal doublet distribution along a curve C extending from A to B. Then the velocity potential is

$$\phi_{\text{doublet}} = \int_A^B \mu \mathbf{n} \cdot \nabla_s \ln \sqrt{(x-x_s)^2 + (y-y_s)^2} ds$$

$$= \int_A^B \mu \frac{(x-x_s)n_x + (y-y_s)n_y}{(x-x_s)^2 + (y-y_s)^2} ds.$$

Defining the tangent vector $\mathbf{t}$ by $\mathbf{t} = \mathbf{n} \times \mathbf{k}$, then the components of the unit tangent are related to those of the unit normal by $t_x = n_y$, $t_y = -n_x$. Note also that

$$\frac{-(y-y_s)}{(x-x_s)^2 + (y-y_s)^2} = \frac{\partial}{\partial x} \left(\tan^{-1} \frac{y-y_s}{x-x_s} \right),$$

$$\frac{x-x_s}{(x-x_s)^2 + (y-y_s)^2} = \frac{\partial}{\partial y} \left(\tan^{-1} \frac{y-y_s}{x-x_s} \right),$$

Then the velocity potential for the doublet can be written as

$$\phi_{\text{doublet}} = -\int_A^B \mu \, \mathbf{t} \cdot \nabla_s \tan^{-1} \frac{y-y_s}{x-x_s} ds$$

$$= -\int_A^B \mu \, \mathbf{t} \cdot \nabla_s \phi_{\text{vortex}} \, ds = -\int_A^B [\mathbf{t} \cdot \nabla_s (\mu \phi_{\text{vortex}}) - \phi_{\text{vortex}} \mathbf{t} \cdot \nabla_s \mu] ds$$

$$= \mu \phi_{\text{vortex}} |_A^B + \int_A^B \phi_{\text{vortex}} \mathbf{t} \cdot \nabla_s \mu \, ds.$$

The first term is the two-dimensional version of the ring vortex; the second is the vortex distribution.

To illustrate the use of equation (2.3.13), we next demonstrate how it can be approximated in two dimensions to give the flow past a body of arbitrary shape. We first elect to use the option of a source distribution. The potential for flow of a uniform stream past the body represented by a source distribution is then, from equation (2.3.16), given by

$$\phi(\mathbf{r}) = xU + \frac{1}{2\pi} \iint_S m(\mathbf{r}_s) \ln \sqrt{(x-x_s)^2 + (y-y_s)^2} \, dS. \qquad (2.3.16)$$

To approximate the integration, the body will be covered with a series of M flat **panels** (in two dimensions, panels become straight lines). On each of these panels the source strength will be taken to be constant. Then equation (2.3.16) is replaced by the approximation

$$\phi(\mathbf{r}) = xU + U \sum_{j=1}^{M} \lambda_j I_j(\mathbf{r}), \qquad (2.3.17)$$

where $\lambda_j = m_j/2\pi U$ is evaluated on the jth panel and

$$I_j(r) = \int_{-L_j/2}^{L_j/2} \ln \sqrt{(x-x_s)^2 + (y-y_s)^2} \, ds_j.$$

Let (X_j, Y_j) and (X_{j+1}, Y_{j+1}) denote the endpoints of the jth panel (sometimes called **nodes**), (x_{cj}, y_{cj}) denote the center of the jth panel, and, referring to Figures 2.3.1 and 2.3.2,

$$x_s = x_{cj} + s_j \sin\beta_j, \, y_s = y_{cj} - s_j \cos\beta_j, \, L_j = \sqrt{(X_{j+1}-X_j)^2 + (Y_{j+1}-Y_j)^2},$$

and $x_{cj} = 0.5(X_j + X_{j+1})$, $y_{cj} = 0.5(Y_j + Y_{j+1})$.

To solve for the λ_j, require that at the center point of the ith panel with coordinates (x_{ci}, y_{ci}) the normal velocity must be zero. Since

$$\frac{\partial\phi}{\partial n} = \cos\beta \frac{\partial\phi}{\partial x} + \sin\beta \frac{\partial\phi}{\partial y},$$

differentiating equation (2.3.17) and forming the normal derivative gives, after carrying out the integration,

$$0 = \cos\beta_i + \pi\lambda_i + \sum_{j \neq i}^{M} \lambda_j I_{ij}, \, i = 1, 2, \ldots M, \qquad (2.3.18)$$

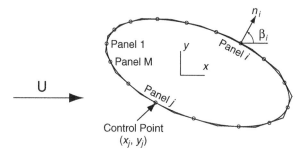

Figure 2.3.1 Paneled body

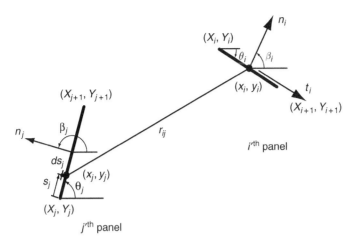

Figure 2.3.2 Panels i and j detail

where

$$I_{ij} = \frac{1}{2}\sin(\beta_j - \beta_i)\ln\frac{0.25L_j^2 + A_{ij}L_j + C_{ij}}{0.25L_j^2 - A_{ij}L_j + C_{ij}}$$
$$+ \cos(\beta_j - \beta_i)\left[\tan^{-1}\frac{0.5L_j + A_{ij}}{B_{ij}} - \tan^{-1}\frac{-0.5L_j + A_{ij}}{B_{ij}}\right],$$
$$A_{ij} = -(x_{ci} - x_{cj})\sin\beta_j + (y_{ci} - y_{cj})\cos\beta_j,$$
$$B_{ij} = (x_{ci} - x_{cj})\cos\beta_j + (y_{ci} - y_{cj})\sin\beta_j,$$

and

$$C_{ij} = (x_{ci} - x_{cj})^2 + (y_{ci} - y_{cj})^2.$$

Equation (2.3.18) represents M algebraic equations in the M unknowns I_{ij} and can readily be solved by most any algebraic solver.

Once the λ_j are known, the velocity at any point can be found. In particular, the tangential velocity on the ith body panel is given by

$$V_i = \frac{\partial\phi}{\partial(\text{tangent})} = \cos\theta_i\frac{\partial\phi}{\partial x} + \sin\theta_i\frac{\partial\phi}{\partial x} = \cos\theta_i + \sum_{j=1}^{M}\lambda_j J_{ij}, \; i = 1, 2, \ldots M, \quad (2.3.19)$$

where

$$J_{ij} = -\frac{1}{2}\cos(\beta_j - \beta_i)\ln\frac{0.25L_j^2 + A_{ij}L_j + C_{ij}}{0.25L_j^2 - A_{ij}L_j + C_{ij}}$$
$$+ \sin(\beta_j - \beta_i)\left[\tan^{-1}\frac{0.5L_j + A_{ij}}{B_{ij}} - \tan^{-1}\frac{-0.5L_j + A_{ij}}{B_{ij}}\right],$$

and

$$D_{ij} = (x_{ci} - x_{cj})\cos\beta_j + (y_{ci} - y_{cj})\sin\beta_j.$$

The tangential velocity is of use in computing the pressure at any point on the body through use of Bernoulli's theorem.

Several arbitrary decisions were made in setting up this approximation. For instance, we took the nodes to be on the body, so our "panelized body" is the polygon inscribed by the body. Alternately, we could have taken the center of the panel on the body, in which case the nodes are outside of the body and the panelized body is a superscribing polygon. Further, our choice of where to put the nodes, as well as the number of nodes, is our engineering decision. Certainly we want the nodes to be placed closer together where the curvature of the body is greatest. Generally, we would expect that the more nodes there are, the better, but factors other than the number of nodes can be more important. For instance, to find the flow past a symmetric body such as a circular cylinder, a panel scheme that represents all symmetries of the problem will generally fare better than a scheme with perhaps more nodes, but that doesn't reflect the symmetries.

Clearly, curved body panels with the source strength varying along the panel in a prescribed fashion is also an option. The result should be more accurate, with necessarily more analysis. For example, with the previous approximation, the velocities at the endpoints of the linear panels are infinite. Curved panels could eliminate these singularities and make the velocity continuous. Thus, there is a certain amount of "art" in actually carrying out the modeling. The present simple approach is sufficient to illustrate the method.

The aerospace industry has developed these programs to a high degree of sophistication, modeling such flows as a Boeing 747 carrying piggyback a space vehicle such as the space shuttle, even including details such as flow into the engines and the like. Panel methods are now an important design tool, particularly in external flows.

We could also have elected to use the option of a normal doublet distribution so that it would be possible to simulate flow past lifting bodies. The potential for flow of a uniform stream past the body would then be given by

$$\phi(\mathbf{r}) = xU - \frac{1}{2\pi} \iint_S \Delta\phi(\mathbf{r}_s)\mathbf{n} \cdot \nabla \ln \sqrt{(x-x_s)^2 + (y-y_s)^2} \, dS. \tag{2.3.20}$$

Again, to carry out the integration, the body is covered with a series of M straight line panels. On these panels the doublet strength is taken to be constant. Then equation (2.3.20) is replaced by the approximation

$$\phi(\mathbf{r}) = xU + U \sum_{j=1}^{M} \mu_j K_j(\mathbf{r}), \tag{2.3.21}$$

where $\mu_j = \Delta\phi/2\pi U$ on the jth panel, and, referring to Figures 2.3.1 and 2.3.2,

$$K_j(\mathbf{r}) = \int_{-L_j/2}^{L_j/2} \frac{(x-x_s)\cos\beta_j + (y-y_s)\sin\beta_j}{(x-x_s)^2 + (y-y_s)^2} \, ds_j,$$

where the notation is as before. For this case the integration can be carried out explicitly, giving

$$K_j(\mathbf{r}) = \tan^{-1} \frac{0.5L_j - P_j}{R_j} - \tan^{-1} \frac{-0.5L_j - P_j}{R_j}, \tag{2.3.22}$$

where

$$P_j(\mathbf{r}) = (x - X_j)\sin\beta_j - (y - Y_j)\cos\beta_j,$$
$$R_j(\mathbf{r}) = (x - X_j)\cos\beta_j + (y - Y_j)\sin\beta_j.$$

The physical interpretation of this solution can be seen by noting that $P_j = (\mathbf{R} - \mathbf{r}_c) \cdot \mathbf{t}_j$, $R_j = (\mathbf{R} - \mathbf{r}_c) \cdot \mathbf{n}_j$, $0.5L_j - P_j = (\mathbf{R}_{j+1} - \mathbf{r}) \cdot \mathbf{t}_j$, and $-0.5L_j - P_j = (\mathbf{R}_j - \mathbf{r}) \cdot \mathbf{t}_j$, where $\mathbf{n}_j$ and $\mathbf{t}_j$ are the unit normal and tangent to the jth panel. Then $K_j(\mathbf{r})$ is the potential of a pair of opposite-signed vortices at the nodes at the panel ends (located at $\mathbf{R}_j$ and $\mathbf{R}_{j+1}$).

Solving for the μ_j, on the ith panel, again requires that the normal velocity at the center point must be zero. Differentiating equation (2.3.22) and then dividing by U, the result is

$$0 = \cos\beta_i + \pi\mu_i + \sum_{j=1}^{M} \mu_j N_{ij}, \quad i = 1, 2, \ldots M, \tag{2.3.23}$$

where

$$N_{ij} = -R_{ji}\sin(\beta_j - \beta_i) - \frac{(0.5L_j - P_{ji})\cos(\beta_j - \beta_i)}{(L_j - P_j)^2 + R_{ji}^2} + R_{ji}\sin(\beta_j - \beta_i)$$
$$- \frac{(0.5L_j - P_{ji})\cos(\beta_j - \beta_i)}{(0.5L_j - P_{ji})^2 + R_{ji}^2 + P_{ji}^2}.$$

The second subscript (i) in the P and R indicates evaluation of the particular quantity at $\mathbf{r}_i$. That is, we have $P_{ji} = P_j(\mathbf{r}_i)$, $R_{ji} = R_j(\mathbf{r}_i)$.

Equation (2.3.23) represents M algebraic equations in the M unknowns μ_j. The solution of this system of equations is not usually as simple as in the case of the source distribution, since the system is poorly conditioned and very sensitive to roundoff errors and the like. In algebraic terms, this means that there is a great disparity in magnitude among the various eigenvalues of the system. Systems that behave in this manner are called *stiff systems*. (Similar problems can arise in systems of differential equations. In the case of differential equations, they are characterized by having solutions that contain terms that are negligible except in specific small regions.) Stiff systems can render standard algebraic and differential equation methods useless. Thus, much greater care must be used in finding its solution than in the case of the source distribution, and special methods suited to dealing with such cases must be used.

Since a pure doublet distribution is difficult to work with, it is sometimes preferable to use a source distribution supplemented by a doublet distribution, where all of the doublet strengths have the same value. This adds one unknown to the set (the doublet strength), and another condition must be added to have a determinate system. One possibility is the *Kutta condition*, discussed in Chapter 3.

2.4   Forces Acting on a Translating Sphere

For a sphere translating in a real fluid, energy will be dissipated by viscous stresses and must be replenished, even for the sphere to maintain a constant velocity. In inviscid flows, a translating sphere will leave the energy unchanged unless the sphere is to be accelerated with respect to the flow, if the sphere sees boundary changes such as an uneven wall or a free surface, or if a cavity is allowed to form in the fluid. In these cases the fluid kinetic energy can change, meaning that the sphere must exert a force on the fluid so the fluid exerts an equal and opposite force on the sphere. These forces are an inviscid drag force on the sphere, exerted by the pressure acting on the sphere.

First consider a sphere of radius b moving at velocity U along the z-axis. In equation (2.2.42) we found that for a stationary sphere in a moving stream, the velocity potential was given by $\phi = zU + \frac{zb^3 U}{2(r^2+z^2)^{3/2}}$. Thus, for a sphere moving in an otherwise stationary fluid the potential is given by subtracting the uniform velocity, giving

$$\phi = \frac{zb^3 U}{2(r^2+z^2)^{3/2}} = \frac{b^3 U \cos\beta}{2R^2}. \tag{2.4.1}$$

The fluid velocity as seen from the perspective of the sphere is then

$$v_R = \frac{\partial\phi}{\partial R} = -\frac{b^3 U \cos\beta}{R^3}, \quad v_\beta = \frac{1}{R}\frac{\partial\phi}{\partial\beta} = -\frac{b^3 U \sin\beta}{2R^3}. \tag{2.4.2}$$

We shall find the force on the sphere by two methods. The first is a global consideration of the rate of work and change of energy. If the sphere is to be accelerated, the fluid kinetic energy has to be changed. Thus, the first task is to compute the kinetic energy of the fluid.

Throughout the region $R \geq b$, the square of the speed is given by equation $v_R^2 + v_\beta^2 = U^2 \left(\frac{b}{R}\right)^6 \left(\cos^2\beta + \frac{1}{4}\sin^2\beta\right)$. Accordingly, by equation (2.4.2) the kinetic energy is given by

$$T = \frac{1}{2}U^2 \int_0^\pi \int_b^\infty \left(\frac{b}{R}\right)^6 (\cos^2\beta + \frac{1}{4}\sin^2\beta)2\pi R^2 \sin\beta\, dR\, d\beta = \frac{\pi b^3}{3}U^2.$$

The rate of work done by the force moving the sphere is $\mathbf{F}\cdot\mathbf{U} = F_z U = dT/dt$; therefore,

$$F_z U = \frac{d}{dt}\left(\frac{\pi b^3 U^2}{3}\right) = \frac{2}{3}\pi b^3 U \frac{dU}{dt},$$

giving finally

$$F_z = \frac{2}{3}\pi b^3 \frac{dU}{dt}. \tag{2.4.3}$$

Note that it was necessary to compute the kinetic energy of a body moving in a quiescent fluid. If the uniform stream had been left in the picture, the kinetic energy would have been infinite.

The force can also be found by direct integration of the pressure over the surface of the sphere. Since the sphere is moving and coordinates relative to the sphere are being used, the pressure is found from the Bernoulli equation in translating coordinates to be given by

$$\frac{p}{\rho} + \frac{\partial\phi}{\partial t} - U\frac{\partial\phi}{\partial z} + \frac{1}{2}\left(v_R^2 + v_\beta^2\right) = \frac{p_\infty}{\rho}, \text{ or}$$

$$\frac{p - p_\infty}{\rho} = -\frac{\partial\phi}{\partial t} + U\frac{\partial\phi}{\partial z} - \frac{1}{2}\left(v_R^2 + v_\beta^2\right)$$

$$= -\frac{dU}{dt}\frac{b^3\cos\beta}{2R^2} + U\frac{b^3\left(1 - 3\cos^2\beta\right)}{2R^3} - \frac{1}{2}U^2\left(\frac{b}{R}\right)^6\left(\cos^2\beta + \frac{1}{4}\sin^2\beta\right).$$

On the surface of the sphere this becomes

$$\frac{p - p_\infty}{\rho} = -\frac{dU}{dt}\frac{b\cos\beta}{2} + U\frac{\left(1 - 3\cos^2\beta\right)}{2} - \frac{1}{2}U^2\left(\cos^2\beta + \frac{1}{4}\sin^2\beta\right).$$

To compute the pressure force on any body due to inviscid effects, it is necessary to carry out the integration

$$\mathbf{F} = -\iint p\mathbf{n} \cdot d\mathbf{A} = -2\pi\mathbf{k}\int_0^\pi p\cos\beta\sin\beta d\beta$$

$$= \pi\rho b\frac{dU}{dt}\mathbf{k}\int_0^\pi \cos^2\beta\sin\beta d\beta = \pi\rho b\frac{dU}{dt}\mathbf{k}\left[-\frac{1}{3}\cos^3\beta\right]_0^\pi = \frac{2}{3}\pi\rho b\frac{dU}{dt}\mathbf{k}, \quad (2.4.4)$$

where $\mathbf{n}$ is the unit outward normal on the body surface and the integration is taken over the entire surface of the body. This second result is, of course, the same as equation (2.4.3), found by energy considerations.

If the sphere is moving at a constant velocity, the drag force due to pressure vanishes. When it is accelerating, the drag force on a sphere is proportional to the acceleration, so the constant of proportionality must have the dimension of mass. Thus, we write $F = m_{\text{added}}dU/dt$, where m_{added} is termed the **added mass**. In the case of the sphere, $m_{\text{added}} = 2/3\rho b^3$, which is one-half the mass of the displaced fluid. Thus, the total force needed to move the sphere is, by virtue of Newton's law, $F = (m_{body} + m_{added})\frac{dU}{dt}$. The combination $m_{\text{body}} + m_{\text{added}}$ is sometimes called the **virtual mass**.

Although both approaches are capable of finding the added mass, the energy method often is the simplest.

2.5 Added Mass and the Lagally Theorem

For inviscid flows, drag forces for translating bodies will always be of the form acceleration times added mass, where added mass will be some fraction of the mass of the fluid displaced by the body. In general, if the body in a quiescent fluid is translating with a velocity U and rotating with an angular velocity $\mathbf{\Omega}$, the velocity potential can be written in terms of axes fixed to the body in the form

$$\phi = \sum_1^6 V_\alpha\phi_\alpha, \quad \text{where } V_\alpha = (U_x, U_y, U_z, \Omega_x, \Omega_y, \Omega_z).$$

Then the kinetic energy will be of the form

$$T = \frac{1}{2}\sum_{\alpha=1}^6\sum_{\beta=1}^6 A_{\alpha\beta}V_\alpha V_\beta, \quad \text{where}$$

$$A_{\alpha\beta} = \rho\iiint_{\text{fluid volume}} \nabla\phi_\alpha \cdot \nabla\phi_\beta dV = \rho\iint_{\text{body surface}} \phi_\alpha\frac{\partial\phi_\beta}{\partial n}dS.$$

$$(2.5.1)$$

The $A_{\alpha\beta}$ are the components of what is referred to as the **added mass tensor**. Of the 36 possible $A_{\alpha\beta}$, only 21 are distinct, since $A_{\alpha\beta} = A_{\beta\alpha}$. A_{11}, A_{22}, A_{33} represent pure translation and have the dimension of mass. A_{44}, A_{55}, A_{66} are for pure rotations and have the dimension of mass times length squared. The remaining terms represent interactions between translations and rotations and have the dimension of mass times length.

The name added mass is clarified if we consider the kinetic energy of the solid body without the fluid being present. That is given by

$$T_B = \frac{1}{2}\iiint_{\text{body volume}} \rho_{\text{body}}\left[(V_1 + V_5x_3 - V_6x_2)^2 + (V_2 + V_6x_1 - V_4x_3)^2\right.$$

$$\left.+ (V_3 + V_4x_2 - V_5x_1)^2\right]dV = \frac{1}{2}\sum_{\alpha=1}^6\sum_{\beta=1}^6 B_{\alpha\beta}V_\alpha V_\beta, \quad (2.5.2)$$

where

$$B_{11} = B_{22} = B_{33} = M,$$

$$B_{44} = I_{11}, \ B_{55} = I_{22}, \ B_{33} = I_{33}, \ B_{12} = -I_{12}, \ B_{13} = -I_{13}, \ B_{23} = -I_{23}, \tag{2.5.3}$$

$$B_{26} = 2M\bar{x}, \ B_{35} = -2M\bar{x}, \ B_{34} = 2M\bar{y}, \ B_{16} = -2M\bar{y}, \ B_{15} = 2M\bar{z}, \ B_{24} = -2M\bar{z}.$$

Here M is the mass of the body, the Is are components of the mass moment of inertia of the body, and the barred coordinates are the position of the body center of mass. The total kinetic energy of the body is then

$$T + T_{\text{body}} = \frac{1}{2} \sum_{\alpha=1}^{6} \sum_{\beta=1}^{6} \left(A_{\alpha\beta} + B_{\alpha\beta} \right) V_\alpha V_\beta. \tag{2.5.4}$$

To move the body through the fluid, then, the energy change required is the same as if its inertia was increased by the added mass.

For bodies made up of source, sink, and doublet distributions, the calculations for $M_{\alpha\beta}$ have been variously evaluated by Taylor (1928), Birkhoff (1953), and Landweber (1956). Their results do not include the possibility of vortex lines or distributed singularities, so using their results are limited to non-lifting cases.

As just mentioned, flows can be unsteady even for bodies that move at constant velocities if their position relative to boundaries varies. The general calculation for such cases was carried out originally by Lagally (1922) and have come to be known as *Lagally forces* and moments. Lagally's results were later generalized by Cummins (1953) and by Landweber and Yih (1956). The derivations and formulae are lengthy and somewhat complicated, so they will not be repeated here. Again, vorticity is not included in their results.

We can see from simple global momentum analysis that for steady flows that involve translating bodies generated solely by source-sink distributions inside or on the body surface, no drag forces are found. It is, however, possible for such bodies to generate forces perpendicular to the direction of translation. These forces are the *lift forces* that normally one would expect to find even with the neglect of viscous effects. This absence of lift in the formulation can be corrected by including vorticity in any model where lift forces are desired. For instance, for the cylinder in the previous example, including a vortex at the center of the cylinder would give the velocity potential and stream function

$$\phi = xU + \frac{xB_x}{x^2 + y^2} + \frac{\Gamma}{2\pi} \tan^{-1} \frac{y}{x}, \quad \psi = yU - \frac{yB_x}{x^2 + y^2} + \frac{\Gamma}{2\pi} \ln\sqrt{x^2 + y^2}. \tag{2.5.5}$$

We can see that the stream function is constant on the cylinder $x^2 + y^2 = a^2$; therefore, the boundary condition on the body is still satisfied. Evaluation of the pressure force now, however, gives a lift force proportional to $\rho U \Gamma$, called the *Magnus effect* after its discoverer.

Where would this vorticity come from in a physical situation? The cylinder could be caused to rotate, and the effects of viscosity then provide the tangential velocity that is provided in our mathematical model by the vortex. This has been attempted in ships (the Flöettner rotor ship, Cousteau's *Alcone*) and experimental airplanes, but it requires an additional power source and is not generally practical. The effect of this rotation is instead provided by having a sharp trailing edge for the lifting surface, or by providing a "flap" on a blunter body. This is done to force the velocity on the body to appear the same as in our model and thereby generate the desired force. The relationship between lift force and vorticity is called the *Joukowski theory of lift*.

For the flow given in equation (2.5.2), the potential can be expressed in cylindrical polar coordinates as

$$\phi = Ur\cos\theta\left(1+\frac{b^2}{r^2}\right)+\frac{\Gamma}{2\pi}\theta, \tag{2.5.6}$$

giving velocity components

$$v_r = \frac{\partial\phi}{\partial r} = U\cos\theta\left(1-\frac{b^2}{r^2}\right), \quad v_\theta = \frac{1}{r}\frac{\partial\phi}{\partial\theta} = -U\sin\theta\left(1+\frac{b^2}{r^2}\right)+\frac{\Gamma}{2\pi r}. \tag{2.5.7}$$

On the boundary $r = b$, the radial component vanishes and the tangential component becomes $v_\theta = -2U\sin\theta + \Gamma/2\pi b$. Stagnation points are thus located at $\sin\theta = \Gamma/4\pi Ub$. The techniques for lift generation mentioned previously amount to control of the locations of the stagnation points, thus generating the circulation necessary for lift. Chapter 3 discusses this in more detail.

2.6 Theorems for Irrotational Flow

Many theorems for irrotational flows exist that are helpful both in analysis and computational methods. They are also helpful for understanding the properties of such flows.

2.6.1 Mean Value and Maximum Modulus Theorems

Theorem: The mean value of ϕ over any spherical surface throughout whose interior $\nabla^2\phi = 0$ is equal to the value of ϕ at the center of the sphere.

   Proof: Let ϕ_c be the value of ϕ at the center of a sphere of radius R, and ϕ_m be the mean value of ϕ over the surface of that sphere. Then

$$\phi_m = \frac{1}{4\pi R^2}\iint_S \phi\, dS = \frac{1}{4\pi}\int \phi\, d\varpi,$$

where $d\varpi = dS/R^2$ is the solid angle subtended by dS. From Green's theorem, it follows that

$$\iiint_V \nabla^2\phi\, dV = 0 = \iiint_V \nabla\cdot(\nabla\phi)dV = \iint_S \mathbf{n}\cdot\nabla\phi\, dS = \iint_S \frac{\partial\phi}{\partial R}dS = R^2\frac{d}{dR}\int\phi\, d\varpi.$$

Therefore, the integral $\int\phi\, d\varpi$ must be independent of the radius R. In particular, if we let R approach zero, $4\pi\phi_m = \int\phi\, d\varpi \to 4\pi\phi_c$, proving the theorem. While we have proven this theorem in three dimensions, a similar theorem can be proved in two dimensions.

   The important concept to be learned from this is that the Laplace equation is a great averager. At any point the solution is the average of its value at all neighboring points. This is the basis for several numerical methods used in the solution of Laplace's equation.

2.6.2 Maximum-Minimum Potential Theorem

Theorem: The potential ϕ cannot take on a maximum or minimum in the interior of any region throughout which $\nabla^2\phi = 0$.

   Proof: Again, taking a sphere around a point in a region where $\nabla^2\phi = 0$, if ϕ_c were a maximum, from the previous theorem it would be greater than the value of ϕ at all points on the surface of the sphere, which contradicts the previous theorem.

2.6.3   Maximum-Minimum Speed Theorem

Theorem: For steady irrotational flows both the maximum and minimum speeds must occur on the boundaries.

Proof: Consider two points P and Q that are close together. Choose axes at P such that $|\mathbf{v}_P|^2 = (\partial\phi/\partial x)_P^2$ and $|\mathbf{v}_Q|^2 = |\nabla\phi|_Q^2$. Since $\nabla^2(\partial\phi/\partial x) = 0$, from the previous theorem we can find a point Q such that $(\partial\phi/\partial x)_P^2 < (\partial\phi/\partial x)_Q^2$, and therefore $|\mathbf{v}_P|^2 < |\mathbf{v}_Q|^2$. Thus, $|\mathbf{v}_P|^2$ cannot be a maximum in the interior. It also follows by the same argument that for steady irrotational flows the speed must also take on a minimum on the boundaries. The previous theorems hold only if $\nabla^2\phi = 0$ *throughout* the region. Generally, if there are interior boundaries,[2] modifications of the above result may be necessary.

2.6.4   Kelvin's Minimum Kinetic Energy Theorem

Theorem: The irrotational motion of a liquid that occupies a simply connected region (that is, a region where all closed surfaces in the region can be shrunk to a point without crossing boundaries) has less kinetic energy than any other motion with the same normal velocity at the boundary. This theorem is credited to Lord Kelvin (born William Thomson, 1824–1907) and published in 1849.

Proof: Let $\mathbf{v}_Q$ represent the velocity field of any motion satisfying $\nabla \cdot \mathbf{v}_Q = 0$ and having kinetic energy T_Q. Let ϕ be the velocity potential of an irrotational flow having kinetic energy T_ϕ. Let the boundary have a normal velocity component $\mathbf{v}_n$ so that $\partial\phi/\partial n = \mathbf{v}_Q \cdot \mathbf{n} = \mathbf{v}_n$ on the boundary (i.e., we are requiring that both velocity fields satisfy the same boundary condition).

Define $\mathbf{v}_{\text{difference}} = \mathbf{v}_Q - \nabla\phi$, and note that $\mathbf{v}_{\text{difference}} = 0$ on the boundary. Then

$$T_Q = \frac{1}{2}\rho \iiint_V \mathbf{v}_Q \cdot \mathbf{v}_Q dV = \frac{1}{2}\rho \iiint_V (\mathbf{v}_{\text{difference}} + \nabla\phi) \cdot (\mathbf{v}_{\text{difference}} + \nabla\phi) dV$$

$$= \frac{1}{2}\rho \iiint_V \mathbf{v}_{\text{difference}} \cdot \mathbf{v}_{\text{difference}} dV + \frac{1}{2}\rho \iiint_V \nabla\phi \cdot \nabla\phi dV + \rho \iiint_V \nabla\phi \cdot \mathbf{v}_{\text{difference}} dV$$

$$= T_{\text{difference}} + T_\phi + \iiint_V \nabla\phi \cdot \mathbf{v}_{\text{difference}} dV.$$

But $\mathbf{v}_{\text{difference}} \cdot \nabla\phi = \nabla \cdot (\phi\mathbf{v}_{\text{difference}}) - \phi\nabla \cdot \mathbf{v}_{\text{difference}} = \nabla \cdot (\phi\mathbf{v}_{\text{difference}})$ because its two components satisfy the continuity equation, the difference velocity also must satisfy the continuity equation. Thus,

$$\iiint_V \nabla\phi \cdot \mathbf{v}_{\text{difference}} dV = \iiint_V \nabla \cdot (\phi\mathbf{v}_{\text{difference}}) dV = \iint_S \phi\mathbf{v}_{\text{difference}} \cdot \mathbf{n} dS = 0.$$

It follows that   (2.6.1)

$$T_Q = T_{\text{difference}} + T_\phi \geq T_\phi \text{ since } T_{\text{difference}} \geq 0.$$

---

[2] Such regions are said to be ***periphractic***.

The equality holds only if $\mathbf{v}_{\text{difference}}$ is zero everywhere in the region.

Note also for future use that

$$
\begin{aligned}
T_\phi &= \frac{1}{2}\rho \iiint_V \nabla\phi \cdot \nabla\phi\, dV = \frac{1}{2}\rho \iiint_V \left[\nabla \cdot (\phi\nabla\phi) - \phi\nabla^2\phi\right] dV \\
&= \frac{1}{2}\rho \iiint_V \nabla \cdot (\phi\nabla\phi)\, dV = \sum_{\substack{\text{all} \\ \text{boundaries}}} \frac{1}{2}\rho \iint_S \phi\frac{\partial\phi}{\partial n}\, dS.
\end{aligned}
\tag{2.6.2}
$$

2.6.5 Maximum Kinetic Energy Theorem

Theorem: The irrotational motion of a liquid occupying a simply connected region (that is, a region where all closed surfaces in the region can be shrunk to a point without crossing boundaries) has more kinetic energy than any other irrotational motion that satisfies the boundary conditions only in the average.

Proof: Let ϕ_Q represent the velocity field of any motion satisfying $\nabla^2\phi_Q = 0$ and having kinetic energy T_Q. Let ϕ be the velocity potential of an irrotational flow having kinetic energy T_ϕ. Let the boundary have a normal velocity component $\mathbf{v}_n$ so that $\partial\phi/\partial n = \mathbf{v}_n$ on the boundary. The potential ϕ_Q satisfies the averaged boundary condition

$$
\iint_S \phi_Q \left(\mathbf{v}_n - \frac{\partial\phi_Q}{\partial n}\right) dS \geq 0.
$$

Define $\phi_{\text{difference}} = \phi - \phi_Q$. Then

$$
\begin{aligned}
T_\phi &= \frac{1}{2}\rho \iiint_V \nabla\phi \cdot \nabla\phi\, dV = \frac{1}{2}\rho \iiint_V \left(\nabla\phi_Q + \nabla\phi_{\text{difference}}\right) \cdot \left(\nabla\phi_Q + \nabla\phi_{\text{difference}}\right) dV \\
&= \frac{1}{2}\rho \iiint_V \nabla\phi_Q \cdot \nabla\phi_Q\, dV + \frac{1}{2}\rho \iiint_V \nabla\phi_{\text{difference}} \cdot \nabla\phi_{\text{difference}}\, dV \\
&\quad + \rho \iiint_V \nabla\phi_Q \cdot \nabla\phi_{\text{difference}}\, dV \\
&= T_Q + T_{\text{difference}} + \rho \iiint_V \nabla\phi_Q \cdot \nabla\phi_{\text{difference}}\, dV.
\end{aligned}
\tag{2.6.3}
$$

But because its two potentials satisfy the Laplace equation, it follows that

$$
\nabla\phi_{\text{difference}} \cdot \nabla\phi_Q = \nabla \cdot \left(\phi_Q\nabla\phi_{\text{difference}}\right) - \phi_Q\nabla \cdot \nabla\phi_{\text{difference}} = \nabla \cdot \left(\phi_Q\nabla\phi_{\text{difference}}\right)
$$

and so

$$
\iiint_V \nabla\phi_Q \cdot \nabla\phi_{\text{difference}}\, dV = \iiint_V \nabla \cdot \left(\phi_Q\nabla\phi_{\text{difference}}\right) dV = \iint_S \phi_Q\left(v_n - \frac{\partial\phi_Q}{\partial n}\right) dS \geq 0.
$$

Thus

$$
T_\phi = T_Q + T_{\text{difference}} \geq T_Q \quad \text{since} \quad T_{\text{difference}} \geq 0.
\tag{2.6.4}
$$

The equality holds only if $\mathbf{v}_{\text{difference}}$ is zero everywhere in the region.

This and the previous theorem give us upper and lower bounds on the kinetic energy. This also is useful for estimating the added mass coefficients and thus the drag forces in accelerating flows.

2.6.6   Uniqueness Theorem

Theorem: There cannot be two different forms of acyclic irrotational motion[3] of a confined mass of a fluid in which the boundaries have prescribed values.

Proof by contradiction: Suppose we have two velocity potentials ϕ_1 and ϕ_2, both satisfying the Laplace equation and the same prescribed boundary conditions. Let $\phi_3 = \phi_1 - \phi_2$. Then the normal derivative of ϕ_3 vanishes on the boundary, and it follows that the kinetic energy is zero. Thus, the velocity associated with ϕ_3 is zero everywhere, and ϕ_3 must be a function of at most time. Notice that this breaks down if the velocity potential is multivalued, which we have already seen is true for flows with concentrated vortices. It also breaks down if there are cavities in the flow.

2.6.7   Kelvin's Persistence of Circulation Theorem

Theorem: For an irrotational flow whose density is either constant or a function of pressure, as we follow a given closed circuit the circulation does not change. This theorem is credited to Lord Kelvin, and was published in 1869.

Proof: In equation (1.11.7) circulation was defined by $\Gamma = \oint_C \mathbf{v} \cdot \mathbf{ds}$. If we take the time derivative following the same fluid particles as they move downstream, we have

$$\frac{D\Gamma}{Dt} = \frac{D}{Dt} \oint_C \mathbf{v} \cdot \mathbf{ds} = \oint_C \left(\frac{D\mathbf{v}}{Dt} \cdot \mathbf{ds} + \mathbf{v} \cdot \frac{D\mathbf{ds}}{Dt} \right) = \oint_C \left[\left(-\frac{1}{\rho} \nabla p + \mathbf{g} \right) \cdot \mathbf{ds} + \mathbf{v} \cdot d\mathbf{v} \right].$$

If $\mathbf{g}$ is a conservative body force such as gravity and the conditions on density stated above are met, then the integrand is an exact differential. Since the path is closed, this means that

$$\frac{D\Gamma}{Dt} = 0. \tag{2.6.5}$$

2.6.8   Weiss and Butler Sphere Theorems

Weiss sphere theorem: For a velocity potential ϕ' having no rigid boundaries and with all singularities at least a distance a from the origin, the flow due to a rigid sphere of radius a introduced into the flow at the origin is characterized by the velocity potential

$$\phi(\mathbf{R}) = \phi'(\mathbf{R}) + \frac{a}{R} \phi' \left(\frac{a^2 \mathbf{R}}{|\mathbf{R}|^2} \right) - \frac{2}{a|\mathbf{R}|} \int_0^a \lambda \phi' \left(\frac{\lambda^2 \mathbf{R}}{|\mathbf{R}|^2} \right) d\lambda. \tag{2.6.6}$$

Butler sphere theorem: If the preceding flow is also axisymmetric and characterized by a stream function $\psi'(R, \theta)$, where R, θ are spherical polar coordinates, then the flow due to introduction of the sphere is given by

$$\psi(R, \beta) = \psi'(R, \beta) - \frac{R}{a} \psi' \left(\frac{a^2}{R}, \beta \right). \tag{2.6.7}$$

The proof of the Weiss theorem is fairly long and will be omitted here.

The proof of the Butler sphere theorem follows from the Weiss theorem quite simply. Substitution of a for R makes the right-hand side of equation (2.6.7) vanish.

---

[3] An *acyclic irrotational motion* is one for which the velocity potential is a single-valued function.

Problems—Chapter 2

2.1 Examine the following functions to see if they could represent inviscid irrotational flows.

    a. $\phi = x + y + z$.

    b. $\phi = x + xy + xyz$.

    c. $\phi = x^2 + y^2 + z^2$.

    d. $\phi = zx^2 - x^2 - z^2$.

    e. $\phi = \sin(x + y + z)$.

    f. $\phi = \ln x$.

2.2 Show that if $\nabla^2 \phi_1 = 0$ and $\nabla^2 \phi_2 = 0$, then $\phi_1 + \phi_2$, $C\phi_1$, and $C + \phi_1$ also satisfy the Laplace equation. Here C is a function of time.

2.3 Given a flow field written in terms of a scalar ϕ, with the velocity defined by $\mathbf{v} = \nabla\phi$, indicate under what conditions the velocity field satisfies each of the following.

    a. irrotationality.

    b. conservation of mass.

    c. the Euler equation.

    d. the Navier-Stokes equation.

2.4 The potential ϕ, the velocity potential, is defined by $\mathbf{v} = \nabla\phi$, stating that the velocity is the gradient of this potential. This means that the velocity is orthogonal to equipotential surfaces. For inviscid flows, find a corresponding potential whose gradient is the acceleration.

2.5 A shear flow has the velocity field $v = (U + yS, 0, 0)$, where U and S are constants.

    a. Find the circulation about a unit square with the bottom side on the x-axis.

    b. Find the circulation about the unit circle centered at the origin.

    c. Find the circulation about the ellipse with major axis of length 2 and minor axis of length 1. The ellipse is centered at the origin.

2.6 Select a proper combination of uniform stream and source strength to generate an axisymmetric Rankine body 1.4 meters in diameter. The pressure difference between ambient and stagnation is 5 kilo-Pascals. The fluid is water ($\rho = 1000 \, \text{kg/meter}^3$)

2.7 a. Determine the stream function for a simple axisymmetric Rankine body that is 3 meters long and 2 meters maximum in diameter. It is in a uniform stream of strength 3 meters/second. The fluid is water.

    b. Find the pressures at the stagnation points and at the place of largest diameter.

2.8 Find the length and breadth of the closed three-dimensional axisymmetric body that is formed by a distributed line sink of strength m per unit length extending from $(-1, 0)$ to $(0, 0)$, a distributed line source of the same strength from $(0, 0)$ to $(1, 0)$, and a uniform flow along the z-axis. Use $m = 20$ and $U = 5$.

2.9 Find the drag force on an accelerating sphere of radius a by integrating the pressure over the body.

2.10 A solid sphere 30 mm in diameter and having a specific gravity of 7 is released in still water. Find its initial acceleration.

2.11 A point doublet of strength μ is located at $r = 0$, $z = b$. Find the stream function if a sphere of radius $a < b$ is inserted at the origin.

2.12 a. Find the equation that the Stokes stream function must satisfy to represent an irrotational flow in three dimensions. Use spherical polar coordinates.

b. Use separation of variables in the form $\psi(R, \theta) = F(R)T(\theta)$, and find the equations that F and T must satisfy.

Chapter 3

Irrotational Two-Dimensional Flows

3.1 Complex Variable Theory Applied to Two-Dimensional Irrotational Flow

The theory of complex variables is ideally suited to solving problems involving two-dimensional flow. The term *complex variable* means that a quantity consists of the sum of a real and an imaginary number. An imaginary number is a real number multiplied by the imaginary number $i = \sqrt{-1}$. (The terms *imaginary* and *complex* distinguish these numbers from real numbers.) A complex number is in fact the sum of two real numbers, the second one being multiplied by the square root of minus one. In many ways complex variable theory is simpler than real variable theory and much more powerful.

Briefly, a complex function F that depends on the coordinates x and y is written in the form

$$F(x, y) = f(x, y) + ig(x, y), \tag{3.1.1}$$

where f and g are real functions. This type of representation has some of the properties of a two-dimensional vector, with the real part standing for the x component and the imaginary part the y component. Thus, the complex number represented by equation (3.1.1) has the directionality properties of the unit vector representation $\boldsymbol{F} = f\mathbf{i} + g\mathbf{j}$, (here $\mathbf{i}$ and $\mathbf{j}$ are Cartesian unit vectors), at least as far as representation and transformation of

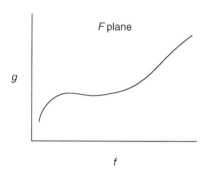

Figure 3.1.1 Complex variable—general complex plane

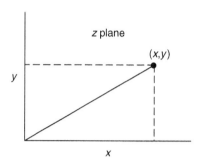

Figure 3.1.2 Complex variable—z plane

coordinates is concerned. The two forms of representation differ considerably, however, in operations like multiplication and division.

A complex function F can be represented in graphical form as in Figure 3.1.1, and a spatial position can be represented as in Figure 3.1.2, where the horizontal axis is the x-axis, and the vertical axis is the y-axis. To denote the position vector, it is traditional to write $z = x + iy$, where z is a complex number (z here is *not* the third-dimensional coordinate) representing the position of a point in space. The plane containing the x- and y-axes is called the ***complex z plane***. Similarly, the complex F plane is shown in Figure 3.1.2, with f and g measured along the horizontal and vertical axes.

The principal interest of complex function theory is in that subclass of complex functions of the form in equation (3.1.1) that have a unique derivative at a point (x, y). Unique derivative means that, if F is differentiated in the x direction ($\Delta z = \Delta x$), obtaining

$$\frac{\partial F}{\partial x} = \frac{\partial f}{\partial x} + i\frac{\partial g}{\partial x}, \tag{3.1.2a}$$

or if F is differentiated in the y direction ($\Delta z = i\Delta y$), obtaining

$$\frac{\partial F}{\partial(iy)} = \frac{\partial f}{\partial(iy)} + i\frac{\partial g}{\partial(iy)} = -i\frac{\partial f}{\partial y} + \frac{\partial g}{\partial y}, \tag{3.1.2b}$$

the results at a given point are the same. For this to be true, the real parts and the imaginary parts of equations (3.1.2a) and (3.1.2b) must be equal to each other. Thus, it must be that

$$\frac{\partial f}{\partial x} = \frac{\partial g}{\partial y}, \quad \frac{\partial g}{\partial x} = -\frac{\partial f}{\partial y}. \tag{3.1.3}$$

The equations in (3.1.3) are called **Cauchy-Riemann conditions**, and functions whose real and imaginary parts satisfy them are called **analytic functions**. Most functions of the complex variable z that involve multiplication, division, exponentiation, trigonometric functions, hyperbolic functions, exponentials, logarithms, and the like are analytic functions. Functions that can be expressed in terms of x only or y only, or involving operations such as magnitude, arguments, or complex conjugations are the few commonly used functions that are not analytic. (Recall that if $F = f + ig$ is a complex number, its complex conjugate is $F^* = f - ig$. If F satisfies the Cauchy-Riemann conditions, F^* will not.) Analytic functions have many useful properties, such as the ability to be expanded in power series, the fact that an analytic function of an analytic function is analytic, and that a transformation of the form $z_1 = F(z_2)$ is angle-preserving. Angle-preserving transformations are said to be **conformal**.

The preceding discussion was phrased in terms of derivatives in x and y. Since the choice of a coordinate system is arbitrary, it should be clear that in fact, at any point in the complex space, derivatives taken in any arbitrary orthogonal directions must satisfy the Cauchy-Riemann conditions.

Comparison of the equations in (3.1.3) with equations (2.2.6a) and (2.2.6b) shows that the complex function

$$w = \phi + i\psi \tag{3.1.4}$$

with ϕ as the velocity potential and ψ as Lagrange's stream function is an analytic function, since we have already seen from the stream function and velocity potential that

$$v_x = \frac{\partial \phi}{\partial x} = \frac{\partial \psi}{\partial y}, \quad v_y = \frac{\partial \phi}{\partial y} = -\frac{\partial \psi}{\partial x},$$

which in fact are the Cauchy-Riemann conditions. The complex function w is termed the **complex velocity potential**, or just the **complex potential**.

From differentiation of w find that

$$\frac{dw}{dz} = \frac{\partial \phi}{\partial x} + i \frac{\partial \psi}{\partial x} = v_x - iv_y. \tag{3.1.5}$$

That is, the derivative of the complex velocity potential is the complex conjugate of the velocity, which is thus an analytic function of z.

Example 3.1.1 Complex variables—analytic functions
For $F(z) = az^3$ with a real, find the real and imaginary parts of F, show that F is an analytic function, and decide whether the mapping from z to F is conformal.

Solution. Putting $z = x + iy$ into F, $F = a(x+iy)^3 = a(x^3 + 3x^2iy + 3xi^2y^2 + i^3y^3)$. Since $i^2 = -1$, this reduces to $F = a(x^3 + 3x^2iy - 3xy^2 - iy^3) = a[x^3 - 3xy^2 + i(3x^2y - y^3)]$. Separation into real and imaginary parts gives $f = a(x^3 - 3xy^2)$, $g = a(3x^2y - y^3)$. To study the analyticity of F, form the partial derivatives of f and g, giving then

$$\frac{\partial f}{\partial x} = 3a(x^2 - y^2) = \frac{\partial g}{\partial y}, \quad \frac{\partial f}{\partial y} = -6axy = -\frac{\partial g}{\partial x}.$$

Thus, F satisfies the Cauchy-Riemann equations, and therefore F is an analytic function of z.

Since $dF/dz = 3az^2$ has no singularities for finite z, and is zero only at $z = 0$, the mapping from the z plane to the F plane is angle preserving except at $z = 0$.

Example 3.1.2 Complex variables—analytic functions
Repeat the previous example, but with $F(z) = a e^z$ with a real.

Solution. Putting $z = x + iy$ into F and using DeMoive's theorem, which states that $e^{iy} = \cos y + i \sin y$, find that $F = ae^{x+iy} = ae^x(\cos y + i \sin y)$. Thus, $f = ae^x \cos y$, $g = ae^x \sin y$. Taking the partial derivatives of f and g,

$$\frac{\partial f}{\partial x} = ae^x \cos y = \frac{\partial g}{\partial y}, \quad \frac{\partial f}{\partial y} = -ae^x \sin y = -\frac{\partial g}{\partial x},$$

and so F is an analytic function of z in the entire finite z plane. Since $dF/dz = ae^z$ is finite and nonzero for all finite z, the mapping from the z plane to the F plane is angle preserving.

––––––––––

Comparison of equation (3.1.4) with our basic flows in Chapter 2 gives the following representations for these flows:

$$\textbf{uniform stream}: \; w = U^* z, \tag{3.1.6}$$

$$\textbf{source/sink at } z_0: \; w = \frac{m}{2\pi} \ln(z - z_0), \tag{3.1.7}$$

$$\textbf{doublet at } z_0: \; w = B/(z - z_0), \tag{3.1.8}$$

$$\textbf{vortex at } z_0: \; w = \frac{-i\Gamma}{2\pi} \ln(z - z_0), \tag{3.1.9}$$

where again an asterisk denotes a complex conjugate. The vortex in equation (3.1.9) is counterclockwise if $\Gamma > 0$, and clockwise if $\Gamma < 0$. The preceding expressions are much more compact and easier to remember and work with than the forms given in Chapter 2, although separating the complex velocity potential into real and imaginary parts can be tedious.

If compactness and ease of use were the only advantages gained from the introduction of complex variables, there would be little justification for introducing them. The power of complex variable theory is that, since an analytic function of an analytic function is analytic, we can solve a flow involving a simple geometry and then use an analytic function to transform, or ***map***, that geometry into a much more complicated one. In the process we concern ourselves only with the geometry, realizing that because we are dealing with analytic functions, the fluid mechanics of the flow (that is, continuity and irrotationality) is automatically satisfied.

Another important use of complex variables is in the integration of functions. Perhaps surprisingly, it is often easier to integrate analytic functions than it is to integrate real functions. The reason is ***Cauchy's integral theorem***. It states that if a function $f(z)$ is analytic and single-valued inside and on a closed contour C, then

$$\oint_C f(z)dz = 0. \tag{3.1.10}$$

Further, for a point z_0 inside C,

$$f^{(n)}(z_0) = \frac{n!}{2\pi i} \oint_C \frac{f(z)}{(z - z_0)^{n+1}} dz. \tag{3.1.11}$$

Here, $f^{(n)}(z_0)$ is the nth derivative of f evaluated at z_0. This theorem is useful in determining forces on bodies.

Note that the theorem is restricted to **single-valued functions**. By that we mean for a given z, there is only one possible value for $f(z)$. An example of a single-valued function is $f(z) = \sin z$. An example of a **multivalued function** is $f(z) = \sin^{-1} z$, which is arbitrary to a multiple of 2π. The stream function for sources and the velocity potential for vortexes are examples of multiple-valued functions. They must be treated with more care than single-valued functions.

3.2   Flow Past a Circular Cylinder with Circulation

We illustrate the mapping process by showing how the flow past a circular cylinder can be transformed into the flow past either an ellipse or an airfoil shape. Recalling from Chapter 2 that the flow of a uniform stream past a circular cylinder is a doublet facing upstream in a uniform stream, write

$$w = zU^* + \frac{a^2}{z}U + \frac{i\Gamma}{2\pi}\ln\frac{z}{a}, \tag{3.2.1}$$

where, for purposes that will become clearer as we proceed, a vortex has been added at the center of the circle. To assure ourselves that this is indeed flow past a circular cylinder, note that, since on the cylinder $z = ae^{i\theta}$, the complex potential on the cylinder is

$$w(ae^{i\theta}) = ae^{i\theta}U^* + \frac{a^2}{ae^{i\theta}}U + \frac{i\Gamma}{2\pi}\ln\left(\frac{ae^{i\theta}}{a}\right) = 2a\,|U|\cos(\theta - \alpha) - \frac{\Gamma\theta}{2\pi}$$

which is real—hence $\psi = 0$ on $|z| = a$. Also, since

$$\frac{dw}{dz} = U^* - \frac{a^2}{z^2}U + \frac{i\Gamma}{2\pi z}, \tag{3.2.2}$$

far away from the cylinder the flow approaches a uniform stream. (Note: In writing the vortex, a constant a was introduced as dividing z in the logarithm. This has the virtue of making ψ zero on the circle rather than a more complicated expression. It has the additional advantage that we are now taking the logarithm of a dimensionless quantity, which is, after all, a necessity. It has no other effect on the flow whatsoever and was only inserted for convenience.)

From equation (3.2.2) we can easily find the stagnation points in the flow. Setting $dw/dz = 0$ and then multiplying by z^2, the result is the quadratic (in z) equation

$$z^2U^* + \frac{i\Gamma}{2\pi}z - a^2U = 0 \quad \text{with roots} \quad z_{\text{separation}} = \frac{\frac{-i\Gamma}{2\pi} \pm \sqrt{4a^2UU^* - \left(\frac{\Gamma}{2\pi}\right)^2}}{2U^*}. \tag{3.2.3}$$

This can be put into a more understandable form by writing $U = |U|e^{-i\alpha}$, $G = \Gamma/4\pi a\,|U|$, giving

$$z_{\text{stagnation}} = ae^{-i\alpha}(-iG \pm \sqrt{1 - G^2}). \tag{3.2.4}$$

When G lies in the range $-1 \le G \le 1$, the stagnation points lie on the cylinder, and equation (3.2.4) can be simplified further by letting $G = \sin\beta$. Then the two stagnation points are at

$$z_{\text{DSP}} = ae^{-i(\alpha+\beta)} \quad \text{and} \quad z_{\text{USP}} = -ae^{-i(\alpha-\beta)} = ae^{-i(\alpha-\beta-\pi)} \tag{3.2.5}$$

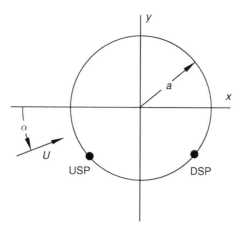

Figure 3.2.1   Cylinder in a uniform stream with circulation

as shown in Figure 3.2.1. In the figure, DSP stands for *downstream stagnation point* and USP for *upstream stagnation point*. The stagnation points on the cylinder are thus oriented at angles $\pm\beta$ with the uniform stream. For the case where $|G|$ is greater than one, equation (3.2.4) shows that the stagnation points move off the circle, with one of them lying inside and the other outside the circle. Both lie on a line that is perpendicular to the uniform stream. Streamlines that show the flow past the cylinder are given in Figure 3.2.2.

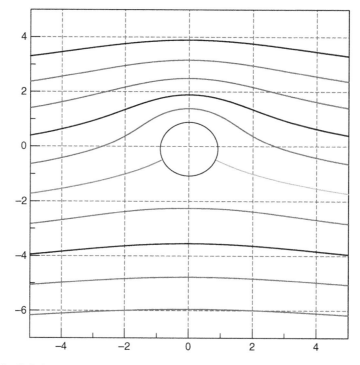

Figure 3.2.2   Cylinder in a uniform stream—streamlines

Before going on, consider the force on the cylinder. From equation (3.2.2) the velocity on the cylinder is given by

$$\frac{dw}{dz}\bigg|_{@z=ae^{i\theta}} = U^* - Ue^{-2i\theta} + \frac{i\Gamma}{2\pi a}e^{-i\theta}.$$

By Bernoulli's equation the pressure on the cylinder is then

$$p = p_0 - \frac{1}{2}\rho\left\{\left[U_x(1-\cos 2\theta) + U_y\sin 2\theta + \frac{\Gamma}{2\pi a}\sin\theta\right]^2 \right.$$
$$\left. + \left[U_x\sin 2\theta + U_y(1+\cos 2\theta) + \frac{\Gamma}{2\pi a}\cos\theta\right]^2\right\},$$

where p_0 is the pressure at the stagnation point. The hydrostatic term has been omitted in the pressure, for it can more easily be included later using Archimede's principle.

The force on the cylinder per unit distance into the paper is then

$$\mathbf{F} = F_x + iF_y = \int_0^{2\pi} p(-\cos\theta - i\sin\theta)a\,d\theta$$

$$= \int_0^{2\pi}\frac{1}{2}\rho a(\cos\theta + i\sin\theta)\left\{\left[U_x(1-\cos 2\theta) + U_y\sin 2\theta + \frac{\Gamma}{2\pi a}\sin\theta\right]^2\right.$$

$$\left. + \left[U_x\sin 2\theta + U_y(1+\cos 2\theta) + \frac{\Gamma}{2\pi a}\cos\theta\right]^2\right\}d\theta$$

$$= \rho\Gamma(U_y + iU_x) = i\rho\Gamma U^*. \tag{3.2.6}$$

Recall from the definition of G and β that

$$\Gamma = 4\pi a\,|U|\sin\beta. \tag{3.2.7}$$

The stagnation pressure has been omitted in the preceding calculation because its integral is zero.

Note three things about equation (3.2.5):

1. The force is independent of the cylinder size, being simply the fluid density times the circulation times the stream speed. (Later it will be seen that the circulation can be affected by the geometry.)
2. The force is always perpendicular to the uniform stream, so it is a lift force.
3. To have this lift force, circulation must be present.

3.3 Flow Past an Elliptical Cylinder with Circulation

To find the flow past an ellipse, introduce the **Joukowski transformation**:

$$z' = z + \frac{b^2}{z}. \tag{3.3.1}$$

This transformation adds to a circle of radius b its **inverse point** to the position z. (An inverse point to a circle is the point z_{inverse} such that $z \cdot z_{\text{inverse}} = b^2$.)

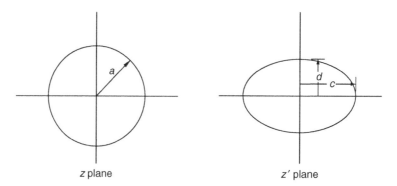

Figure 3.3.1   Ellipse in a uniform stream—transformations

To see what the Joukowski transformation does, let $z = ae^{i\theta}$ in equation (3.3.1) and then divide the resulting expression into real and imaginary parts. We then have

$$x' = \left(a + \frac{b^2}{a}\right)\cos\theta, \quad y' = \left(a - \frac{b^2}{a}\right)\sin\theta, \tag{3.3.2}$$

and the circle of radius a in the z plane is seen to have been transformed into an ellipse in the z' plane with semimajor axis $c = a + b^2/a$ and semiminor axis $d = a - b^2/a$ (Figure 3.3.1). Upon elimination of θ the equation of the ellipse is found to be given by

$$\left(\frac{x'}{a + b^2/a}\right)^2 + \left(\frac{y'}{a - b^2/a}\right)^2 = 1. \tag{3.3.3}$$

With this knowledge in hand, the flow past an ellipse can be computed.

The direct approach would be to solve equation (3.3.1) for z, obtaining

$$z = \frac{1}{2}\left(z' + \sqrt{(z')^2 - 4b^2}\right), \tag{3.3.4}$$

and then substituting this into equation (3.2.1), obtaining

$$w = \frac{1}{2}U^*\left(z' + \sqrt{z'^2 - 4b^2}\right) + \frac{2Ua^2}{z' + \sqrt{(z')^2 - 4b^2}} + \frac{i\Gamma}{2\pi}\ln\frac{z' + \sqrt{z'^2 - 4b^2}}{2a}. \tag{3.3.5}$$

The velocity is then found by forming dw/dz' from equation (3.3.5).

A slightly less direct approach—but with some advantages—would be to compute the z coordinate corresponding to a point z' by use of equation (3.3.4). The velocity can then be found using this with equation (3.2.2) and the chain rule of calculus to be

$$\frac{dw}{dz'} = \frac{dw}{dz}\frac{dz}{dz'} = \frac{1}{1 - b^2/z^2}\frac{dw}{dz}. \tag{3.3.6}$$

From equation (3.3.6) it is easy to see that the velocity is infinite at the points $z = \pm b$. Using the expressions for the semimajor and -minor axes, the parameter b in the transformation is given by

$$b = a\sqrt{(c - d)/(c + d)}. \tag{3.3.7}$$

The points of infinite velocity in equation (3.3.6) lie inside the ellipse unless $d = 0$, in which case they lie on the ellipse. For $d = 0$ the ellipse degenerates to a flat plate, and the velocity is infinite at both endpoints of the plate, unless of course the plate is aligned parallel to the uniform stream.

Notice that the possibility exists in the preceding analysis that, at a point where the denominator in equation (3.3.6) becomes zero, dw/dz also is zero. We will use this when we discuss airfoils.

3.4   The Joukowski Airfoil

The flow past an ellipse provides the clue needed to generate an airfoil shape. The principle characteristic of an airfoil is that it has a sharp trailing edge. At that point, the airfoil has a discontinuity in slope, and the mapping will not be conformal. To obtain such a shape, Joukowski proposed the following sequence of transformations.

The circle in the z plane is first translated in a new coordinate system with origin at z'_c (Figure 3.4.1) as given by

$$z' = z + z'_c, \quad \text{where } z'_c = x'_c + iy'_c \text{ and } x'^2_c + y'^2_c \le a^2. \tag{3.4.1}$$

Under this translation the circle $z = ae^{i\theta}$ in the z plane will still be a circle in the z' plane, where it now satisfies the equation $(x' - x'_c)^2 + (y' - y'_c)^2 = a^2$ and has its center at z'_c. The stagnation points originally at $z_{\text{DSP}} = ae^{i(\alpha - \beta)}$ and $z_{\text{USP}} = ae^{i(\alpha + \beta + \pi)}$ in the z plane are now at

$$z'_{\text{DSP}} = z'_c + ae^{i(\alpha - \beta)} \quad \text{and} \quad z'_{\text{USP}} = z'_c + ae^{i(\alpha + \beta + \pi)} \tag{3.4.2}$$

in the z' plane, as shown in Figure 3.4.2.

The z' transformation is next followed by the Joukowski transformation

$$z'' = z' + b^2/z'. \tag{3.4.3}$$

Here b is a real number. Since

$$\frac{dw}{dz''} = \frac{dw}{dz}\frac{dz}{dz'}\frac{dz'}{dz''} = \frac{dw}{dz}\frac{z'^2}{z'^2 - b^2}, \tag{3.4.4}$$

it is seen that the velocity in the z'' plane will be infinite at the points $z' = \pm b$ *unless* one of these points coincides with a stagnation point in the z plane. Therefore, to make the transformed circle into an airfoil shape, the first stagnation point in equation (3.4.2)

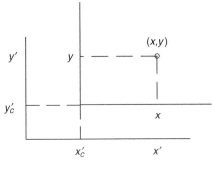

z and z' planes

Figure 3.4.1   Joukowsky transformation—z and z' planes

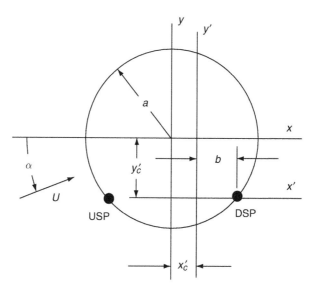

Figure 3.4.2 Joukowsky airfoil—z and z' planes

must lie on the circle and be where $z' = +b$. This will give a sharp trailing edge, and the mapping of the cylinder to the airfoil will not be conformal at that point. This results in

$$y'_c = -a\sin(\alpha - \beta) \quad \text{and} \quad b = x'_c + a\cos(\alpha - \beta) = x'_c + \sqrt{a^2 - y'^2_c}. \qquad (3.4.5)$$

Further, require for later purposes that the point $z' = (-b, 0)$ does not lie outside the circle, this condition being satisfied provided that

$$(b + x'_c)^2 + y'^2_c < a^2. \qquad (3.4.6)$$

This means that, except inside the body, the only possible place where the velocity can be singular is at $z' = b$. Satisfaction of both parts of equation (3.4.5) means that $x'_c \le 0$.

Investigating the transformation further, note that on the circle in the z'' plane, at the trailing edge/stagnation point the angle the surface turns through is 2π (or zero), whereas on the airfoil the corresponding angle is π. The reason that the requirement was made that $z' = -b$ must not lie outside the circle is that we wanted to ensure that no singularities were introduced into the flow, only inside the circle.

The shape of the airfoils that this family of transformations has generated can now be found. The following outlines the approach:

1. Specify $|U|$, the attack angle α and the coordinates of the center of the translated circle x'_c and y'_c. The latter is to be done such that $x'_c < 0$ and $x'^2_c + y'^2_c \le a^2$. (The choice of a is in fact arbitrary, representing a scaling of the coordinate system, and could just as well be taken as unity. The same is true for $|U|$, which scales the velocity.) The parameter x'_c controls the airfoil thickness, while y'_c controls its camber.

2. Compute b, β, and Γ.: From equation (3.4.5), $\beta = \alpha + \sin^{-1} y'_c/a$, $b = x'_c + \sqrt{a^2 - y'^2_c}$. From equation (3.2.7)

$$\Gamma = 4\pi a |U| \sin \beta.$$

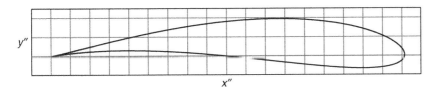

Figure 3.4.3 Joukowsky airfoil shape

3. Choose an arbitrary point $z = ae^{i\theta}$ on the circle by selecting a value for θ. Compute the corresponding transformed position of this point on the airfoil using equations (3.4.1) and (3.4.3).

4. To find the velocity and pressure on the airfoil, use equation (3.4.4) for the velocity and then compute the pressure coefficient according to $C_p = \frac{1}{2}\left(1 - |U/U_\infty|^2\right)$.

5. Repeat steps 3 and 4 for a number of values of θ to generate the airfoil shape and the pressure distribution.

6. Repeat steps 1 through 5 for various values of x'_c and y'_c to see the effect of these parameters on airfoil shape.

Figure 3.4.3 shows a typical airfoil generated by this procedure. Here

$$|\mathbf{U}| = 1, \quad \alpha = 0, \quad A = 1, \quad x'_c = -0.1, \quad y'_c = 0.1.$$

In the beginning of the analysis, the vortex was somewhat arbitrarily added to the flow. Its presence is an absolute necessity if a lift force is to be present. How, then, does the flow introduce circulation?

In the case of the circular cylinder, circulation would have to be added artificially, either by rotating the cylinder (the **Magnus effect**) and letting viscosity and the no-slip condition introduce the swirl or by forcing the position of the stagnation points either by the addition of a small flap where the trailing edge would be (**Thwaite's flap**) or by some other means to control the pressure on the boundary. In the case of the airfoil, note that our theory predicts an infinite velocity at the trailing edge *unless $dw/dz = 0$* at that point. In fact, unless we impose the condition $dw/dz = 0$ at the trailing edge, the velocity there is not only infinite, but it turns through 180 degrees in zero distance. A real fluid cannot do this because it implies infinite acceleration requiring infinite force.

Kutta proposed that the circulation will adjust itself so that both the infinite velocity and the sharp turn in the flow at the trailing edge are avoided. At this stage both numerator and denominator in equation (3.4.4) are zero, meaning that dw/dz'' is of the form zero over zero. To resolve the value of the true speed at the trailing edge, use L'Hospital's rule, in the form

$$\frac{dw}{dz''} = \lim_{z' \to b}\left[z'^2 \frac{\dfrac{d^2w}{dz^2}}{\dfrac{dz'}{dz}\dfrac{d}{dz'}(z'^2 - b^2)} \right] = \frac{b^2}{2b}\left[\frac{2a^2 U}{z^3} - \frac{i\Gamma}{2\pi z^2} \right]_{z = ae^{-i(\alpha+\beta)}} \tag{3.4.7}$$

$$= \frac{b}{a}|U| e^{2i(\alpha+\beta)}\cos\beta.$$

This tells us that the body slope at the trailing edge is turned an angle of 2β from the attack angle α.

Kutta's hypothesis (usually referred to as a **Kutta condition**) gives a very good value for the circulation on a Joukowski airfoil, as has been verified in a series of tests (National Advisory Committee for Aeronautics Technical Report number 391 and Technical Memorandums 422 and 768). Joukowski airfoils do have very good stalling behavior and can achieve angles of attack (α) as high as 30 degrees with only a slight drop in lift. They have the disadvantage of having a rather high drag coefficient, and the very thin, sharp trailing edge makes their construction very difficult.

3.5   Kármán-Trefftz and Jones-McWilliams Airfoils

There have been several variations of the Joukowski airfoil that add several helpful features. One, proposed by von Kármán and Trefftz, eliminates the disadvantage of the thin trailing edge.

Transformation equation (3.4.1) can be written in the equivalent forms

$$z'' + 2b = (z' + b)^2/z', \quad z'' - 2b = (z' - b)^2/z'.$$

Taking the ratio of these, the Joukowski transformation can therefore be written in the form

$$\frac{z'' + 2b}{z'' - 2b} = \left(\frac{z' + b}{z' - b}\right)^2. \tag{3.5.1}$$

Von Kármán and Trefftz suggested replacing the Joukowski transformation with the alternate transformation

$$\frac{z'' + 2b}{z'' - 2b} = \left(\frac{z' + b}{z' - b}\right)^n. \tag{3.5.2}$$

The trailing edge then has an inside angle of $(2 - n)\pi$, rather than zero. The details of the shape can be carried out in a manner similar to the Joukowski airfoil, with one more variable (n) available to the designer.

A second variation of the Joukowski airfoil was proposed by R. T. Jones and R. McWilliams in a pamphlet distributed at an Oshkosh Air Show. They also start with equation (3.4.1) but then follow it with the two transformations

$$z'' = z' - \frac{\varepsilon}{z' - \Delta} \tag{3.5.3}$$

and

$$z''' = z'' + \frac{b^2}{z''}, \tag{3.5.4}$$

where ε is a complex number and Δ is real. Carrying through an analysis similar to what we did for the Joukowski profile, it can be shown using the same general analysis as for the Joukowski transformation that

$$\varepsilon = (x'_T - b)(x'_T - \Delta) - (y'_T)^2 + iy'_T(2x'_T - \Delta - b), \tag{3.5.5}$$

where $(x'_T, \ y'_T)$ is the location of the trailing edge in the z' plane.

It is convenient to set b, a scale factor, arbitrarily to 1. The parameters $\Delta, x'_c, y'_c, x'_T, y'_T$ are then set by the designer. The parameter ε is determined by equation (3.5.5) subject to the inequality

$$\left|\Delta - b \pm \sqrt{(\Delta + b)^2 + 4\varepsilon} - 2(x'_c + iy'_c)\right| < 2a, \tag{3.5.6}$$

and the parameter a is determined by

$$a = \sqrt{(x_T' - x_c')^2 + (y_T' - y_c')^2}. \qquad (3.5.7)$$

The circulation is then found from

$$\Gamma = 4\pi a |U| B, \qquad (3.5.8a)$$

with B given by

$$B = \frac{(x_T' - x_c') \sin\theta - (y_T' - y') \cos\theta}{a}. \qquad (3.5.8b)$$

With a careful selection of the parameters, Jones and McWilliams have generated airfoils with the properties of the NACA 6 series, the 747 series, the Clark Y, and the G-387.

3.6   NACA Airfoils

Early airfoil designs by Eiffel, the Wright brothers, and many others were largely the result of intuition and trial and error, although the Wright brothers did utilize a crude wind tunnel and were well read as to the aerodynamic theory of their day. During World War I, Prandtl at Gottingen, Germany, started to put these designs on a firmer basis, and efficient airfoils such as the Gottingen 398 and the Clark Y were developed. These airfoil shapes, or extensions of them, were utilized almost exclusively during the 1920s and 1930s. In the 1930s, the NACA (National Advisory Committee for Aeronautics, which later became NASA, the National Aeronautics and Space Administration) started a comprehensive and systematic development program that used both theory and experiment. They were able to separate out the effects of camber and thickness distribution and also carried out experiments at higher Reynolds numbers than had previously been possible.

Airfoil shapes are generally given by specifying many coordinate points on the surface of the foil. The effects of camber and thickness can be separated by writing

$$x_U = x_c - y_t \sin\theta, \qquad (3.6.1a)$$

$$y_U = y_c + y_t \cos\theta \qquad (3.6.1b)$$

on the upper surface of the foil, and

$$x_L = x_c + y_t \sin\theta, \qquad (3.6.1c)$$

$$y_L = y_c - y_t \cos\theta \qquad (3.6.1d)$$

on the lower surface. Here (x_c, y_c) are the coordinates of a point on the ***mean line***, $\tan\theta$ is the slope of the mean line, and y_t is the half-thickness. Mean lines are frequently designed to have specific load distributions. For example, the following mean line equation has constant loading over the chord (Abbott and von Doenhoff, pages 74 and 75, 1959):

$$y = -\frac{c_{Li}}{4\pi} \left[\left(1 - \frac{x}{c}\right) \ln\left(1 - \frac{x}{c}\right) + \frac{x}{c} \ln\frac{x}{c} \right], \qquad (3.6.2)$$

where $c_{Li} = 2\gamma/U$, γ is the circulation per unit chord length, and U is the free stream velocity.

For two famous series of airfoils, the NACA four-digit and five-digit series, the shapes are defined completely by formulae. In the four-digit series, the thickness is given by

$$y_t = 5y_{max}\left[0.2969\sqrt{\frac{x}{c}} - 0.126\frac{x}{c} + 0.3537\left(\frac{x}{c}\right)^2 + 0.28431\left(\frac{x}{c}\right)^3 - 0.1015\left(\frac{x}{c}\right)^4\right],$$

(3.6.3)

where y_{max} is the maximum value of the half-thickness, and c is the chord length. Near the leading edge, neglecting all terms on the right-hand side of equation (3.6.3) beyond the first, the foil is approximately given by the parabola $(y_t/y_{max})^2 \approx 2.20374(x/c)$. The local radius of curvature at the leading edge is thus

$$R_{LE} \approx 1.10187\, y_{max}^2/c.$$

(3.6.4)

The mean lines for the airfoils of this series are given by

$$y_c = \frac{m(2px - x^2)}{p^2} \quad \text{for } 0 \le x \le p,$$
$$= \frac{m(c-x)(x-2p+c)}{(c-p)^2} \quad \text{for } p \le x \le c,$$

(3.6.5)

where m is the maximum camber (distance of the mean line from the chord line), and p is the chordwise location of the maximum camber point.

The numbering system for this series is based on the foil shape. The first integer is $100\,m/c$, the second integer is $10\,p/c$, and the third and fourth are $100\,y_{max}/c$. Thus, for an NACA 2415 foil, we have

$$m/c = 0.02, \quad p/c = 0.4, \quad y_{max}/c = 0.15.$$

(3.6.6)

Foils with numbers NACA 00xx are symmetrical—that is, their mean and chord lines coincide.

For the four-digit series, tests show that the maximum lift coefficient increases as the position of maximum camber is shifted forward from midchord. The mean lines for the four-digit series were not suitable for forward positions of maximum camber, so the five-digit series was introduced. For the five-digit series, the thickness is given by equation (3.6.3) and the mean lines by

$$y_c = k[x^3 - 3mx^2 + m^2 x(3 - m/c)] \quad \text{for } 0 \le x \le m,$$
$$= \frac{km^3(c-x)}{c} \quad \text{for } m \le x \le c.$$

(3.6.7)

where k originally was calculated to give a design lift coefficient of 0.3, and m determines where the maximum camber points occur. Differentiating equation (3.6.7), the maximum camber points are found to occur at

$$x_{MC} = m(1 - \sqrt{m/3c}).$$

(3.6.8)

The numbering system for the five-digit series is based on a combination of foil shape and aerodynamic characteristics. The first integer is $6.66667 \cdot C_L$, where C_L is the design lift coefficient. The second and third integers are $200 \cdot x_{MC}/c$, and the fourth and fifth are $100 \cdot y_{max}/c$, where y_{max} is the maximum thickness. Thus, an NACA 23012 wing section has a design lift coefficient of 0.3, a maximum camber at 0.15 of the chord length, and $y_{max}/c = 0.12$.

Details of these series and others are given in Abbott and van Doenhoff (1959). This book includes a good discussion of airfoil theory along with data on lift, drag, and moment coefficients. Modern airfoils generally have more complicated shapes than given by the NACA four- and five-digit series. Their shapes are usually given in tabular form rather than by formulae.

3.7   Lifting Line Theory

At this point in the discussion there is a good bit of theory on foils of infinite span and also information on test data on a large number of foil shapes. All good things come to an end, however, including airplane wings, turbine blades, and vorticity lines (recall that the last, however, do not end in the interior of the fluid). How, then, does the body of two-dimensional knowledge fit into a three-dimensional world?

Suppose for example that a wing is to be designed and that the lift coefficient distribution on the wing is given. This is a good starting point in the design process, as the local value of the lift coefficient gives the load on the wing at that point, and this load is intimately connected to the structural requirements of the wing. From the lift coefficient, previous results for the Joukowski airfoil give the circulation distribution.

If a very simple wing design is used where the circulation is constant along the span, in the interior of the wing, the portion of the vortex line along the span would simply be a straight line. This portion of the vortex is called the **bound vortex**, since it is bound to the wing. When this vortex line reaches the wing tips, it leaves the wing and is convected rearward with the flow. Considering the situation where the flow starts from rest, the vortex line for this case will be (approximately) a flat, closed rectangle, consisting of the bound vortex, the **tip vortices**, and the **starting vortices**, as shown in Figure 3.7.1. The length of this rectangle increases with time, as the starting vortices move away from the wing. (The tip vortices can frequently be seen from an airplane window. Because of the high rotational speeds at their core, the pressure there is lowered. Water vapor can then condense, and flow visualization occurs.) The vortex line is thus closed, so the requirement that a vortex line can neither originate nor end in the fluid is met. When the starting vortex is very far away (at infinity), the pattern we have described is called a **horseshoe vortex**.

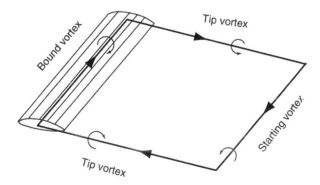

Figure 3.7.1   Wingtip, boundary, and starting vortices

Recalling that a doublet distribution is equivalent to a vortex sheet, with the local vorticity being proportional to the change of the doublet strength, the velocity potential for the vorticity associated with a finite wing can be found readily. Since we want vorticity only on the boundaries of our rectangle, a constant-strength doublet facing normal to the x-y plane is sufficient. This results in

$$
\begin{aligned}
\phi &= \frac{\Gamma}{4\pi}\int_{-s}^{s} d\eta \int_{x_a}^{L} \frac{z\,d\xi}{[(x-\xi)^2+(y-\eta)^2+z^2]^{3/2}} \\
&= \frac{\Gamma}{4\pi}\int_{-s}^{s} \left(\frac{z(\xi-x)}{[(\eta-y)^2+z^2]\sqrt{(\xi-x)^2+(y-\eta)^2+z^2}}\right)_{x_a}^{L} d\eta \\
&= \frac{\Gamma}{4\pi}\left(\tan^{-1}\frac{(\eta-y)(x-x_a)}{z\sqrt{(x-x_a)^2+(\eta-y)^2+z^2}} - \tan^{-1}\frac{(\eta-y)(x-L)}{z\sqrt{(x-L)^2+(\eta-y)^2+z^2}}\right)_{-s}^{s}.
\end{aligned}
$$

$$(3.7.1)$$

From Figure 3.7.1, and with the help of the right-hand rule of vectors, it is seen that the effect of this vortex rectangle is to induce a downward velocity in the interior, called the ***downwash***. The downwash in turn contributes a drag to the wing, called the ***induced drag***.

If the circulation distribution is not constant, such as for tapered wings, this pattern is repeated, with each point in change in the circulation acting (weakly) as a wing tip. Thus, the vorticity pattern might be as in Figure 3.7.2. The velocity potential for such a pattern could be generated by including the effects of variable Γ inside the original integral in equation (3.7.1). The integration involved in such a procedure would very quickly overtax our calculus skills!

The preceding procedure reduces the wing to a line. It was originated by Prandtl (1921) and is called ***lifting-line*** theory. The ***vortex-lattice*** method is a more detailed procedure than a lifting line theory. It gives some geometry to the wing and is similar to the panel method introduced earlier. The constant-strength doublet panels with which we might expect to cover our wing (with some mathematical difficulties) are instead replaced by a series of horseshoe vortices as we just saw. Experience suggests that the best results are found by placing the bound vortex portion of the horseshoe at the quarter-chord point. Computer programs to calculate the resulting velocities are closely related to our panel programs. (See, for instance, Moran (1984) for more details.)

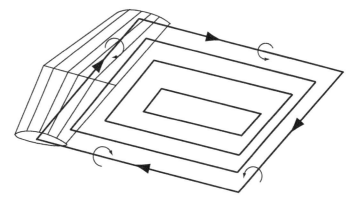

Figure 3.7.2   Vortex pattern for a tapered wing

It was recognized as early as the 1890s that to increase wing lift and decrease drag, it was important to keep the flow over the wings as two-dimensional as possible. Frederick W. Lanchester[1] in England patented the first concept of ***winglets***, vertical surfaces at the tips of a wing, in 1897. His version was essentially an endplate that, by making the flow more two-dimensional, did reduce induced drag but had the unfortunate consequence of increasing viscous drag, negating the benefits. In the 1970s, Richard T. Whitcomb at NASA Langley Research Laboratory reexamined the concept and found that the airflow above the wing tip of a typical airfoil is directed inward, while below the wing tip it is directed outward. By careful design of the ***cant*** (upward angle) and ***toe*** (inward angle) of the winglet the small lift force developed by the winglet could be directed forward so that the trailing vortex contributes to thrust rather than drag. This has strongly influenced design procedures used by general aviation and business jet manufacturers and led to the introduction of winglets on many new planes and to the retrofit of many older ones.

3.8 Kármán Vortex Street

For flows past two-dimensional bodies, wake vortices are found to alternately shed from the two sides of the body. Their pattern tends to be initially symmetrical (providing the body is symmetrical) but soon changes to an alternating pattern, as seen in Figure 3.8.1. Von Kármán questioned whether this alternating pattern (called a ***vortex street***) was necessarily the only pattern that could exist; in other words, was the pattern stable? To idealize the problem, he omitted the body and considered two rows of vortices, each consisting of vortices extending to infinity for both positive and negative x. The upper row consists of counterclockwise vortices at $x = na$, $n = 1, 2, \ldots$, $y = b/2$, with circulation Γ, and the lower row of clockwise vortices at $x = (n + c)a$, $n = 1, 2, \ldots$, $-1 < c < 1$, $y = -b/2$ with circulation $-\Gamma$. The complex velocity potential for this flow (the calculations become unmanageable if we do not use complex variables here) is then

$$
\begin{aligned}
w &= \frac{-i\Gamma}{2\pi} \sum_{n=-\infty}^{\infty} \left[\ln\left(z - na - \frac{ib}{2} \right) - \ln\left(z - na - ca + \frac{ib}{2} \right) \right] \\
&= \frac{-i\Gamma}{2\pi} \left\{ \ln\left[\sin \frac{\pi\left(z - \dfrac{ib}{2} \right)}{a} \right] - \ln\left[\sin \frac{\pi\left(z - ca + \dfrac{ib}{2} \right)}{a} \right] \right\}.
\end{aligned}
\tag{3.8.1}
$$

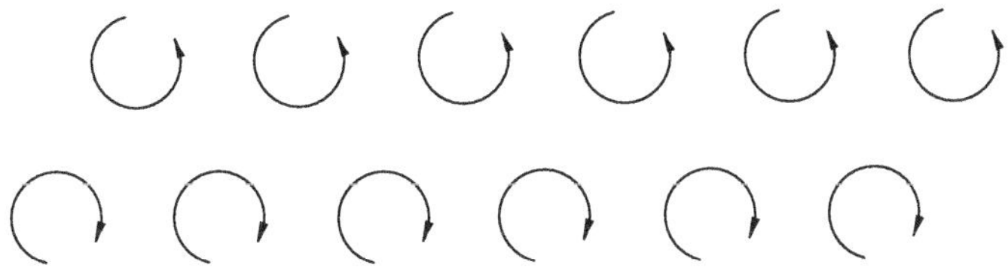

Figure 3.8.1   Kármán vortex street

---

[1] Frederick William Lanchester (1868–1946) made many contributions to aerodynamics and automotive engineering. His work was influential in the study of aircraft stability, and his 1919 book *Aviation in Warfare: The Dawn of the Fourth Arm* eventually led to the development of the study of logistics and the field of operations research.

The second form follows from

$$\sum_{n=-\infty}^{\infty} \ln(p-na) = \ln p + \sum_{n=1}^{\infty} \ln(p^2 - n^2)$$

$$= \ln p(1-p^2)\left(1 - \frac{p^2}{2^2}\right) \cdots \left(1 - \frac{p^2}{n^2}\right) \cdots + \text{constant}$$

$$= \pi \sin \pi p,$$

where the constant is unimportant for our purposes.

In the analysis that follows, we will be using several formulas that come from Fourier series. Three particularly helpful expansions are the following:

Expansion #1. $\frac{\cosh k(\pi-\beta)}{\sinh k\pi} = \frac{1}{\pi} + \frac{2k}{\pi} \sum_{n=1}^{\infty} \frac{\cos m\beta}{n^2+k^2}$. Evaluating this at $\beta = 0$ gives

$$\coth k\pi = \frac{1}{k\pi} + \frac{2}{\pi} \sum_{n=1}^{\infty} \frac{k\cos n\beta}{n^2 + k^2}.$$

Expansion #2. $\frac{\sinh k(\pi-\beta)}{\cosh k\pi} = \frac{2}{\pi} \sum_{n=0}^{\infty} \frac{k\cos(n+1/2)\beta}{(n+1/2)^2+k^2}$. Evaluating this at $\beta = 0$ gives

$$\tanh k\pi = \frac{2}{\pi} \sum_{n=1}^{\infty} \frac{k}{(n+1/2)^2 + k^2}.$$

Differentiating this with respect to k gives

$$\frac{d(\tanh k\pi)}{dk} = \frac{\pi}{\cosh^2 k\pi} = \frac{2}{\pi} \sum_{n=0}^{\infty} \frac{(n+1/2)^2 - k^2}{[(n+1/2)^2 + k^2]^2}.$$

Expansion #3. $\frac{\beta}{4}(2\pi - \beta) = \sum_{n=1}^{\infty} \frac{1 - \cos n\beta}{n^2}$.

The complex conjugate velocity corresponding to the complex potential (equation (3.8.1)) is

$$u - iv = \frac{dw}{dz} = \frac{-i\Gamma}{2a}\left[\cot \frac{\pi\left(z - \frac{ib}{2}\right)}{a} - \cot \frac{\pi\left(z - ca + \frac{ib}{2}\right)}{a} \right]. \tag{3.8.2}$$

The first cotangent represents the velocity due to vortices in the upper row, and the second the velocities due to the vortices in the lower row.

Considering the velocity induced at an arbitrary vortex in the upper row, notice that the other vortices in the same row induce no net velocities, as the induced velocities due to vortices at equal distances right and left from our chosen vortex will cancel out each another. Thus, the net induced velocity at any vortex in the upper row (for computational convenience consider the vortex to be at $z = ib/2$) is found from the vortices in the lower row to be

$$V_u^* = \frac{i\Gamma}{2a} \cot \pi\left(\frac{ib}{a} - c\right), \tag{3.8.3}$$

the asterisk denoting the complex conjugate induced velocity. A similar computation for the lower vortices gives $V_l^* = V_u^*$, as should be expected.

To a good approximation, the vortices in a vortex street behind a bluff body do not drift either up or down. Equation (3.8.3), on the other hand, states that there will be a vertical component to the induced velocity unless either $c = 0$ (when the two vortex rows are aligned), for which

$$V^* = \frac{\Gamma}{2a} \coth \frac{\pi b}{a}, \tag{3.8.4}$$

or when $c = 1/2$ (the vortices in the lower row are placed below the midpoints of the vortices in the upper row), when

$$V^* = \frac{\Gamma}{2a} \tanh \frac{\pi b}{a}. \tag{3.8.5}$$

Thus, the two cases we will consider are for c either 0 or 1/2, and the vortices will move with velocity V^* given by either equation (3.8.4) or (3.8.5).

To examine the stability of this arrangement, notice that at time t the vortices in the upper row will be at $z_{n0} = na + V^*t + ib/2$, $-\infty < n < \infty$, and the vortices in the lower row will be at $z'_{n0} = (n+c)a + V^*t - ib/2$, $-\infty < n < \infty$.

For the flow disturbance, displace the vortex originally at z_{n0} in the upper row to the location $z_{n0} + z_n$ and displace the vortex in the lower row originally at z'_{n0} to the location $z'_{n0} + z'_n$. Further assume that our disturbance displacement is of the form $z_n = \gamma \cos n\theta$ and $z'_n = \gamma' \cos(n+c)\theta$, where γ and γ' are small time-dependent complex numbers, so small that equations can be linearized. This corresponds to a wavy displacement of each vortex row. Of course, this is not the most general displacement that is possible. If, however, we use this displacement and find that γ and γ' grow with time, then, because the flow is unstable to at least one disturbance, we can conclude that the flow is unstable. If, however, we find that for this displacement γ and γ' do not grow with time, since our choice of disturbance was not a general one, we cannot draw any definite conclusions other than to say that it is a possible configuration.

The new velocity potential is now

$$w = \frac{-i\Gamma}{2\pi} \sum_{n=-\infty}^{\infty} [\ln(z - z_{n0} - z_n) - \ln(z - z'_{n0} - z'_n)] \tag{3.8.6}$$

with a complex velocity

$$\frac{dw}{dz} = \frac{-i\Gamma}{2\pi} \sum_{n=-\infty}^{\infty} \left(\frac{1}{z - z_{n0} - z_n} - \frac{1}{z - z'_{n0} - z'_n} \right). \tag{3.8.7}$$

If we consider a vortex in the upper row—say, the one corresponding to $n = 0$—then it moves with a velocity $V^* + d\gamma^*/dt$ and is at $z_0 + Vt + ib/2$. Setting this equal to equation (3.8.7) and linearizing the right-hand side, we have with the help of the binomial theorem

$$V^* + \frac{d\gamma^*}{dz} = \frac{-i\Gamma}{2\pi} \left(\sum_{\substack{n=-\infty \\ n \neq 0}}^{\infty} \frac{1}{z_0 - z_{n0} - z_n} - \sum_{n=-\infty}^{\infty} \frac{1}{z_0 - z'_{n0} - z'_n} \right)$$

$$= \frac{-i\Gamma}{2\pi} \sum_{n=1}^{\infty} \left(\frac{1}{z_0 - z_{n0} - z_n} + \frac{1}{z_0 - z_{-n0} - z_{-n}} \right) + \frac{i\Gamma}{2\pi} \sum_{n=-\infty}^{\infty} \frac{1}{z_0 - z'_{n0} - z'_n}$$

$$= \frac{-i\Gamma}{2\pi} \sum_{n=1}^{\infty} \left(\frac{1}{-na + \gamma - \gamma \cos n\theta} + \frac{1}{na + \gamma - \gamma \cos n\theta} \right)$$

$$+ \frac{i\Gamma}{2\pi} \sum_{n=-\infty}^{\infty} \frac{1}{\gamma + ib - (n+c)a - \gamma' \cos(n+c)\theta}$$

$$\simeq \frac{-i\Gamma}{2\pi} \sum_{n=1}^{\infty} \left[\frac{-1}{na} \left(1 + \frac{\gamma - \gamma \cos n\theta}{na} \right) + \frac{1}{na} \left(1 - \frac{\gamma - \gamma \cos n\theta}{na} \right) \right]$$

$$+ \frac{di\Gamma}{2\pi} \sum_{n=0}^{\infty} \frac{1 - \dfrac{\gamma - \gamma' \cos(n+c)\theta}{ib - (n+c)a}}{ib - (n+c)a} + \frac{i\Gamma}{2\pi} \sum_{n=1}^{\infty} \frac{1 - \dfrac{\gamma - \gamma' \cos(-n+c)\theta}{ib - (-n+c)a}}{ib - (-n+c)a}$$

$$= \frac{i\Gamma\gamma}{\pi} \sum_{n=1}^{\infty} \frac{1 - \cos n\theta}{n^2 a^2} + \frac{i\Gamma}{2\pi} \sum_{n=0}^{\infty} \left[\frac{1}{ib - (n+c)a} \left(1 - \frac{\gamma - \gamma' \cos(n+c)\theta}{ib - (n+c)a} \right) \right.$$

$$\left. + \frac{1}{ib + (n+c)a} \left(1 - \frac{\gamma - \gamma' \cos(n+c)\theta}{ib + (n+c)a} \right) \right]$$

$$= \frac{i\Gamma}{2\pi} \left(\frac{2}{ib - ca} + \sum_{n=1}^{\infty} \frac{2(ib - ca)}{(ib - ca)^2 - n^2 a^2} \right) + \frac{i\Gamma\gamma}{\pi} \sum_{n=1}^{\infty} \frac{1 - \cos n\theta}{n^2 a^2}$$

$$- \frac{i\Gamma}{2\pi} \sum_{n=0}^{\infty} \left(\frac{\gamma - \gamma' \cos(n+c)\theta}{[ib - (n+c)a]^2} + \frac{\gamma - \gamma' \cos(n+c)\theta}{[ib + (n+c)a]^2} \right). \tag{3.8.8}$$

Canceling V^* from both sides of the equation, we are left with

$$\frac{d\gamma^*}{dt} = \frac{i\Gamma\gamma}{\pi} \sum_{n=1}^{\infty} \frac{1 - \cos n\theta}{n^2 a^2} - \frac{i\Gamma}{2\pi} \sum_{n=0}^{\infty} \left(\frac{\gamma - \gamma' \cos(n+c)\theta}{[ib - (n+c)a]^2} + \frac{\gamma - \gamma' \cos(n+c)\theta}{[ib + (n+c)a]^2} \right)$$

$$= \frac{i\Gamma}{\pi a^2} (A\gamma + C\gamma'), \tag{3.8.9}$$

where

$$A = \sum_{n=1}^{\infty} \frac{1 - \cos n\theta}{n^2} - \frac{1}{2} \sum_{n=0}^{\infty} \left(\frac{1}{(n+c - ib/a)^2} + \frac{1}{(n+c + ib/a)^2} \right)$$

$$= \sum_{n=1}^{\infty} \frac{1 - \cos n\theta}{n^2} - \sum_{n=0}^{\infty} \frac{(n+c)^2 - (b/a)^2}{[(n+c)^2 + (b/a)^2 +]^2},$$

and

$$C = \sum_{n=0}^{\infty} \frac{(n+c)^2 - (b/a)^2}{[(n+c)^2 + (b/a)^2]^2} \cos(n+c)\theta$$

are both real.

A similar calculation for the lower row gives

$$\frac{d\gamma'^*}{dt} = \frac{-i\Gamma}{\pi a^2} (A\gamma' + C\gamma). \tag{3.8.10}$$

To solve the set of equations for γ and γ', differentiate the complex conjugate of equation (3.8.9) with respect to time and use equation (3.8.10) to eliminate $d\gamma'/dt$. The result is

$$\frac{d^2\gamma}{dt^2} + \left(\frac{\Gamma}{\pi a^2}\right)^2 (C^2 - A^2)\gamma = 0. \tag{3.8.11}$$

Consider first the case where $c = 0$—that is, the vortices are not staggered. Then

$$A = \sum_{n=1}^{\infty} \frac{1 - \cos n\theta}{n^2} - \sum_{n=0}^{\infty} \frac{n^2 - (b/a)^2}{[n^2 + (b/a)^2]^2} = \frac{\theta}{2}\left(\pi - \frac{\theta}{2}\right) - \frac{\pi^2}{2\sinh^2\dfrac{b\pi}{a}},$$

$$C = \sum_{n=0}^{\infty} \frac{n^2 - (b/a)^2}{[n^2 + (b/a)^2]^2}\cos n\theta = -\frac{\pi^2 \cosh\dfrac{b\pi}{a}}{\sinh^2\dfrac{b\pi}{a}} - \frac{\pi\theta\sinh\dfrac{b}{a}(\pi - \theta)}{\sinh\dfrac{b\pi}{a}}.$$

Note that for $\theta = \pi$

$$A = -\frac{\pi^2}{4} - \frac{\pi^2}{2\sinh^2\dfrac{b\pi}{a}} = -\frac{\pi^2\left(\cosh^2\dfrac{b\pi}{a} + 1\right)}{4\sinh^2\dfrac{b\pi}{a}} \quad \text{and} \quad C = -\frac{\pi^2\cosh\dfrac{b\pi}{a}}{\sinh^2\dfrac{b\pi}{a}},$$

so that

$$C^2 - A^2 = -\frac{\pi^4\left(\cosh^4\dfrac{b\pi}{a} + \cosh^2\dfrac{b\pi}{a} + 1\right)}{16\sinh^4\dfrac{b\pi}{a}} < 0.$$

Therefore, there is at least one value of θ for which this arrangement is unstable.

For the staggered configuration $c = 1/2$,

$$A = \sum_{n=1}^{\infty} \frac{1 - \cos n\theta}{n^2} - \sum_{n=0}^{\infty} \frac{(n+1/2)^2 - (b/a)^2}{[(n+1/2)^2 + (b/a)^2]^2} = \theta\left(\pi - \frac{\theta}{2}\right) + \frac{\pi^2}{\cosh^2\dfrac{b\pi}{a}},$$

$$C - \sum_{n=0}^{\infty} \frac{(n+1/2)^2 - (b/a)^2}{[(n+1/2)^2 + (b/a)^2]^2}\cos(n+1/2)\theta.$$

When $\theta = \pi$, $C = 0$ and $A = \pi^2(1/2 - 1/\cosh^2 b\pi/a)$. Thus, a necessary condition for stability is that $A = 0$, giving $\cosh^2 b\pi/a = 2$, or $b/a = 0.281$. For that value of θ

$$V^* = \frac{\Gamma}{2a}\tanh 2 = \frac{\Gamma}{2a\sqrt{2}} = \frac{0.3536\Gamma}{a}. \tag{3.8.12}$$

In practice, the $c = 1/2$ case is usually found to be a good model several vortices downstream from the body. The spacing parameter b is determined by the body shape and size, as is the circulation.

Vortex shedding can be frequently seen in everyday life, from the waving of tree branches in the wind to the torsional oscillations of stop signs on windy corners. Ancient Greeks called the sounds made by wind passing tree branches **Aeolian tones**. Open automobile windows can cause low-frequency oscillations in the automobile interior, with subsequent discomfort to passengers. In open areas with frequent crosswinds, power and telephone lines can be seen to "gallop" on windy days. (Simple weights hung at the quarter-wave points can alter the natural frequency of the line and thus prevent catastrophic motion.) Perhaps the most famous case of motion due to vortex shedding is the Tacoma Narrows Bridge, dubbed "galloping Gertie," that self-destructed months after it was built. (Von Kármán was on the committee that was formed to investigate the collapse.) The final word on this bridge has not yet been written, as scientific articles are still appearing with additions and further explanations for the collapse. Pictures can be found by searching the Internet.

Our model for a vortex street here was the most simple one, where the vortices are concentrated. In three dimensions no such simple models are found, and the analysis becomes more complicated. More details are given in the chapter on numerical calculations and in the book by Saffman (1992).

3.9 Conformal Mapping and the Schwarz-Christoffel Transformation

In performing the transformation of the circle to the ellipse by means of the Joukowski transform, we say that we have **mapped** the circle in the z plane into an ellipse and that the mapping is conformal, or angle-preserving. This means that if two lines cross at some point in the z plane with an angle α between them, at the corresponding point in the z' plane the angle between the mapped lines is still α. The relation between w and z can be thought of as a mapping of one plane into another.

There are a few mappings that have been found to be generally useful, particularly for dealing with **free streamline flows**.[2] The simplest one is the logarithm function used in mapping the **hodograph plane** dw/dz. Letting $dw/dz = u - iv = qe^{-i\theta}$, we see that $\ln dw/dz = \ln q - i\theta$. Thus, regions of constant speed are vertical line segments in the hodograph plane, whereas regions of constant direction of velocity are vertical line segments.

The **Schwarz-Christoffel transformation** is used to map the interior of a closed polygon into either the upper or lower half plane. The definition of polygon in this case includes cases where the length of one or more sides can be infinite.

To illustrate the Schwarz-Christoffel transformation, consider for the sake of argument a five-sided polygon in the z plane with corners A, B, C, D, and E. We wish to transform this into a half-plane in the ζ plane. The angles one turns through in passing a corner of the polygon are defined in Figure 3.9.1. By virtue of the polygon being closed the angles must satisfy

$$\alpha + \beta + \gamma + \delta + \varepsilon = 2\pi. \tag{3.9.1}$$

This mapping can be carried out by the transformation

$$\frac{dz}{d\varsigma} = R(\varsigma - a)^{-\alpha/\pi}(\varsigma - b)^{-\beta/\pi}(\varsigma - c)^{-\gamma/\pi}(\varsigma - d)^{-\delta/\pi}(\varsigma - e)^{-\varepsilon/\pi}, \tag{3.9.2}$$

where a, b, c, d and e are real numbers. Usually three of these numbers may be chosen arbitrarily. Notice that it is necessary to traverse the polygon in a counterclockwise

---

[2] A free streamline or free surface is a line or surface of constant pressure.

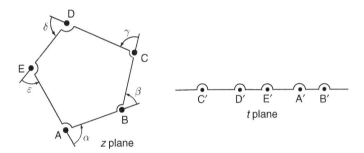

Figure 3.9.1   Schwarz-Christoffel transformation

direction to map to the upper-half t plane. If the polygon has n corners rather than five, all of the preceding statements hold with the obvious addition (or subtraction if $n < 5$) of terms to equations (3.9.1) and (3.9.2).

A variation of the Schwarz-Christoffel transformation allows the polygon to be made up of circular arcs instead of straight lines. To simplify matters, it often is convenient to map one point into the point at infinity, making one of the constants a, b, and so on be infinite. To accomplish this, simply drop the corresponding term in equations (3.9.1) and (3.9.2).

As an example, consider a rectangle in the z plane of semi-infinite length with the points c and d at infinity in the t plane. Since $\alpha = \beta = \pi/2$, we have $dz/d\varsigma = R/\sqrt{(\varsigma - a)(\varsigma - b)}$. If we exercise our freedom to choose, let $a = 1$ and $b = -1$. Then

$$z = R \int \frac{dt}{\sqrt{\varsigma^2 - 1}} = R \cosh^{-1} \varsigma + S.$$

Exercising our freedom of choice once more, let $S = 0$. Then

$$\varsigma = \cosh \frac{z}{R}. \tag{3.9.3}$$

Then $z = 0$ maps to $\varsigma = +1$, and $z = i\pi R$ maps to $\varsigma = -1$. In other words, our semi-infinite rectangle is bounded by the y-axis and horizontal lines starting at $z = 0$ and $z = i\pi R$. See Figure 3.9.2.

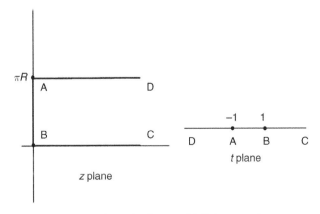

Figure 3.9.2   Schwarz-Christoffel transformation for a semi-infinite rectangle

3.10   Cavity Flows

To illustrate further the application of the Schwarz-Christoffel transformation, we consider the two-dimensional case of a uniform stream impinging on a flat plate. We could use the result for an ellipse and let one of the axes be of zero length, collapsing the ellipse into a plate perpendicular to the uniform stream. We would find that at the corners of the plate the speed would be infinite due to the flow turning through 180 degrees at those points. This is not a physically realizable situation.

Instead, allow the fluid to leave the plate tangentially and have a constant pressure region, or *cavity*, develop behind the plate. Neglecting gravity, the condition that the pressure in the cavity be constant means that the Bernoulli equation says the fluid speed must be constant on the free surface.

Notice that for steady flows this theory requires that the cavity never close. For it to do so, one of two things would have to happen: Either the closure point where the top and bottom flows meet would have be a stagnation point (ruled out by the constant speed condition), or the two streams would have to form a cusp at closure (not easily realizable).

The following presentation is given in Lamb (1932, pp. 99–102). The original solution is credited to Kirchhoff. Since for symmetric flow the flows above and below the streamline going to the stagnation point are the same, we need to consider only the upper flow. For convenience let the just described streamline have the value zero, and let the velocity potential be zero at the stagnation point. We will use the following planes in our discussion. They are shown in Figure 3.10.1.

- *z* **plane**: This is the physical plane. The plate lies on the line segment DAB, A being the stagnation point, D and B the edges of the plate. BC and DC are the free streamline parts of the streamline $\psi = 0$ going from the plate edges to infinity, while AC is the portion of that streamline going from the stagnation point to

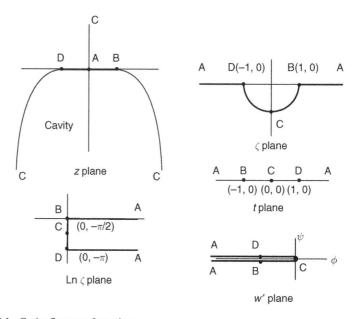

Figure 3.10.1   Cavity flow transformations

far upstream at infinity. Generally, in mapping theory infinity is regarded as a single point.

- $t = (dw/dz)^{-1}$ **plane**: This is the inverse of the ***hodograph plane***.[3] The stagnation point is set at infinity. Since the free streamlines require pressure to be constant, in the absence of gravity this means that the speed is constant. Thus, the free streamlines make up a semicircle BCD, while the flow along the plate are horizontal lines going from B and D to A. The flow lies in the region below the curve ADCBA. If we let $dw/dz = qe^{-i\theta}$ where q is the speed (real), then $t = \frac{1}{q}e^{i\theta}$.

- **ln *t* plane**: This transformation maps the flow region into a rectangle with corners at A, B, and D. Notice that $\ln t = -\ln q + i\theta$.

- **ζ plane**: The Schwarz-Christoffel transformation is used to transfer the rectangle in the ln t plane into the upper-half ζ plane. The boundary is ABCDA.

- **w plane**: This is the complex potential plane used to relate the velocity potential and stream function to the above planes. It consists of a slit ABCDA in the entire plane.

- **w' plane**: Here $w' = 1/w$. This inverse w plane looks much like the w plane but with some rearranging. The slit is now at CDABC.

The analysis proceeds much as that used in transforming the rectangle in the previous section. Using the Schwarz-Christoffel transformation between the ζ and ln t planes, we have

$$\ln t = \cosh^{-1}\varsigma - i\pi, \quad \text{or} \quad t = \cosh(\ln\varsigma + i\pi) = -\frac{1}{2}\left(\varsigma + \frac{1}{\varsigma}\right). \tag{3.10.1}$$

The latter form of the expression follows from the definition of the hyperbolic function.

Using the Schwarz-Christoffel transformation between the ζ and w' planes, we have $dw'/d\varsigma = -R\varsigma$, so $w' = -1/2R\varsigma^2 + S$. At C we choose to have $\varsigma = 0$, $w^{-1} = 0$, which makes $S = 0$, leaving

$$w' = -\frac{1}{2}R\varsigma^2. \tag{3.10.2}$$

To determine R, notice that so far the length L of the plate has not been used. Along the plate t is real, so

$$\varsigma = -\frac{1}{2}\left(q + \frac{1}{q}\right),$$

giving $q = -\varsigma - \sqrt{\varsigma^2 - 1}$. Here, the sign of the radical was chosen so as to make q approach zero as ς approaches minus infinity.

Along the plate $\partial\phi/\partial x = q$ so that $dx/d\phi = 1/q$. Also, $w = 1/w' = -2/R\varsigma^2 = \phi + i\psi$, so that $d\phi/d\varsigma = 4/R\varsigma^3$.

Then

$$L = 2\int_{-\infty}^{-1}\frac{dx}{d\phi}\frac{d\phi}{d\varsigma}d\varsigma = 4R\int_{-\infty}^{-1}\frac{1}{q}\frac{1}{\varsigma^3}d\varsigma = 2R\int_{-\infty}^{-1}\left(-\varsigma + \sqrt{\varsigma^2 - 1}\right)\frac{1}{\varsigma^3}dv. \tag{3.10.3}$$

---

[3] In fluid mechanics the hodograph plane usually refers to the complex velocity plane dw/dz.

Letting $\varsigma' = -1/\varsigma$, this becomes

$$L = \frac{8}{R} \int_0^1 \left(1 + \sqrt{1 - \varsigma'^2}\right) d\varsigma' = \frac{2}{R}(4 + \pi), \quad \text{giving } R = \frac{2(\pi + 4)}{L}. \qquad (3.10.4)$$

Along the free boundary BC, $\ln t = i\theta$. Then $\varsigma = -\cos\theta$ and $\phi = 2R \sec^2\theta$, giving the intrinsic equation of the curve in the form $s = \frac{L}{\pi + 4} \sec^2\theta$ with θ varying between zero and $-\frac{\pi}{2}$. In terms of x and y, the coordinates of the free surface are then

$$x = \frac{2L}{\pi + 4}\left(\sec\theta + \frac{\pi}{4}\right), \quad y = \frac{L}{\pi + 4}\left[\sec\theta\tan\theta - \ln\left(\frac{\pi}{4} + \frac{\theta}{2}\right)\right]. \qquad (3.10.5)$$

The origin is taken at the center of the plate.

To find the force on the plate notice that the difference in pressure between front and back is given by $\frac{1}{2}\rho\left(1 - q^2\right)$, so the net force is given

$$F = \rho \int_{-\infty}^{-1} (1 - q^2)\frac{dx}{d\varsigma} d\varsigma = -\rho R \int_{-\infty}^{-1} \left(\frac{1}{q} - q\right)\frac{d\varsigma}{\varsigma^3} = -2\rho R \int_{-\infty}^{-1} \sqrt{\varsigma^2 - 1}\frac{d\varsigma}{\varsigma^3} = \frac{1}{2}\pi\rho R.$$

Using equation (3.10.4), the net force is found to be

$$F = \frac{\pi}{\pi + 4}\rho U^2 L = 0.440\rho U^2 L. \qquad (3.10.6)$$

3.11   Added Mass and Forces and Moments for Two-Dimensional Bodies

For two-dimensional flows, the 21 independent added masses for three-dimensional flows discussed in Chapter 2 reduce to six: $A_{11}, A_{12}, A_{22}, A_{61}, A_{62}, A_{66}$. The remaining A_{12}, A_{16}, A_{26} follow from these by symmetry. Of these six, a simplifying formula can be found for all but A_{66}. While it is possible to derive the two-dimensional results from the three-dimensional, it is easier to start afresh.

The added masses are going to be generated by the $p = -\rho\frac{\partial\phi}{\partial t}$ term. Thus, the force on the body is given by $\mathbf{F} = \rho\frac{d}{dt}\oint \phi\mathbf{n}\,ds$, where the integration is around the body. Letting $\phi = \sum_{\alpha=1,2,6} V_\alpha\phi_\alpha$ as before, then

$$A_{\alpha 1} + iA_{\alpha 2} = \rho\oint \phi_\alpha n_1\,ds + i\rho\oint \phi_\alpha n_2\,ds = i\rho\oint \phi_\alpha\,dz$$

$$= i\rho\oint w_\alpha\,dz + \oint \psi_\alpha\,dz, \qquad (3.11.1)$$

since $n_1\,ds + in_2\,ds = dy - idx = -idz$. Also, $\frac{\partial\psi_\alpha}{\partial s} = n_\alpha$, and so $\oint \psi_\alpha\,dz = -\oint z\frac{\partial\psi_\alpha}{\partial z}\,dz = -\oint z\frac{\partial\psi_\alpha}{\partial s}\,ds = -\oint zn_\alpha\,ds$. Let

$$B_{\alpha 1} + iB_{\alpha 2} = \rho\oint x_1 n_\alpha\,ds + i\rho\oint x_2 n_\alpha\,ds, \qquad (3.11.2)$$

so equation (3.11.1) becomes

$$(A_{\alpha 1} + B_{\alpha 1}) + i(A_{\alpha 2} + B_{\alpha 2}) = \rho\oint w_\alpha\,dz. \qquad (3.11.3)$$

Note from the definitions in equation (3.11.2), $B_{11} = B_{22} = B$, $B_{12} = B_{21} = 0$, $B_{61} = -B\bar{y}$, $B_{62} = B\bar{x}$, where B is the mass of the fluid displaced by the body and $\bar{x}$ and $\bar{y}$ are the coordinates of the mass center of the displaced fluid. For a body made up of distributed and isolated sources and sinks as well as doublets, application of Cauchy's theorem gives

$$(A_{\alpha 1} + B_{\alpha 1}) + i(A_{\alpha 2} + B_{\alpha 2}) = 2\pi\rho \left[\int \upsilon_\alpha z_s dA + \sum (m_\alpha z_s + \mu_\alpha) \right], \qquad (3.11.4)$$

where σ_α is the distributed source/sink strength, m_α the isolated source/sink strength and z_s its location, and μ_α is the doublet strength. For a circular cylinder, the doublet strength is Ub^2, so $A_{11} + B_{11} = A_{11} + \pi b^2 \rho = 2\pi b^2 \rho$. Thus the added mass of a cylinder is $A_{11} = m_{added} = \pi b^2 \rho$.

For computation of other forces and moments on the body, again recourse could be made to the general three-dimensional Lagally theorem. However, an approach introduced by Blasius (1910) preceded the Lagally work and is in fact easier to use and understand.

The force on an infinitesimal piece of the body surface is

$$dF = dX + idY = ipe^{i\theta} ds = ipdz,$$

where θ is the inclination of the body surface. The complex conjugate of dF is then

$$dF^* = dX - idY = -ipe^{-i\theta} ds = -ipdz^* = -ipe^{-2i\theta} dz. \qquad (3.11.5)$$

Similarly, the moment of dF about the origin is

$$dM = -ydX + xdY = pds(x\sin\theta + y\cos\theta), \qquad (3.11.6)$$

the right-hand side being the real part of $izdF^*$. Thus, add the imaginary part of $izdF^*$ to equation (3.11.6), obtaining

$$dM + idN = izdF^* = pzdze^{-i2\theta}, \qquad (3.11.7)$$

where idN is introduced strictly for mathematical convenience and has no particular physical meaning.

Since for steady irrotational flows $p = p_0 - \frac{1}{2}\rho q^2 = p_0 - \frac{1}{2}\rho\left(\frac{dw}{dz}\right)^2 e^{2i\theta}$, then integration of equations (3.11.5) and (3.11.7) gives us Blasius's results—namely,

$$X - iY = \frac{i}{2}\rho \oint \left(\frac{dw}{dz}\right)^2 dz \qquad (3.11.8)$$

and

$$M + iN = -\frac{1}{2}\rho \oint z\left(\frac{dw}{dz}\right)^2 dz. \qquad (3.11.9)$$

Recalling the Cauchy integral theorems from earlier in this chapter, particularly equations (3.1.10) and (3.1.11), the following infinite series representation of analytic functions can be made.

Taylor series: If $f(z)$ is analytic on and inside a simple closed contour C, and if z_0 is a point inside C, then

$$f(z) = f(z_0) + (z - z_0)f'(z_0) + \cdots + \frac{(z - z_0)^n}{n!}f^{(n)}(z_0) + \cdots$$

$$= \sum_{n=0}^{\infty} \frac{(z - z_0)^n}{n!}f^{(n)}(z_0) \qquad (3.11.10)$$

is convergent everywhere inside C.

Laurent series: If $f(z)$ is analytic on and between two concentric circles C and C′ with center at z_0 is a point inside C, then $f(z)$ can be expanded in positive and negative powers of $z - z_0$. It is convergent everywhere inside the ring-shaped region between C and C′.

$$f(z) = \sum_{n=-N}^{\infty} a_n (z - z_0)^n. \tag{3.11.11}$$

Here, $a_n = \frac{1}{2\pi i} \oint \frac{f(z)\,dz}{(z-z_0)^{n+1}}$.

If $N = \infty$, the function f is said to have an essential singularity at z_0. If N is a positive nonzero number, f is said to have a singularity called a pole of order N at z_0. An example of an essential singularity is $e^{1/z}$. Examples of poles of order one and two are $1/z$ and $1/z^2$, respectively. Essential singularities are seldom encountered in flow problems, while poles of orders one to four are frequently encountered in computing forces and moments due to sources, sinks, vortices, and doublets.

You have no doubt already encountered the Taylor/Laurent series for real functions in calculus classes. You may have wondered at that time why the expansion of $1/(1-x)$ about the origin failed at both $x = +1$ and $x = -1$. The circle domains required by these theorems affords an explanation of that question.

Example 3.11.1   Find the force exerted on a circular cylinder, center at the origin, and with a radius a, in a uniform stream with circulation

Solution. From our earlier work, the complex potential is

$$w = U\left(z + \frac{a^2}{z}\right) + \frac{i\Gamma}{2\pi}\ln z.$$

Thus,

$$\left(\frac{dw}{dz}\right)^2 = \left[U\left(1 - \frac{a^2}{z^2}\right) + \frac{i\Gamma}{2\pi z}\right]^2 = U^2\left(1 - 2\frac{a^2}{z^2} + \frac{a^4}{z^4}\right) + \frac{i\Gamma U}{\pi z}\left(1 - \frac{a^2}{z^2}\right) - \left(\frac{\Gamma}{2\pi z}\right)^2.$$

When this is inserted into equations (3.11.8) and (3.11.9), Cauchy's theorems tell us that the only contribution is from the first order pole. Thus,

$$X = 0, \quad Y = \rho\Gamma U, \quad M = 0.$$

Problems—Chapter 3

3.1 a. Find the equation that the stream function must satisfy in order to represent an irrotational flow in two dimensions. Use cylindrical polar coordinates.

   b. Use separation of variables in the form $\psi(r, \theta) = R(r)T(\theta)$, and find the equations that R and T must satisfy.

3.2 Show that for steady two-dimensional incompressible flow the acceleration can be written as $\left[\left(\frac{dw}{dz}\right)^* \frac{d^2w}{dz^2}\right]^*$.

3.3 It is sometimes convenient to regard a uniform flow as a source at minus infinity and a sink at plus infinity, To demonstrate this, consider the potential for a source of strength m at $(-a, 0)$ and a sink of equal strength at $(a, 0)$. Fix z, and expand the complex potential in a Taylor series of z/a. Show that as a and m each go to infinity such that their ratio remains constant, the result is a uniform stream of speed $U = m/\pi a$.

3.4 Show that, if $w_1(z)$ represents the complex potential of a two-dimensional irrotational flow and has no singularities in $|z| < a$, then $w(z) = w_1(z) + w_1^*\left(a^2/z\right)$ represents the same flow with a circle of radius a introduced at the origin. The star represents that the complex conjugate of $w_1(z)$ is to be taken. This is known as the **circle theorem**. It is a two-dimensional counterpart of the Weiss sphere theorem.

3.5 Use the circle theorem to find the complex potential for a circular cylinder of radius a in a flow field produced by a counterclockwise vortex of strength Γ located at the point $z = b$, where $b > a$. Also find the location of the stagnation points on the circle.

3.6 a. Write the complex potential for a uniform stream, a source of strength $2m$ at $(-a, 0)$, and sinks of strength m at $(0, 0)$ and $(a, 0)$.

   b. Find the equation of the closed streamline.

   c. What is the length of the closed body when $m/2\pi aU = 2/3$, $a = 2$?

3.7 A line vortex with circulation 10 is placed in a corner at $(2, 3)$. There are walls at $x = 0$ and at $y = 0$.

   a. Write the complex potential for the vortex and two corners.

   b. Find the induced velocity at the vortex.

3.8 Find the complex velocity potential for a line vortex at the origin and between two parallel walls at $x = 0$, $y \pm a$. Hint: Multiple mirrors are frequently encountered in clothing stores, generating multiple reflections.

3.9 Show that the flow past a flat plate on the real z-axis is given by $w(z) = Ua\cosh(\varsigma - i\alpha) + \frac{i\Gamma}{2\pi}\varsigma$, where $z = a\cosh\varsigma$ and U represents a uniform stream. The angle α is the angle between the plate and the uniform stream. The plate is at $0 \leq \text{Imag}(\varsigma) \leq 1$ in the ς plane.

3.10 Show that the flow of fluid inside a rotating elliptical cylinder with semimajor axes a and b is given by $w(z) = iAz^2$. Find A in terms of the geometry and angular rate of rotating. Hint: The boundary condition is $\psi = \frac{1}{2}\Omega\left(x^2 + y^2\right) + \text{constant}$.

3.11 Show that the flow of fluid outside a rotating elliptical cylinder with semimajor axes a and b is given by $w(z) = iAe^{-2\varsigma}$, where $z = c\cosh\varsigma$. Find A in terms of the geometry and angular rate of rotating. Hint: The boundary condition is $\psi = \frac{1}{2}\Omega\left(x^2 + y^2\right) + \text{constant}$.

3.12 a. Write the complex potential for a line vortex of circulation Γ located at $z = ih$ and a line vortex of circulation $-\Gamma$ located at $z = -ih$. Show that the x-axis is a streamline and so can be considered an infinite flat plate.

   b. Calculate the pressure on the surface of the plate from the Bernoulli equation, letting the pressure at infinity be zero. Integrate this pressure over the entire length of the plate in order to find the force acting on the plate due to the vortex.

    c. If a pair of sources, each of strength m, had been used, what would the force be?

3.13 Using the method of images, find the complex potential for a vortex located along the centerline of a channel of width a. The vortex is at point $(b, 0)$, and the channel is in the region $x > 0$, $-a/2 \leq y \leq a/2$.

3.14 Show that the flow about a circular arc can be found using the Joukowski transformation with an intermediate translation of axes that puts the origin on the vertical axes. This means that both the leading and trailing edges lie on the surface of the arc.

3.15 Show that the transformation that maps the interior of the sector $0 \leq \theta \leq \pi/n$, n an integer, in the z plane onto the upper half of the ς plane is $\varsigma = z^n$. Then if $w = A\varsigma$, A real, represents a uniform stream in the ζ plane, find the corresponding flow in the z plane.

3.16 The infinite strip shown can be regarded as a two-sided polygon and hence the Schwarz-Christoffel transformation can be used to transform it into the upper-half ζ plane. Find the transformation that puts A and D at infinity and B and C at the origin in the upper-half ζ plane. Take the width of the strip to be h.

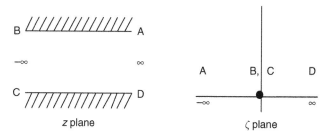

Figure P3.16   Problem 3.16—Infinitely long strip

3.17 The semi-infinite strip shown can be regarded as a three-sided polygon, hence the Schwarz-Christoffel transformation can be used to transform it into the upper-half ζ plane. Find the transformation that puts A and D at infinity and B and C at plus and minus one in the upper-half ζ plane. Take the width of the strip to be h.

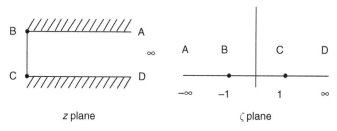

Figure P3.17   Problem 3.17—Semi-infinitely long strip

3.18 The Schwarz-Christoffel transformation can be used to transform the right-angle 90-degree bend shown into the upper-half ζ plane. The bend can be thought of as a four-sided polygon. Find the transformation that places the corner A at 1, B at the origin, C at -1, and D at infinity in the ζ plane.

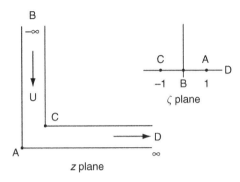

Figure P3.18   Problem 3.18—Elbow

Surface and Interfacial Waves

Many different mechanisms for wave propagation in fluids exist. Compressibility of both liquids and gases allows for compression waves such as sound waves and shock waves. These generally have quite large velocities of propagation. Normally, they cannot be seen by the human eye without suitable instrumentation or lighting. Waves on surfaces and interfaces, on the other hand, are generally slower and can be observed easily. Gravity, inertial forces, surface tension, and viscosity are also important mechanisms in generating waves.

Much of the work presented in this chapter was originated by three English scientists: Lord Kelvin (born William Thomson, 1824–1907), Lord Rayleigh (born William Strutt, 1842–1919), and Sir Horace Lamb (1848–1934). The names in the literature for the first two can be confusing, because, for example, both Kelvin and Thomson are used, depending on whether the date of publication preceded or followed the bestowing of honors. Much of the work is summarized in the books by Rayleigh (1945) and Lamb (1932), and in the lengthy review in Wehausen and Laitone (1960).

4.1 Linearized Free Surface Wave Theory

4.1.1 Infinitely Long Channel

Consider the case of small amplitude waves in a two-dimensional channel of constant depth h, an infinite length in the x direction, and having a uniform stream of velocity

U. It is convenient to choose a coordinate system with an origin on the undisturbed free surface. When dealing with an unsteady flow problem with a free surface, it is necessary to find the pressure by using the Bernoulli equation. Therefore, working with the velocity potential, rather than the stream function, is indicated.

Write the total velocity potential in the form $\phi = Ux + \phi'$, where ϕ' represents the velocity field due to the waves. The equations to be solved are then

$$\nabla^2 \phi = 0, \tag{4.1.1}$$

such that

$$\frac{\partial \phi}{\partial y} = v_y = 0 \quad \text{on } y = -h. \tag{4.1.2}$$

Also, if the free surface is elevated an amount $\eta(x, t)$ from the static position, then

$$\frac{p}{\rho} = -\left(\frac{\partial \phi}{\partial t} + \frac{|\nabla \phi|^2}{2} + g\eta\right) + \frac{\sigma}{\rho}\left(\frac{1}{R_1} + \frac{1}{R_2}\right) = \text{constant on } y = \eta, \tag{4.1.3}$$

and

$$\frac{D\eta}{Dt} = \frac{\partial \phi}{\partial y} = v_y \quad \text{on } y = \eta \tag{4.1.4}$$

(see Figure 4.1.1). Here, σ is the surface tension and R_1 and R_2 are the principal radii of curvature. Since the pressure above the free surface is constant and can be taken as gauge pressure, it can usually be set to zero.

Equation (4.1.4) states that the free surface moves up and down with a velocity equal to the vertical component of the fluid velocity. If this problem were to be solved with all nonlinearities included, it would be necessary to work instead with the vector $\boldsymbol{\eta}$, which is the total vector displacement of a fluid particle on the free surface, and to replace equation (4.1.4) by

$$\frac{D\boldsymbol{\eta}}{Dt} = \nabla \phi, \quad \frac{D\phi}{Dt} = \frac{1}{2}|\nabla \phi|^2 - g\eta_y$$

on the free surface $D\boldsymbol{\eta}/Dt = \mathbf{v}$. Here, η is the vertical component of $\boldsymbol{\eta}$.

Even though equation (4.1.1) is linear, the boundary conditions equations (4.1.3) and (4.1.4), together with the fact that the free surface position is a priori unknown make the problem highly nonlinear. Only in a few very special cases can a closed form expression be found. Instead, consider the case of waves whose amplitude is small

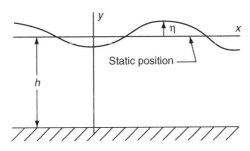

Figure 4.1.1 Wave definitions—two dimensions

compared to their wavelength, and replace the exact free surface conditions in equations (4.1.3) and (4.1.4) by the linearized conditions

$$\frac{\partial \phi'}{\partial t} + U \frac{\partial \phi'}{\partial x} + g\eta = \frac{\sigma}{\rho} \frac{\partial^2 \eta}{\partial x^2} \quad \text{on } y = 0, \tag{4.1.5}$$

$$\frac{\partial \eta}{\partial t} + U \frac{\partial \eta}{\partial x} = \frac{\partial \phi'}{\partial y} \quad \text{on } y = 0. \tag{4.1.6}$$

Here, the problem has been simplified still further by imposing equations (4.1.5) and (4.1.6) at the static position $y = 0$ rather than on the actual surface. The problem is now amenable to analysis by the method of separation of variables.

Start by assuming that ϕ can be written as the superposition of terms like

$$\phi'(x, y, t) = T(t) \, X(x) \, Y(y), \tag{4.1.7}$$

and insert this form into equation (4.2.1). The result is

$$TX_{xx}Y + TXY_{yy} = 0.$$

Divided by ϕ, this becomes

$$X_{xx}/X + Y_{yy}/Y = 0.$$

Since the first term depends only on x and the second depends only on y, this relationship can hold throughout the flow region only if each term is itself a constant. Thus,

$$Y_{yy}/Y = -X_{xx}/X = \text{constant} = k^2 \text{ (say)}.$$

The form of the constant k^2 has been selected by "looking ahead" at the solution. It is expected that the disturbance will be oscillatory in the x direction and also that it will die out away from the free surface. Making this decision now is not necessary, but it does save a lot of cumbersome and confusing notation later on. Experience with separation of variables problems, and a little foresight, can be very useful in avoiding a morass of symbols.

The X solutions are then trigonometric functions of kx, and the Y solutions are hyperbolic functions of ky. Taking equation (4.1.2) into account, using DeMoive's theorem that $e^{i\theta} = \cos\theta + i\sin\theta$, write

$$X = A \, e^{ikx} + B \, e^{-ikx}, \tag{4.1.8}$$

$$Y = D \cosh k(y + h). \tag{4.1.9}$$

Since both D and T multiply the still unknown constants A and B in X, at this point D and T can be combined with A and B, letting

$$\phi' = \cosh k(y + h) \left(A(t)e^{ikx} + B(t)e^{-ikx} \right).$$

Guided by this form for the velocity potential, choose

$$\eta' = F(t)e^{ikx} + G(t)e^{-ikx}.$$

Substituting these expressions into equations (4.1.5) and (4.1.6), it is found that for these boundary conditions to be satisfied for all x it is necessary that

$$A = \frac{1}{k \sinh kh} \left(\frac{dF}{dt} + ikUF \right), \quad B = \frac{1}{k \sinh kh} \left(\frac{dG}{dt} - ikUG \right), \tag{4.1.10}$$

and

$$F = -\frac{\cosh kh}{g + k^2\frac{\sigma}{\rho}}\left(\frac{dA}{dt} + ikUA\right) \quad \text{and} \quad G = -\frac{\cosh kh}{g + k^2\frac{\sigma}{\rho}}\left(\frac{dB}{dt} - ikUB\right). \quad (4.1.11)$$

Combining these equations and letting

$$\omega^2 = k \tanh kh \left(g + k^2\frac{\sigma}{\rho}\right) \quad (4.1.12)$$

results in

$$A = -\frac{1}{\omega^2}\left(\frac{d^2A}{dt^2} + 2ik\frac{dA}{dt} - k^2U^2A\right) \quad \text{and} \quad B = -\frac{1}{\omega^2}\left(\frac{d^2B}{dt^2} - 2ik\frac{dB}{dt} - k^2U^2B\right).$$

These have solutions like $A = e^{-ikUt}(\alpha(k)e^{i\omega t} + \beta(k)e^{-i\omega t})$, $B = e^{ikUt}(\gamma(k)e^{i\omega t} + \delta(k)e^{-i\omega t})$.

Combining these results gives finally

$$\phi' = \cosh k(y+h)\left[e^{ik(x-Ut)}\left(\alpha(k)e^{i\omega t} + \beta(k)e^{-i\omega t}\right) + e^{-ik(x-Ut)}\left(\gamma(k)e^{i\omega t} + \delta(k)e^{-i\omega t}\right)\right],$$
$$(4.1.13)$$

$$\eta' = i\frac{\sinh kh}{\omega}\left[e^{ik(x-Ut)}\left(\alpha(k)e^{i\omega t} - \beta(k)e^{-i\omega t}\right) + e^{-ik(x+Ut)}\left(\gamma(k)e^{i\omega t} - \delta(k)e^{-i\omega t}\right)\right].$$
$$(4.1.14)$$

These represent a wave traveling upstream with a velocity $U \pm \omega/k$ and are called *traveling waves*. The parameters k and ω are called the *wavenumber* and *circular frequency*, respectively. The *wavelength* λ of the wave is given by $\lambda = 2\pi/k$ and the wave speed relative to the uniform speed by

$$c = \frac{\omega}{k} = \sqrt{\frac{1}{k}\left(g + k^2\frac{\sigma}{\rho}\right)\tanh kh}. \quad (4.1.15)$$

For depths large compared to the wavelength (the long wave case) $\tanh kh \approx 1$ and $c \approx \sqrt{(g + k^2\frac{\sigma}{\rho})\frac{\lambda}{2\pi}}$. For shallow depths, $\tanh kh \approx kh$ and $c = \frac{\omega}{k} = \sqrt{h(g + k^2\frac{\sigma}{\rho})}$. If surface tension effects are not important, the wave speed reduces to $c \approx \sqrt{gh}$ and could have been found by elementary momentum considerations.

The fact that the wave speed depends on the wave number means that the wave is *dispersive*. That is, if a wave consists of two or more wave numbers, as the wave propagates, each component moves at a different speed, so the wave disperses, or changes shape, as it travels.

Since waves on a very large body of water are being considered, with x extending to infinity in both directions, there are no boundary conditions left to determine k. Thus, there can be a continuous spectrum of k. The linearity of the problem allows for superposition of wave numbers, leading to a solution in the form of a *Fourier integral*. This is conveniently written as

$$\phi''(x,y,t) = \int_{-\infty}^{\infty} \phi'(x,y,t,k)\,dk, \quad \eta''(x,y,t) = \int_{-\infty}^{\infty} \eta'(x,y,t,k)\,dk, \quad (4.1.16)$$

where ω is given by equation (4.1.12). Information on the initial shape and velocity of the disturbance is necessary to determine α, β, γ, and δ, and to be able to carry out the integrations.

Undersea earthquakes generate waves of very long wavelengths, giving rise to waves called **tsunamis**, which can travel at very high speeds across the ocean. In deep water these waves are of small amplitude, and ships encountering such waves at sea may not experience much if any of a disturbance as the waves pass. However, as the waves approach shallow water near the shore, the previous results for wave speed show that they slow down and, by exchanging their kinetic energy for potential energy, steepen drastically, frequently causing catastrophic damage and loss of life. This explanation also can be used to understand wave amplification and breaking of waves on beaches. Waves caused by high winds such as hurricanes can be of large amplitude near the driving force but do not tend to steepen as much.

4.1.2 Waves in a Container of Finite Size

If there are walls at the ends of our channel, reflections will occur and the traveling wave form of the solution is no longer convenient for describing the behavior of the flow. Instead of the $e^{\pm ikx}$ in equation (4.1.8) a better selection is to use $\cos kx$ and $\sin kx$. If the channel is bounded at $x = 0$ and at $x = L$, then $\partial\phi/\partial x = 0$ at those locations. This rules out the use of $\sin kx$ and requires that $\sin kL = 0$. This gives $k = n\pi/L$, where n is any positive integer greater than zero. Our solution now becomes

$$\phi' = \sum_{n=0}^{\infty} A_n(t) \cos \frac{n\pi x}{L} \cosh k(y+h). \qquad (4.1.17)$$

The boundary conditions still to be satisfied are

$$\frac{\partial\phi'}{\partial t} + g\eta = \frac{\sigma}{\rho}\frac{\partial^2\eta}{\partial x^2} \quad \text{on } y = 0, \qquad (4.1.18)$$

and

$$\frac{\partial\eta}{\partial t} = \frac{\partial\phi'}{\partial y} \quad \text{on } y = 0. \qquad (4.1.19)$$

The U term has been dropped because the boundaries do not allow a uniform stream.

Taking the displacement of the form $\eta = \sum_{n=0}^{\infty} F_n(t)\cos n\pi x/L$ and applying equations (4.1.18) and (4.1.19) results in

$$\sum_{n=0}^{\infty} \frac{dA_n}{dt} \cos \frac{n\pi x}{L} \cosh kh = -\sum_{n=0}^{\infty} \left(g + \frac{n^2\pi^2}{L^2}\frac{\sigma}{\rho}\right) F_n \cos \frac{n\pi x}{L},$$

$$\sum_{n=0}^{\infty} \frac{dF_n}{dt} \cos \frac{n\pi x}{L} = \sum_{n=0}^{\infty} k A_n \sinh kh \cos \frac{n\pi x}{L}.$$

Using the orthogonality properties of the cosines by multiplying each side of the above by $\cos m\pi x/L$ and then integrating over the channel length gives

$$\frac{dA_n}{dt} \cosh kh = -\left(g + \frac{n^2\pi^2}{L^2}\frac{\sigma}{\rho}\right) F_n, \quad \frac{dF_n}{dt} = k A_n \sinh kh.$$

Solving these for A_n and F_n results in

$$\phi' = \sum_{n=0}^{\infty} (a_n \sin \omega t + b_n \cos \omega t) \cos \frac{n\pi x}{L} \cosh k(y+h), \qquad (4.1.20)$$

$$\eta = -\sum_{n=0}^{\infty} \frac{\omega \cosh kh}{g + \frac{n^2 \pi^2}{L^2} \frac{\sigma}{\rho}} (a_n \cos \omega t - b_n \sin \omega t) \cos \frac{n\pi x}{L}, \qquad (4.1.21)$$

with

$$\omega_n = \sqrt{\left(g + \frac{\sigma}{\rho} \frac{n^2 \pi^2}{L^2}\right) \frac{n\pi}{L} \tanh \frac{n\pi h}{L}}. \qquad (4.1.22)$$

The solutions in equations (4.1.21) and (4.1.22) are in the form of Fourier series, whose coefficients can be found by imposing initial conditions on η and $\partial\eta/\partial t$. The solution given by equations (4.1.20) and (4.1.21) is referred to as a ***standing wave***.

It is important to understand how the physics and mathematics have interacted. Laplace's equation is said to be of ***elliptic type***. This type of equation governs phenomena such as steady-state heat transfer, deformation of membranes, and flow in porous media. Normally, waves are not anticipated when dealing with such equations. The free surface conditions have altered the physics of the situation, giving solutions that in mathematics are more usual to equations of ***hyperbolic type***—equations that permit waves.

Wave behavior was clearly seen in the case where the channel extended to infinity in the x direction. When the x extent was finite, this wave nature was concealed by the form of the solution. Consider, for instance, a wave that starts out traveling to the right. It will have a solution of the form $e^{i(mx-\omega t)}$. When it encounters a wall, it is reflected and becomes a left-traveling wave with the form $e^{-i(mx+\omega t)}$. This wave is reflected at the left wall, and the process repeats endlessly. The presence of the wall ***selects*** waves—that is, it allows only certain wave numbers to be present. The form of the solution in turn breaks down the exponential $e^{i(mx\pm\omega t)}$ into $\sin mx$, $\cos mx$, $\sin \omega t$, and $\cos \omega t$—thus disguising the traveling wave nature of the solution. The infinite domain case with no walls present and solutions like $e^{i(mx\pm\omega t)}$ results in traveling waves with the general solution of the form $f(x-ct)+g(x+ct)$. In the finite domain case, because of reflections from walls, solutions like $\sin mx \ \sin \omega t$ or $\cos mx \ \cos \omega t$ result in standing waves after startup conditions have occurred.

4.2   *Group Velocity*

In the previous section it was stated that because the wave speed depended on the wave number, the waves were dispersive. That means that if a wave shape is broken down into Fourier components and more than one wave number is present, the wave shape would change as the wave proceeds. If one observes an isolated group of waves traveling with nearly the same wavelength, some of the waves are observed to proceed through the group until it approaches the front of the group. At that point they appear to die out. The process is then repeated by other waves.

To analyze this situation, consider the sum of two sine waves of equal amplitude and nearly equal wave length, and observe the change in the envelope of the resulting wave. Let

$$
\begin{aligned}
\eta &= A\sin(kx - \omega t) + A\sin\left[(k + \Delta k)x - (\omega + \Delta\omega)t\right] \\
&= 2A\cos 0.5(\Delta kx - \Delta\omega t)\sin\left[(k + 0.5\Delta k)x - (\omega + 0.5\Delta\omega)t\right].
\end{aligned}
\tag{4.2.1}
$$

This has the appearance of an envelope that looks much like a slowly traveling wave of wavelength $\Delta\lambda = \pi/\Delta k$ that encloses a much more rapidly traveling wave of wavelength $\lambda = 2\pi/k$. In the limit, as the wave numbers approach each other, the speed of propagation of the envelope is then

$$
c_g = \frac{d\omega}{dk} = \frac{d(kc)}{dk} = c + k\frac{dc}{dk} = c - \lambda\frac{dc}{d\lambda}.
\tag{4.2.2}
$$

The quantity c_g is called the **group velocity**.

In the previous section we saw that the general formula for c as a function of wave number was given by equation (4.1.15) as

$$
c = \frac{\omega}{k} = \sqrt{\frac{1}{k}\left(g + k^2\frac{\sigma}{\rho}\right)\tanh kh}.
$$

Thus, for these waves the group velocity is given by

$$
c_g = \frac{\left(g + 3k^2\dfrac{\sigma}{\rho}\right)\tanh kh + kh\left(g + k^2\dfrac{\sigma}{\rho}\right)\operatorname{sech}^2 kh}{\sqrt[2]{k\left(g + k^2\dfrac{\sigma}{\rho}\right)\tanh kh}}.
\tag{4.2.3}
$$

For large wavelengths compared to the depth this is approximated by

$$
c_g \approx \frac{g + 3k^2\dfrac{\sigma}{\rho}}{\sqrt[2]{k\left(g + k^2\dfrac{\sigma}{\rho}\right)}},
\tag{4.2.4}
$$

whereas for long waves, where surface tension effects are small ($\lambda \gg 2\pi\sqrt{\sigma/g\rho}$),

$$
c_g \approx \frac{1}{2}\sqrt{\frac{g}{k}} \simeq \frac{1}{2}c.
\tag{4.2.5}
$$

There are at least two ways for explaining the concept of group velocity. Since the group velocity is smaller than the wave velocity, the waves advance within their envelopes until they approach the neck, or nodal point, at the right of the envelope. As they approach that point, they are gradually extinguished and replaced by their successors that are formed on the right side of the following node. The waves are thus "grouped" within the necks of their envelope.

Another explanation of group velocity has to do with the speed at which the energy of the wave travels. The rate at which work is done on a section of the fluid is $\int_{-\infty}^{0} p\frac{\partial\phi}{\partial x}\,dy$. The only part of the Bernoulli equation that contributes to the work in the linearized theory is $\rho \frac{\partial \phi}{\partial t}$. Thus, taking the velocity potential from equation (4.1.15) in the abbreviated form $\phi = \beta e^{ky+ik(x-ct)}$ (no uniform stream) results in

$$\int_{-\infty}^{0} p \frac{\partial \psi}{\partial x} \, dy = \rho \int_{-\infty}^{0} \frac{\partial \psi}{\partial t} \frac{\partial \psi}{\partial x} \, dy = \frac{1}{2} \rho c k \beta^2 \cos^2 k(x - ct), \qquad (4.2.6)$$

where the real part of the integrand has been used. Averaging this over a wavelength gives $\frac{1}{2} \rho c k \beta^2 \frac{\lambda}{2} = \frac{1}{2} \pi \rho c \beta^2$. The kinetic energy passing through the section is

$$\frac{1}{2} \rho \int_{-\infty}^{0} \left(\frac{\partial \phi}{\partial x} \right)^2 dy = \frac{1}{2} \rho \int_{-\infty}^{0} \left(ik\beta e^{ky+ik(x-ct)} \right)^2 dy = \frac{1}{4} \rho k \beta^2 \cos^2 k(x - ct), \quad (4.2.7)$$

which when averaged over the same period gives $\frac{1}{4} \rho k \beta^2 \frac{\lambda}{2} = \frac{1}{4} \pi \rho \beta^2$. Thus, the kinetic energy is moved through the fluid at only one-half the wave speed c, which corresponds to the group velocity.

4.3   Waves at the Interface of Two Dissimilar Fluids

Following the procedure described in the previous section, the theory of waves at an interface can also be developed. This time consider two fluids: the upper one ($y > 0$) moving with a velocity U_1 and the lower one ($y \leq 0$) moving with a velocity U_2. The investigation will be to determine whether a small disturbance introduced at the interface will either grow or decay. To accomplish this, choose a flow plus a disturbance in the form of a progressive wave such that

$$\phi = \begin{cases} \phi_1 = xU_1 + C_1 e^{-ky+ik(x-ct)} & \text{for } y > 0, \\ \phi_2 = xU_2 + C_2 e^{ky+ik(x-ct)} & \text{for } y \leq 0. \end{cases} \qquad (4.3.1)$$

Notice that both of these velocity potentials satisfy the Laplace equation. They represent waves that are traveling in the positive x direction whose amplitude dies out away from the interface. The (real) parameter k is the wave number and is twice pi divided by the wave length. The real part of c is the wave speed, and the imaginary part of kc is the growth rate. If $\text{Imag}(kc)$ is positive, the wave will grow, whereas if it is negative the wave decays. Neutral stability is obtained if c is real.

The form of the interface disturbance that is suited to these potentials is $\eta = Ae^{ik(x-ct)}$ and the appropriate boundary conditions to be imposed at $y = 0$ are

$$\frac{D\eta}{Dt} = \frac{\partial \phi}{\partial n} \quad \text{and} \quad p_2 - p_1 = \frac{\sigma}{R}, \qquad (4.3.2)$$

where σ is the surface tension and R is the radius of curvature of the interface.

As we saw in Section 4.1, imposing nonlinear boundary conditions on an unknown boundary is a very difficult task. Instead, again assume that the disturbance is very small—so small, in fact, that nonlinear terms can be neglected—and also that these boundary conditions can be applied at the undisturbed interface. This will tell us whether or not a small disturbance will grow. It does not say what happens if growth occurs. In fact, it says the growth will be unbounded. Obviously, the nonlinear terms will become important long before that, and in that situation a nonlinear analysis must be used.

Our linearized boundary conditions to be applied then on $y = 0$ are

$$\frac{\partial \eta}{\partial t} + U_1 \frac{\partial \eta}{\partial x} = \frac{\partial \phi_1}{\partial y}, \quad \frac{\partial \eta}{\partial t} + U_2 \frac{\partial \eta}{\partial x} = \frac{\partial \phi_2}{\partial y}, \quad \text{and}$$

$$p_2 - p_1 = \rho_2 \left(\frac{\partial \phi_2}{\partial t} + U_2 \frac{\partial \phi_2}{\partial x} + g\eta \right) - \rho_1 \left(\frac{\partial \phi_1}{\partial t} + U_1 \frac{\partial \phi_1}{\partial x} + g\eta \right) = \sigma \frac{\partial^2 \eta}{\partial x^2}. \tag{4.3.3}$$

The radius of curvature has been linearized here according to

$$\frac{1}{R} = \frac{\dfrac{d^2 y}{dx^2}}{\left[1 + \left(\dfrac{dy}{dx} \right)^2 \right]^{3/2}} \approx \frac{d^2 y}{dx^2}.$$

Substitution of our velocity potentials into this and removing multiplying terms gives

$$ik(U_1 - c)A = -kC_1, \quad ik(U_2 - c)A = kC_2,$$

$$\rho_2 \left[ik\,(U_2 - c)\,C_2 + gA \right] - \rho_1 \left[ik\,(U_1 - c)\,C_1 + gA \right] = -\sigma k^2 A. \tag{4.3.4}$$

Eliminating the C_1, C_2, and A and solving for c gives

$$\rho_2 [-k\,(U_2 - c)^2 + g] - \rho_1 [k\,(U_1 - c)^2 + g] = -\sigma k^2, \quad \text{or finally}$$

$$c = \frac{\rho_1 U_1 + \rho_2 U_2}{\rho_1 + \rho_2} \pm \left\{ \frac{g\,(\rho_2 - \rho_1) + \sigma k^2}{k\,(\rho_1 + \rho_2)} - \frac{\rho_1 \rho_2\,(U_1 - U_2)^2}{(\rho_1 + \rho_2)^2} \right\}^{1/2}. \tag{4.3.5a}$$

We can also write this as

$$c = c_0 \pm c_1,$$

$$\text{where} \quad c_0 = \frac{\rho_1 U_1 + \rho_2 U_2}{\rho_1 + \rho_2} \quad \text{and} \quad c_1 = \left\{ \frac{g\,(\rho_2 - \rho_1) + \sigma k^2}{k\,(\rho_1 + \rho_2)} - \frac{\rho_1 \rho_2\,(U_1 - U_2)^2}{(\rho_1 + \rho_2)^2} \right\}^{1/2}.$$

$$\tag{4.3.5b}$$

To decide on the stability of this flow note that c_0 is always real and positive. If the flow is to be unstable, it must be c_1 that is imaginary. That would occur provided

$$(U_1 - U_2)^2 > \frac{g\,(\rho_2^2 - \rho_1^2) + \sigma k^2\,(\rho_1 + \rho_2)}{\rho_1 \rho_2 k}. \tag{4.3.6}$$

We can see that surface tension is always a stabilizing influence, but for long waves (small values of k) its effect loses significance. Having a denser fluid on top of a lighter fluid is of course always destabilizing.

Helmholtz pointed out that where $U_1 = U_2$, $\rho_1 = \rho_2$ can be used to explain the flapping of sails and flags. In this case c_1 vanishes, and our solution procedure fails because of the double root. Repeating the analysis for this special case would be the same except for the expression for η, which now would have to include a time to the power one multiplier. A disturbance introduced at the edge of a sail or flag would then grow as it travels along the flag, a frequently observed phenomenon.

The difference in velocities on either side of the interface means that the interface can be considered as a concentrated vortex sheet. To see this, suppose that rather than having a discrete jump, the undisturbed flow was given by a velocity profile such as

$U = \frac{U_1+U_2}{2} + \frac{U_1-U_2}{2}\tanh\frac{y}{\delta}$. In the limit as δ gets small this approaches the discontinuous profile. It has the vorticity component $\frac{dU}{dy} = \frac{U_1-U_2}{2\delta\cosh^2\frac{y}{\delta}}$. Vortices have a tendency to roll up, as exhibited in the nature of the displacement of the interface.

4.4   Waves in an Accelerating Container

Transporting a liquid in a moving container is a very effective way of causing surface waves, as anyone who has carried a cup of coffee across a room has experienced. In transporting oil across the oceans in large tankers, the breaking of these waves against the bulkheads can easily create substantial damage. Even in airplanes, as fuel tanks empty the sloshing of the fuel can create problems. One solution is to break up the surface into smaller regions by floating dividers, but this leads to difficulties in cleaning the tanks.

Different types of container motion lead to different mechanisms of wave generation. One of these—waves due to an acceleration in the vertical direction—is examined next.

Consider a rectangular tank of liquid subjected to a translational acceleration in the vertical direction. Neglect surface tension effects. The liquid will be taken to originally be in the rectangular region $0 \le x \le a, 0 \le y \le b, -h \le z \le 0$, where the coordinate system is fixed to the tank. The velocity potential then satisfies the Laplace equation together with the boundary conditions

$$\left.\frac{\partial\phi}{\partial x}\right|_{x=0} = \left.\frac{\partial\phi}{\partial x}\right|_{x=a} = 0, \quad \left.\frac{\partial\phi}{\partial y}\right|_{y=0} = \left.\frac{\partial\phi}{\partial y}\right|_{y=b} = 0, \quad \left.\frac{\partial\phi}{\partial z}\right|_{z=-h} = 0$$

$$\frac{\partial\eta}{\partial t} = \left.\frac{\partial\phi}{\partial z}\right|_{z=0}, \quad \left.\frac{\partial\phi}{\partial t}\right|_{z=0} + \left(g + \frac{dV_z}{dt}\right)\eta = 0. \tag{4.4.1}$$

The solution of the Laplace equations that satisfies the boundary conditions at the walls is of the form

$$\phi = \sum_{m=0}^{\infty}\sum_{n=0}^{\infty}\frac{dA_{mn}}{dt}\cos\frac{m\pi x}{a}\cos\frac{m\pi y}{b}\cosh\pi r(z+h), \tag{4.4.2}$$

and

$$\eta = \pi r\sum_{m=0}^{\infty}\sum_{n=0}^{\infty}A_{mn}\cos\frac{m\pi x}{a}\cos\frac{m\pi y}{b}\sinh\pi rh, \tag{4.4.3}$$

where $r = \sqrt{\frac{m^2}{a^2} + \frac{n^2}{b^2}}$. This solution also satisfies the kinematic condition at the free surface.

The pressure condition at the free surface requires that

$$\frac{d^2A_{mn}}{dt^2} + \pi r\left(g + \frac{dV_z}{dt}\right)A_{mn}\tanh\pi rh = 0. \tag{4.4.4}$$

Next introduce the circular frequency $\omega^2 = \pi rg\tanh\pi rh$ and take $V_z = -V_0\sin\Omega t$. Make time dimensionless by choosing $T = \frac{1}{2}\Omega t$ and introduce dimensionless parameters

$$a = \left(\frac{2\omega}{\Omega}\right)^2, \quad 2q = \frac{V_0\Omega}{g}\left(\frac{2\omega}{\Omega}\right)^2.$$

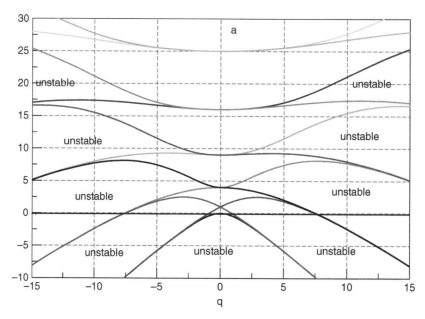

Figure 4.4.1 Strutt diagram

Equation (4.4.4) then becomes

$$\frac{d^2 A_{mn}}{dT^2} + (a - 2q\cos 2T)\, A_{mn} = 0. \qquad (4.4.5)$$

This is the standard form of the **Mathieu equation**,[1] which has been well studied, particularly by Rayleigh. He showed that regions of stability were governed by combinations of the parameters p and α as shown in Figure 4.4.1. This figure is called the **Strutt diagram**. If the parameter pair *(a, q)* lie in a region not labeled, the solution is oscillatory but not periodic. This nonperiodicity of the wave is an example of the "weakness" of this type of wave. If the pair (p, α) lie in a region labeled unstable, at least one of the solutions will increase without limit with increasing time.

The lines separating the stable-unstable regions in Figure 4.4.1 are lines along which one of the solutions is periodic with period $2\pi/\nu$, where ν is the largest integer with p as an upper bound. The other solution increases in time without bound. Thus, these dividing lines are considered part of the unstable region. Further information on the computation of these lines is given in the Appendix.

4.5 Stability of a Round Jet

Liquid jets have a number of industrial applications, including the production of a stream of droplets. In the early days of the development of the inkjet printer two technologies were considered. The first one used jet breakup to produce a steady stream of electrically charged droplets, which then were steered by electric fields to either write on a

---

[1] This equation is also encountered in analyzing certain mechanical vibrations, including the inverted pendulum and the pendulum of varying length.

page or to be collected and subsequently recirculated. The second approach, the "drop-on-demand," produced individual droplets all of which struck the page. It is the technology most frequently used today for home printers.

Modern "airless" paint sprayers used for painting automobiles depend on a rotating cone to throw off a series of jets, each of which rapidly break up into droplets. The water-based paint is electrically charged as it enters the cone, and an electric potential between the cone and the item to be painted steers the paint to the target.

The breakup of a liquid jet was first studied by Raleigh in 1876. For a jet of radius a traveling at a speed U, the velocity potential for the disturbed flow is taken as $\phi = \phi_1(r)e^{i(kz+s\theta+\omega t)}$. This is in the form of a wave traveling along the jet moving at a speed c. The θ dependence has been selected so that traveling around the jet circumference returns us to the starting point. The z dependence gives a wavelike shape in the z direction.

Inserting this form into the Laplace equation gives

$$\nabla^2\phi = \left(\frac{\partial^2}{\partial r^2} + \frac{1}{r}\frac{\partial^2}{\partial r^2} + \frac{1}{r^2}\frac{\partial^2}{\partial \theta^2} + \frac{\partial^2}{\partial z^2}\right)\phi = e^{i(kz+n\theta+\omega t)}\left(\frac{d^2}{dr^2} + \frac{1}{r}\frac{d^2}{dr^2} - \frac{n^2}{r^2} - k^2\right)\phi_1 = 0.$$
(4.5.1)

The operator acting on ϕ_1 is of the Bessel type. The solution that is finite within the jet is $\phi_1 = AI_n(kr)$, where $I_n(kr)$ is a modified Bessel function given by the series

$$I_n(kr) = i^{-n}J_n(ikr) = i^{-s}\sum_{m=0}^{\infty}\frac{(-1)^m}{m!\Gamma(m+n+1)}\left(\frac{ikr}{2}\right)^{s+2m}$$

$$= \left(\frac{kz}{2}\right)^n\sum_{m=0}^{\infty}\frac{1}{m!\Gamma(m+n+1)}\left(\frac{kr}{2}\right)^{2m}.$$
(4.5.2)

The boundary conditions to be satisfied on the free surface are

$$\frac{\partial \eta}{\partial t} + U\frac{\partial \eta}{\partial z} = \frac{\partial \phi}{\partial r} \quad \text{on } r = a$$

and

$$\frac{p-p_0}{\rho} = -\frac{\partial \phi}{\partial t} - U\frac{\partial \phi}{\partial z} = \frac{\sigma}{\rho}\left(\frac{1}{R_1} + \frac{1}{R_2}\right) - \frac{\sigma}{\rho a} = \frac{\sigma}{\rho}\left[-\frac{1}{a^2}\left(\eta + \frac{\partial^2 \eta}{\partial \theta^2}\right) - \frac{\partial^2 \eta}{\partial z^2}\right] \quad \text{on } r = a$$

The kinematic condition gives

$$\eta = \frac{-i}{kU+\omega}\frac{d\phi_1}{dr}\bigg|_{r=a}e^{i(kz+n\theta+\omega t)} = \frac{-ikA}{kU+\omega}\frac{dI_s(\varsigma)}{d\varsigma}\bigg|_{\varsigma=ka}e^{i(kz+n\theta+\omega t)}.$$
(4.5.3)

The pressure condition gives

$$-iA(kU+\omega)I_s(ka)\, e^{i(kz+n\theta+\omega t)} = \frac{i\sigma k A}{\rho(kU+\omega)}\frac{1}{a^2}\frac{dI_s(\varsigma)}{d\varsigma}\bigg|_{\varsigma=ka}\left(n^2+a^2k^2-1\right)e^{i(kz+n\theta+\omega t)},$$

or

$$(kU+\omega)^2 = \frac{\sigma}{a^3\rho}\frac{kaI_s'(ka)}{I_s(ka)}\left(n^2+a^2k^2-1\right).$$
(4.5.4)

There is thus a possibility that the right-hand side of equation (4.5.4) can be zero in the range $0 < ka < 1$. This is a case where the wavelength is greater than the circumference of the jet. The right-hand side takes its largest negative value when

$k^2 a^2 = 0.4858$, or when the wavelength equals $\lambda = 2\pi/k = 4.508 \times 2a$. Thus, the jet takes on a series of enlargements and contractions with continually increasing amplitude, until it breaks up into a series of separated drops.

4.6 Local Surface Disturbance on a Large Body of Fluid—Kelvin's Ship Wave

The waves caused by creating a local disturbance to the surface of a large body of fluid is of interest in studying the wave pattern associated with ships. Considering the fluid to be very deep and linearizing the boundary conditions as before, the basic equations are

$$\nabla^2 \phi = 0 \quad \text{for } -\infty \leq z \leq 0,$$

$$\frac{p}{\rho} = -\frac{\partial \phi}{\partial t} - g\eta \quad \text{on } z = 0, \tag{4.6.1}$$

$$\frac{\partial \eta}{\partial t} = \frac{\partial \phi}{\partial z} \quad \text{on } z = 0.$$

Working in cylindrical polar coordinates and assuming independence of the angle θ, separation of variables gives

$$\phi(r, z, t) = A(t) e^{kz} J_0(kr), \quad J_0(kr) = \sum_{j=0}^{\infty} \frac{(-1)^j}{(j!)^2} \left(\frac{kr}{2}\right)^{2j}, \tag{4.6.2}$$

where $J_0(kr)$ is the Bessel function of order zero. Inserting equation (4.6.2) into the boundary conditions gives

$$\eta = -\frac{1}{g} \frac{\partial \phi}{\partial t}\bigg|_{z=0} = \frac{1}{g} \frac{dA}{dt} J_0(kr) \quad \text{and} \quad \frac{\partial \eta}{\partial t} = \frac{\partial \phi}{\partial z}\bigg|_{z=0} = kA J_0(kr). \tag{4.6.3}$$

Having in mind a problem where there is an initial displacement given to the free surface at $r = 0$, choose as a solution the forms

$$\phi(r, z, t) = f(k) \sqrt{\frac{g}{k}} e^{kz} \sin \sqrt{gk}\, t J_0(kr), \quad \eta(r, t) = f(k) \cos \sqrt{gk}\, t J_0(kr). \tag{4.6.4}$$

To find the nature of this solution, first consider a very basic initial disturbance where the solution is concentrated near the origin. This is a disturbance much like our Green's functions of Chapter 2, or like concentrated loads in elementary beam theory, and is akin to a Dirac delta function. Use the property of integral transforms that

$$f(k) = \int_0^\infty f(r) J_0(kr) 2\pi r\, dr \quad \text{and} \quad f(r) = \frac{1}{2\pi} \int_0^\infty f(a) J_0(ar) a\, da \tag{4.6.5}$$

and choose the initial disturbance to be

$$f(r) = \begin{cases} \lim_{a \to 0} \dfrac{1}{\pi a^2} & \text{for } 0 \leq r \leq a, \\[2mm] 0 & \text{for } a < r. \end{cases}$$

Then $\lim\limits_{a\to 0}\int_0^\infty f(r)2\pi r dr = 1$ and

$$\phi(r,\, z,\, t) = \int_0^\infty \sqrt{\frac{g}{k}} e^{kz} \sin\sqrt{gk}t J_0(kr) k dk \int_0^\infty f(a) J_0(ka) a da$$

$$= \int_0^\infty \sqrt{\frac{g}{k}} e^{kz} \sin\sqrt{gk}t J_0(kr) k dk, \qquad (4.6.6)$$

$$\eta(r,\, t) = \int_0^\infty \cos\sqrt{gk}t J_0(kr) k dk \int_0^\infty f(a) J_0(ka) a da$$

$$= \int_0^\infty \cos\sqrt{gk}t J_0(kr) k dk. \qquad (4.6.7)$$

It is not possible to evaluate these integrals in closed form. However, if in the expression for the displacement the cosine is expanded in a Taylor series, it is found that

$$\eta(r,\, t) = \frac{1}{2\pi r^2}\left\{ \frac{1^2}{2!}\frac{gt^2}{r} - \frac{1^2\cdot 3^2}{6!}\left(\frac{gt^2}{r}\right)^3 + \frac{1^2\cdot 3^2\cdot 5^2}{10!}\left(\frac{gt^2}{r}\right)^5 - \cdots\right\}. \qquad (4.6.8)$$

This tells us that for any particular phase of the displacement, the quantity gt^2/r is constant. Thus, the waves travel radially out from where they originated, each phase traveling with a constant acceleration.

The more important problem would be to determine what the wave pattern would be as the source moves, emitting wave pulses along the way. That would involve superposition of the preceding solution over a time integral. The problem was first solved by Kelvin (1891) using the method of stationary phase that he had in fact originated. The paper, however, leaves out many details and does not tell how he performed the analysis. Lamb (1932, §256) provides a figure, but, again, the presentation is sketchy and difficult to follow. Later papers by Peters (1949) and Ursell (1960) clarified the details omitted by the previous authors. In any case, the mathematics at this point involves quite complicated use of asymptotic methods and becomes very involved. Details are left to the discussions in Lamb, Ursell, and Peters.

Using the notation of Peters, the results can be summarized as follows:

- The waves are contained within an angle of $2\cdot\sin^{-1}\frac{1}{3} = 38°56'$.

- Two wave patterns exist within this triangle: a set of transverse waves and a set of divergent waves. The patterns of constant phase are shown in Figure 4.6.1. They are given by the following:

$$\textbf{Transverse waves: } \frac{g}{c^2} r h_1(\theta) = \left(2n+\frac{1}{4}\right)\pi,\ n > 0, \qquad (4.6.9)$$

$$\textbf{Divergent waves: } \frac{g}{c^2} r h_2(\theta) = \left(2n+\frac{3}{4}\right)\pi,\ n > 0. \qquad (4.6.10)$$

Here, r and θ are the usual cylindrical coordinates centered at the present location of the disturbance, and c is the speed at which it travels. The various functions are given by

$$h_i = \sin\theta \coth p_i \cosh^2 p_i, \qquad (4.6.11)$$

with $\sinh p_1 = \frac{1}{4}(\cot\theta - \sqrt{\cot^2\theta - 8})$, $\sinh p_2 = \frac{1}{4}(\cot\theta + \sqrt{\cot^2\theta - 8})$.

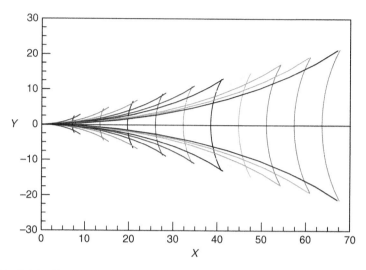

Figure 4.6.1 Kelvin ship wave—waves of constant phase

The p_i's are real if $\cot^2 \theta \geq 8$—that is, $-19.47° \leq \theta \leq 19.47°$. Outside of this region the disturbance is greatly diminished.

The preceding pattern is a reasonable depiction of the bow wave on a slender ship or the pattern due to a tree branch or fish line in the water. Understanding the pattern also has a practical use. Resistance to the motion of ships is about equally divided between viscous shear and wave drag. (The waves have a considerable amount of potential energy, all provided by the ship.) Simple models of a ship use a single source-sink pair—the source modeling the ship's bow—to generate the pattern given above. In the 1950s Inui (1957) and others discovered that if somehow another "source" could be placed in front of the ship's bow, it could create waves out of phase with the bow wave and thus cancel them. Experiments conducted in the towing tank at The University of Michigan by Inui and others confirmed this theory, leading to the now-famous ***bulbous bow***. This is an underwater rounded protuberance extended from the bow. This idea was embraced by the world's tanker fleet, which added bulbous bows and midsections to existing tankers, enabling greater capacity using the original power supplies. Similar wave-cancellation ideas are used in noise-cancelling headphones.

4.7   Shallow-Depth Free Surface Waves—Cnoidal and Solitary Waves

Many years ago, Russell (1838, 1894) pointed out that for wavelengths that were large with respect to the depth, waves are observed for which shape changes only occur very gradually as the wave proceeds along a shallow channel. The wave height of these waves is not necessarily small. To investigate this situation, consider a two-dimensional case where the wave pattern varies along the channel but not with the direction transverse to the channel walls. To make the wave pattern steady, the undisturbed flow in the channel is taken to move at a speed c, this speed being whatever is necessary to make the wave pattern stationary.

First, consider a periodic wave pattern in an infinitely long channel. The x-axis is taken as being the direction of wave propagation, and the coordinate origin is at the bottom of the channel. The governing equations and boundary conditions are

$$\frac{\partial^2 \phi}{\partial x^2} + \frac{\partial^2 \phi}{\partial y^2} = 0,$$

$$\frac{\partial \phi}{\partial y} = 0 \quad \text{on } y = 0, \tag{4.7.1}$$

$$\frac{\partial \phi}{\partial x}\frac{\partial \eta}{\partial x} = \frac{\partial \phi}{\partial y} \quad \text{and} \quad \frac{1}{2}\left(\frac{\partial \phi}{\partial x}\right)^2 + \frac{1}{2}\left(\frac{\partial \phi}{\partial y}\right)^2 + g\eta = C \quad \text{on } y = \eta.$$

Here, C is a constant that represents the total energy at a point, to be determined later.

Next, following Lamb (1932, §252), assume that a solution can be found for the velocity potential in the form of a Taylor series in y. Then to satisfy the Laplace equation, let

$$\phi(x, y) = G - \frac{y^2}{2!}G'' + \frac{y^4}{4!}G'''' \mp \cdots, \tag{4.7.2}$$

where G is a function of x and primes denote x derivatives.

Substituting equation (4.7.2) into the boundary conditions gives

$$\left(G' - \frac{\eta^2}{2!}G''' + \frac{\eta^4}{4!}G''''' \mp \cdots\right)\eta' = -\eta G'' + \frac{\eta^3}{3!}G'''' \mp \cdots,$$

$$\left(G' - \frac{\eta^2}{2!}G''' + \frac{\eta^4}{4!}G''''' \mp \cdots\right)^2 + \left(-\eta G'' + \frac{\eta^3}{3!}G'''' \mp \cdots\right)^2 = 2C - 2g\eta. \tag{4.7.3}$$

At this point, since long waves are being considered, assume that the more G is differentiated, the smaller its effect becomes. Then, under this assumption, equation (4.7.3) becomes after rearrangement

$$\eta'G' + \eta G'' - \eta'\frac{\eta^2}{2!}G''' - \eta'\frac{\eta^3}{3!}G'''' \mp \cdots = 0,$$

$$-2C + 2g\eta + (G')^2 + (-\eta G'')^2 - \eta^2 G'G''' \pm \cdots = 0. \tag{4.7.4}$$

The first of these can be integrated, giving

$$\eta G' - \frac{\eta^3}{3!}G''' \pm \cdots = D, \quad \text{or} \quad GF' = \frac{D}{\eta} + \frac{\eta^2}{3!}G''' \mp \cdots, \tag{4.7.5}$$

where D is a constant of integration. Inserting this into the pressure condition and simplifying gives

$$\frac{-2C + 2g\eta}{D^2} + \frac{1}{\eta^2} + \frac{2}{3}\frac{\eta''}{\eta} - \frac{\eta'^2}{3\eta^2} = 0. \tag{4.7.6}$$

Equation (4.7.6) is a form of the ***Korteweg-deVries equation*** (1895). It can be integrated by first multiplying by η'. The result is

$$\frac{-2C\eta + g\eta^2}{D^2} - \frac{1}{\eta} + \frac{\eta'^2}{3\eta} = H, \tag{4.7.7}$$

where H is still another constant of integration.

The constants of integration can be evaluated by noting that at both the crests and the troughs of the waves the slope is zero. Thus,

$$\eta'^2 = \frac{3(\eta - h_{\text{trough}})(h_{\text{crest}} - \eta)(\eta - L)}{h_{\text{crest}} h_{\text{trough}} L},$$

(4.7.8)

where $L = \dfrac{2C}{g} - (h_{\text{crest}} + h_{\text{trough}})$ and $L \le h_{\text{trough}} \le \eta \le h_{\text{crest}}$.

The magnitude restriction on L comes about because the right-hand side must always be positive to match the left-hand side.

Applying the conditions at the crest and trough of the waves, the constants D and H are found to be given by

$$D^2 = gL h_{\text{trough}} h_{\text{crest}}, \quad H = -\frac{h_{\text{trough}} + h_{\text{crest}}}{h_{\text{trough}} h_{\text{crest}}} - \frac{1}{L}.$$

(4.7.9)

The solution can be advanced further by letting $\eta = h_{\text{trough}} + (h_{\text{crest}} - h_{\text{trough}}) \cos^2 \varsigma$. Then the equation for the displacement becomes

$$\left(\frac{d\varsigma}{dx}\right)^2 = \frac{3(h_{\text{crest}} - L)}{h_{\text{crest}} h_{\text{trough}} L}\left(1 - \frac{h_{\text{crest}} - h_{\text{trough}}}{h_{\text{crest}} - L} \sin^2 \varsigma\right).$$

This can be integrated to give x as a function of ζ and also the wavelength. The result is

$$x = \beta \int_0^\phi \frac{dp}{\sqrt{1 - k^2 \sin^2 p}} = \beta F(k, \varsigma),$$

(4.7.10)

where $\lambda = \text{wavelength} = 2\beta \int_0^{\pi/2} \frac{dp}{\sqrt{1 - k^2 \sin^2 p}} = 2\beta F(k, \pi/2),$

and $k^2 = \dfrac{h_{\text{crest}} - h_{\text{trough}}}{h_{\text{crest}} - L}$ and $\beta = \sqrt{\dfrac{h_{\text{crest}} h_{\text{trough}} L}{3(h_{\text{crest}} - L)}}.$

Here, x is measured from the wave crest, and $F(k, \varsigma)$ is the *incomplete elliptic function of the first kind*. (When $\varsigma = \pi/2$, it becomes the *complete elliptic integral of the first kind*.) The function that is essentially the inverse of F, giving ζ as a function of x, is

$$\eta = h_{\text{trough}} + (h_{\text{crest}} - h_{\text{trough}}) cn^2\left(\frac{x}{\beta}\right),$$

(4.7.11)

where $cn(x/\beta) = \cos \varsigma$ is called the *cnoidal function*, and the wave described by equation (4.7.11) is the *cnoidal wave*.

The remaining task is to determine the constant C. From equation (4.7.1) it is seen that C can be expressed in the form $C = c^2/2 + gh$, where h is the average depth and c is the speed necessary to hold the wave steady. The amount of fluid in one wavelength is

$$h\lambda = \int_0^\lambda \eta \, dx = \int_0^{\pi/2} \eta \frac{dx}{dp} \, dp = 2\beta \int_0^{\pi/2} \frac{h_{\text{crest}} - (h_{\text{crest}} - h_{\text{trough}}) \sin^2 p}{\sqrt{1 - k^2 \sin^2 p}} \, dp,$$

or

$$h = 2\beta \left[LF(k, \pi/2) + (h_{\text{crest}} - L)E(k, \pi/2) \right] / \lambda, \qquad (4.7.12)$$

where $E(k, \pi/2) = \int_0^{\pi/2} \sqrt{1 - k^2 \sin^2 p}\, dp$ is the **complete elliptic integral of the second kind**.

The procedure for completing the solution for a specific case is as follows. First, choose three of the variables—say, h_{trough}, h_{crest}, and L—since their relative magnitudes are known. From equation (4.7.10) and (4.7.12) k can be found, as well as x as a function of the local wave elevation. Since $C = c^2/2 + gh$, it follows from equation (4.7.8) that $c = \sqrt{g(L + h_{\text{trough}} + h_{\text{crest}} - h)}$. Thus, all of the parameters of the problem have been determined. Plots of the elliptic functions are given in Figures 4.7.1 and 4.7.2, where $\theta = \sin^{-1} k$ is expressed in degrees. Values of these functions are shown in tables in the Appendix.

A more elaborate presentation of this material is given in Wehausen and Laitone (1960, §31). Their procedure is to make x dimensionless using the wavelength (x/λ) and y dimensionless by a representative depth h (y/h) and then writing equation (4.7.2) as

$$\phi(x, y) = G(x/\lambda) + \sum_{n=1}^{\infty} \frac{(-1)^n \varepsilon^{2n}}{(2n)!} \left(\frac{y}{h} \right)^{2n} \frac{d^{(2n)} G(x/\lambda)}{d(x/\lambda)^{(2n)}}, \qquad \varepsilon = \frac{h}{\lambda}.$$

This form clarifies what is meant by the statement that the more G is differentiated, the smaller the terms become, and also allows better assessment of the order of accuracy of the results.

As α approaches unity, equation (4.7.10) can be integrated in a simpler form. This value of α gives a wave whose wavelength approaches infinity. It is called the **solitary wave**, with $L \to h_{\text{trough}} = h_{\infty}$. The result is

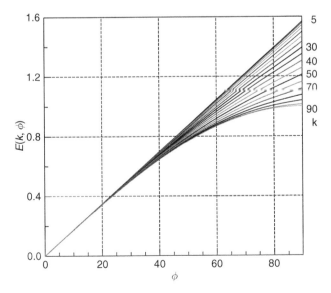

Figure 4.7.1   Elliptic function $E(k, \phi)$

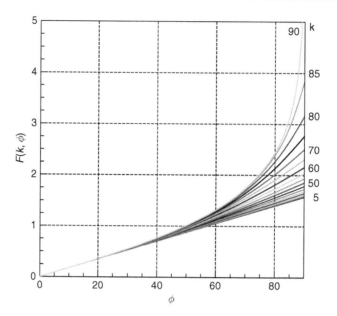

Figure 4.7.2 Elliptic function $F(k, \phi)$

$$\eta'^2 = \frac{3(\eta - h_\infty)^2(h_{\text{crest}} - \eta)}{h_{\text{crest}} h_\infty^2}, \text{ which can be integrated to give}$$

$$\eta = h_\infty + \frac{h_{\text{crest}} - h_\infty}{\cosh^2 0.5\kappa x} \quad \text{with } \kappa = \sqrt{\frac{3(h_{\text{crest}} - h_\infty)}{h_\infty^2 h_{\text{crest}}}} \tag{4.7.13}$$

The speed necessary to keep this wave stationary can be found in this case by realizing that far upstream the constant C in equation (4.6.13) is $C = c^2/2 + gh_\infty$. Putting this in the expression for L in equation (4.6.18) gives $c = \sqrt{gh_\infty}$. Conversely, this would be the speed at which the wave travels in otherwise still water.

Wehausen and Laitone (1960, §31) show that there is a theoretical limit to the height of finite amplitude waves, given by

$$\left(\frac{h_{\text{crest}} - h_{\text{trough}}}{h}\right)_{\text{max}} = \frac{8\alpha^2}{9\alpha^2 + 2}. \tag{4.7.14}$$

At this amplitude theory shows that there is a reversal in the vertical velocity near the crest. For the solitary wave this has a value of $8/11 = 0.7273$. Experiments with these waves show that as the wave amplitude approaches this limit, the wave starts to break.

4.8 Ray Theory of Gravity Waves for Nonuniform Depths

When the depth of the flow is variable, particularly in shallow regions where depth is comparable to wavelength, observations of waves from the shoreline show waves both steepening and changing direction. To understand this phenomena, the approach must be altered a bit from the previous.

Since a three-dimensional problem is now being considered, choose z as being the direction perpendicular to the free surface. The waves will be considered small so that the linearized theory can be used. The governing equations are then

$$\nabla^2 \phi = 0 \quad \text{in } 0 \leq z \leq -h(x, y), \tag{4.8.1}$$

$$\frac{\partial \phi}{\partial t} + g\eta = 0 \quad \text{on } z = 0, \tag{4.8.2}$$

$$\frac{\partial \eta}{\partial t} = \frac{\partial \phi}{\partial z} \quad \text{on } z = 0, \tag{4.8.3}$$

$$\frac{\partial \phi}{\partial z} = v_z = -\left(\frac{\partial \phi}{\partial x} \frac{\partial h}{\partial x} + \frac{\partial \phi}{\partial y} \frac{\partial h}{\partial y} \right) \quad \text{on } z = -h(x, y). \tag{4.8.4}$$

Since equation (4.8.4) involves products of function of x and y, methods such as separation of variables cannot be used for general functions $h(x,y)$.

To simplify the equations, certain assumptions must be made concerning the relative values of the various quantities. The easiest way to do this is to deal with dimensionless quantities. L will be taken to be an appropriate length, possibly a wavelength, measure in the x-y plane, and changes in the z direction will be taken to occur at a faster rate than in the x-y directions. Thus, εL will be the appropriate scale in the z direction. Our dimensionless coordinates will then be $X = x/L, Y = y/L, Z = z/\varepsilon L$, and $H = \varepsilon h/L$. The appropriate time dimensionalization is $T = t\sqrt{\varepsilon g/L}$.

As a result of this,

$$\frac{\partial}{\partial x} = \frac{1}{L} \frac{\partial}{\partial X}, \quad \frac{\partial}{\partial y} = \frac{1}{L} \frac{\partial}{\partial Y}, \quad \frac{\partial}{\partial z} = \frac{1}{\varepsilon L} \frac{\partial}{\partial Z}, \quad \text{and} \quad \frac{\partial}{\partial t} = \sqrt{\frac{\varepsilon g}{L}} \frac{\partial}{\partial T}.$$

Thus, our equations become

$$\nabla_2^2 \phi + \frac{1}{\varepsilon^2} \frac{\partial^2 \phi}{\partial Z^2} = 0 \quad \text{in } 0 \leq Z \leq -H(x, y), \tag{4.8.5}$$

$$\text{where } \nabla_2 = \mathbf{i} \frac{\partial}{\partial X} + \mathbf{j} \frac{\partial}{\partial Y}, \quad \nabla_2^2 = \frac{\partial^2}{\partial X^2} + \frac{\partial^2}{\partial Y^2}.$$

Also,

$$\sqrt{\frac{\varepsilon g}{L}} \frac{\partial \phi}{\partial T} + g\eta = 0, \quad \sqrt{\frac{\varepsilon g}{L}} \frac{\partial \eta}{\partial T} = \frac{1}{\varepsilon L} \frac{\partial \phi}{\partial Z} \quad \text{on } Z = 0, \tag{4.8.6}$$

$$\frac{1}{\varepsilon} \frac{\partial \phi}{\partial Z} = -\varepsilon \left(\frac{\partial \phi}{\partial X} \frac{\partial H}{\partial X} + \frac{\partial \phi}{\partial Y} \frac{\partial H}{\partial Y} \right) \text{ on } Z = -H. \tag{4.8.7}$$

If these are rearranged and η is solved for, the result is

$$\varepsilon^2 \nabla_2^2 \phi + \frac{\partial^2 \phi}{\partial Z^2} = 0 \quad \text{in } 0 \leq Z \leq -H(x, y), \tag{4.8.8}$$

$$\eta = -\sqrt{\frac{\varepsilon}{gL}} \frac{\partial \phi}{\partial T}, \quad \frac{\partial \eta}{\partial T} = \sqrt{\frac{1}{\varepsilon^3 gL}} \frac{\partial \phi}{\partial Z} \text{ on } Z = 0. \tag{4.8.9}$$

The displacement η can be eliminated from consideration by differentiating the first result in equation (4.8.9) and using it in the second. Thus,

$$\frac{\partial \phi}{\partial Z} = -\varepsilon^2 \left(\frac{\partial \phi}{\partial X} \frac{\partial H}{\partial X} + \frac{\partial \phi}{\partial Y} \frac{\partial H}{\partial Y} \right) \quad \text{on } Z = -H, \tag{4.8.10}$$

$$\varepsilon^2 \frac{\partial^2 \phi}{\partial T^2} + \frac{\partial \phi}{\partial Z} = 0 \quad \text{on } Z = 0. \tag{4.8.11}$$

To proceed with a solution, recall that in the constant-depth case there were solutions like $e^{i(kx-\varpi t)}$ to represent traveling waves. Here, in correspondence with this, write $\phi(X, Y, Z, T) = f(X, Y, Z)e^{is(X,Y,T)/\varepsilon}$. Substituting this into our equations yields

$$\varepsilon^2 \nabla_2^2 f + i\varepsilon\beta(2\nabla_2 s \cdot \nabla_2 f + f \nabla_2^2 s) + \left(\frac{\partial^2 f}{\partial Z^2} - |\nabla_2 s|^2 f \right) = 0 \text{ in } 0 \leq Z \leq -H(x, y),$$

$$\left[\frac{\partial f}{\partial Z} - \left(\frac{\partial s}{\partial T} \right)^2 f \right] + i\varepsilon \frac{\partial^2 s}{\partial T^2} f = 0 \quad \text{on } Z = 0,$$

$$\frac{\partial f}{\partial Z} + i\varepsilon f \nabla_2 s \cdot \nabla_2 H + \varepsilon^2 \nabla_2 f \cdot \nabla_2 H = 0 \quad \text{on } Z = -H.$$

As yet neither L nor ε have been defined. In the constant depth case, it was seen that the quantity $kx - \omega t$ corresponds to our $s(X, Y, T)/\varepsilon$, where k is the wave number and ω the circular frequency. In analogy with this, it is customary to choose the scaling so that $\partial s/\partial T = 1$, which assumes the circular frequency $\sqrt{g/\varepsilon L}$ is independent of time, and to write $k = |\nabla_2 s|$ as a wave number that depends on X and Y. This equation for k is called the ***eikonal equation*** in geometric optics. (*Eikonal* is the Greek word for "ray." It sounds much more impressive than the short word *ray*.) The surfaces $s = constant$ are ***wave fronts***, or lines of constant phase, and $\nabla_2 s$ is a vector normal to the wave front and represents a ***ray***.

If everything has been scaled properly, the terms not multiplied by ε should be handled first. From them it is found as the lowest approximation to f that

$$f_0 = A_0(x, y) \cosh k(Z + H) \quad \text{with } k \tanh kh = 1. \tag{4.8.12}$$

Checking back with the constant depth case, the results are the same, but now the wave number k is a function of the horizontal coordinates.

The amplitude solution can be continued by writing

$$f = f_0 + \sum_{n=1}^{\infty} \varepsilon^n \cosh k(Z + H) A_n(X, Y, Z)$$

and substituting into the equations and collecting terms on powers of ε. The process is tedious but straightforward. The approach is discussed in Meyer (1979).

Of main interest, however, is how a wave front moves. To carry out a solution numerically, one would start with a given wave front (i.e., a given value of s). At any given depth, H is known, so k can be found. The eikonal equation plus the known directionality of $\nabla_2 s$ at any point (perpendicular to the wave front) is then integrated to find the ray direction and establish a new wave front. The process repeats until a boundary is encountered.

Note that it is possible that rays can cross one another, just as light rays cross when focusing a beam of light. The intersection of two rays is called a ***caustic***. Where they intersect, our theory predicts infinite amplitude of the wave. In fact, our theory breaks down there, just as it does at the envelopes of the rays. The same happens in geometric optics theory. In both cases a different scaling is needed.

To summarize, in dimensional terms the various quantities in terms of dimensional units are given by

$$\text{dimensional wave number} \to k_D = \frac{1}{\varepsilon}|\nabla_2 s| = \frac{k}{\varepsilon L},$$

$$\text{dimensional circular frequency} \to \omega_D = \frac{1}{\varepsilon}\frac{\partial s}{\partial T} = \sqrt{\frac{g}{\varepsilon L}},$$

$$\text{dimensional wave speed} \to c_D = \frac{\omega_D}{k_D} = \frac{\sqrt{\varepsilon g L}}{k},$$

$$\text{dimensional wave length} \to \lambda_D = \frac{2\pi}{k_D} = \frac{2\pi\varepsilon L}{k}.$$

The monograph by Mader (1988) presents more details as well as numerical programs for calculating the progress of such waves.

Problems—Chapter 4

4.1 For a monochromatic (a single wavelength) traveling wave on a deep layer of fluid, the potential and the elevation are given by $\phi(x,t) = -ace^{ky}\cos k(x-ct)$, $\eta(x,t) = a\sin k(x-ct)$. The kinetic and potential energies contained in one wavelength are

$$KE = \frac{1}{2}\rho\int_{-\infty}^{0}\int_{0}^{\lambda}\left[\left(\frac{\partial\phi}{\partial x}\right)^2 + \left(\frac{\partial\phi}{\partial y}\right)^2\right]d(x-ct)dy \text{ and } PE = \frac{1}{2}\rho g\int_{0}^{\lambda}\eta^2 d(x-ct).$$

Show that the two energies are equal. This is called the ***principle of equi-partition of energy***.

4.2 Find the wavelengths to be expected for standing waves on the free surface of a rectangular tank a by b, with depth d.

4.3 Find the wavelengths to be expected for standing waves on the free surface of a cylindrical tank of radius a and depth d.

4.4 Two fluids of different densities are separated by an interface at $y = 0$. The fluids can each be considered to have infinite extent. Find the speed at which interfacial waves travel, including the effects of surface tension. The linearized surface tension condition can be taken as $\Delta p = \sigma\frac{\partial^2\eta}{\partial x^2}$.

4.5 Find the possible interfacial waves in a canal, where the bottom layer of depth a has a higher density than the top layer of thickness b. The top layer is open to the atmosphere. Neglect surface tension. (Ships moving in fjords where the top layer is brackish and the lower layer is salty can notice increased ship drag due to waves generated at the interface of the two fluids.)

Chapter 5

Exact Solutions of the Navier-Stokes Equations

5.1 Solutions to the Steady-State Navier-Stokes Equations When Convective Acceleration Is Absent

Because of the mathematical nonlinearities of the convective acceleration terms in the Navier-Stokes equations when viscosity is included, and also because the order of the Navier-Stokes equations is higher than the order of the Euler equations, finding solutions is generally difficult, and the methods and techniques used in the study of inviscid flows are generally not applicable. In this and the following chapters, a number of cases where exact and approximate solutions of the Navier-Stokes equations can be found are discussed.

In particular, for flows where the velocity gradients are perpendicular to the velocity, the convective acceleration terms vanish. The results are then independent of the Reynolds number. Several such cases will be considered.

5.1.1   Two-Dimensional Flow Between Parallel Plates

Consider flow between parallel plates as shown in Figure 5.1.1. Write the velocity in the form $\mathbf{v} = (u(y), 0, 0)$, which automatically satisfies the continuity equation. With gravity acting in the plane of the figure, the Navier-Stokes equations then become

$$\mu \frac{d^2 u}{dy^2} = \frac{\partial p}{\partial x} - \rho g \cos \theta,$$

$$0 = \frac{\partial p}{\partial y} - \rho g \sin \theta, \tag{5.1.1}$$

$$0 = \frac{\partial p}{\partial y},$$

where θ is the angle between the x-axis and the direction of gravity. Since the viscous terms are functions of at most y, and since from the second and third equation p must

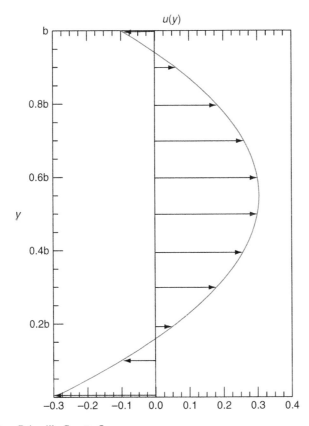

Figure 5.1.1   Plane Poiseuille-Couette flow

be linear in y and independent of z, $\partial p/\partial x$ must be a constant. Thus, integration of equation (5.1.1) gives

$$u = \frac{1}{2\mu}\left(\frac{\partial p}{\partial x} - \rho g\cos\theta\right)(y^2 + c_1 y + c_2),$$

where c_1 and c_2 are constants of integration.

If there is an upper plate at $y = b$ moving with velocity U_b, and a lower plate at $y = 0$ moving with velocity U_0, then applying these conditions to our expression for u yields

$$U_0 = \frac{c_2}{2\mu}\left(\frac{\partial p}{\partial x} - \rho g\cos\theta\right)$$

and

$$U_b = \frac{1}{2\mu}\left(\frac{\partial p}{\partial x} - \rho g\cos\theta\right)(b^2 + c_1 b + c_2).$$

Solving for c_1 and c_2 gives

$$u(y) = \frac{1}{2\mu}\left(\frac{\partial p}{\partial x} - \rho g\cos\theta\right)(y^2 - by) + U_0 + (U_b - U_0)\frac{y}{b} \qquad (5.1.2)$$

with a shear stress given by

$$\tau_{xy}(y) = \mu\frac{du}{dy} = \frac{1}{2}\left(\frac{\partial p}{\partial x} - \rho g\cos\theta\right)(2y - b) + \frac{\mu(U_b - U_0)}{b}, \qquad (5.1.3)$$

a pressure by

$$p(x, y) = \frac{\partial p}{\partial x}x + \rho g y\sin\theta + p_0, \qquad (5.1.4)$$

and a volumetric discharge per unit width of

$$Q = a\int_0^b u\,dy = a\int_0^b\left[\frac{1}{2\mu}\left(\frac{\partial p}{\partial x} - \rho g\cos\theta\right)(y^2 - by) + U_0 + (U_b - U_0)\frac{y}{b}\right]dy$$

$$= \frac{-ab^3}{12\mu}\left(\frac{\partial p}{\partial x} - \rho g\cos\theta\right) + ab\frac{(U_b - U_0)}{2}. \qquad (5.1.5)$$

The terms in the velocity that are associated with the pressure gradient and body forces are called the **Poiseuille flow** terms. The terms due to the motion of the boundaries are called the **Couette flow** terms.

5.1.2 Poiseuille Flow in a Rectangular Conduit

For flows in a rectangular conduit, the analysis is similar to that in the previous section with of course one more space dimension, since the velocity depends on both y and z. Letting $\mathbf{v} = (u(y, z), 0, 0)$, the Navier-Stokes equations for this flow are

$$0 = -\frac{\partial p}{\partial x} + \rho g\cos\theta + \mu\left(\frac{\partial^2 u}{\partial y^2} + \frac{\partial^2 u}{\partial z^2}\right),$$

$$0 = -\frac{\partial p}{\partial y} + \rho g\sin\theta, \qquad (5.1.6)$$

$$0 = -\frac{\partial p}{\partial y},$$

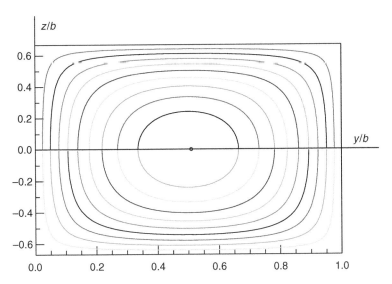

Figure 5.1.2   Rectangular conduit—stream surfaces

with $u = 0$ on the (stationary) walls at $0 \leq y \leq b, -0.5a \leq z \leq 0.5a$ as shown in Figure 5.1.2.

Separation of variables suggests taking

$$u_s = \frac{1}{2\mu}\left(\frac{\partial p}{\partial x} - \rho g \cos \theta\right)(y^2 - by) + Y(y)Z(z).$$

The terms quadratic in y are needed to make the equation homogeneous so that the method of separation of variables works. Inserting this into equation (5.1.6) gives

$$0 = \frac{d^2Y}{dy^2}Z + Y\frac{d^2Z}{dz^2}, \text{ which after dividing by } YZ \text{ becomes}$$

$$\frac{1}{Y}\frac{d^2Y}{dy^2} = -\frac{1}{Z}\frac{d^2Z}{dz^2} = \text{a constant—say,} - k^2.$$

Solving these equations leads to Y being made up of sines and cosines of ky and Z being hyperbolic sines and hyperbolic cosines of kx. Applying the no-slip boundary conditions and symmetry in z eliminates the cosine and sinh functions. To meet the remaining condition, it is necessary that $k = \frac{n\pi}{b}$, where n is any positive integer. After summing these in a Fourier series, the result is

$$u_x = \frac{1}{2\mu}\left(\frac{\partial p}{\partial x} - \rho g \cos \theta\right)\left[(y^2 - by) + \sum_{n=1}^{\infty} A_n \sin \frac{n\pi y}{b} \cosh \frac{n\pi z}{b}\right], \qquad (5.1.7)$$

with shear stress components

$$\tau_{xy} = \mu\frac{\partial u_x}{\partial y} = \frac{1}{2}\left(\frac{\partial p}{\partial x} - \rho g \cos \theta\right)\left[(2y - b) + \sum_{n=1}^{\infty} A_n \frac{n\pi}{b} \cos \frac{n\pi y}{b} \cosh \frac{n\pi z}{b}\right],$$

$$\tau_{xz} = \mu\frac{\partial u_x}{\partial z} = \frac{1}{2}\left(\frac{\partial p}{\partial x} - \rho g \cos \theta\right)\sum_{n=1}^{\infty} A_n \frac{n\pi}{b} \sin \frac{n\pi y}{b} \sinh \frac{n\pi z}{b}. \qquad (5.1.8)$$

Use of the orthogonality conditions for the sine gives

$$A_n = \frac{\int_0^b (yb - y^2) \sin \dfrac{n\pi y}{b} dy}{\cosh \dfrac{n\pi a}{b} \int_0^b \sin^2 \dfrac{n\pi y}{b} dy} = \frac{4 \dfrac{b^2}{n^3 \pi^3} (1 - (-1)^n)}{\cosh \dfrac{n\pi a}{b}}. \tag{5.1.9}$$

The volumetric discharge is given by

$$Q = \int_0^b dy \int_{-0.5a}^{0.5b} u_x dz = \frac{1}{2\mu} \left(\frac{\partial p}{\partial x} - \rho g \cos \theta \right)$$

$$\times \left[\frac{-ab^3}{6} + \sum_{n=1}^{\infty} A_n \frac{2b^2}{n^2 \pi^2} [1 - (-1)^n] \sinh \frac{n\pi a}{2b} \right]. \tag{5.1.10}$$

5.1.3 Poiseuille Flow in a Round Conduit or Annulus

For Poiseuille flow in a circular annulus as shown in Figure 5.1.3, adapting the previous forms to cylindrical polar coordinates suggests letting $\mathbf{v} = (0, 0, w(r))$. Again, this form for the velocity automatically satisfies the incompressible continuity equation. Then the Navier-Stokes equations in cylindrical polar coordinates (see Appendix) in the z direction (along the tube) reduces to

$$\frac{\partial p}{\partial z} - \rho g \cos \theta = \mu \left(\frac{d^2 w}{dr^2} + \frac{1}{r} \frac{dw}{dr} \right).$$

Since by use of the product rule of calculus

$$\frac{d^2 w}{dr^2} + \frac{1}{r} \frac{dw}{dr} = \frac{1}{r} \frac{d}{dr} \left(r \frac{dw}{dr} \right),$$

this can be integrated twice to obtain

$$w = \frac{1}{4\mu} \left(\frac{\partial p}{\partial z} - \rho g \cos \theta \right) (r^2 + c_1 \ln r + c_2). \tag{5.1.11}$$

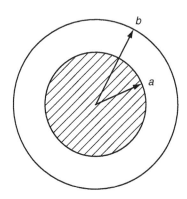

Figure 5.1.3 Rotating Couette flow—inner and outer cylinders

If $w = W_a$ at $r = a$ and $w = W_b$ at $r = b$, proceeding as in the plane Poiseuille flow case, we have

$$w = \frac{1}{4\mu} \left(\frac{\partial p}{\partial z} - \rho g \cos \theta \right) \left[r^2 - b^2 - (b^2 - a^2) \frac{\ln r/b}{\ln a/b} \right]$$
$$+ W_b + (W_b - W_a) \frac{\ln r/b}{\ln a/b}, \tag{5.1.12}$$

with a shear stress

$$\tau_{rz} = \mu \frac{dw}{dr} = \frac{1}{4\mu} \left(\frac{\partial p}{\partial z} - \rho g \cos \theta \right) \left[2r - \frac{b^2 - a^2}{r \ln a/b} \right] + \frac{W_b - W_a}{r \ln a/b}. \tag{5.1.13}$$

If there is no inner cylinder ($a = 0$), all logarithmic terms in equation (5.1.12) must be ruled out, since they would make both the velocity and the shear stress infinite at $r = 0$. In that case the solution reduces to

$$w = \frac{1}{4\mu} \frac{\partial p}{\partial z} (r^2 - b^2) + W_b, \tag{5.1.14}$$

with

$$\tau_{zr} = \mu \frac{dw}{dr} = \frac{r}{2} \left(\frac{\partial p}{\partial z} - \rho g \cos \theta \right). \tag{5.1.15}$$

If, instead, the outer cylinder is removed, no solution exists that is bounded at large r. Thus, if a long cylinder is moved longitudinally through a fluid, one-dimensional motion, depending only on r is impossible.

5.1.4   Poiseuille Flow in Conduits of Arbitrarily Shaped Cross-Section

Except for conduits of a few simple geometries, such as the circle, ellipse, rectangle, and equilateral triangle, exact solutions of equation (5.1.6) are impossible to find. However, the discharge can still be estimated very accurately. From the definition of Q we have

$$Q = \iint_S u \, dS. \tag{5.1.16}$$

Rewrite equation (5.1.6) as

$$\frac{\partial p}{\partial x} - \rho g \cos \theta = \mu \left(\frac{\partial^2 u}{\partial y^2} + \frac{\partial^2 u}{\partial z^2} \right).$$

For convenience, let $K = -\mu / (\frac{\partial p}{\partial x} - \rho g \cos \theta)$, where K is a positive constant. Dividing this equation by the negative of the pressure gradient and gravity term, we have

$$1 = -K \left(\frac{\partial^2 u}{\partial y^2} + \frac{\partial^2 u}{\partial z^2} \right) = -K \nabla^2 u. \tag{5.1.17}$$

Multiply equation (5.1.17) by u, and integrate over the conduit area, giving

$$Q = \iint_S u \, dS = -K \iint_S u \nabla^2 u \, dS = -K \iint_S [\nabla \cdot (u \nabla u) - |u|^2] \, dS = K \iint_S |u|^2 \, dS, \tag{5.1.18}$$

where we have used the product rule of calculus, then the divergence theorem, and finally the no-slip boundary condition.

To get the first estimate on Q, choose an approximation to the velocity that satisfies the continuity equation and the no-slip boundary conditions but not necessarily the momentum equation (5.1.17). Then, by some elementary algebra and rearrangements,

$$\frac{1}{2}K \iint_S [|\nabla v|^2 - |\nabla u|^2]dS = \frac{1}{2}K \iint_S [|\nabla(v-u)|^2 + 2\nabla v \cdot \nabla u - 2|\nabla u|^2]dS$$

$$= \frac{1}{2}K \iint_S [|\nabla(v-u)|^2 + 2\nabla u \cdot \nabla(v-u)]dS$$

$$= \frac{1}{2}K \iint_S [|\nabla(v-u)|^2 + 2\nabla(v-u) \cdot \nabla u - 2(v-u)\nabla^2 u]dS$$

$$= \frac{1}{2}K \iint_S |\nabla(v-u)|^2 dS + \iint_S (v-u)dS \geq \iint_S (v-u)dS.$$

Rewriting this and using equation (5.1.17), we have

$$\frac{1}{2}K \iint_S [|\nabla v|^2 - |\nabla u|^2]dS - \frac{1}{2}Q \geq \iint_S vdS - Q,$$

or after rearrangement,

$$Q \geq 2\iint_S vdS - K\iint_S |\nabla v|^2 dS, \tag{5.1.19}$$

the equal sign holding only if $v = u$ everywhere.

A different bound on Q can be found by choosing an approximation w for the velocity that satisfies equation (5.1.6) but not necessarily the no-slip boundary conditions. Proceeding as before,

$$\frac{1}{2}K \iint_S [|\nabla w|^2 - |\nabla u|^2]dS = \frac{1}{2}K \iint_S [|\nabla(w-u)|^2 + 2\nabla u \cdot \nabla(w-u)]dS$$

$$= \frac{1}{2}K \iint_S [|\nabla(w-u)|^2 + 2\nabla u \cdot \nabla(w-u) - 2u\nabla^2(w-u)]dS$$

$$= \frac{1}{2}K \iint_S [|\nabla(w-u)|^2 dS \geq 0$$

by virtue of both u and w satisfying equation (5.1.6) and u also satisfying the no-slip conditions. Expanding the term on the left of the inequality, and using equations (5.1.16) and (5.1.18), the result is

$$\frac{1}{2}K \iint_S |\nabla w|^2 dS - \frac{1}{2}Q \geq 0,$$

or finally,

$$Q \leq K\iint_S |\nabla w|^2 dS. \tag{5.1.20}$$

Again, the equal sign holds only if w equals u everywhere. Thus, from equations (5.1.19) and (5.1.20) Q is bounded on both sides according to

$$K\iint_S |\nabla w|^2 dS \leq Q \leq \iint_S (2v - K|\nabla v|^2)dS. \tag{5.1.21}$$

By refining the approximating functions v and w, we can come as close to u, or at least to Q, as we desire.

5.1.5   Couette Flow Between Concentric Circular Cylinders

For flow between concentric circular cylinders, again use cylindrical polar coordinates and let $\mathbf{v} = (0, v(r), 0)$. This satisfies continuity automatically. Then the Navier-Stokes equations reduce to

$$0 = \mu \left(\frac{\partial^2 v}{\partial r^2} + \frac{1}{r}\frac{\partial v}{\partial r} - \frac{v}{r^2} \right),$$

$$\rho\frac{v^2}{r} = \frac{\partial p}{\partial r}. \tag{5.1.22}$$

The first equation is of Euler type, and its solutions are therefore of the form r^n. Substituting this form into equation (5.1.22a), we find that $(n-1)(n+1) = 0$. Thus, the solution is

$$v = Ar + B/r. \tag{5.1.23}$$

The term multiplied by the constant A represents a rotational rigid body rotation. The term multiplied by B is an irrotational vortex, which we saw in Chapter 2. If $v = b\Omega_b$ at $r = b$ and $v = a\Omega_a$ at $r = a$, applying these boundary conditions to equation (5.1.23) gives for A and B

$$A = \frac{b^2\Omega_b - a^2\Omega_a}{b^2 - a^2},$$

$$B = \frac{a^2 b^2 (\Omega_a - \Omega_b)}{b^2 - a^2}. \tag{5.1.24}$$

Notice that if $a^2\Omega_a = b^2\Omega_b$, the flow is irrotational, so viscous flows *can* in rare cases be irrotational. Generally, however, they are rotational because, except for very special boundary conditions such as we have here, vorticity is generated at boundaries by the no-slip condition and then is diffused throughout the flow by viscosity.

   If the outer cylinder is absent, $A = 0$, which again is an irrotational flow. If the inner cylinder is absent, $B = 0$, and the motion is a rigid body rotation.

5.2   Unsteady Flows When Convective Acceleration Is Absent

Next consider two important cases of unsteady flow that were first solved by Stokes. Both of the problems have a velocity of the form $\mathbf{v} = (u(y, t), 0, 0)$ and an infinite plate located at $y = 0$. For these flows, since the flow is due solely to the motion of the plate and no other forcing is imposed, the pressure gradient is absent, and the x momentum equation becomes

$$\rho\frac{\partial u}{\partial t} = \mu\frac{\partial^2 u}{\partial y^2}. \tag{5.2.1}$$

5.2.1   Impulsive Motion of a Plate—Stokes's First Problem

Stokes considered the case where at an initial time a plate is suddenly caused to move in the x direction with a velocity U. Considering the various parameters in the problem, we see that u must depend only on y, t, U, ρ, and μ. Thus, we have $u = u(y, t, U, \rho, \mu)$. In dimensionless form this can be written as

$$\frac{u}{U} = f\left(y\sqrt{\frac{\rho}{\mu t}}, \frac{\rho U y}{\mu} \right). \tag{5.2.2}$$

Since the momentum equation (5.2.1) in this case is linear, and since U appears only in the boundary condition, expect that U will appear only as a multiplying factor, so the dimensionless form (5.2.2) can be reduced to

$$\frac{u}{U} = f\left(y\sqrt{\frac{\rho}{\mu t}}\right). \tag{5.2.3}$$

To verify this, note that $\frac{\partial u}{\partial t} = -\frac{Uy}{2t}\sqrt{\frac{\rho}{\mu t}}f'$, $\frac{\partial u}{\partial y} = U\sqrt{\frac{\rho}{\mu t}}f'$, and $\frac{\partial^2 u}{\partial y^2} = \frac{U\rho}{\mu t}f''$, where a prime denotes differentiation with respect to the dimensionless variable $\eta = y\sqrt{\rho/\mu t}$. Inserting the preceding into equation (5.2.1), the result is $-\rho\frac{Uy}{2t}\sqrt{\frac{\rho}{\mu t}}f' = \mu\frac{U\rho}{\mu t}f''$, or

$$f'' + 0.5\,\eta f' = 0. \tag{5.2.4}$$

The boundary conditions on f are as follows: Because of the no-slip condition, $u = U$ on the flat plate, which requires that $f(0) = 1$. At very large values of y, u must go to zero, since the velocity is transported away from the plate only by viscous diffusion, and it will take a very large amount of time for the effect to be noticed away from the plate. Therefore, $f(\infty) = 0$.

Equation (5.2.4) can be integrated once to give $f' = A\,exp(-\eta^2/4)$, where A is a constant of integration. Integrating this once again gives $f = A\int_0^\eta e^{-s^2/4}ds + B$, where B is a second constant of integration. Since $f(0) = 1$, $B = 1$. Since $f(\infty) = 0$,

$$0 = A\int_0^\infty e^{-s^2/4}ds + 1, \text{giving } A = \frac{-1}{\int_0^\infty e^{-s^2/4}ds}.$$

The integral appearing in the denominator of A has the value $\sqrt{\pi}$; thus, $A = -1/\sqrt{\pi}$ and finally

$$u = U\,(1 - Err(\eta)) = U * Errc(\eta), \tag{5.2.5}$$

where $Err(\eta) = 1/\sqrt{\pi}\int_0^\eta e^{-s^2/4}ds$. The integral appearing in equation (5.2.5) is a form of the well-known *error function*, and 1 minus the error function is the *complementary error function*. Typical values for the solution are seen in Table 5.2.1 and in Figure 5.2.1. The combination $1 - u/U$ corresponds to the case where the plate is at rest and the outer fluid is impulsively started in motion.

TABLE 5.2.1   The Error Function

η	u/U	$1-u/U$	η	u/U	$1-u/U$	η	u/U	$1-u/U$
0.0	1.0000	0.0000	1.3	0.3580	0.6420	2.6	0.0660	0.9340
0.1	0.9436	0.0564	1.4	0.3222	0.6778	2.8	0.0477	0.9523
0.2	0.8875	0.1125	1.5	0.2888	0.7112	3.0	0.0339	0.9661
0.3	0.832	0.1680	1.6	0.2579	0.7421	3.2	0.0236	0.9764
0.4	0.7773	0.2227	1.7	0.2293	0.7707	3.4	0.0162	0.9838
0.5	0.7237	0.2763	1.8	0.2031	0.7969	3.6	0.0109	0.9891
0.6	0.6714	0.3286	1.9	0.1791	0.8209	3.8	0.0072	0.9928
0.7	0.6206	0.3794	2.0	0.1573	0.8427	4.0	0.0047	0.9953
0.8	0.5716	0.4284	2.1	0.1376	0.8624	4.2	0.0030	0.9970
0.9	0.5245	0.4755	2.2	0.1198	0.8802	4.4	0.0018	0.9982
1.0	0.4795	0.5205	2.3	0.1039	0.8961	4.6	0.0011	0.9989
1.1	0.4367	0.5633	2.4	0.09	0.9103	4.8	0.0007	0.9993
1.2	0.3961	0.6039	2.5	0.077	0.9229	5.0	0.0004	0.9996

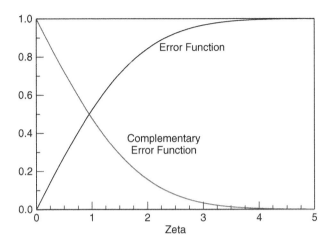

Figure 5.2.1   Error and complementary error functions

The variable η that appears in this solution is called a ***similarity variable***, and f is called a ***similarity solution***. By *similarity* we mean that at two different times, t_1 and t_2, we have geometric similarity in our flow (one velocity profile is merely an enlargement of the other) at locations y_1 and y_2, providing that

$$y_1\sqrt{\frac{\rho}{\mu t_1}} = y_2\sqrt{\frac{\rho}{\mu t_2}}, \quad \text{or equivalently,} \quad \eta_1 = \eta_2. \tag{5.2.6}$$

Similarity solutions play an important role in fluid mechanics. It reduces the Navier-Stokes partial differential equations to ordinary differential equations that are much easier to solve, with a solution that is much easier to understand physically. The existence of a similarity solution depends on simplifying conditions such as the domain having infinite extent, which allows the number of dimensionless parameters to be small. Thus, the similarity solutions are very special solutions and exist only in a relatively few flow cases. We will see more of similarity solutions in later sections.

This Stokes problem is a reasonably good approximation of the flow on a flat plate if we consider Ut to be a "distance" from the leading edge of the plate. To see how "deep" the viscous effects penetrate, note that for $\eta = 3$ the fluid velocity is reduced to less than 4% of the plate velocity. Consequently, we can consider this to be approximately the outer edge of the effects of viscosity. This thickness is termed the ***boundary layer thickness***. In terms of distance from the leading edge of the plate, we have

$$\text{boundary layer thickness} = 3\sqrt{\mu t/\rho} = 3\sqrt{\mu x/\rho U} = 3x/\sqrt{\rho Ux/\mu}, \tag{5.2.7}$$

We will consider the problem of flow on a flat plate again in Chapter 6 and find it to be in general agreement with those results.

5.2.2   Oscillation of a Plate—Stokes's Second Problem

A second problem considered by Stokes is one where an infinite plate oscillates back and forth with a circular frequency Ω, the motion having taken place sufficiently long that the starting conditions can be ignored. Then, $u(0, t) = U \cos \Omega t$, with U being the maximum speed of the plate.

If we were to try to solve this problem using separation of variables, the "traditional" method of solving partial differential equations, we would find phase differences in time between the boundary velocity and the fluid velocities, and we would need both the sine and cosine of Ωt. The result would be that we would end up having to solve two coupled second order equations. An easier approach is to let $u(0, t) = Ue^{i\Omega t}$, where i is the imaginary number $\sqrt{-1}$, and by DeMoivre's theorem $e^{i\Omega t} = \cos(\Omega t) + i \sin(\Omega t)$. Since the real part of $e^{i\Omega t}$ is $\cos \Omega t$, our solution is found by taking only the real part of u at the very end of our analysis.

Proceeding in this manner, let

$$u = f(y)e^{i\Omega t} \tag{5.2.8}$$

and substitute this into equation (5.7.8). The result is

$$i\rho\Omega f = \mu f'' \tag{5.2.9}$$

with the solution

$$f = Ae^{-(1+i)y/a} + Be^{(1+i)y/a}, \tag{5.2.10}$$

where $a = \sqrt{12\mu/\rho\Omega}$ is a characteristic length for the problem. The parameter a tells us how far the viscous effects penetrate from the boundary into the fluid.

Because of the condition at $y = 0$, we have $A + B = U$. If we consider the only solid boundary to be at $y = 0$, then u must vanish as y becomes large. Since the term that B multiplies grows exponentially as y increases, then the requirement is met if $B = 0$ is to be everywhere finite. Then it follows that $A = U$ and $u(y, t) = Ue^{-(1+i)y/a+i\Omega t}$. Taking the real part of this gives

$$u(y, t) = Ue^{-y/a} \cos \Omega(t - y/a). \tag{5.2.11}$$

If there were a solid boundary at $y = b$ instead, then $u(b, t) = 0$ and $0 = Ae^{-(1+i)b/a} + Be^{(1+i)b/a}$. Solving for A and B yields

$$f(y) = U\frac{e^{(1+i)(b-y)/a} - e^{-(1+i)(b-y)/a}}{e^{(1+i)b/a} - e^{-(1+i)b/a}} = U\frac{\sinh\dfrac{(1+i)(b-y)}{a}}{\sinh\dfrac{(1+i)b}{a}}. \tag{5.2.12}$$

Notice that if a is small, the limit of equation (5.2.12) is $f \approx Ue^{-(1+i)y/a}$, which says that the fluid motion is confined to a thin region near the wall with thickness of order a. In fact, equation (5.2.12) shows us that when $b \geq 2a$, and since $e^{-2} \approx 0.1$, for practical purposes the upper plate can be thought of as being at infinity for such values of a.

The results of Stokes's second problem are important to the study of pulsatile flows. There are few analytical tools available to deal with these flows, and Stokes's solution at least indicates some of the primary effects that can occur.

Example 5.2.1   Work done in an oscillating flow
An acoustic pressure $p_0 \sin (\Omega x/c) \sin (\Omega t)$ is set up in a gas that has a speed of sound c. This in turn induces a velocity field $U_0 \cos (\Omega x/c) \cos (\Omega t)$, where $U_0 = p_0/\rho c$. No useful work is produced by this, since the pressure and velocity are 90 degrees out of phase. Show that by introducing a flat plate into the flow and parallel to the velocity field, useful work is produced.

Solution. Assume the following: The plate is long enough that edge effects can be ignored; the viscous terms in the Navier-Stokes equations that involve second derivatives with respect to x can be neglected compared to the terms involving second derivatives with respect to y. The justification of this is that a boundary layer exists. The equation that will govern the flow and use the preceding assumption on the viscous terms is

$$\rho \frac{\partial u}{\partial t} = -\frac{\partial p}{\partial x} + \mu \frac{\partial^2 u}{\partial x^2}.$$

This is the counterpart of equation (5.2.1) that, however, also includes the pressure gradient term. As suggested by Stokes's second solution and the far field, take

$$u(x, y, t) = U_0[1 + f(y)]\cos(\Omega x/c)e^{i\Omega t}, \quad p(x, t) = -ip_0\sin(\Omega x/c)e^{i\Omega t},$$

the $-i$ in front of the pressure taking care of the time phase difference. Again, real values will be taken after solving for f.

Substituting these forms into the governing equation, we have, after cancelling out the common x and t factors, that $i\Omega\rho U_0(1 + f) = -\Omega p_0/c + \mu U_0 f''$. By virtue of the relation between U_0 and p_0, the constant terms cancel. Thus, we are left with $i\rho\Omega f = \mu f''$ the same as equation (5.2.9). Again, take the solution that decays as y increases. Thus, $f = Ae^{-(1+i)y/a}$, where a is defined as in Stokes's second problem.

Applying the boundary condition gives $A = -1$. Thus,

$$u(x, y, t) = U_0(1 - e^{-(1+i)y/a})\cos(\Omega x/c)e^{i\Omega t},$$

which satisfies all boundary conditions. Taking the real part, we have for the final form of the velocity

$$u(x, y, t) = U_0\{[1 - e^{-y/a}\cos(y/a)]\cos(\Omega t) - e^{-y/a}\sin(y/a)\sin(\Omega t)\}\cos(\Omega x/c).$$

The useful local work LW performed over a period $T = 2\pi/\Omega$ is

$$LW = \frac{1}{T}\int_0^T pu \, dt.$$

Putting our expressions for pressure and velocity, we have

$$LW = \frac{1}{T}\int_0^T p_0\sin(\Omega x/c)\sin(\Omega t)U_0\{[1 - e^{-y/a}\cos(y/a)]\cos(\Omega t)$$
$$- e^{-y/a}\sin(y/a)\sin(\Omega t)\}\cos(\Omega x/c)dt.$$

The time terms that involve the product of the sine and cosine will average to zero over the interval, while the average of the sine squared terms is $T/2$. Thus, we have

$$LW = -0.5e^{-y/a}\sin(y/a)\cos(\Omega x/c).$$

Notice that the term independent of y, which represents the work done in the far field, dropped out of LW. The total work TW per unit width and length is

$$TW = \int_0^\infty LW \, dy = -0.5p_0U_0\cos(\Omega x/c)\int_0^\infty e^{-y/a}\sin(y/a) \, dy = -0.25p_0U_0\cos(\Omega x/c).$$

Thus useful work is done due to the phase shift introduced in the velocity field by the no-slip condition and the presence of the plate. The results given here are a much simplified version of a device known as the "acoustic refrigerator."

5.3 Other Unsteady Flows When Convective Acceleration Is Absent

5.3.1 Impulsive Plane Poiseuille and Couette Flows

Several of the flows in the first section of this chapter can also be solved for if they are unsteady flows. For the plane Poiseuille flow, suppose that the fluid is initially at rest, and then suddenly the pressure gradients and plate motions are applied with the walls held still. Then $\mathbf{v} = (u(y, t), 0, 0)$, and a $\rho \partial u / \partial t$ term must be included in the acceleration. It is convenient now to let $u = u_s + u_t$, where u_s is given by equation (5.1.2) and u_t is the transient solution. Then, by the linearity of the equations, and getting $\nu = \mu / \rho$,

$$\frac{\partial u_t}{\partial t} = \nu \frac{\partial^2 u_t}{\partial y^2} \tag{5.3.1}$$

and $u_t = -u_s$ at $t = 0$, with $u_t = 0$ at $y = 0$ and $y = b$.

Separation of variables along with application of the boundary conditions yields

$$u_t = \sum_{n=1}^{\infty} A_n e^{-n^2 \pi^2 \nu t / b^2} \sin \frac{n \pi y}{b}. \tag{5.3.2}$$

Applying the initial condition and using the orthogonality of the sine function gives

$$A_n = \frac{-\int_0^b u_s \sin \dfrac{n \pi y}{b} dy}{\int_0^b \sin^2 \dfrac{n \pi y}{b} dy}. \tag{5.3.3}$$

For stationary walls $u_x = \dfrac{1}{2\mu} \left(\dfrac{\partial p}{\partial x} - \rho g \cos \theta \right) (y^2 - by)$, giving

$$A_n = -\frac{2}{b\mu} \left(\frac{\partial p}{\partial x} - \rho g \cos \theta \right) \left(\frac{b}{n\pi} \right)^3 (1 - (-1)^n).$$

A plot of the developing flow showing velocity profiles as time passes is shown in Figure 5.3.1.

The solution given by equation (5.3.2) is for a step function increase in the driving force $\partial p / \partial x - \rho g \sin \theta$. From this solution the velocity for a driving force that is an arbitrary function of time can easily be found. Let $u_u = u_{su}(y) + u_{tu}(y, t)$, where the subscript u stands for our solution (5.3.2) with $\partial p / \partial x - \rho g \sin \theta$ set to unity. For an arbitrary forcing function $\partial p / \partial x - \rho g \sin \theta = f(t)$, **Duhamel's superposition theorem**[1] can be used to give the velocity

$$u(y, t) = u_u(t_0) + \int_{t_0}^t \frac{\partial u_u(y, \tau)}{\partial \tau} f(t - \tau) d\tau \tag{5.3.4}$$

Thus, only a single time integration is needed to find the velocity for the arbitrary forcing function.

---

[1] Duhamel's theorem is similar to the **convolution theorem** encountered in the use of Fourier integrals.

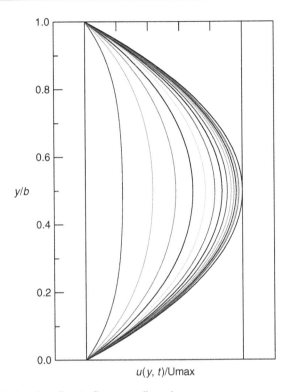

Figure 5.3.1   Developing plane Couette flow—two dimensions

5.3.2   Impulsive Circular Couette Flow

For impulsively started transient Couette flow in a circular annulus $a \leq r \leq b$, take
$\mathbf{v} = (0, v_s(r) + v_t(r, t), 0)$ with v_s given by equation (5.2.1). Using the unsteady Navier-
Stokes equations in cylindrical polar coordinates together with separation of variables,
a procedure similar to the preceding yields

$$v_t = \sum_k e^{-k^2 \nu t} [A_k J_1(kr) + B_k Y_1(kr)],$$

where J_1 and Y_1 are Bessel functions of the first and second kind. Application of the
boundary conditions yields

$$A_k J_1(ka) + B_k Y_1(ka) = 0,$$

$$A_k J_1(kb) + B_k Y_1(kb) = 0.$$

Solving these for A and B, find that

$$B_k = -A_k J_1(ka)/Y_1(ka) \tag{5.3.5}$$

and

$$J_1(kb)Y_1(ka) - J_1(ka)Y_1(kb) = 0. \tag{5.3.6}$$

Equation (5.3.6) determines the characteristic values of k that are to be used in the
summation.

For the case where no inner wall is present, $0 \leq r \leq b$, and as a approaches zero in equation (5.3.5), B_k also approaches zero and equation (5.3.6) becomes $J_1(kb) = 0$. A_k is determined as before, using $v_t(r, 0) = -v_s(r)$.

For flow outside a cylinder and extending to infinity ($b \leq r \leq \infty$), the problem is not as simple. Since both Bessel functions J_1 and Y_1 go to zero as r goes to infinity, the condition at infinity is automatically satisfied. Thus, only the condition at the wall need be imposed, which gives equation (5.3.5).

In this case, rather than a Fourier series solution as in equation (5.3.4), there is a continuous spectrum of k rather than a discrete one, and the summation over the finite spectrum is replaced by an integral over a continuous spectrum of ks. The result is

$$v_t = e^{-k^2 \nu t} \int_0^\infty A(k) \left[Y_1(ka) J_1(kr) - Y_1(kr) J_1(ka) \right] dk, \qquad (5.3.7)$$

where we have implicitly assumed that k is real—that is, that the solution decays exponentially in time with no oscillations. $A(k)$ is again determined by the initial condition, which this time requires solution of an integral equation. Some zeros of the Bessel functions have been listed in tables. Others can be found by numerical integration.

5.4 Steady Flows When Convective Acceleration Is Present

There are a few special flows that allow similarity solutions of the full Navier-Stokes equations even when the convective acceleration terms are present. Generally these problems involve specialized geometry and are of an idealized nature. These similarity solutions, even though they are for idealized geometries, do, however, provide much of our basic understanding of laminar viscous flows. They also serve both as a starting point, as well as a validation, of approximate and numerical solutions.

Our analysis of these flows starts with a stream function in all cases. To have a similarity solution, the stream function will generally be of the form $\psi = \psi(x, y, U, \nu)$. Putting this into dimensionless form, it becomes

$$\psi = \frac{\mu}{\rho} x^n F \left(\frac{\rho x U}{\mu}, \frac{\rho y U}{\mu} \right), \qquad (5.4.1)$$

where $n = 0$ for two-dimensional problems (Lagrange's stream function) and $n = 1$ for three-dimensional axisymmetric problems (Stokes stream function). Here, $U = U(x)$ is a characteristic velocity of the flow and is generally given by the tangential slip velocity at the boundary found from an inviscid solution of the problem. In the flows for which similarity solutions hold, F is generally taken to be of the form

$$F = \sqrt{\frac{\rho x U}{\mu}} f(\eta), \qquad (5.4.2)$$

where, reminiscent of Stokes's first problem, the dimensionless variable η is defined by

$$\eta = \left(\frac{\rho y U}{\mu} \right) \Big/ \sqrt{\frac{\rho x U}{\mu}} = \frac{y}{x} \sqrt{\frac{\rho x U}{\mu}}. \qquad (5.4.3)$$

This form will reduce the nonlinear partial differential Navier-Stokes equations to a nonlinear ordinary differential equation, which then usually must be solved numerically. We can show the application of this with several examples.

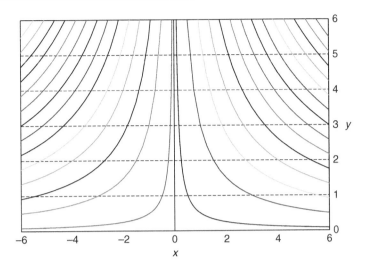

Figure 5.4.1   Streamlines for plane stagnation point flow

5.4.1   Plane Stagnation Line Flow

The flow impinging normally against a plane located at $y = 0$, with axes set so that there is a stagnation point at (0,0) as in Figure 5.4.1, was first solved by Hiemenz in 1911. The simplest irrotational flow that corresponds to this case is $\psi = axy$, with a slip velocity $U = ax$ on the boundary. The constant a has dimensions of reciprocal time.

The irrotational flow $\psi = axy$ by itself is not particularly interesting. However, it is a good model for the stream function in the vicinity of a stagnation point. For instance, it can be shown to be the local value of the stream function for any flow past an obstacle—for example, a uniform flow past a circular cylinder.

Taking $U = ax$ as the representative velocity from the preceding, we have from our general form (5.4.1)

$$\psi = \frac{\mu}{\rho}\sqrt{\frac{\rho x U}{\mu}}\,f(\eta) = \sqrt{\frac{\mu a x^2}{\rho}}\,f(\eta), \quad \text{or } \psi = x\sqrt{\frac{\mu a}{\rho}}f(\eta), \qquad (5.4.4)$$

where $\eta = y\sqrt{\rho a/\mu}$. We see then that $v_x = \partial\psi/\partial y = axf'$, $v_y = -\partial\psi/\partial x = -\sqrt{a\mu/\rho}f$, where primes denote differentiation with respect to η.

The no-slip boundary conditions are that $v_x = v_y = 0$ at $y = 0$, requiring that $f(0) = f'(0) = 0$. We will also require that v_x approach U far from the plate, so that f' must approach 1 as η approaches infinity.

The following quantities are needed for use in the Navier-Stokes equations:

$$\frac{\partial v_x}{\partial x} = af', \quad \frac{\partial v_x}{\partial y} = xa\sqrt{\rho a/\mu}f'', \quad \frac{\partial^2 v_x}{\partial x^2} = 0, \quad \frac{\partial^2 v_x}{\partial y^2} = \frac{\rho x a^2}{\mu}f''',$$

$$\frac{\partial v_y}{\partial x} = 0, \quad \frac{\partial v_y}{\partial y} = -af', \quad \frac{\partial^2 v_y}{\partial x^2} = 0, \quad \frac{\partial^2 v_y}{\partial y^2} = -a\sqrt{\rho a/\mu}f''.$$

Notice the following order of magnitude relations for later reference:

$$O\left(\frac{v_x}{v_y}\right) = \sqrt{\rho x^2 a/\mu} = \sqrt{\rho x U/\mu} \quad \text{and} \quad O\left(\frac{\frac{\partial^2}{\partial x^2}}{\frac{\partial^2}{\partial y^2}}\right) = 0. \qquad (5.4.5)$$

Here, $\sqrt{\rho x U/\mu}$ is the local Reynolds number for the flow. These orderings will be used in Chapter 6 in developing the boundary layer approximation.

Substituting the above into the two-dimensional Navier-Stokes equations, we have

$$\rho\left[(axf')af' + \left(-\sqrt{a\mu/\rho}f\right)xa\sqrt{a\rho/\mu}f''\right] = -\frac{\partial p}{\partial x} + xa^2\mu f'''$$

and

$$\rho\left[(axf')\cdot 0 + (-\sqrt{a\mu/\rho}f)(-af')\right] = -\frac{\partial p}{\partial y} + \mu\left(-a\sqrt{a\mu/\rho}f''\right).$$

Since we started with a stream function, continuity is automatically satisfied.

Solving for the pressure gradients from the preceding equations, we find that

$$\frac{1}{\rho}\frac{\partial p}{\partial x} = xa^2[-(f')^2 + ff'' + f'''] \qquad (5.4.6)$$

and

$$\frac{1}{\rho}\frac{\partial p}{\partial y} = \sqrt{a/\nu}\frac{1}{\rho}\frac{\partial p}{\partial \eta} = -a\sqrt{a\nu}(ff' + f'') \qquad (5.4.7)$$

Again, note for later reference that

$$O\left(\frac{\frac{\partial p}{\partial x}}{\frac{\partial p}{\partial \eta}}\right) = \sqrt{Ux\rho/\mu}.$$

The second of these equations can be integrated exactly, giving $p = -a\mu(0.5f^2 + f') +$ function of x. For now, let the function of x be written as $\rho h(x)$. Substituting the pressure into equation (5.4.6), the result is

$$\frac{dh}{dx} = xa^2[-(f')^2 + ff'' + f'''].$$

Comparing this with equation (5.4.6), we see that the only possibility of having the variables x and η separate is for

$$h = -0.5x^2a^2A,$$

giving

$$f''' + ff'' - (f')^2 = -A = \text{integration constant.} \qquad (5.4.8)$$

For the potential flow corresponding to $\psi = axy$, Bernoulli's theorem gives the pressure as

$$p = -0.5\rho(v_x^2 + v_y^2) = -0.5\rho a^2(x^2 + y^2). \qquad (5.4.9)$$

TABLE 5.4.1 Two-dimensional stagnation point flow against a plate (Hiemenz)

η	f	f'	f''	η	f	f'	f''
0.0	0.0000	0.0000	1.2326	2.0	1.3620	0.9732	0.0658
0.1	0.0060	0.1183	1.1328	2.2	1.5578	0.9839	0.0420
0.3	0.0510	0.3252	0.9386	2.6	1.9538	0.9946	0.0156
0.4	0.0881	0.4145	0.8463	2.8	2.1530	0.9970	0.0090
0.5	0.1336	0.4946	0.7583	3.0	2.3526	0.9984	0.0051
0.6	0.1867	0.5663	0.6752	3.2	2.5523	0.9992	0.0028
0.7	0.2466	0.6299	0.5973	3.4	2.7522	0.9996	0.0014
0.8	0.3124	0.6859	0.5251	3.6	2.9521	0.9998	0.0007
0.9	0.3835	0.7351	0.4587	3.8	3.1521	0.9999	0.0004
1.0	0.4592	0.7779	0.3980	4.0	3.3521	1.0000	0.0002
1.1	0.5389	0.8149	0.3431	4.2	3.5521	1.0000	0.0001
1.2	0.6220	0.8467	0.2938	4.4	3.7521	1.0000	0.0000
1.3	0.7081	0.8738	0.2498	4.6	3.9521	1.0000	0.0000
1.4	0.7967	0.8968	0.2110	4.8	4.1521	1.0000	0.0000
1.5	0.8873	0.9162	0.1770	5.0	4.3521	1.0000	0.0000
1.6	0.9798	0.9323	0.1474	5.5	4.8521	1.0000	0.0000
1.7	1.0737	0.9458	0.1218	6.0	5.3521	1.0000	0.0000
1.8	1.1689	0.9568	0.1000	6.5	5.8520	1.0000	0.0000
1.9	1.2650	0.9659	0.0814	7.0	6.3520	1.0000	0.0000

Since far from the wall the condition f' approaches zero implies that f approaches η plus a constant, Bernoulli's equation and our viscous result for the pressure are seen to agree far from the wall, providing A is taken to be unity. Then we are left with

$$\tau_{xy\,\text{wall}} = \mu \frac{\partial v_x}{\partial y}\Big|_{y=0} = xa\mu\sqrt{a\rho/\mu}\,f''(0) = 1.2326\rho U^2/\sqrt{\rho x U/\mu}. \qquad (5.4.10)$$

with

$$f''' + ff'' - (f')^2 = -1. \qquad (5.4.11)$$

The numerical solution of equation (5.4.8) subject to the boundary conditions was originally obtained by Hiemenz in 1911 and is given in Table 5.4.1 and Figure 5.4.2.

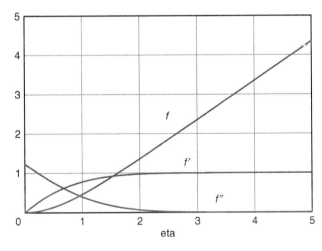

Figure 5.4.2 Hiemenz's solution for plane stagnation point flow

The table shows that for y greater than $2.4\sqrt{\mu/\rho a}$, the difference between the irrotational solution and the Navier-Stokes solution is less than 1 percent. Thus, when we required our viscous solution to approach the inviscid solution as η approached infinity, this limit was, for practical purposes, achieved at $\eta = 2.4$. It follows that a constant thickness boundary layer exists in this case, with the boundary layer thickness in the form

$$\delta = 2.4x/\sqrt{\rho Ux/\mu}. \qquad (5.4.12)$$

Inside this boundary layer, viscosity affects the flow in an important manner. Outside of this region, viscous effects have a negligible effect on our flow.

5.4.2   Three-Dimensional Axisymmetric Stagnation Point Flow

In 1936 Homann provided a three-dimensional analog of the previous two-dimensional stagnation point case. This time use

$$v_r = -\frac{1}{r}\frac{\partial \psi}{\partial z}, \quad v_z = \frac{1}{r}\frac{\partial \psi}{\partial r}.$$

The axisymmetric potential flow $\psi = -ar^2 z$ will be used as our starting point, since it has a downward velocity toward the plate of $v_z = -2'ar$ and a radial velocity of $v_r = az$. Thus, we have $U = ar$ for the slip velocity along the plate and by Bernoulli's equation a pressure of

$$p = -\rho a^2 \frac{4r^2 + z^2}{2}.$$

Again, a is a constant with dimensions of reciprocal time. Letting

$$\psi = -r^2\sqrt{a\nu}f(\eta), \quad \eta = z\sqrt{a/\nu},$$

we can carry out the calculations in the same manner as in the two-dimensional case. The final result of these manipulations is the equation

$$f''' + 2ff'' - (f')^2 = -1, \qquad (5.4.13)$$

which differs from equation (5.4.8) only by the number 2 multiplying the term containing the second derivative. The numerical results satisfying the boundary conditions $f(0) = f'(0) = 0$, f' approaching 1 as η gets large, differ little from the preceding case, as can be seen from Table 5.4.2 and Figure 5.4.3.

5.4.3   Flow into Convergent or Divergent Channels

One interesting flow that shows a form of flow separation is that between two nonparallel plates an angle α apart, as shown in Figure 5.4.4. This flow was first studied by Jeffrey (1915) and elaborated further by Hamel (1917). A source or a sink is placed where the two planes meet. To use our standard form (5.4.2) for the stream function, we work in cylindrical polar form and recognize that the inviscid flow in this case is like a source or sink, with streamlines being straight radial lines passing through the source/sink. From continuity considerations our radial velocity component then is proportional to $1/r$, and the stream function for the channel must therefore be of the form

$$\psi = \frac{\mu}{\rho}F(\theta, \alpha, \rho Q/\mu), \qquad (5.4.14)$$

TABLE 5.4.2   Three-dimensional stagnation point flow against a plate (Homann)

η	f	f'	f''	η	f	f'	f''
0.0	0.0000	0.0000	1.3121	1.4	0.8495	0.9401	0.1736
0.1	0.0064	0.1262	1.2121	1.5	0.9443	0.9555	0.1359
0.2	0.0249	0.2424	1.1123	1.6	1.0405	0.9675	0.1046
0.3	0.0545	0.3487	1.0129	1.7	1.1377	0.9766	0.0793
0.4	0.0943	0.4451	0.9147	1.8	1.2357	0.9835	0.0591
0.5	0.1432	0.5317	0.8183	1.9	1.3343	0.9886	0.0433
0.6	0.2003	0.6088	0.7247	2.0	1.4334	0.9923	0.0221
0.7	0.2647	0.6767	0.6347	2.2	1.6323	0.9968	0.0154
0.8	0.3354	0.7359	0.5494	2.4	1.8319	0.9989	0.0072
0.9	0.4116	0.7868	0.4696	2.6	2.0318	0.9999	0.0032
1.0	0.4925	0.8300	0.3961	2.8	2.2318	0.9999	0.0014
1.1	0.5774	0.8663	0.3295	3.0	2.4319	0.9999	0.0007
1.2	0.6655	0.8962	0.2702	3.2	2.6321	0.9999	0.0004
1.3	0.7564	0.9205	0.2182				

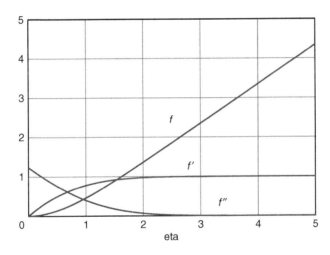

Figure 5.4.3   Homann's solution for axisymmetric stagnation point flow

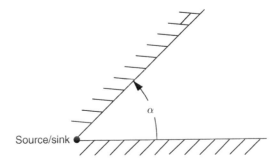

Figure 5.4.4   Flow in a converging–diverging channel

the radial coordinate r dropping out of the list, since it cannot be combined with the other variables to form a dimensionless parameter. Thus, only a radial velocity component is present.

This velocity is of the form $v_r = \frac{\mu}{\rho r} f(\theta)$, with $f(\theta) = \frac{dF}{d\theta}$, which automatically satisfies continuity. Using primes to denote differentiation with respect to θ, find then that

$$\frac{\partial v_r}{\partial r} = -\frac{\mu}{\rho r^2} f, \quad \frac{\partial v_r}{r \partial \theta} = \frac{\mu}{\rho r} f', \quad \frac{\partial^2 v_r}{\partial r^2} = \frac{2\mu}{\rho r^3} f, \quad \frac{\partial^2 v_r}{r \partial \theta^2} = \frac{\mu}{\rho r} f''.$$

The Navier-Stokes equations in cylindrical polar form in this case are

$$\rho v_r \frac{\partial v_r}{\partial r} = -\frac{\partial p}{\partial r} + \mu \left(\frac{\partial^2 v_r}{\partial r^2} + \frac{1}{r} \frac{\partial v_r}{\partial r} + \frac{1}{r^2} \frac{\partial^2 v_r}{\partial \theta^2} - \frac{v_r}{r^2} \right), \quad 0 = -\frac{1}{r} \frac{\partial p}{\partial \theta} + \mu \left(\frac{2}{r^2} \frac{\partial v_r}{\partial \theta} \right),$$

giving for the pressure gradients

$$\frac{1}{\rho} \frac{\partial p}{\partial r} = \frac{\mu^2}{\rho^2 r^3} (f^2 + f''), \quad \frac{1}{\rho} \frac{\partial p}{r \partial \theta} = \frac{2\mu^2}{\rho^2 r^3} f'.$$

Integration of these gives

$$\frac{p}{\rho} = \frac{2\nu^2}{r^2} f + h(r) = -\frac{\nu^2}{2r^2} \left(f^2 + f'' \right) + H(\theta). \tag{5.4.15}$$

Comparing terms on the left and right sides of this equation, we can see that $h(r)$ has to be of the form $\mu^2 A / \rho r^2$, where A is a constant, and $H(\theta)$ has to be a constant that can arbitrarily be set to zero. The $1/r^2$ dependency of p is consistent with the pressure field for an inviscid flow far from a source/sink. Thus,

$$p = \frac{\mu^2}{\rho r^2} (2f + A) = -\frac{\mu^2}{2\rho r^2} \left(f^2 + f'' \right) \tag{5.4.16}$$

and

$$f'' + f^2 + 4f + 2A = 0. \tag{5.4.17}$$

Equation (5.4.17) can be integrated once if we first multiply it by f'. The result is

$$0.5(f')^2 + (1/3)f^3 + 2f^2 + 2Af + B = 0, \tag{5.4.18}$$

where B is a second constant of integration.

Before going further with the solution, consider the boundary conditions. From the no-slip conditions, f must vanish at both walls of the wedge. Thus, from equations (5.4.16) and (5.4.18), $A = -f''_{wall}$ and $B = -0.5(f'_{wall})^2$. Unfortunately neither f'_{wall} nor f''_{wall} are known.

We could try to express our constants in terms of the discharge per unit length into the paper, $Q = \int r v_r d\theta = \nu \int f d\theta$, the integral being over the angle of the wedge. Q would normally be a given quantity but we would first have to know f to be able to carry out the integration.

Alternately, we could interpret A in terms of the pressure along a wall, so $A = \rho r^2 p_{wall} / \mu^2$, where p_{wall} varies like $1/r^2$. This perhaps makes the most sense, as it is the pressure gradient that drives the flow.

Traditional procedure has been to note that equation (5.4.30) is a first-order equation and is one where the variables can be separated. This gives

$$\theta = \int \frac{df}{\sqrt{-2B - 4Af - 4f^2 - \frac{2}{3}f^3}} + C, \tag{5.4.19}$$

where C is still another constant of integration. The three constants A, B, and C are to be determined such that the no-slip and discharge conditions

$$f(\pm\alpha) = 0, \quad Q = \int_{-\alpha}^{\alpha} rv_r\,d\theta, \tag{5.4.20}$$

where Q would normally be a given quantity.

This solution can be rewritten in the form of elliptic integrals, but the interpretation of the result is still complicated, since it gives θ as a function of f rather than vice versa, and either numerical integration or tables are necessary to interpret the elliptic integrals. Further, at a point of maximum f the integrand become singular and the sign of the square root changes.

Two separate approaches have been used to carry out computations of equation (5.4.19). In 1940 Rosenhead made the substitution

$$A = -(ab + ac + bc)/6, \quad B = -abc/3, \quad \text{where } a + b + c = -6.$$

This enabled him to factor the cubic terms under the square root in equation (5.4.3). The following conclusions have been reached for various ranges of these constants:

- If the constant a is real and positive, the constants b and c are complex conjugates of each other. In this case, the flow will have a positive velocity (sink flow) throughout.

- If the constants a, b, and c are real, with $a > b > c$, and $a > 0$, $c < b < 0$, there can be one or more regions of reversed flow.

- Rosenhead also found that for a given wedge angle and value of $\rho Q/\mu$, there is an infinite number of solutions. Also, for a given number of inflow-outflow regions and wedge angle there is a critical value of $\rho Q/\mu$ above which the solution doesn't exist.

- Millsaps and Pohlhausen (1953), using a slightly different approach, made further calculations and found that for pure outflow problems as $\rho Q/\mu$ is increased, the flow concentrates more and more in the center of the wedge.

- If $Q < 0$ (sink flow), f is symmetric about a center line and always negative. There is no flow reversal (Figure 5.4.5a).

- If $Q > 0$ (source flow), too large a value of $\rho Q/\mu$ for a given wedge angle can result in unsymmetric flow with reversed flow near one or more walls. The nature of the flow has to do with the question of whether the roots of the cubic term under the square root sign in equation (5.4.30) are real or imaginary equation (Figure 5.4.5b).

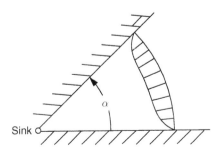

Figure 5.4.5a Velocity profile for a converging channel

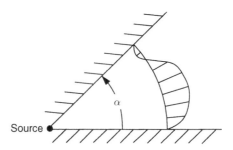

Figure 5.4.5b   Velocity profile for a diverging channel

- Most of the simpler two-dimensional flows have three-dimensional counterparts. Interestingly, this one does not! (Try a radial velocity inversely proportional to R^2 in spherical coordinates to convince yourself.) This suggests that flow into an ideal conical funnel must be much more complicated than the simple radial flow one might expect.

This seemingly simple problem with an exact solution (if you are willing to accept that elliptic integrals are "well-known functions") illustrates the complexities that can arise from solutions of the Navier-Stokes equations.

An alternate approach to the solution of this equation for those who have access to personal computers involves straightforward numerical integration of equation (5.4.18) after suitable choice of the constants A and B. Select values for A and B, and then integrate equation (5.4.18) numerically, using, say, a Runge-Kutta scheme, continuing the integration until f' becomes zero. You may consider the corresponding angle to be an acceptable value of the wedge angle, or you may continue if you wish to allow reversed flow. Numerical integration of f according to equation (5.4.31) will give you the discharge for your choice of parameters. Since there are no square roots in this procedure, it is not necessary to test for signs or patch solutions together.

If you want your constants to give a pure source flow, the integration should proceed until f' becomes zero, which would be at the location of the center line of the wedge. The flow past the center line is a mirror image of this flow. After you have done many computations for values of the constants, you should have a better understanding of these flows.

5.4.4   Flow in a Spiral Channel

Hamel also considered flow into a channel, which is in the shape of a logarithmic spiral whose shape is given in cylindrical polar coordinates by $A\theta + B \ln r = $ constant. Like the approach in the previous section, he let

$$\psi = \frac{\mu}{\rho} f(\beta) \quad \text{and} \quad p = p_0 + \rho \left(\frac{\nu}{r}\right)^2 P(\beta), \quad \text{where } \beta = A\theta + B \ln r. \qquad (5.4.21)$$

Letting primes denote differentiation with respect to β, the radial and theta velocity components are

$$v_r = \frac{1}{r}\frac{\partial\psi}{\partial\theta} = \frac{A\nu}{r}f', \quad v_\theta = -\frac{\partial\psi}{\partial r} = -\frac{B\nu}{r}f' = -\frac{B}{A}v_r.$$

Substitution of these into the Navier-Stokes equations leads to

$$-2P + BP' = \left(A^2 + B^2\right)\left[\left(f'\right)^2 + Af'''\right],$$
$$AP' = \left(A^2 + B^2\right)\left(2f'' - Bf'''\right). \tag{5.1.22}$$

From these the pressure is found to be

$$P = \frac{A^2 + B^2}{2A}\left[-A\left(f'\right)^2 + 2Bf'' - \left(A^2 + B^2\right)f'''\right], \tag{5.4.23}$$

and the governing equation for the stream function is

$$\left(A^2 + B^2\right)f^{iv} - 4Bf''' + \left(4 + 2Af'\right)f'' = 0.$$

This can be integrated once to give

$$\left(A^2 + B^2\right)f''' - 4Bf'' + \left(4 + Af'\right)f' = c_1, \tag{5.4.24}$$

which is second order in f'. The boundary conditions are that f' vanish on the walls—say, $\beta = 0$ and β_1. The remainder of the solution procedure is much the same as in the previous section.

5.4.5   Flow Due to a Round Laminar Jet

The solution for the round laminar axisymmetric jet was first presented by Landau (1944) and later rediscovered by Squire (1951). Working in spherical coordinates they considered a (vanishing) thin tube injecting fluid at the origin with a stream function

$$\psi = \frac{\mu}{\rho} R f(\varsigma), \quad \text{where } \varsigma = \cos\beta. \tag{5.4.25}$$

(The use of $\cos\beta$ rather than β itself simplifies the following equations considerably.) Then the velocity components are given by

$$v_R = \frac{1}{R^2 \sin\beta}\frac{\partial\psi}{\partial\beta} = -\frac{\mu}{\rho R}\frac{df}{d\varsigma},$$
$$v_\beta = -\frac{1}{R \sin\beta}\frac{\partial\psi}{\partial R} = -\frac{\mu f}{\rho R \sin\beta}. \tag{5.4.26}$$

For axisymmetric flow the Navier-Stokes equations are

$$\rho\left(v_R\frac{\partial v_R}{\partial R} + \frac{v_\beta}{R}\frac{\partial v_R}{\partial\beta} - \frac{v_\beta^2}{R}\right) = -\frac{\partial p}{\partial R} + \mu\left(\nabla^2 v_R - \frac{2v_R}{R^2} - \frac{2}{R^2}\frac{\partial v_\beta}{\partial\beta} - \frac{2v_\beta\cot\beta}{R^2}\right),$$
$$\rho\left(v_R\frac{\partial v_\beta}{\partial R} + \frac{v_\beta}{R}\frac{\partial v_\beta}{\partial\beta} + \frac{v_R v_\beta}{R}\right) = -\frac{1}{R}\frac{\partial p}{\partial\beta} + \mu\left(\nabla^2 v_\beta + \frac{2}{R^2}\frac{\partial v_R}{\partial\beta} - \frac{v_\beta}{R^2\sin^2\beta}\right),$$
$$\frac{1}{R^2}\frac{\partial\left(R^2 v_R\right)}{\partial R} + \frac{1}{R\sin\beta}\frac{\partial\left(v_\beta\sin\beta\right)}{\partial\beta} = 0, \tag{5.4.27}$$

$$\text{where } \nabla^2 = \frac{1}{R^2}\frac{\partial}{\partial R}\left(R^2\frac{\partial}{\partial R}\right) + \frac{1}{R^2\sin\beta}\frac{\partial}{\partial\beta}\left(\sin\beta\frac{\partial}{\partial\beta}\right).$$

These equations can be considerably simplified by recognizing for this problem that from the form of the stream function, the following holds:

$$\frac{\partial v_R}{\partial R} = -\frac{v_R}{R}, \qquad \frac{\partial v_\beta}{\partial R} = -\frac{v_\beta}{R}.$$

Using these and some tedious but straightforward calculus, the Navier-Stokes equations become

$$\rho\left(-\frac{v_R^2 + v_\beta^2}{R} - \frac{v_\beta \sin\beta}{R}\frac{\partial v_R}{\partial \beta}\right) = -\frac{\partial p}{\partial R} + \frac{\mu}{R^2}\frac{\partial^2 v_R}{\partial \beta^2},$$

$$-\rho v_\beta \frac{\partial v_\beta}{\partial \beta} = -\frac{\partial p}{\partial \beta} - \frac{\mu}{R}\frac{\partial v_R}{\partial \beta}. \tag{5.4.28}$$

The second of these can be integrated to obtain

$$p - p_0 = \rho\left(-\frac{v_\beta^2}{2} + \frac{v_R v_\beta}{R} + \frac{c_1}{R^2}\right). \tag{5.4.29}$$

Here, p_0 is the constant pressure at infinity. Since R was held constant during the integration, the "constant of integration" c_1 is in fact a function of R. The radial momentum equation dictates that it be proportional to R^{-2}.

Using the form for the pressure, inserting it into the radial momentum equation, and using the stream function, the result is

$$(f')^2 + ff'' = 2f' + \left[(1 - s^2)f''\right]' - 2c_1.$$

Primes here denote differentiation with respect to ζ. Integrating this once gives

$$ff' = 2f + (1 - s^2)f'' - 2c_1 s - c_2,$$

and once more gives

$$\frac{1}{2}f^2 = 2sf + (1 - s^2)f' - c_1 s^2 - c_2 s - c_3 \tag{5.4.30}$$

with boundary conditions $f(0) = f(\pi) = 0$.

Squire solved this for the special case $c_1 = c_2 = c_3 = 0$, obtaining

$$f = \frac{2(1 - s^2)}{a + 1 - s}, \tag{5.4.31}$$

where a is a parameter representing the strength of the jet.

The momentum flux M in the x direction that passes through the surface of a sphere centered at the origin is

$$M = \int_0^\pi \rho v_R \left(v_R \cos\beta - v_\beta \sin\beta\right) 2\pi R^2 \sin\beta\, d\beta$$

$$= 2\pi\frac{\mu^2}{\rho}\left[\frac{32(a+1)}{3a(a+2)} - 16(a+1) + 8a(a+2)\ln\frac{a+2}{a}\right], \tag{5.4.32}$$

and the force component for the same sphere is

$$F_z = -\int_0^\pi \left\{ \left(-p + 2\mu \frac{\partial v_R}{\partial R} \right) \cos\beta - \mu \left[R \frac{\partial}{\partial R} \left(\frac{v_\beta}{R} \right) + \frac{1}{R} \frac{\partial v_R}{\partial \beta} \right] \sin\beta \right\} 2\pi R^2 \sin\beta \, d\beta$$

(5.4.33)

$$= \frac{2\pi\mu^2}{\rho} \left[24(a+1) - (12a^2 + 24a + 4) \ln \frac{a+2}{a} \right].$$

The ratio

$$\frac{M + F_z}{\mu^2/\rho} = 2\pi \left[\frac{32(a+1)}{3a(a+2)} + 8(a+1) - 4(a+1)^2 \ln \frac{a+2}{a} \right]$$

(5.4.34)

gives a measure of the momentum strength of the jet. For values of $a = 1, 0.1$, and 0.01, this ratio takes on the values 34.76, 314.0, and 3,282. Whether there are other important combinations of the three constants of integration is not resolved by this analysis.

5.4.6 Flow Due to a Rotating Disk

One of the very few similarity solutions that involve all three velocity components was found by von Kármán in 1921 for flow caused by a large rotating disk in a stagnant fluid. We alter our previous "general" form for the stream function by replacing U by $r\Omega$, where Ω is the angular speed of the disk. Following our form for the general similarity solution and remembering that for axisymmetric flows the stream function has dimensions of length cubed over time, one length dimension higher than in the two-dimensional case, we have

$$\psi = -r\frac{\mu}{\rho}\sqrt{\rho r U/\mu} f(\eta) = -r^2 \sqrt{\Omega\mu/\rho} f(\eta),$$

(5.4.35)

where

$$\eta = (z\rho U/\mu)/\sqrt{\rho r U/\mu} = z\sqrt{\rho\Omega/\mu}.$$

(5.4.36)

(We have also altered our general solution in one other trivial detail: We have put a minus sign in ψ because we expect to have a downward vertical velocity component directed toward the disk, and it is convenient to have f positive.) The appropriate form for the swirl velocity component to accompany the stream function was found by von Kármán to be

$$v_\theta = r\Omega g(\eta).$$

(5.4.37)

Differentiation of the velocity components gives

$$v_r = r\Omega f', \quad v_\theta = r\Omega g, \quad v_z = -2\sqrt{\Omega\mu/\rho} f,$$

$$\frac{\partial v_r}{\partial r} = \Omega f', \quad \frac{\partial v_\theta}{\partial r} = \Omega g, \quad \frac{\partial v_z}{\partial r} = 0,$$

$$\frac{\partial^2 v_r}{\partial r^2} = 0, \quad \frac{\partial^2 v_\theta}{\partial r^2} = 0, \quad \frac{\partial^2 v_z}{\partial r^2} = 0,$$

$$\frac{\partial v_r}{\partial z} = r\Omega\sqrt{\rho\Omega/\mu} f'', \quad \frac{\partial v_\theta}{\partial z} = r\Omega\sqrt{\rho\Omega/\mu} g', \quad \frac{\partial v_z}{\partial z} = -2\Omega f',$$

$$\frac{\partial^2 v_r}{\partial z^2} = \frac{\rho r\Omega^2}{\mu} f''', \quad \frac{\partial^2 v_\theta}{\partial z^2} = \frac{\rho r\Omega^2}{\mu} g'', \quad \frac{\partial^2 v_z}{\partial z^2} = -2\Omega\sqrt{\rho\Omega/\mu} f''.$$

Substituting the preceding results into the Navier-Stokes equations in cylindrical polar form gives

$$\frac{\partial p}{\partial r} = \rho r \Omega^2 \left[-(f')^2 + 2ff'' + g^2 + f''' \right],$$

$$0 = -2f'g + 2fg' + g'',$$

$$\frac{\partial p}{\partial z} = \rho \sqrt{\frac{\rho \Omega}{\mu}} \frac{1}{\rho} \frac{\partial p}{\partial \eta} = -2\rho \Omega \sqrt{\Omega \mu / \rho} \left(2ff' + f'' \right).$$
$$(5.4.38)$$

The no-slip boundary conditions at the wall give $v_z(0) = v_r(0) = 0$, or equivalently $f(0) = f'(0) = 0$, $g(0) = 1$. As η becomes large (far away from the wall), the viscous solution must approach the inviscid flow in having v_r and v_θ approach zero (equivalently, f' and g approach zero). Notice that the relative scalings of the various velocities and the velocity and pressure gradients are as previously noticed. Since the flow is entirely driven by the disk, the flow away from the disk is very weak. The disk is in fact acting as a rather inefficient centrifugal pump.

Elimination of the pressure in the preceding forms results in

$$f''' + 2ff'' - (f')^2 + g^2 = 0, \qquad (5.4.39)$$

$$g'' + 2fg' - 2gf' = 0, \qquad (5.4.40)$$

and

$$p = -2\rho \Omega (f^2 + f'). \qquad (5.4.41)$$

The numerical solution of this system of equations (5.4.39) and (5.4.40) subject to the boundary conditions is shown in Figure 5.4.6. and Table 5.4.3 The unknown derivatives at the wall are found to have the values $g'(0) = -0.6159$ and $f''(0) = 0.510$.

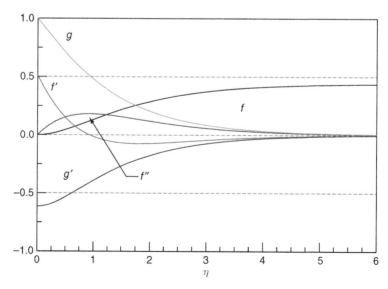

Figure 5.4.6 Von Kármán's solution for a rotating disk

TABLE 5.4.3   Rotating disk boundary layer (von Kármán)

η	f	f'	f''	g	$-g'$
0.0	0.0000	0.0000	0.5102	1.0000	0.6159
0.1	0.0024	0.0462	0.4163	0.9386	0.6112
0.2	0.0090	0.0836	0.3338	0.8780	0.5987
0.3	0.0189	0.1133	0.2620	0.8190	0.5803
0.4	0.0314	0.1364	0.1999	0.7621	0.5577
0.5	0.0459	0.1536	0.1467	0.7076	0.5321
0.6	0.0620	0.1660	0.1015	0.6557	0.5047
0.7	0.0790	0.1742	0.0635	0.6067	0.4763
0.8	0.0967	0.1789	0.0317	0.5605	0.4476
0.9	0.1147	0.1807	0.0056	0.5171	0.4191
1.0	0.1327	0.1802	−0.0157	0.4766	0.3911
1.1	0.1506	0.1777	−0.0327	0.4389	0.3641
1.2	0.1682	0.1737	−0.0461	0.4038	0.3381
1.3	0.1853	0.1686	−0.0564	0.3712	0.3133
1.4	0.2019	0.1625	−0.0640	0.3411	0.2898
1.5	0.2178	0.1559	−0.0693	0.3132	0.2677
1.6	0.2331	0.1487	−0.0728	0.2875	0.2470
1.7	0.2476	0.1413	−0.0747	0.2638	0.2276
1.8	0.2613	0.1338	−0.0754	0.2419	0.2095
1.9	0.2743	0.1263	−0.0751	0.2218	0.1927
2.0	0.2866	0.1188	−0.0739	0.2033	0.1771
2.2	0.3089	0.1044	−0.0698	0.1708	0.1494
2.4	0.3284	0.0910	−0.0643	0.1433	0.1258
2.6	0.3454	0.0788	−0.0580	0.1202	0.1057
2.8	0.3600	0.0678	−0.0517	0.1008	0.0888
3.0	0.3726	0.0581	−0.0455	0.0845	0.0745
3.0	0.3726	0.0581	−0.0455	0.0845	0.0745
3.2	0.3834	0.0496	−0.0397	0.0708	0.0625
3.4	0.3925	0.0422	−0.0343	0.0594	0.0525
3.6	0.4003	0.0358	−0.0296	0.0498	0.0440
3.8	0.4069	0.0303	−0.0253	0.0417	0.0369
4.0	0.4125	0.0257	−0.0216	0.0349	0.0309
4.2	0.4172	0.0217	−0.0184	0.0293	0.0259
4.4	0.4212	0.0183	−0.0156	0.0245	0.0217
4.6	0.4246	0.0154	−0.0132	0.0205	0.0182
4.8	0.4274	0.0129	−0.0112	0.0172	0.0152
5.0	0.4298	0.0109	−0.0095	0.0144	0.0128

Since $f(\infty) = 0.4422$, the flow far above the plate is $v_z = -0.8845\sqrt{\nu\Omega}$. The boundary layer thickness, again constant since η is independent of r, is approximately

$$\delta = 5\sqrt{\nu/\Omega}. \tag{5.4.42}$$

Even though our result is for an infinite disk, we can find the moment over a finite radius R of one side of the disk by

$$M = -\int_0^R r\tau_{z\theta}|_{z=0} 2\pi r\, dr = -2\pi\Omega\sqrt{\rho\Omega/\mu}\, g'(0)\int_0^R r^3 dr = -0.5\pi\Omega R^4\sqrt{\rho\Omega/\mu}\, g'(0).$$

Since $-\pi g'(0) = 1.935$, we have for a moment coefficient

$$C_M = \frac{M}{\frac{1}{2}\rho\Omega^2 R^5} = \frac{1.935}{\sqrt{\rho R^2 \Omega/\mu}}. \tag{5.4.43}$$

In practice, the flow is found to be laminar for values of the Reynolds number $\rho R^2 \Omega/\mu$ less than 10^5, and turbulent for larger Reynolds numbers.

Problems—Chapter 5

5.1 Two fluids of different densities and viscosities are flowing down a plane with slope θ. The lower layer of fluid occupies the region $0 \leq y \leq a$ and has density and viscosity ρ_1 and μ_1, while the upper fluid occupies the region $a \leq y \leq b$ and has density and viscosity ρ_2 and μ_2. The lower fluid has the higher density. The surface at $y = b$ is open to the atmosphere. Find the pressure and velocity distributions.

5.2 A layer of viscous fluid flows down a vertical plate under the action of gravity. The density of the fluid varies linearly from ρ_1 at the wall to $\rho_2 < \rho_1$ at $y = b$, the free surface. Find the velocity distribution and the velocity at the free surface.

5.3 Find the flow in a rotating pipe of radius a. A pressure gradient is present. Take the flow to be axially symmetric and steady.

5.4 Verify that the velocity in a conduit with an elliptic cross-section with semimajor and semiminor axes a and b is given by

$$v_x = K\left(\frac{y^2}{a^2} + \frac{z^2}{b^2} - 1\right), \quad \text{where } K = \frac{1}{2\mu}\frac{\partial p}{\partial x}\frac{a^2 b^2}{a^2 + b^2}.$$

5.5 For flow in the conduit with an elliptic cross-section (problem 5.4), find the ratio b/a that gives the maximum flow rate for a given pressure gradient and cross-sectional area.

5.6 Verify that the laminar velocity in a pipe with the cross-section of an equilateral triangular of side b is given by

$$v_x = -\frac{\sqrt{3}}{6\mu b}\frac{\partial p}{\partial x} z\left(z + \sqrt{3}y - \frac{\sqrt{3}b}{2}\right)\left(z - \sqrt{3}y - \frac{\sqrt{3}b}{2}\right).$$

5.7 Using the solution for flow between concentric rotating cylinders, find the velocity distribution caused by a circular cylinder that is rotating in an infinite fluid. Show that this is the same as that due to a line vortex along the z-axis. What is the strength of this vortex?

5.8 Find the flow of a layer of viscous fluid on an oscillating plate ($U_0 \cos \Omega t$) at $y = 0$. The fluid is of constant thickness a; the surface at $y = a$ is stress-free.

5.9 A large disk is rotated sinusoidally at a frequency Ω in a semi-infinite fluid. Find the steady-state fluid velocity, assuming $\mathbf{v}(r, z, t) = (0, rf(z, t), 0)$.

5.10 A concentrated line vortex of the type discussed in Chapter 2 is suddenly introduced into a viscous fluid. The action of viscosity is to diffuse the vorticity. The problem to be solved for is thus the following:

$$\frac{\partial \omega}{\partial t} = \nu \left(\frac{\partial^2 \omega}{\partial r^2} + \frac{1}{r} \frac{\partial \omega}{\partial r} \right), \quad \omega(r, 0) = \begin{cases} 0 & \text{for } r > 0, \\ \Gamma & \text{for } r = 0, \end{cases} \quad \int_0^\infty \omega 2\pi r \, dr = \Gamma.$$

    a. Show the form of a similarity solution $\omega = \omega(r, t, \mu, \rho, \Gamma)$ that is a possible solution for this problem.

    b. Solve for the vorticity.

5.11 Show that the form $\omega(r, t) = A t^\alpha f(\eta)$, where $\eta = r \sqrt{\dfrac{\rho}{\mu t}}$, when inserted into the vorticity equation $\rho \dfrac{\partial \omega}{\partial t} = \mu \left(\dfrac{\partial^2 \omega}{\partial r^2} + \dfrac{1}{r} \dfrac{\partial \omega}{\partial r} \right)$ leads to an ordinary differential equation.

Chapter 6

The Boundary Layer Approximation

6.1 *Introduction to Boundary Layers*

As we saw in Chapter 5, solutions to the full Navier-Stokes equations are few in number and difficult to obtain. In the exact solutions of the Navier-Stokes equations, it was repeatedly seen that when a local Reynolds number was large, viscous effects are felt mainly in the immediate vicinity of a solid boundary. In 1904 Prandtl introduced an approximate form of the Navier-Stokes equations that holds in the thin boundary layer near the wall, where viscous effects are comparable to the inertia effects. Here, the relationship between the boundary layer equations and the full Navier-Stokes equations is investigated by demonstrating how the boundary layer equations can be derived as a limiting form of the Navier-Stokes equations.

To understand how inviscid flow theory and boundary layer theory fit with the Navier-Stokes equations, consider the following example, first presented by Friedrichs in 1942. His model equation is

$$\varepsilon \frac{d^2 f}{dx^2} + \frac{df}{dx} = a, \quad f(0) = 0, \quad f(1) = 1. \tag{6.1.1}$$

The second-order derivative can be thought of as the viscous terms with a small viscosity ε, the first derivative as a "momentum," and the constant a as a pressure force. When the highest-order derivative is neglected ("inviscid" approximation), the first-order equation that remains has the solution $f = ax + c$, c being a constant of integration. Clearly, only one of the boundary conditions can be satisfied. Our version of relaxing the "no-slip" condition would be to impose $f(1) = 1$, giving $c = 1 - a$. Thus, our "inviscid" solution is

$$f_1(x) = 1 + a(x - 1). \tag{6.1.2}$$

To take care of the unsatisfied boundary condition, recall that in all of the similarity solutions the coordinate was "stretched" by dividing by the small viscosity, giving a thin layer at the wall. For this example, let $\eta = x/\varepsilon$, so that our equation becomes

$$\frac{1}{\varepsilon} \frac{d^2 f}{d\eta^2} + \frac{1}{\varepsilon} \frac{df}{d\eta} = a.$$

The temptation now is to disregard the right-hand side and find

$$f = b + ce^{-\eta}.$$

To satisfy the boundary condition at $x = 0$, select $c = -b$, so

$$f_2 = b\left(1 - e^{-\eta}\right). \tag{6.1.3}$$

To "match" the two solutions, require next that there is some region where the two solutions overlap. In that region, write both solutions in one of the two variables (x and η), and then take a limit, the limit being either $x \to 0$ *or* $\eta \to \infty$, depending on which variable was chosen.

Frequently, it is simpler to choose η as the variable of choice. In that case our first solution is $f_1(x) = 1 + a(\varepsilon\eta - 1)$. To the lowest-order, $b = 1 - a$ and

$$f_2 = (1 - a)\left(1 - e^{-\eta}\right) \times \begin{cases} 1 + a(x - 1) & \text{away from the wall,} \\ (1 - a)\left(1 - e^{-\eta}\right) & \text{near the wall.} \end{cases} \tag{6.1.4}$$

Putting this in the context of fluid mechanics, to the lowest-order approximation as the Reynolds number becomes large ($\varepsilon = 1/\text{Re} \to 0$), a good approximation to the solution of the Navier-Stokes equations is to solve the Euler equations for the slip velocity on the boundary and use Prandtl's boundary layer approximation for the flow near the wall.

Other forms of boundary layers exist as well. As we will see in Chapter 8, slow flows at low Reynolds numbers also require similar handling. Boundary layers can also exist as shear layers in the interior of a region—for example, the Gulf Stream where it departs the U.S. coast and crosses the Atlantic Ocean. Other areas of physics exhibit similar phenomena, such as the behavior of the free sides of a thin elastic plate when it is bent and current flowing in a solid wire at high frequencies. For more detailed explanations of the matching process, see Lagerstrom and Cole (1955), Goldstein (1960), and Van Dyke (1964).

6.2   The Boundary Layer Equations

In all of our exact solutions of the Navier-Stokes equations it was seen that the pressure gradient along a wall was of greater magnitude than that of the pressure gradient perpendicular to the wall, and that the viscous terms involving second derivatives along the wall were smaller than those involving derivatives taken perpendicular to the wall. The continuity equation was always satisfied in full, whereas the momentum equation in the direction perpendicular to the wall introduced only very low orders of magnitude terms.

To help in deriving the boundary layer equations, scale the various terms in the Navier-Stokes equations so that these orders of magnitude hold true, particularly when the Reynolds number is large. As a bookkeeping scheme it is convenient to build these notions as to the orders of magnitude of the various terms into our dimensionalization so that these orderings appear automatically. To do this, introduce dimensionless coordinates

$$x = x_D/L, \quad y = y_D\sqrt{\mathrm{Re}}/L, \quad t = t_D/T, \tag{6.2.1}$$

with the Reynolds and Strouhal numbers given by

$$\mathrm{Re} = U_D L/\nu \quad \text{and} \quad \mathrm{St} = L/U_D T. \tag{6.2.2}$$

Here, x_D, y_D are the dimensional lengths along and perpendicular to a wall, U_D is a representative constant body or stream velocity, T is a time scale for unsteady effects, and L is a body length along the wall. The existence of constant L and U_D, and hence of a constant Re, is vital to what follows, as will be discussed later.

Since changes along the body occur on a length scale L, whereas those perpendicular to the body occur over a distance of the order of the (thin) thickness of the boundary layer $L/\sqrt{\mathrm{Re}}$, the biased dimensionalization introduced in equation (6.2.1) "stretches" the thin y_D by multiplying it by $\sqrt{\mathrm{Re}}$ to make the partial derivatives reflect their true orders of magnitude.

Introducing a dimensionless stream function (as done in the similarity solutions of the full Navier-Stokes equations) by

$$\psi = \rho\psi_D/\mu\sqrt{\mathrm{Re}}, \tag{6.2.3}$$

gives dimensionless velocities

$$v_x = \frac{\partial\psi}{\partial y} = \frac{1}{U_D}\frac{\partial\psi}{\partial y} = \frac{u_D}{U_D}, \tag{6.2.4}$$

$$v_y = -\frac{\partial\psi}{\partial x} = -\frac{\sqrt{\mathrm{Re}}}{U_D}\frac{\partial\psi}{\partial x} = \sqrt{\mathrm{Re}}\frac{v_D}{U_D}. \tag{6.2.5}$$

These dimensionless velocities bring out the physical fact that, at the outer edge of the boundary layer u_D will be of magnitude U_D, and v_D will be of the (small) magnitude $U_D/\sqrt{\mathrm{Re}}$. Our stretching of the coordinates, then, along with the nondimensionalization of ψ, has stretched the velocity components appropriately as well.

Making pressure dimensionless by

$$p = p_D/\rho U_D^2 \tag{6.2.6}$$

the dimensionless Navier-Stokes equations for incompressible flow with constant density and viscosity in two-dimensional Cartesian coordinates become

$$\mathrm{St}\frac{\partial v_x}{\partial t} + v_x\frac{\partial v_x}{\partial x} + v_y\frac{\partial v_x}{\partial y} = -\frac{\partial p}{\partial x} + \frac{\partial^2 v_x}{\partial y^2} + \frac{1}{\mathrm{Re}}\frac{\partial^2 v_x}{\partial x^2}, \tag{6.2.7}$$

and

$$\text{St}\frac{\partial v_y}{\partial t} + v_x\frac{\partial v_y}{\partial x} + v_y\frac{\partial v_y}{\partial y} = -\text{Re}\frac{\partial p}{\partial x} + \frac{\partial^2 v_y}{\partial y^2} + \frac{1}{\text{Re}}\frac{\partial^2 v_y}{\partial x^2}. \tag{6.2.8}$$

For large values of the Reynolds number, the limit of equations (6.2.7) and (6.2.8) is

$$\text{St}\frac{\partial v_x}{\partial t} + v_x\frac{\partial v_x}{\partial x} + v_y\frac{\partial v_x}{\partial y} = -\frac{\partial p}{\partial x} + \frac{\partial^2 v_x}{\partial y^2} \tag{6.2.9}$$

and

$$0 = -\frac{\partial p}{\partial y}. \tag{6.2.10}$$

Equation (6.2.10) states that the pressure gradient across the thin boundary layer is negligible, and thus the pressure gradient at the outer edge of the boundary layer, as obtained from the Euler equations, can be used. Since the inviscid velocity component perpendicular to a wall vanishes, this means that the Euler equations at the wall are

$$-\frac{\partial p}{\partial x} = \text{St}\frac{\partial U}{\partial t} + U\frac{\partial U}{\partial x}, \tag{6.2.11}$$

where U is the velocity along the wall (made dimensionless by U_D) as predicted from inviscid theory. Thus, in equation (6.2.9) the pressure gradient is known, and, since v is related to u through continuity, there is only one equation in one unknown to be solved. The appropriate boundary conditions are the no-slip conditions at the wall,

$$v_x = v_y = 0 \quad \text{at} \quad y = 0, \quad u = v = 0 \quad \text{at} \quad y = 0, \tag{6.2.12}$$

and a joining to the inviscid velocity—that is,

$$v_x \to U \tag{6.2.13}$$

at the outer edge of the boundary layer.

Having developed the boundary layer equations in dimensionless form, usually it is more convenient when doing problems to start with the dimensional form. This is, in two dimensions with $\mathbf{v} = (u, v, 0)$,

$$\frac{\partial u}{\partial t} + u\frac{\partial u}{\partial x} + v\frac{\partial u}{\partial y} = \frac{\partial U}{\partial t} + U\frac{\partial U}{\partial x} + \nu\frac{\partial^2 u}{\partial y^2}, \tag{6.2.14}$$

$$\frac{\partial p}{\partial y} = 0, \tag{6.2.15}$$

and

$$\frac{\partial u}{\partial x} + \frac{\partial v}{\partial y} = 0, \tag{6.2.16}$$

with $u = v = 0$ on the wall and u approaches U at the outer edge of the boundary layer.

Mathematically, the outer edge of the boundary layer can be defined as being at an infinite value of dimensionless y, since in the stretching of the y coordinate the Reynolds number has been made large. In practice, a value for dimensionless y of five or so can often be regarded as infinity for practical purposes, since, as has been seen in the previous similarity solutions, u will not vary much for values of y beyond that point.

Our boundary layer equations have been derived for a flat boundary. For curved boundaries, providing x is regarded as being locally tangent to the boundary and y as being locally normal to the boundary, the equations still hold, since curvature effects are of higher order unless the curvature is extreme.

In approximating the Navier-Stokes equations by the boundary layer equations, the mathematical character of the equations have changed. The original Navier-Stokes equations are of elliptic nature, while the boundary layer equations are parabolic. (The boundary layer equations are sometimes referred to as the ***parabolized Navier-Stokes equations***.) Equations of parabolic type have solutions that can be "marched" in the *x* direction, and the numerical methods appropriate for parabolic equations are quite different from those used for elliptic equations. The implication now is that upstream conditions completely determine downstream behavior. When this is not true, for example, where a flow starts to separate, the boundary layer assumption breaks down.

6.3 *Boundary Layer Thickness*

In Chapter 5 where exact solutions were being considered, boundary layer thickness was established by first finding the solution and then considering for what value of η the horizontal velocity component approached some percentage of the outer velocity. For both theoretical and experimental purposes, it is preferable to have a less subjective definition of "thickness." Two such definitions have been found to be useful.

The ***displacement thickness*** of the boundary layer is defined by

$$\delta_D = \frac{1}{U} \int_0^\delta (U - u)dy = \text{displacement thickness,} \tag{6.3.1}$$

where δ is a value "large enough" so that choosing a slightly larger value would not significantly affect the value of δ_D. Looking at Figure 6.4.1 (on page 177), we can see that the integral represents the area $U\delta$ minus the actual discharge in the layer of thickness δ. As the value of δ is increased, there comes a point where no significant area is added to the integral, so whether 90%, 99%, or evaluation by eye of experimental data is used, the value of the displacement thickness remains virtually unchanged.

Another interpretation of the displacement thickness is that because the volumetric flow in the boundary layer is actually $\int_0^\delta u\,dy$, the same flow would be achieved if the wall was displaced upward into the flow in an amount δ_D, since the definition can be written in the form

$$(\delta - \delta_D)U = \int_0^\delta u\,dy.$$

Similarly, a ***momentum thickness*** definition has been found to be useful. It is given by

$$\delta_M = \frac{1}{U^2} \int_0^\delta u(U - u)dy = \text{momentum thickness.} \tag{6.3.2}$$

Again, its value is relatively insensitive to increases in δ as long as a large enough δ is chosen. Its physical interpretation is a bit more difficult to explain than the previous definition, as it involves the momentum difference between the actual flow and an inviscid one with the same discharge, as follows from

$$U \int_0^\delta u\,dy - U^2 \delta_M = \int_0^\delta u^2 dy.$$

As we will see in Section 6.5, both these quantities arise naturally when the boundary layers are integrated over the boundary layer thickness.

6.4   Falkner-Skan Solutions for Flow Past a Wedge

To illustrate the behavior of the boundary layer equations, next consider a solution of similarity type first credited to Falkner and Skan (1930). This is a solution where U is proportional to x^m. The irrotational flow that corresponds to this is the flow past a wedge of angle radians with a complex potential and velocity given by

$$w = \phi + i\psi = \frac{Az^{m+1}}{m+1} = \frac{Ar^{m+1}\left[\cos(m+1)\,\theta + i\sin(m+1)\,\theta\right]}{m+1}, \tag{6.4.1}$$

$$\frac{dw}{dz} = u - iv = Ar^m\left(\cos m\theta + i\sin m\theta\right). \tag{6.4.2}$$

Evaluating equation (6.4.1) on the surface of the wedge ($\theta = 0$) gives $U = Ax^m$. From the Euler equations,

$$-\frac{\partial p}{\partial x} = \rho U\,\frac{dU}{dx} = \rho m A^2 x^{2m-1}.$$

For the similarity form of the solution use

$$\psi = \frac{\mu}{\rho}\sqrt{\rho Ux/\mu}f(\eta), \quad \text{with } \eta = y\sqrt{\rho U/x\mu}$$

as in the full Navier-Stokes equations. Then with

$$\frac{\partial}{\partial x} = \frac{m-1}{2}\frac{\eta}{x}\frac{d}{d\eta}, \quad \frac{\partial}{\partial y} = \sqrt{\frac{U}{\nu x}}\frac{d}{d\eta},$$

from differentiation of ψ find that

$$u = Uf', \quad v = -\frac{1}{2}\sqrt{\frac{U\mu}{\rho x}}\left[(m+1)f + \eta(m-1)f'\right],$$

$$\frac{\partial u}{\partial x} = \frac{U}{x}\left[mf' + 0.5(m-1)\eta f''\right], \quad \frac{\partial u}{\partial y} = U\sqrt{\rho U/\mu x}f'', \quad \frac{\partial^2 u}{\partial y^2} = \frac{\rho U^2}{\mu x}f'''.$$

Substituting all of these into the boundary layer equations, the result is

$$m\,(f')^2 - 0.5(m+1)ff'' = m + f'''.$$

The terms on the left-hand side result from the convected acceleration, the constant on the right side results from the pressure gradient, and the third derivative is the viscous stress gradient. Rearranging to put all terms on the left hand side, the result is

$$f''' + 0.5(m+1)ff'' + m[1 - (f')^2] = 0. \tag{6.4.3}$$

Equation (6.4.3) is to be solved together with the no-slip boundary conditions

$$f(0) = f'(0) = 0 \tag{6.4.4}$$

and the outer condition

$$f' \text{ approaches 1 as } \eta \text{ becomes large.} \tag{6.4.5}$$

The shear stress at the wall is

$$\tau_{\text{wall}} = \mu\left.\frac{\partial u}{\partial y}\right|_{y=0} = \mu U\sqrt{\rho U/x\mu}f''(0). \tag{6.4.6}$$

Consider next several special cases of equation (6.4.6).

6.4.1 Boundary Layer on a Flat Plate

The first solution of equation (6.4.3) was carried out by Blasius (1908) for a semi-infinite flat plate, where $m = 0$. Then equation (6.4.3) reduces to

$$f''' + 0.5ff'' = 0. \tag{6.4.7}$$

Equation (6.4.7) has the minimum number of terms that equation (6.4.3) can have for a boundary layer flow, but still no exact solution is possible. A numerical solution must be resorted to, and since boundary conditions must be applied at both the wall and the outer edge of the boundary layer, generally our choice is either to use a shooting method (first guess $f''(0)$, then integrate out from the wall to see if f' approaches unity, then iterate on this procedure) or use a two-point method that has the boundary conditions at both ends included. Blasius found that $f''(0) = 0.332$.

For this special problem only, numerical integration using a shooting method can be simplified. First, make the change of variables $\zeta = c\eta$ and $f = cF(\zeta)$ so that equation (6.4.7) is replaced by

$$F''' + 0.5FF'' = 0, \tag{6.4.8}$$

where this time a prime is a derivative with respect to ζ.

Set $F''(0)$ arbitrarily to unity, and integrate to find that F' approaches some value b at large ζ. Then, letting $c = 1/\sqrt{b}$ will make f' approach unity as ζ becomes large, completing the desired solution. The results of such a numerical integration are shown in Table 6.4.1 and Figure 6.4.1.

Several interesting features of boundary layer flows can be seen from the results of this problem.

1. Since the outer edge of the boundary layer is at some constant value of η, the exact value depending on how the boundary layer thickness has been defined, then $\delta \sim x/\sqrt{\rho U x/\mu}$.

2. As η becomes large, f approaches the asymptotic value $\eta - 1.7208$. Thus, the displacement thickness is

$$\delta_D = \frac{1}{U}\int_0^\infty (U - u)dy = 1.7208x\sqrt{\mu/\rho U x}.$$

3. The shear stress is proportional to $x^{-1/2}$, so the shear stress decreases going downstream from the plate leading edge. This decrease is due to the thickening of the boundary layer, with a resulting decrease in velocity gradient.

4. The drag force per unit width over a length L of the plate is

$$F = \int_0^L \tau_{\text{wall}}dx = 2\mu U f''(0)\sqrt{\rho U L/\mu}.$$

5. From the asymptotic value of f, the stream function is seen to approach

$$\psi = Uy - 1.7208\sqrt{Ux\mu/\rho} + \cdots.$$

Thus, the displacement thickness effect does not result in an irrotational flow away from the wall.

TABLE 6.4.1   Flat plate boundary layer (Blasius)

η	f	f'	f''	η	f	f'	f''
0.0	0.0000	0.0000	0.3320	2.0	0.6500	0.6298	0.2668
0.1	0.0017	0.0332	0.3320	2.2	0.7812	0.6813	0.2484
0.2	0.0066	0.0664	0.3320	2.4	0.9223	0.7290	0.2281
0.3	0.0149	0.0996	0.3318	2.6	1.0725	0.7725	0.2065
0.4	0.0266	0.1328	0.3315	2.8	1.2310	0.8115	0.1840
0.5	0.0415	0.1659	0.3309	3.0	1.3968	0.8460	0.1614
0.6	0.0597	0.1989	0.3301	3.2	1.5691	0.8761	0.1391
0.7	0.0813	0.2319	0.3289	3.4	1.7469	0.9018	0.1179
0.8	0.1061	0.2647	0.3274	3.6	1.9295	0.9233	0.0981
0.9	0.1342	0.2974	0.3254	3.8	2.1160	0.9411	0.0801
1.0	0.1656	0.3298	0.3230	4.0	2.3057	0.9555	0.0642
1.1	0.2002	0.3619	0.3201	4.2	0.9670	2.4980	0.0505
1.2	0.2379	0.3938	0.3166	4.4	2.6924	0.9759	0.0390
1.3	0.2789	0.4252	0.3125	4.6	2.8882	0.9827	0.0295
1.4	0.3230	0.4563	0.3079	4.8	3.0853	0.9878	0.0219
1.5	0.3701	0.4868	0.3026	5.0	3.2833	0.9915	0.0159
1.6	0.4203	0.5168	0.2967	5.5	3.7806	0.9969	0.0066
1.7	0.4735	0.5461	0.2901	6.0	4.2796	0.9990	0.0024
1.8	0.5295	0.5748	0.2829	6.5	4.7793	0.9997	0.0008
1.9	0.5884	0.6027	0.2751	7.0	5.2792	0.9999	0.0002

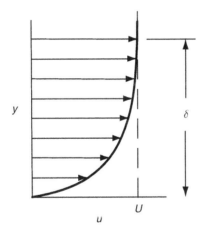

Figure 6.4.1   Blasius's solution for boundary layer flow on a semi-infinite plate

6. The shear stress is infinite for *all* values of y at $x = 0$. We could be content with this if it were infinite only at the leading edge, but it is perhaps surprising that it should be singular along a line perpendicular to the plate. Also, a solution does not exist for negative x, for in that case η becomes imaginary.

The criticisms raised in the last two points may seen to be minor ones, but they are needless errors that have in fact been introduced by our choice of a coordinate system. In performing any mathematical solution, it is always best to use a ***natural coordinate system*** (Kaplun, 1954)—that is, one where setting one of the coordinates to a constant value yields the desired boundary. For our choice of a coordinate system, the boundary is at $y = 0$ for $x > 0$.

Having to qualify the value of x by saying that it must be positive means that this is not a natural coordinate system for this problem. A coordinate system that is a natural coordinate system for the semi-infinite flat plate is a parabolic system, given by

$$\xi^2 = x + \sqrt{x^2 + y^2}, \quad \eta^2 = -x + \sqrt{x^2 + y^2}, \tag{6.4.9}$$

with the inverse relations

$$x = \frac{\xi^2 - \eta^2}{2}, \quad y = \xi\eta. \tag{6.4.10}$$

Use of these coordinates eliminates both problems raised in points e and f, leaves equation (6.4.7) unchanged so the problem does not have to be re-solved, and the parabolic η behaves as the η originally introduced, since far away from the leading edge along the plate it is seen that $\xi \simeq \sqrt{2x}$, $\eta \simeq y/\sqrt{2x}$.

You may have noticed another point that involves general principles in what has been done. In deriving the boundary layer approximation in the first place, it was stated that there was a definite boundary length scale L upon which our Reynolds number was based. In the Falkner-Skan problems, however, there is no such L, and our results come out with a Reynolds number based on the local distance x. The distinction is perhaps subtle, but it is important as far as an understanding of the implications is concerned. The question was not answered until the 1960s, and the lack of attention to the problem resulted in some confusing and erroneous publications prior to that time.

When there is a fixed-length scale L in a problem, what is being done in the boundary layer approximation is expanding the solution to the Navier-Stokes equations in terms of negative powers (and perhaps logarithms) of the Reynolds number. This is then an expansion in the constant parameter Re. Near a boundary, the first term in the expansion is the boundary layer equation.

Away from boundaries, the first term in the expansion is the Euler equation. Presumably, it is possible to go to higher-order terms in the expansion (albeit with increasing difficulty and work) and still have the solutions match at the overlap of the Euler and boundary layer regions.

When there is no geometric length scale L, the expansion is in terms of a coordinate such as x, but it may not be able to find solutions in the various regions that mathematically match together properly, although they might numerically join together satisfactorily. The question of how to obtain higher-order approximations is thus left unanswered, and we may have to be satisfied with just the lowest-order approximation. (We usually settle for this in any case, but it would be nice to at least know that higher-order approximations *could* be carried out if we wanted them.) A more thorough discussion of this matter can be found in Van Dyke (1964).

6.4.2 Stagnation Point Boundary Layer Flow

When $\beta = \pi$, the inviscid flow becomes flow perpendicular to a flat plate, and our equation reduces to that found in the full solution of the Navier-Stokes equations. In this case, then, the boundary layer solution is an exact solution of the Navier-Stokes equations for the velocity (but not the pressure).

6.4.3 General Case

Values for $f''(0)$ are given in Table 6.4.2 for a number of wedge angles. The last 18 sets of values in were computed by Hartree in 1937. He found that if $\beta > 0$, the

TABLE 6.4.2 $f''(0)$ as a function of β and m

β		m	$f''(0)$
radians	degrees		
−0.56549	−32.4	−0.08257	−0.0657
−0.47124	−27	−0.06977	−0.09002
−0.31416	−18	−0.04762	−0.0973
−0.15708	−9	−0.02439	−0.07543
−0.07854	−4.5	−0.01235	−0.052
−0.62202	−35.64	−0.09008	0
−0.59692	−34.2	−0.08676	0.05811
−0.5655	−32.4	−0.08257	0.0871
−0.50262	−28.8	−0.07407	0.12928
−0.43982	−25.2	−0.06542	0.16338
−0.31417	−18	−0.04762	0.22013
0	0	0	0.33206
0.31415	18	0.05263	0.42585
0.62831	36	0.11111	0.51131
0.94248	54	0.17647	0.59363
1.25664	72	0.25	0.67586
1.57078	90	0.33333	0.75689
1.88495	108	0.42857	0.84177
2.51312	144	0.66667	1.0224
3.14159	180	1	1.2326
3.76991	216	1.5	1.49669
5.02655	288	4	2.40491
6.28318	360	∞	∞

solution of equation (6.4.7) was unique and that for $\beta > 2.0$, the velocity components are imaginary. He carried the solution out for $\beta = 2.4$ to facilitate interpolation and for values of $\beta < 0$ for which f' approached one exponentially from below.

Stewartson (1954) computed the first five values in the table. These flows show reversed flow regions near the wall and also demonstrate that for a given value of m, the solution of equation (6.4.3) is not necessarily unique.

The fact that solutions of the Falkner-Skan equation may not be unique is interesting in itself. It also is a caution that simply formulating a similarity solution in terms of a nonlinear ordinary differential equation along with appropriate boundary conditions is not the end of the story. At least one solution must be found and then investigated over at least some useful parameter range so that there is reasonable cause to believe that there are no other solutions for those parameters.

6.5   The Integral Form of the Boundary Layer Equations

Next, an integral form for the boundary layer approximation will be derived by integrating the boundary layer equation. First, rearrange equation (6.2.14) in the form

$$\rho\left[\frac{\partial(U-u)}{\partial t} + U\frac{\partial U}{\partial x} - u\frac{\partial u}{\partial x} - v\frac{\partial u}{\partial y}\right] = -\mu\frac{\partial^2 u}{\partial y^2}. \qquad (6.5.1)$$

From integration of the continuity relation equation (6.2.16), find that

$$v_\delta = -\int_0^\delta \frac{\partial u}{\partial x} dy. \tag{6.5.2}$$

Integrating equation (6.5.1) with respect to y and using the continuity equation and the product rule of calculus, the result is

$$\frac{\tau_{\text{wall}}}{\rho} = \int_0^\delta \left[\frac{\partial (U-u)}{\partial t} + U\frac{\partial U}{\partial x} - u\frac{\partial u}{\partial x} - v\frac{\partial u}{\partial y} \right] dy$$

$$= \frac{\partial (U\delta_D)}{\partial t} + \int_0^\delta \left(U\frac{\partial U}{\partial x} - u\frac{\partial u}{\partial x} - \frac{\partial (uv)}{\partial y} + u\frac{\partial v}{\partial y} \right) dy$$

$$= \frac{\partial (U\delta_D)}{\partial t} + \int_0^\delta \left[U\frac{\partial U}{\partial x} - 2u\frac{\partial u}{\partial x} + u\left(\frac{\partial u}{\partial x} + \frac{\partial v}{\partial y} \right) \right] dy - Uv|_\delta .$$

Using equation (6.5.2) to bring the last term back under the integral sign gives

$$\frac{\tau_{\text{wall}}}{\rho} = \frac{\partial (U\delta_D)}{\partial t} + \int_0^\delta \left[U\frac{\partial U}{\partial x} - \frac{\partial u^2}{\partial x} + U\frac{\partial u}{\partial x} \right] dy$$

$$= \frac{\partial (U\delta_D)}{\partial t} + \int_0^\delta \left[U\frac{\partial U}{\partial x} - \frac{\partial u^2}{\partial x} + \frac{\partial (Uu)}{\partial x} - u\frac{\partial U}{\partial x} \right] dy \tag{6.5.3}$$

$$= \frac{\partial (U\delta_D)}{\partial t} + \int_0^\delta \left\{ \frac{\partial U}{\partial x} (U-u) + \frac{\partial}{\partial x} [u(U-u)] \right\} dy$$

$$= \frac{\partial (U\delta_D)}{\partial t} + U\frac{\partial U}{\partial x}\delta_D + \frac{\partial (U^2\delta_M)}{\partial x}.$$

To use equation (6.5.3) to estimate the nature of the boundary layer a form for $u(y)$ must be chosen. For illustration of the method, choose first a linear approximation,

$$u(y) = \begin{cases} U\dfrac{y}{\delta}, & 0 \le y \le \delta, \\ U, & y \ge \delta. \end{cases}$$

Then, $\delta_D = 0.5\delta$, $\delta_D = 0.167\delta$, and $\tau_{\text{wall}} = \mu U/\delta$. Inserting these into equation (6.5.3) gives

$$\frac{\mu U}{\rho\delta} = \frac{\partial}{\partial t} (0.5U\delta) + 0.5U\delta\frac{\partial U}{\partial x} + \frac{\partial}{\partial x} \left(0.167U^2\delta \right).$$

Multiplying this by δ gives

$$\frac{\mu U}{\rho} = \frac{\partial}{\partial t} \left(0.25U\delta^2 \right) + 0.5U\delta^2\frac{\partial U}{\partial x} + \frac{\partial}{\partial x} \left(0.0835U^2\delta^2 \right).$$

Provided U is known, this is a linear equation in δ^2. For the case of the steady flow past a flat plate it reduces to $\mu U/\rho = d/dx \left(0.0835U^2\delta^2 \right)$, giving $\delta = \sqrt{\mu x/0.0835\rho U} = 3.46x\sqrt{\mu/\rho xU}$.

We can see that this has the correct Reynolds number dependency. Using this to determine the shear stress gives

$$\tau_{\text{wall}} = \frac{\mu U}{3.46\sqrt{\mu x/\rho U}} = 0.289\rho U^2 \sqrt{\frac{\mu}{\rho Ux}}.$$

Again, the Reynolds number behavior is correct, and the 0.289 is only 13% apart from the Blasius value of 0.332.

The linear approximation used here is, of course, unnecessarily crude. Other profiles that could have been used follow, where $\eta = y/\delta$:

Parabolic: $u = \begin{cases} U\left(2\eta - \eta^2\right), & 0 \leq \eta \leq 1, \\ U, & \eta \geq 1, \end{cases}$

Sinusoidal: $u = \begin{cases} U \sin \dfrac{\pi \eta}{2}, & 0 \leq \eta \leq 1, \\ U, & \eta \geq 1, \end{cases}$ (6.5.4)

Pohlhausen: $u = \begin{cases} U\left[2\eta - 2\eta^3 + \eta^4 + 2\Lambda\left(\eta - 3\eta^2 + 3\eta^3 - \eta^4\right)\right], & 0 \leq \eta \leq 1, \\ U, & \eta \geq 1. \end{cases}$

The results for the flat plate are given in Table 6.5.1.

The Pohlhausen profile[1] (1921) was selected to agree with all solutions that were known by 1921. Pohlhausen used the following boundary conditions in the selection process:

$$\text{At } \eta = 0, \quad u = 0, \quad \mu \frac{\partial^2 u}{\partial y^2} = -\rho U \frac{dU}{dx}. \quad \text{At } \eta = 1, \quad u = U, \quad \frac{\partial u}{\partial y} = 0, \quad \frac{\partial^2 u}{\partial y^2} = 0.$$

The parameter $\Lambda = \frac{\rho \delta^2}{\mu} \frac{dU}{dx}$ must lie in the range $-12 \leq \Lambda \leq 12$, since at $\Lambda = -12$ the shear stress becomes zero at the wall, implying flow separation, and at $\Lambda = 12$ becomes greater than U within the boundary layer.

If the Kármán-Pohlhausen approximation is inserted into our equations, the result is

$$\delta_D = \delta \left(\frac{3}{10} - \frac{\Lambda}{120}\right) \delta_M = \frac{\delta}{63}\left(\frac{37}{5} - \frac{\Lambda}{15} - \frac{\Lambda^2}{144}\right) \text{ and } \tau_{\text{wall}} = 2\mu \frac{U}{\delta}\left(1 + \frac{\Lambda}{12}\right). \quad (6.5.5)$$

TABLE 6.5.1   Comparison of various boundary layer profiles with the Blasius results

Velocity Profile	$\dfrac{\delta_D}{\delta}$	$\dfrac{\delta_M}{\delta}$	$\dfrac{\delta}{\sqrt{\nu x/U}}$	$\dfrac{\tau_{\text{wall}}}{\rho U \sqrt{U\nu/x}}$	Error in τ_{wall}
Linear	$1/2 = 0.5$	$1/3 = 0.167$	$\sqrt{12} = 3.46$	$1/\sqrt{12} = 0.288$	13%
Parabolic	$1/3 = 0.333$	$2/15 = 0.133$	$\sqrt{30} = 5.48$	$\sqrt{2/15} = 0.365$	9.9%
Sinusoidal	$1 - 2/\pi = 0.363$	$2/\pi - 1/2 = 0.137$	$\pi\sqrt{2/(4-\pi)} = 4.8$	$\sqrt{2 - \pi/2}/2 = 0.328$	1.5%
Kármán-Pohlhausen	0.3	$37/315 = 0.117$	$\sqrt{1260/37} = 5.836$	$\sqrt{37/315} = 0.343$	3.3%
Blasius	0.344	0.1328	5.0	0.332	0%

---

[1] Pohlhausen was acting on a suggestion by von Kármán (1921), so the method is usually referred to as the Kármán-Pohlhausen method.

Putting these into the integrated Navier-Stokes equations results (after a reasonable amount of algebraic manipulation) in

$$\frac{U}{\nu}\frac{d\delta_M^2}{dx} = H(K), \text{ where}$$

$$F(K) = \frac{\delta_D}{\delta_M} = \frac{63\,(3 - \Lambda/12)}{10\,(37/5 - \Lambda/15 - \Lambda^2/144)},$$

$$G(K) = \frac{\delta_M}{U}\frac{\partial u}{\partial y}\bigg|_{\text{wall}} = \frac{2}{63}\left(1 + \frac{\Lambda}{12}\right)\left(\frac{37}{5} - \frac{\Lambda}{15} - \frac{\Lambda^2}{144}\right), \qquad (6.5.6)$$

$$H(K) = 2G(K) - 2K\left[2 + F(K)\right],$$

$$K = \frac{\delta_M^2}{\nu}\frac{dU}{dx} = \frac{1}{3,969}\left(\frac{37}{5} - \frac{\Lambda}{15} - \frac{\Lambda^2}{144}\right)^2 \Lambda.$$

While this looks formidable, plotting H versus K shows that over a good range the relationship is practically linear and given by $H(K) = 0.47 - 6K$. Thus,

$$\frac{\rho U}{\mu}\frac{d\delta_M^2}{dx} \approx 0.47 - 6K, \quad \text{or} \quad \frac{\rho}{\mu U^5}\frac{d\left(U^6\delta_M^2\right)}{dx} \approx 0.47.$$

After integration of this it becomes

$$\delta_M^2 = \delta_M^2\big|_0 + \frac{0.47}{U^6}\int_0^x U^5 dx. \qquad (6.5.7)$$

As we will see in the next section, this result is useful in predicting separation.

Because of the present day availability of ready access to powerful personal computers, the integral formulation of the boundary layer equations is no longer as important as it once was. However, the integrated form of the boundary layer is still convenient for providing an approximate solution of the boundary layer equations with a minimum amount of labor with reasonable accuracy. It is also useful for providing starting values for the wall shear when more accurate numerical methods are used.

6.6 Axisymmetric Laminar Jet

Using the boundary layer equations, it is also possible to solve for a laminar jet, similar to Squire's solution in Section 5.4.5. Working in cylindrical polar coordinates, let

$$\psi = \frac{4\mu z}{\rho}f(\eta) \quad \text{with } \eta = \frac{1}{8}\frac{r}{z}\left(\frac{3\rho M}{\pi\mu^2}\right)^{1/2}.$$

Here, M is the momentum of the jet given by

$$M = \rho\int_0^\infty w^2 2\pi r\,dr. \qquad (6.6.1)$$

The boundary layer momentum equation in the z direction then becomes, after a bit of rearranging,

$$\frac{d}{d\eta}\left(\frac{4f}{\eta}\frac{df}{d\eta}\right) = \frac{d}{d\eta}\left(\frac{1}{\eta}\frac{df}{d\eta} - \frac{d^2f}{d\eta^2}\right) \qquad (6.6.2)$$

with the boundary conditions

$$f(0) = 0, \quad \left.\frac{df}{d\eta}\right|_0 = 0, \quad \left.\frac{df}{d\eta}\right|_\infty = 0. \tag{6.6.3}$$

Integrating equation (6.6.2) gives $4f\frac{df}{d\eta} = \frac{df}{d\eta} - \eta\frac{d^2f}{d\eta^2}$. Since $4f\frac{df}{d\eta} = 2\frac{df^2}{d\eta}$, this can be integrated to give

$$2f^2 = 2f - \eta\frac{df}{d\eta} + C_1.$$

From the boundary conditions the constant of integration $C_1 = 0$. Then $\frac{df}{-f^2+f} = 2\frac{d\eta}{\eta}$. This can be integrated to give $f(\eta) = \frac{C_2\eta^2}{1+C_2\eta^2}$. Using equation (6.6.1) find $C_2 = 1$, so finally

$$f(\eta) = \frac{\eta^2}{1+\eta^2} \tag{6.6.4}$$

and

$$u = \frac{1}{2}\left(\frac{3M}{\pi\rho z^2}\right)^{1/2}\frac{\eta\left(1-\eta^2\right)}{\left(1+\eta^2\right)^2}, \quad w = \frac{3M}{8\pi\rho\nu z}\frac{1}{\left(1+\eta^2\right)^2}. \tag{6.6.5), \quad (6.6.6}$$

6.7   *Flow Separation*

When the pressure gradient along a body surface is negative (***favorable pressure gradient***), the pressure acts to locally accelerate the flow along the body and to overcome the viscous forces that act to decelerate. However, when the pressure gradient becomes positive (***adverse pressure gradient***), both pressure and viscous stresses act to decelerate the flow, and the boundary layer thickness rapidly increases. The flow soon reverses $\left(\left.\frac{\partial u}{\partial y}\right|_{\text{wall}} < 0\right)$, and ***flow separation*** is said to occur.

The point at which the flow reversal starts is called the separation point and is usually a short distance downstream from the point where the pressure gradient changes sign. Boundary layer theory is capable of predicting where the separation point is located but is not able to predict the flow past this point.

Thwaites (1949) correlated a number of solutions of the integrated boundary layer equations and found that separation could be closely predicted by the result $\delta_M^2 = -\frac{0.09\mu}{\rho dU/dx}$. Inserting this into equation (6.5.7) gives

$$-\frac{0.09\mu}{\rho dU/dx} = \frac{0.47}{U^6}\int_0^x U^5 dx, \tag{6.7.1}$$

where the integration has been started from a stagnation point.

A second criteria suggested by Stratford (1954) suggests that separation will occur at the point where

$$(x - x_0)^2 c_p \left(\frac{dc_p}{dx}\right)^2 = 0.0104, \quad \text{where } c_p = 1 - U^2/U_{\text{max}}^2. \tag{6.7.2}$$

Both criteria require knowledge of U at the boundary. The Stratford criterion is somewhat easier to use in that it doesn't involve an integration.

Even though the Thwaites and Stratford criteria are for two-dimensional flows, they are frequently used in three-dimensions in situations where the flow is predominately two-dimensional in behavior.

Example 6.7.1 Calculation of the separation point on a circular cylinder in a uniform stream.

Schmidt and Wenner (1941) found experimentally that a good approximation to the laminar surface flow on a circular cylinder of radius a was given by

$$U = U_0 \left(2s - 0.451s^3 - 0.00578s^5\right), \quad \text{where}$$

$s = x/a$ and x is the distance along the circumference of the cylinder.

Find the predicted separation point using the Stratford and Thwaite criteria.

Solution using Stratford's criterion. Taking the derivative of U, find that U has a maximum where $2 - 1.353s^2 - 0.0289s^4 = 0$. By the quadratic formula, find that root to be $s_{max} = 1.19765$, giving $U_{max} = 1.6063U_0$. This is 68.62 degrees from the stagnation point. Then

$$(s - s_{max})^2 \left[1 - (U/U_{max})^2\right] \left[\frac{-2U}{U_{max}^2} \frac{dU}{dx}\right]^2 - 0.0104 = 0.$$

By successively substituting values of s greater than s_{max} find the value $s = 1.447$, which corresponds to a separation point at 82.97 degrees from the stagnation point.

Solution using Thwaite's criterion. Multiplying U to the fifth power and then carrying out the integration in equation (6.7.1) is tedious, to say the least. Instead the integration is best carried out numerically using something like the Newton-Cotes formula. A program for doing this is given in Chapter 10. Separation is predicted at $s = 1.576$, corresponding to an angle of 90.30 degrees from the stagnation point.

These criteria both assume that the separation point does not move in time. In reality, the region beyond the separation points usually develop vortices and a wake that contains a vortex street as modeled by von Kármán (Section 3.8). If the flow is started gradually from rest, the vortices may first align side by side. Soon, however, they switch to the staggered position, as found in the von Kármán vortex street. As a vortex grows and is shed, the separation point can move back and forth. As yet, no simple method is available for predicting this analytically. For the most part, description of wakes appears to be best done by numerical methods.

Even wakes with minimal large-scale vortices have proven difficult for analytical methods. In 1930 Goldstein attempted an expansion solution of the wake behind a flat plate. Some progress was made, but the formidable computations required had to be done by hand, and progress was small. In two papers in 1968 and 1969, Stewartson elaborated further on the structure of the flow near the trailing edge of the plate and found that while Goldstein's viscous solution held in a very thin layer in the rear of the body, there was also a thin inviscid layer containing vorticity outside the Goldstein layer before inviscid irrotational flow was achieved. Complexity thus reigns in wake flow!

6.8 Transformations for Nonsimilar Boundary Layer Solutions

The solutions of the boundary layer equations presented so far are limited to very special external flows—namely, those outer-velocity profiles given in the Falkner-Skan problem. Other outer pressure gradients require different methodology and numerical techniques for their solution. Several transformations have been introduced over the years in attempts to condition the boundary layers to simplify solution methods. These transformations offer a possible starting point for the utilization of numerical methods.

6.8.1   Falkner Transformation

The transformations used in the solution for flow past a wedge can also be used for nonsimilar solutions. Start with the steady form of equation (6.2.14)—namely,

$$\rho \left(u \frac{\partial u}{\partial x} + v \frac{\partial u}{\partial y} \right) = \rho U \frac{\partial U}{\partial x} + \mu \frac{\partial^2 u}{\partial y^2}. \tag{6.8.1}$$

It is convenient to first suppress as much as is easily possible the growth of the boundary layer thickness in the x direction. Introduce variables suggested by our Falkner-Skan solution according to

$$\xi = x, \quad \eta = y \sqrt{\frac{\rho U x}{\mu}}, \quad \psi = \frac{\mu}{\rho} \sqrt{\frac{\rho U x}{\mu}} f(x, \eta), \quad P = \frac{x}{U} \frac{dU}{dx}. \tag{6.8.2}$$

Then, since

$$\frac{\partial}{\partial \xi} = \frac{\partial}{\partial x} + \frac{\eta}{2} \left(-\frac{1}{x} + \frac{1}{U} \frac{dU}{dx} \right) \frac{\partial}{\partial \eta} \quad \text{and} \quad \frac{\partial}{\partial y} = \sqrt{\frac{\rho U}{\mu x}} \frac{\partial}{\partial \eta},$$

it follows that

$$u = \frac{\partial \psi}{\partial y} = \sqrt{\frac{\rho U}{x \mu}} \frac{\partial}{\partial \eta} \left(\frac{\mu}{\rho} \sqrt{\frac{\rho U x}{\mu}} f \right) = U \frac{\partial f}{\partial \eta} \tag{6.8.3}$$

and

$$v = -\frac{\partial \psi}{\partial x} = -\sqrt{U \mu x / \rho} \left[\frac{1}{2x} (1 + P) f + \frac{\partial f}{\partial \xi} + \frac{\eta}{2x} (-1 + P) \frac{\partial f}{\partial \eta} \right]. \tag{6.8.4}$$

Substituting these into equation (6.8.1), we find, after some arranging, that

$$U \frac{\partial f}{\partial \eta} \left[\frac{dU}{dx} \frac{\partial f}{\partial \eta} + U \frac{\partial^2 f}{\partial \eta \partial \xi} + \frac{\eta U}{2x} (-1 + P) \frac{\partial^2 f}{\partial \eta^2} \right]$$

$$- U^2 \frac{\partial^2 f}{\partial \eta^2} \left[\frac{1}{2x} (1 + P) f + \frac{\partial f}{\partial \xi} + \frac{\eta}{2x} (-1 + P) \frac{\partial f}{\partial \eta} \right] = \frac{U^2}{x} \frac{\partial^3 f}{\partial \eta^3} + U \frac{dU}{dx}.$$

Multiplying the above equation by x/U^2 and simplifying, the result is

$$\frac{\partial^3 f}{\partial \eta^3} + \frac{1+P}{2} f \frac{\partial^2 f}{\partial \eta^2} + P \left[1 - \left(\frac{\partial f}{\partial \eta} \right)^2 \right] = \xi \left(\frac{\partial f}{\partial \eta} \frac{\partial^2 f}{\partial \eta \partial \xi} - \frac{\partial f}{\partial \xi} \frac{\partial^2 f}{\partial \eta^2} \right). \tag{6.8.5}$$

The left-hand side of the equation is essentially the equation seen in the similarity solution, except now P can depend on $\xi(x)$. Since a stream function is the starting point, continuity is automatically satisfied.

It is seen that equation (6.8.5) is a third-order partial differential equation. In numerical solutions, it is somewhat easier to deal with first-order partial differential equations. To get to that point here, make the change of variables

$$F = \frac{\partial f}{\partial \eta}, \quad V = \frac{\mu}{\rho} \sqrt{\frac{\rho \xi}{U \mu}},$$

where F is then the dimensionless u velocity component, and V is a scaled dimensionless version of the v velocity component. Then equation (6.8.3) becomes

$$V = -\left(\frac{1+P}{2} f + \xi \frac{\partial f}{\partial \xi} + \frac{\eta(P-1)}{2} \frac{\partial f}{\partial \eta} \right).$$

Differentiating this with respect to η gives

$$\frac{\partial V}{\partial \eta} = -\left(PF + \xi \frac{\partial F}{\partial \xi} + \frac{\eta(P-1)}{2} \frac{\partial F}{\partial \eta} \right). \tag{6.8.6}$$

The momentum equation (6.8.4) can now be written as

$$\frac{\partial^2 F}{\partial \eta^2} = -P + F\left(PF + \xi \frac{\partial F}{\partial \xi} \right) + \frac{\partial F}{\partial \eta}\left(V + \frac{\eta}{2}(P-1)F \right). \tag{6.8.7}$$

6.8.2 von Mises Transformation

Von Mises (1927) suggested the transformation $\chi = U^2 - u^2$ and used χ as the dependent variable and x and ψ as the independent variables. Then

$$d\psi = -v\,dx + u\,dy, \tag{6.8.8}$$

$$d\chi = 2\left(U\frac{dU}{dx} - u\frac{\partial u}{\partial y} - v\frac{\partial u}{\partial x} \right)dx - 2\frac{\partial u}{\partial y}d\psi. \tag{6.8.9}$$

It is seen from this that

$$\frac{\partial \chi}{\partial x} = 2\left(U\frac{dU}{dx} - u\frac{\partial u}{\partial y} - v\frac{\partial u}{\partial x} \right) = -2\nu\frac{\partial^2 u}{\partial y^2}, \tag{6.8.10}$$

$$\frac{\partial \chi}{\partial \psi} = -2\frac{\partial u}{\partial y}d\psi, \tag{6.8.11}$$

where the term on the far right of equation (6.8.10) comes from the momentum equation. Taking further differentials gives

$$d\left(\frac{\partial \chi}{\partial \psi} \right) = -2\left(\frac{\partial^2 u}{\partial x \partial y} + \frac{v}{u}\frac{\partial^2 u}{\partial y^2} \right) - 2\frac{\partial^2 u}{\partial y^2}d\psi.$$

From this it follows that

$$\frac{\partial^2 \chi}{\partial \psi^2} = -\frac{2}{u}\frac{\partial^2 u}{\partial y^2}. \tag{6.8.12}$$

Combining equations (6.8.10) and (6.8.12) gives the final form of the transformed equation as

$$\frac{\mu}{\rho}\sqrt{U^2 - \chi}\frac{\partial^2 \chi}{\partial \psi^2} = \frac{\partial \chi}{\partial x}. \tag{6.8.13}$$

This is to be solved subject to the conditions $\chi = U^2$ at $\psi = 0$, $\chi = 0$ at $\psi = \infty$.

Once equation (6.8.13) is integrated, u can be found from $u = \sqrt{U^2 - \chi}$. Since $dy = d\psi + udx/u$, y is found from $y = \int_0 d\psi/u$, the integration being performed on a line of constant x. The v velocity component is found from $v = \int_0 \frac{\partial v}{\partial y}dy = -\int \frac{\partial u}{\partial x}dy = -\int \frac{1}{u}\frac{\partial u}{\partial x}d\psi$ on a line of constant x. Since there is a singularity in the integrand at the wall, care must be used in carrying out the integration.

6.8.3   Combined Mises-Falkner Transformation

The previous two transformations can be combined into one step by the following:

$$\xi = \frac{1}{U_\infty} \int^\Lambda U ds, \quad \eta = f(\xi)\left(\frac{\psi}{\sqrt{\mu U_\infty/\rho}}\right)^m, \quad u = U\sqrt{w}. \tag{6.8.14}$$

At this point m is a constant and f is a function, both to be chosen. For convenience, note that

$$\frac{\partial \xi}{\partial x} = \frac{U}{U_\infty}, \quad \frac{\partial \xi}{\partial y} = 0, \quad \frac{\partial \eta}{\partial x} = \frac{\eta}{f}\frac{\partial f}{\partial \xi} - \frac{m\mu\eta}{\rho\psi}, \quad \frac{\partial \eta}{\partial y} = \frac{m\eta u}{\psi},$$

$$\frac{\partial}{\partial x} = \frac{U}{U_\infty}\frac{\partial}{\partial \xi} + \eta\left(\frac{1}{f}\frac{\partial f}{\partial \xi} - \frac{mv}{\psi}\right)\frac{\partial}{\partial \eta}, \quad \frac{\partial}{\partial y} = \frac{m u \eta}{\psi}\frac{\partial}{\partial \eta}.$$

Using these in the boundary layer equation yields

$$\frac{\partial w}{\partial \xi} = \frac{2}{U}\frac{dU}{d\xi}(1-w) - \frac{\eta}{f}\frac{df}{d\xi}\frac{\partial w}{\partial \eta} + m^2\eta^{2-2/m}f^{2/m}\sqrt{w}\left(\frac{\partial^2 w}{\partial \eta^2} + \frac{1-1/m}{\eta}\frac{\partial w}{\partial \eta}\right). \tag{6.8.15}$$

There are two choices for m and f that have proven to be useful in the past.

1. $m = 1/2$, $f = \xi^{-1/4}$. In this case equation (6.8.15) reduces to

$$\xi\frac{\partial w}{\partial \xi} = \frac{2\xi}{U}\frac{dU}{d\xi}(1-w) - \frac{\eta}{4}\frac{\partial w}{\partial \eta} + \frac{\sqrt{w}}{4\eta^2}\left(\frac{\partial^2 w}{\partial \eta^2} - \frac{1}{\eta}\frac{\partial w}{\partial \eta}\right). \tag{6.8.16}$$

2. $m = 1$, $f = 1$. In this case equation (6.8.15) reduces to

$$\frac{\partial w}{\partial \xi} = \frac{2}{U}\frac{dU}{d\xi}(1-w) + \sqrt{w}\frac{\partial^2 w}{\partial \eta^2}. \tag{6.8.17}$$

Integration in both cases proceeds as in the previous two cases.

6.8.4   Crocco's Transformation

Crocco's transformation (1939) is unusual in that it uses shear stress rather than a velocity function as a dependent variable. Letting $\tau = \mu\frac{\partial u}{\partial y}$, since

$$du = \frac{\partial u}{\partial x}dx + \frac{\partial u}{\partial y}dy = \frac{\partial u}{\partial x}dx + \frac{\tau}{\mu}dy, \text{ then}$$

$$dy = \frac{\mu}{\tau}\left(du - \frac{\partial u}{\partial x}dx\right). \text{ Also}$$

$$d\tau = \frac{\partial \tau}{\partial x}dx + \frac{\partial \tau}{\partial y}dy = \left(\frac{\partial \tau}{\partial x} - \frac{\mu}{\tau}\frac{\partial u}{\partial x}\frac{\partial \tau}{\partial y}\right)dx + \frac{\mu}{\tau}\frac{\partial \tau}{\partial y}du, \text{ so that}$$

$$\frac{\partial \tau}{\partial x} = \frac{\partial \tau}{\partial x} - \frac{\mu}{\tau}\frac{\partial u}{\partial x}\frac{\partial \tau}{\partial y}, \quad \frac{\partial \tau}{\partial u} = \frac{\mu}{\tau}\frac{\partial \tau}{\partial y}.$$

Then

$$d\left(\frac{\partial \tau}{\partial u}\right) = \mu\frac{\partial}{\partial x}\left(\frac{1}{\tau}\frac{\partial \tau}{\partial y}\right)dx + \mu\frac{\partial}{\partial y}\left(\frac{1}{\tau}\frac{\partial \tau}{\partial y}\right)\left(\frac{\mu}{\tau}du - \frac{\mu}{\tau}\frac{\partial u}{\partial x}dx\right), \text{ so that}$$

$$\frac{\partial^2 \tau}{\partial u^2} = \frac{\mu^2}{\tau^2}\left[\frac{\partial^2 \tau}{\partial y^2} - \frac{1}{\tau}\left(\frac{\partial \tau}{\partial y}\right)^2\right]. \text{ Similarly } \frac{\partial^2 \tau}{\partial y^2} = \rho\left(u\frac{\partial \tau}{\partial x} + v\frac{\partial \tau}{\partial y}\right).$$

With further algebra, the preceding can finally be reduced to

$$\frac{\partial^2 \tau}{\partial u^2} + \rho\mu \left[u \frac{\partial}{\partial x}\left(\frac{1}{\tau}\right) + U \frac{dU}{dx}\frac{\partial}{\partial u}\left(\frac{1}{\tau}\right)\right] = 0. \tag{6.8.18}$$

The accompanying boundary conditions are $\partial\tau/\partial u = 0$ at $u = 0$ and $\tau = 0$ at $u = U$.

In using any of the previous transformations it is necessary to first decide whether the dependent variable is the one best suited to the method. The Falkner transformation results in an equation for $\psi(x,\eta)$, the von Mises an equation for essentially $u(\psi,y)$, the combined method an equation for $u(\xi,\eta)$, and Crocco's method for an equation for $\tau(x,u)$. The equations are of about the same order of complexity, and all require a guess of either the dependent variable or its derivative on the wall. The choice of method to be used then depends on the end result sought and the suitability of the method to the boundary conditions.

6.8.5 Mangler's Transformation for Bodies of Revolution

Mangler's transformation differs in intent from the four previous transformations in that its purpose is to change an axisymmetric flow problem into a two-dimensional flow problem rather than to condition the equations.

Consider the flow on a body generated by revolving the curve $y = r(x)$ about the x-axis. Providing the body radius r is much greater than the boundary layer thickness, the boundary layer equations are

$$u\frac{\partial u}{\partial x} + v\frac{\partial u}{\partial y} = U\frac{dU}{dx} + \frac{\mu}{\rho}\frac{\partial^2 u}{\partial y^2}, \quad \frac{\partial(ru)}{\partial x} + \frac{\partial(rv)}{\partial y} = 0.$$

Using the transformation $x' = \int_0^x r^2 dx$, $y' = ry$, then $u' = u$, $v' = \frac{1}{r}\left(v + \frac{yu}{r}\frac{dr}{dx}\right)$, so that $u = u'$, $v = -\frac{yu'}{r}\frac{dr}{dx} + rv'$, then

$$\rho\left(u'\frac{\partial u'}{\partial x'} + v'\frac{\partial u'}{\partial y'}\right) = \rho U\frac{dU}{dx'} + \mu\frac{\partial^2 u'}{\partial y'^2},$$

$$\frac{\partial u'}{\partial x} + \frac{\partial v'}{\partial y} = 0. \tag{6.8.19}$$

Thus, any solution procedure suited to two-dimensional boundary layer problems can readily be transferred to flow past axisymmetric bodies.

6.9 Boundary Layers in Rotating Flows

The flows previously studied are caused by either pressure gradients, body forces, or moving boundaries. In these cases, the resulting flow is in the direction of the gradient, or body force, or boundary motion. In the atmosphere, however, as is apparent from weather maps that show isobars, the flow tends to be along the isobars, due to Earth's rotation.

Studying flow on a sphere is difficult, so to simplify matters, adopt the viewpoint of the meteorologists and oceanographers and use the beta plane approach. That is, look at a relatively small segment of Earth—say, a square region 1,000 kilometers on a side—and consider that portion of Earth to be flat, rotating at the rate $\Omega_0 \sin\theta$, where θ is the latitude of the center of the region and Ω_0 is Earth's rotation speed of 2π radians per day. Introducing this into equation (1.17.4), the principal term of interest is the Coriolis acceleration term $2\Omega_0 \sin\theta \times \mathbf{v}$, where the angular velocity vector is perpendicular to Earth's surface, and the velocity vector is in plane of the tangent to the Earth's surface. The neglect of the centripetal acceleration is justified, since it adds essentially a small body force, and the remaining terms are considered small compared to the Coriolis terms.

If the vertical velocity component is neglected, the simplified Navier-Stokes equations, first presented by Ekman (1905), are then

$$-2\rho v \Omega_0 \sin\theta = -\frac{\partial p}{\partial x} + \mu \nabla^2 u, \quad 2\rho u \Omega_0 \sin\theta = -\frac{\partial p}{\partial y} + \mu \nabla^2 v. \qquad (6.9.1)$$

When viscosity is neglected, it is seen that the flow is along the lines of constant pressure.

Ekman considered the problem where the wind shear at the surface of the ocean was in the x direction, no pressure gradients were present, and the velocity components only depended on the vertical coordinate. Then equations (6.9.1) reduce to

$$-2\rho v \Omega_0 \sin\theta = \mu \frac{d^2 u}{dz^2}, \quad 2\rho u \Omega_0 \sin\theta = \mu \frac{d^2 v}{dz^2}. \qquad (6.9.2)$$

Multiplying the second equation by i and adding it to the first gives

$$2i\rho(u + iv)\Omega_0 \sin\theta = \mu \frac{d^2(u+iv)}{dz^2}.$$

With z negative downward, this has the simple solution with

$u + iv = (U + iV)e^{(1+i)z\sqrt{\rho\Omega_0 \sin\theta/\mu}}$, or

$$u = e^{z\sqrt{\rho\Omega_0 \sin\theta/\mu}}\left(U\cos z\sqrt{\rho\Omega_0 \sin\theta/\mu} - V\sin z\sqrt{\rho\Omega_0 \sin\theta/\mu}\right), \qquad (6.9.3)$$

$$v = e^{z\sqrt{\rho\Omega_0 \sin\theta/\mu}}\left(V\cos z\sqrt{\rho\Omega_0 \sin\theta/\mu} + U\sin z\sqrt{\rho\Omega_0 \sin\theta/\mu}\right).$$

In particular, if the wind shear is such that the surface water is in the x direction, then $V = 0$ and

$$u = Ue^{z\sqrt{\rho\Omega_0 \sin\theta/\mu}}\cos z\sqrt{\rho\Omega_0 \sin\theta/\mu},$$
$$v = Ue^{z\sqrt{\rho\Omega_0 \sin\theta/\mu}}\sin z\sqrt{\rho\Omega_0 \sin\theta/\mu}. \qquad (6.9.4)$$

It is seen from equations (6.9.4) that the path of the tip of the velocity vector traces a spiral as we move down from the free surface. In the northern hemisphere it starts pointing eastward, say, then turns north, then west, then southward, and repeating the process and decreasing in magnitude as we proceed downward. Computing the discharge by integrating the first of equations (6.9.3), find that

$$Q_x + iQ_y = \int_0^{-\infty}(u+iv)\,dz = U\int_0^{-\infty} e^{\sqrt{\rho\Omega_0 \sin\theta/\mu}(1+i)z}\,dz = \frac{U(1-i)}{2\sqrt{\rho\Omega_0 \sin\theta/\mu}}. \qquad (6.9.5)$$

Thus, equal quantities of fluid are transported perpendicular to the wind shear stress as in the direction of the wind stress.

Boundary layers associated with rotating flows are referred to as **Ekman layers**. From equations (6.9.4) we see that the thickness of the layer is of the order $\sqrt{\mu/\rho\Omega_0 \sin\theta}$.

Another example of rotating boundary layers that does not necessarily need Earth-magnitude dimensions is for a fluid in weak rotation above a stationary plate. A stirred cup of tea with a few loose tea leaves would be an example of such a flow. In this case there is a radial pressure gradient $\partial p/\partial r = \rho r \Omega^2$ outside of the boundary layer. Taking radial and circumferential velocity components in the form $u = r\Omega f(z)$, $v = r\Omega h(z)$, the equations are

$$-2\rho r h\Omega^2 - \rho r\Omega^2 = \mu r\Omega \frac{d^2 f}{dz^2},$$

$$2\rho r f\Omega^2 = \mu r\Omega \frac{d^2 h}{dz^2}.$$

(6.9.6)

Again, multiplying the second of equations (6.9.6) by i and adding it to the first, the result is

$$\rho\left[2i(f+ih) - \Omega\right] = \mu \frac{d^2(f+ih)}{dz^2}.$$

With $z = 0$ at the bottom of the cup, the solution is

$$\mu \frac{d^2(f+ih)}{dz^2} - 2\rho\Omega i(f+ih) = -\rho\Omega,$$

therefore,

$$f + ih = -i\left(1 - e^{-(1+i)z\sqrt{\rho\Omega/\mu}}\right)$$

$$= -i\left[1 - e^{-z\sqrt{\rho\Omega/\mu}}\left(\cos z\sqrt{\rho\Omega/\mu} - i\sin z\sqrt{\rho\Omega/\mu}\right)\right].$$

Splitting into real and imaginary parts gives

$$h = -1 + e^{-z\sqrt{\rho\Omega/\mu}}\cos z\sqrt{\rho\Omega/\mu},$$

$$f = e^{-z\sqrt{\rho\Omega/\mu}}\sin z\sqrt{\rho\Omega/\mu}.$$

(6.9.7)

Above the boundary layer the almost neutrally buoyant tea leaves move in concentric circles, the centripetal force balanced by the pressure gradient. As they move into the boundary layer, the centripetal force diminishes, while the pressure gradient remains unchanged, pushing the tea leaves to the center of the cup.

There is a popular belief that when draining a bathtub the water swirls one way in the northern hemisphere and the opposite way in the southern hemisphere. Since Earth's rotation is small, its effects are mostly felt when lengths of hundreds of kilometers are involved. For something like a bathtub, the length scale is so small that the Coriolis acceleration is negligible. The behavior instead depends strongly on what swirl was introduced when the tub was filled and on the geometry of the tub. Experiments with well-designed circular containers with carefully centered outlets show that it is necessary to wait many hours—perhaps even an entire day—for initial effects to die out. After that, the swirl is hardly noticeable until the depth is of the order of the Ekman layer. When that happens, the direction of swirl reverses several times as the layer becomes thinner and thinner.

Problems—Chapter 6

6.1 The similarity solution for flow into a sink between two inclined plates can be obtained from the Falkner-Skan equations by putting

$$\alpha = \frac{\rho\xi}{\mu}\frac{d(U\xi)}{dx} = 0, \quad \beta = \frac{\rho\xi^2}{\mu}\frac{dU}{dx} = 1, \quad \xi(x) = y/\eta = \sqrt{\mu x/\rho}$$

giving the equation

$$f''' + 1 - (f')^2 = 0, \quad f(0) = f'(0) = 0, \quad f' \to 1 \text{ as } \eta \to \infty.$$

Show that this equation may be integrated to give

$$f'(\eta) = 3\tanh^2\left(\frac{\eta}{\sqrt{2}} + 1.146\right) - 2,$$

where the prime denotes differentiation with respect to r.

6.2 The equation for source flow between inclined plates $f'' + f^2 + 4f + 4a = 0$ can be linearized if in some sense f can be considered to be small. In that case, the equation becomes $f'' + 4f + 4a \approx 0$. Solve this equation subject to f vanishing at θ_1 and at θ_2.

6.3 Find the similarity solution for a two-dimensional submerged jet emitting at the origin. Show that the form $\psi(x, y) = \psi(\eta) = \frac{\mu}{\rho}f(\eta)$, $\eta = y\left(\rho M/\mu^2 x^2\right)^{1/3}$, reduces the Navier-Stokes equations to a third-order ordinary differential equation. This equation can be integrated twice, reducing it to w first-order differential equation.

6.4 The following form has been suggested for flow in the wake of a flat plate:

$$\psi(x, y) = \left(\Omega\mu^2 x^2/\rho^2\right)^{1/3} f(\eta), \quad \eta = y\left(\Omega\rho/\mu x\right)^{1/3},$$

with $U = \sqrt{3C}\left(\frac{\Omega^2\mu x}{\rho}\right)^{1/3}$. Here, Ω represents the constant vorticity at the edge of the boundary layer, and C is also a constant. Show that this is a suitable similarity solution of the boundary layer equations. Do not attempt to solve the equation.

6.5 The boundary layer equations place severe restrictions on the possibility of having a similarity solution. To make a general search for similarity solutions, start with the form $\psi(x, y) = \frac{\mu}{\rho}[e(x) + g(x)f(\eta)]$, $\eta = \frac{y}{h(x)}$, and substitute it into the boundary layer equations. Find the restrictions that must be placed on $e(x)$, $g(x)$, and $h(x)$ to obtain a differential equation solely in terms of f and η.

6.6 Show that a solution of the boundary layer equation (in fact, the full Navier-Stokes equation) can be found for an infinite flat plate in a uniform stream (speed U) with constant suction at the wall. The suction induces a downward velocity of strength V_0. Such suctions have been used as a method for controlling boundary layer separation. Use a velocity form of the type $\mathbf{v}(y) = (u(y), V)$.

6.7 Fluid enters a channel of width $2b$ at a uniform speed U_0. Because of the no-slip condition, the wall shear changes gradually from the uniform flow to the standard Poiseuille parabolic flow. The flow will consist of two wall boundary layers plus an inner core where the speed changes with x but is uniform across its domain. Since the flow can be expected to be symmetric with respect to the centerline, choose y as the distance measured from the bottom wall.

Using the profile $v_x = \begin{cases} U(2\eta - \eta^2), & 0 \le \eta \le 1, \\ U, & 1 \le \eta \le h(x)/\delta, \quad \eta = y/\delta, \end{cases}$

with U being the speed outside of the boundary layer, find U and the distance needed to establish the flow.

6.8 The potential flow against an infinite plate is described by the velocity potential $\phi = A(x^2 - y^2)$, the plate being at $y = 0$. Using a parabolic velocity profile, solve for the boundary layer thickness and wall shear.

6.9 The form $u = \begin{cases} U(A + B\cos C\eta), & 0 \le \eta \le 1, \\ U, & \eta > 1, \quad \eta = \frac{y}{\delta}, \end{cases}$

is proposed as an approximation for the velocity profile in a boundary layer. Find appropriate choices for $A, B,$ and C, and calculate the displacement and momentum thicknesses.

6.10 A thin viscous layer of liquid flows down a vertical wall under the force of gravity. The flow at the outer edge of the layer has the speed $U = \sqrt{2gx}$. Assuming the layer in the liquid to be of the form $v_x = \begin{cases} U(2\eta - \eta^2), & 0 \le \eta \le 1, \\ U, & 1 \le \eta \le h(x)/\delta, \quad \eta = y/\delta, \end{cases}$

find $h(x)$ and $\delta(x)$.

6.11 Viscosity does play a small role in the damping of free surface gravity waves. A simple model for computing this is to say $\frac{d}{dt}(KE + PE) = -\Phi$, where Φ is the dissipation function discussed in Chapter 1. For the two-dimensional traveling wave given by $\phi(x, z, t) = cA \sin k(x - ct)e^{kz}$, $\eta(x, t) = A\cos k(x - ct)^i$, $c = \sqrt{g/k}$, compute the potential and kinetic energies and the dissipation function to find the rate of decay of the wave. Assume A depends exponentially on time.

6.12 The profile $U = U_0\left(1 - \frac{x}{L}\right)$ has been used to study separated flows. Calculate the location of the separation point using both the Thwaites and Stratford criteria.

Thermal Effects

7.1 Thermal Boundary Layers

If temperature gradients exist in a flow, the momentum boundary layer for large Reynolds numbers discussed in Chapter 6 can be accompanied by a thermal boundary layer. Recall equation (1.14.7), where it was found from the first law of thermodynamics that

$$\rho \frac{Du}{Dt} = -\nabla \cdot \mathbf{q} + \frac{dr}{dt} + \Phi. \tag{1.14.7}$$

Taking $u = c_p T$ and $\mathbf{q} = -k\nabla T$ for the internal energy and heat flux, and considering only cases where there is no internal heat generation and dissipation and compressibility effects are secondary, this becomes

$$\rho c_p \frac{DT}{Dt} = k\nabla^2 T. \tag{7.1.1}$$

Here, c_p is the heat capacity at constant pressure $(J/kg \cdot K)$, and k is the thermal conductivity $(W/(m \cdot K))$.

Choose dimensionless parameters in the manner used for the flat plate in steady flow. Using the form

$$x = x_D/L, \quad y = y_D\sqrt{Re}/L,$$

$$v_x = u_D/U_0, \quad v_y = v_D\sqrt{Re}/L,$$

$$T = (T_D - T_{wall})/(T_\infty - T_{wall}),$$

$$Pr = \mu c_p/k.$$

then equation (7.1.1) becomes

$$v_x \frac{\partial \Theta}{\partial x} + v_y \frac{\partial \Theta}{\partial y} = \frac{1}{Pr} \frac{\partial^2 \Theta}{\partial y^2}. \tag{7.1.2}$$

The resemblance to the form of equation (7.1.2) and the momentum equations of Chapter 6 suggest that similarity solutions are possible.

An important dimensionless parameter that comes up often in thermal flows is the *Prandtl number*, defined as $Pr = \mu c_p/k$. It represents the ratio of viscous diffusivity to thermal diffusivity. The Prandtl number is strictly a function of the properties of the fluid and not on flow properties and plays a strong role in determining the ratio of the thermal boundary layer to the momentum boundary layer. Representative Prandtl numbers are shown for various fluids in Tables 7.1.1, 7.1.2, and 7.1.3.

TABLE 7.1.1   Prandtl numbers for liquid metals

Metal	Temperature (K)	Pr	Metal	Temperature (K)	Pr
Bismuth	589	0.014	Mercury	273	0.029
	1,033	0.0083		589	0.0084
Lead	644	0.024	Potassium	422	0.0066
	977	0.016		977	0.0030
Lithium	477	0.065	Sodium	366	0.0011
	1,255	0.027		400	0.0163

TABLE 7.1.2   Prandtl numbers for gases at 300 K

Gas	Pr	Gas	Pr	Gas	Pr
Air	0.707	Carbon monoxide (CO)	0.73	Nitrogen (N_2)	0.716
Ammonia (NH_3)	0.887	Helium (He)	0.68	Oxygen (O_2)	0.711
Carbon dioxide (CO_2)	0.766	Hydrogen (H_2)	0.701		

TABLE 7.1.3   Prandtl numbers for liquids at 300 K

Liquid	Pr	Liquid	Pr
Engine oil	6,400	Glycerin [$C_3H_5(OH)_3$]	6,780
Ethylene glycol [$C_2H_4(OH)_2$]	151	Water (H_2O)	5.83

7.2 *Forced Convection on a Horizontal Flat Plate*

7.2.1 Falkner-Skan Wedge Thermal Boundary Layer

Consider again the Falkner-Skan solution for flow past a wedge. Take the wedge surface to be at a temperature T_{wall}, and the ambient temperature to be T_∞. If buoyancy effects are negligible, the momentum equation is unaffected by the presence of thermal gradients. Then, with η as before and letting $\Theta = \frac{T - T_{\text{wall}}}{T_\infty - T_{\text{wall}}}$, equation (7.1.2) becomes

$$\frac{d^2\Theta}{d\eta^2} + \frac{m+1}{2} \Pr f \frac{d\Theta}{d\eta} = 0. \tag{7.2.1}$$

This is to be solved subject to the isothermal plate condition $\Theta = 0$ at $\eta = 0$, $\Theta \to 1$ as $\eta \to \infty$.

The momentum boundary layer is unchanged from what was discussed in Chapter 6, and the solution found there for the stream function is valid here. The energy equation again requires a numerical solution, with Runge-Kutta-type methods among the easiest to use. They require guessing $\Theta(0)$ and then integrating after choosing values for m and the Prandtl number.

7.2.2 Isothermal Flat Plate

A formal solution of equation (7.2.1) was found by Pohlhausen (1921) for the case of the flat plate ($m = 0$) by choosing Θ as a function of η and the Prandtl number. Then, from equation (6.4.7), $f = -2f'''/f''$. Substitution of this into equation (7.2.1), integrating and applying the thermal boundary conditions gives

$$\Theta(\eta, \Pr) = \frac{F(\eta)}{F(\infty)}, \quad \text{where } F(\eta) = \int_0^\eta (f'')^{\Pr} d\eta. \tag{7.2.2}$$

For the special case $\Pr = 1$, this gives $\Theta = f' = u/U$.

The rate of heat transfer per unit area from the plate is given by

$$q''_{\text{wall}} = -\left. k \frac{\partial T}{\partial y}\right|_{\text{wall}} = k(T_{\text{wall}} - T_\infty)\sqrt{\frac{U_\infty}{\nu x}} \left.\frac{d\Theta}{d\eta}\right|_{\eta=0}. \tag{7.2.3}$$

Detailed information of the heat transfer rate thus requires the integration of equation (7.2.2) for a number of Prandtl numbers. Results show that to a good approximation

$$\left.\frac{d\Theta}{d\eta}\right|_{\eta=0} = \begin{cases} 0.564\sqrt{\Pr} & \text{as } \Pr \to 0, \\ 0.332\,\Pr^{1/3} & \text{for } 0.6 < \Pr < 15, \\ 0.339\,\Pr^{1/3} & \text{as } \Pr \to \infty. \end{cases}$$

The thickness of the thermal boundary layer can be defined analogous to the displacement and momentum thicknesses as

$$\delta_T = \frac{1}{U_\infty(T_{\text{wall}} - T_\infty)} \int_0^h u(T - T_\infty) dy. \tag{7.2.4}$$

Since the Prandtl number represents the rate of momentum diffusivity to thermal diffusivity,[1] for low Prandtl numbers (e.g., liquid metals) momentum transfer in the boundary layer is more efficient than heat transfer. The converse is true if the Prandtl number is greater than unity (e.g., water). Therefore, in general expect that

$$\delta_T > \delta \quad \text{if} \quad \text{Pr} < 1,$$

$$\delta_T = \delta \quad \text{if} \quad \text{Pr} = 1,$$

$$\delta_T < \delta \quad \text{if} \quad \text{Pr} > 1.$$

The previous solution was extended by Pohlhausen (1921) to include the effects of dissipation. The dissipation function Φ in the case of the boundary layer reduces to $\mu \left(\partial u / \partial y \right)^2$, so equation (7.2.1) becomes

$$\frac{d^2\Theta}{d\eta^2} + \frac{m+1}{2} \text{Pr} f \frac{d\Theta}{d\eta} = A \text{Pr} \left(\frac{\partial u}{\partial y} \right)^2, \tag{7.2.5}$$

with $A = \frac{2U^2}{c_p(T_{\text{wall}} - T_\infty)}$. Using our knowledge of the solution of the homogeneous equation, the particular integral can easily be found. The solution now becomes

$$\Theta(\eta) = \frac{F(\eta)}{F(\eta)} \left[1 - KF_1(\infty) \right] + KF_1(\eta), \quad \text{where } F_1(\eta) = \int_0^\eta \left[(f'')^{\text{Pr}} \int_\infty^\eta (f'')^{2-\text{Pr}} \, d\eta \right] d\eta.$$

$$(7.2.6)$$

The function F is the same as that found in the case where dissipation was neglected. Equation (7.2.6) satisfies the conditions $\Theta(0) = 0$, $\Theta(\infty) = 1$. The heat flux also goes to zero at the outer edge of the boundary layer.

Note that if K is large compared with unity, the temperature inside the boundary layer can be higher than either the wall or exterior temperatures.

7.2.3   Flat Plate with Constant Heat Flux

The case of constant heat flux emanating from the plate is also of some interest. For the isothermal plate the correct choice for temperature dependency was a function of η. To meet the constant heat flux condition, the proper choice of boundary condition is

$$T = T_\infty - q''_{\text{wall}} \sqrt{\frac{\nu x}{U}} \Theta(\eta). \tag{7.2.7}$$

This makes $\frac{\partial T}{\partial y} = -\frac{q''_{\text{wall}}}{k} \frac{d\Theta}{d\eta}$, so the boundary condition at the wall is $\frac{d\Theta}{d\eta}\Big|_{\text{wall}} = 1$. Insertion of equation (7.2.7) into the energy boundary layer equation gives

$$\frac{d^2\Theta}{d\eta^2} + \frac{\text{Pr}}{2} \left(f \frac{d\Theta}{d\eta} - \theta \frac{df}{d\eta} \right) = 0, \tag{7.2.8}$$

along with the boundary conditions $\frac{d\Theta}{d\eta}\Big|_{\text{wall}} = 1$, $\Theta \to 0$ as $\eta \to \infty$.

---

[1] Thermal diffusivity is defined as $\alpha = k/\rho c_p$.

7.3   The Integral Method for Thermal Convection

The integral method introduced in Chapter 6 can be easily extended to thermal problems. In using it the temperature dependence of the density is often taken in the simplified form

$$\rho \simeq \rho_\infty [1 - \beta (T - T_\infty)], \quad \text{where} \quad \beta = -\frac{1}{\rho} \left(\frac{d\rho}{dT} \right)_p . \tag{7.3.1}$$

Density variations are usually neglected in the multiplier of the acceleration. Then, the integrated momentum equation including buoyancy force becomes

$$\rho_\infty \frac{d}{dx} \int_0^h v_x (U_\infty - v_x) dy + \rho_\infty \frac{dU_\infty}{dx} \int_0^h (U_\infty - v_x) dy$$
$$= \rho_\infty g \beta \cos \gamma \int_0^h (T - T_\infty) dy + \mu \frac{\partial v_x}{\partial y} \bigg|_{\text{wall}} . \tag{7.3.2}$$

Here, γ is the angle the x-axis makes with gravity. The constant part of the gravity force has been included with the pressure gradient as the driving force for U outside the boundary layer—that is,

$$\rho_\infty U \frac{dU}{dx} = -\frac{\partial p}{\partial x} + \rho_\infty g \cos \gamma.$$

Carrying out a procedure in the manner used in Chapter 6 to obtain the integrated momentum equation, find for the integrated first law equation

$$\frac{d}{dx} \int_0^h \rho c_p v_x (T_\infty - T) dy = k \frac{\partial T}{\partial y} \bigg|_{\text{wall}} . \tag{7.3.3}$$

In using the integral methods for thermal problems, the methodology employed in the previous chapters work well as long as they are adopted to the circumstances at hand. For example, if there is an external flow, the velocity profiles from the previous chapter are satisfactory, but if the external flow is stagnant, the velocity profiles used must vanish at the edge of the momentum boundary layer. An example of such a profile is

$$v_x(x, y) = U \frac{y}{\delta} \left(1 - \frac{y}{\delta} \right)^2 . \tag{7.3.4}$$

This satisfies the conditions $u(x, 0) = 0$, $\frac{\partial^2 u}{\partial y^2} = -\frac{4u_L}{\delta^2}$, $u(x, \delta) = 0$, $\frac{\partial u}{\partial y}(x, \delta) = 0$. The quantity U can be a function of x and is usually related to buoyancy terms.

For problems where wall temperatures are prescribed, forms such as

$$T(x, y) = T_\infty + (T_{\text{wall}} - T_\infty) \left(1 - \frac{y}{\delta} \right)^2 \tag{7.3.5}$$

are suitable since this ensures that $T(x, 0) = T_{\text{wall}}$, whereas if heat flux is specified, the form

$$T(x, y) = T_\infty + \frac{q''_{\text{wall}} \delta}{2k} \left(1 - \frac{y}{\delta} \right)^2 \tag{7.3.6}$$

is a proper choice, since $k \frac{\partial T}{\partial y}(x, 0) = -q''_{\text{wall}}$. As we saw in the similarity solutions, the choice of appropriate velocity and temperature profiles first requires that the approximating functions satisfy the most important boundary conditions.

7.3.1 Flat Plate with a Constant Temperature Region

Suppose that a region of the flat plate of length x_0 extending from the leading edge is at temperature T_∞. The remainder of the plate is at temperature T_{wall}. For the velocity and temperature profiles in the thermal boundary layer use

$$v_x(x, y) = U\left[\frac{3}{2}\frac{y}{\delta} - \frac{1}{2}\left(\frac{y}{\delta}\right)^2\right], \tag{7.3.7}$$

$$T(x, y) = T_{\text{wall}} + (T_\infty - T_{\text{wall}})\left[\frac{3}{2}\frac{y}{\delta_T} - \frac{1}{2}\left(\frac{y}{\delta_T}\right)^2\right]. \tag{7.3.8}$$

Then

$$\rho c_p \frac{d}{dx}\int_0^{\delta_T}(T_\infty - T)v_x\,dy = \rho c_p U(T_\infty - T_{\text{wall}})\frac{d}{dx}\left[\delta_T\int_0^1\left(1 - \frac{3}{2}\eta_T + \frac{1}{2}\eta_T^3\right)\left(\frac{3}{2}\eta - \frac{1}{2}\eta^3\right)\right]d\eta_T$$

$$= \rho c_p U(T_\infty - T_{\text{wall}})\frac{d}{dx}\left[\varsigma\delta_T\left(\frac{3}{20} - \frac{3}{280}\varsigma^2\right)\right] = k\frac{\partial T}{\partial y}\bigg|_{\text{wall}} = \frac{3}{2}\frac{k}{\delta_T}(T_\infty - T_{\text{wall}}),$$

where $\varsigma = \delta_T/\delta$.

Since at least in the front portion of the plate the thermal boundary is thinner than the momentum layer, the squared term in ς can be neglected, leaving

$$\frac{d}{dx}\left(\varsigma^2\delta\right) = \frac{10k}{\rho U c_p \varsigma\delta}.$$

Since $\delta = 4.64\sqrt{\nu x/U}$, with some manipulation this can be rewritten as

$$\varsigma^3 + \frac{4}{3}x\frac{d\varsigma^3}{dx} = \frac{0.929}{\text{Pr}}.$$

The solution of this first-order differential equation in ς^3 is then

$$\varsigma^3 = Cx^{-3/4} + \frac{0.929}{\text{Pr}}.$$

C is a constant determined from $\varsigma(0) = 0$ at $x = x_0$, so $C = -\frac{0.929}{\text{Pr}}x_0^{3/4}$.

Hence, the boundary layer thickness ratio is given by

$$\varsigma = 0.976\,\text{Pr}^{-1/3}\left[1 - \left(\frac{x_0}{x}\right)^{3/4}\right]^{1/3}. \tag{7.3.9}$$

The local heat flux is given as

$$q_{\text{wall}}'' = -k\frac{\partial T}{\partial y}\bigg|_{\text{wall}} = 0.331\frac{k(T_{\text{wall}} - T_\infty)}{x}\text{Pr}^{1/3}\frac{Ux}{\nu}\left[1 - \left(\frac{x}{x_0}\right)^{-3/4}\right]^{-1/3}. \tag{7.3.10}$$

General wall temperature distributions can be handled using Duhamel's superposition method. Let $\phi(\xi, x, y)$ be the solution for

$$v_x\frac{\partial\phi}{\partial x} + v_y\frac{\partial\phi}{\partial y} = \frac{k}{\rho c_p}\frac{\partial^2\phi}{\partial y^2}$$

with boundary conditions

$$\text{at } y = 0, \quad \phi(\xi, x, y) = \begin{cases} 0, & 0 < x < \xi, \\ 1, & x > \xi, \end{cases}$$

$$\text{as } y \to \infty, \quad \phi \to 0, \quad \text{at } x = \xi, \quad \phi = 0.$$

Then

$$T(x, y) = T_\infty + [T_{\text{wall}}(0) - T_\infty]\,\phi(0, x, y) + \int_0^x \phi(\xi, x, y)\frac{dT_{\text{wall}}}{d\xi}d\xi. \tag{7.3.11}$$

7.3.2   Flat Plate with a Constant Heat Flux

The case of a plate with constant heat flux, solved using similarity methods in the previous section, can also be solved by the integral method. With the integral method, however, it is possible to have the heat flux start at a distance of x_0 from the leading edge of the plate. Using

$$v_x(x, y) = U\left[\frac{3}{2}\frac{y}{\delta} - \frac{1}{2}\left(\frac{y}{\delta}\right)^2\right], \tag{7.3.12}$$

$$T(x, y) = T_{\text{wall}} + (T_\infty - T_{\text{wall}})\left[\frac{3}{2}\frac{y}{\delta_T} - \frac{1}{2}\left(\frac{y}{\delta_T}\right)^2\right]. \tag{7.3.13}$$

Proceeding as before, find that

$$\rho c_p \frac{d}{dx}\int_0^{\delta_T}(T_\infty - T)v_x\,dy = \frac{1}{10}\rho c_p U q''_{\text{wall}}\frac{d}{dx}\left[\delta_T{}^2\left(\varsigma - \frac{1}{14}\varsigma^2\right)\right] = k\left.\frac{\partial T}{\partial y}\right|_{\text{wall}} = -q''_{\text{wall}},$$

giving

$$\frac{d}{dx}\left[\delta_T^2\left(\varsigma - \frac{1}{14}\varsigma^2\right)\right] = \frac{10k}{\rho c p U}.$$

Again, since this is in the thermal boundary layer and, at least in the front portion of it, the thermal layer is thinner than the momentum layer, the cubic term in ς can be neglected, leaving

$$\frac{d}{dx}\left(\varsigma^3\delta\right) = \frac{10k}{\rho U c_p}.$$

Integration of this gives $\delta^2\varsigma^3 = C + \frac{10k}{\rho c_p U}$, where C is a constant determined from $\varsigma(0) = 0$ at $x = x_0$, so $C = -\frac{10k}{\rho c_p U}x_0$. Using our previous knowledge of δ, the boundary layer thickness ratio is given by

$$\varsigma = 0.774\,\text{Pr}^{-1/3}\left(1 - \frac{x_0}{x}\right)^{1/3}. \tag{7.3.14}$$

The local temperature is given as

$$T_{\text{wall}} = 2.394\frac{q''_{\text{wall}}}{k}\text{Pr}^{-1/3}\left(\frac{Ux}{\nu}\right)^{-1/2}\left[1 - \frac{x_0}{x}\right]^{1/3}. \tag{7.3.15}$$

Superposition as used in arriving at equation (7.3.10) can be used for heat flux that varies along the plate.

7.4 Heat Transfer Near the Stagnation Point of an Isothermal Cylinder

The irrotational flow past a circular cylinder predicts a velocity $U = 2U_\infty \sin\gamma$ along the surface of a cylinder of radius a. Here, γ is the angle measured from the stagnation point. As long as we don't venture too far from the stagnation point, a good approximation is $\sin\gamma \simeq x/a$. The Kármán–Pohlhausen velocity profile discussed in Chapter 6 was of the form $v_x = U\left[F(\eta) + \Lambda G(\eta)\right]$, with $F(\eta) = 2\eta - 2\eta^3 + \eta^4$ and $G(\eta) = \frac{1}{6}\eta(1-\eta)^3$, giving $\delta_D = \frac{\delta}{10}\left(3 - \frac{\Lambda}{12}\right)$ and $\delta_M = \frac{\delta}{63}\left(\frac{37}{5} - \frac{\Lambda}{15} - \frac{\Lambda^2}{144}\right)$.

Recalling from equation (6.5.6) that $\frac{U}{\nu}\frac{d\delta_M^2}{dx} = H$, since U is zero at the stagnation point and $d\delta_M^2/dx$ must be finite, H must be zero at the stagnation point. This will be true if

$$\Lambda^3 + 147.4\Lambda^2 - 1670.4\Lambda + 9{,}072 = 0. \tag{7.4.1}$$

The three roots of this equation are 7.052, 17.75, and −70. Only the first of these is in the range $-12 < \Lambda < 12$, which is required for the flow to be nonseparated. From the definition of G in equation (6.5.6) it follows that

$$\delta^2 = \frac{\mu\Lambda}{\rho\, dU/dx} = \frac{\mu\Lambda}{2U_\infty\rho}. \tag{7.4.2}$$

Hence,

$$\delta^2_{\text{stagnation}} = \frac{3.526a\mu}{\rho U_\infty}. \tag{7.4.3}$$

In the thermal boundary layer, using the cubic temperature profile, we again find that

$$\theta = \frac{3}{2}\frac{y}{\delta_T} - \frac{1}{2}\left(\frac{y}{\delta_T}\right)^3. \tag{7.4.4}$$

From the integrated energy equation find

$$\varsigma\delta\rho c_p \frac{d}{dx}\left[\varsigma\delta U\left(M + \Lambda N\right)\right] = \frac{3k}{2},$$

$$\text{where} \quad \varsigma = \delta_T/\delta, \quad M = \frac{1}{5}\varsigma - \frac{3}{70}\varsigma^3 + \frac{1}{80}\varsigma^4,$$

$$N = \frac{1}{6}\left[\frac{1}{10}\varsigma - \frac{1}{8}\varsigma^2 + \frac{9}{140}\varsigma^3 - +\frac{1}{80}\varsigma^4\right].$$

Again, neglecting higher powers of ς and using $U = 2U_\infty x/a$ along with $\delta^2 = \Lambda a\mu/2\rho U_\infty$, this simplifies to $\varsigma\sqrt{\Lambda}\frac{d}{dx}[x\varsigma^2\sqrt{\Lambda}(12 + \Lambda)] = \frac{90}{\text{Pr}}$. Using the stagnation value of 7.052 for Λ reduces this to $x\frac{d\varsigma^3}{dx} + \frac{3}{2}\varsigma^3 = \frac{1.0048}{\text{Pr}}$. The solution of this that keeps ς finite at $x = 0$ is

$$\varsigma = \frac{0.875}{\text{Pr}^{1/3}}. \tag{7.4.5}$$

Then

$$q''_{\text{wall}} = -k \left. \frac{\partial T}{\partial y} \right|_{\text{wall}} = \frac{3}{2} \frac{k}{s\delta} = 0.645 \, (T_{\text{wall}} - T_\infty) \frac{k}{a} \text{Pr}^{1/3} \text{Re}^{1/2}, \qquad (7.4.6)$$

where $\text{Re} = 2U_\infty \rho a / \mu$.

7.5   Natural Convection on an Isothermal Vertical Plate

In natural convection (also referred to as free convection) flow is due solely to buoyancy forces. The boundary layer momentum equation for these flows is

$$\rho \left(v_x \frac{\partial v_x}{\partial x} + v_y \frac{\partial v_x}{\partial y} \right) = -\frac{\partial p}{\partial x} + \rho g + \mu \frac{\partial^2 v_x}{\partial y^2}. \qquad (7.5.1)$$

The pressure gradient here is solely hydrostatic and balances the average buoyancy force. The coordinate x is directed upward in the vertical direction.

Using the density form of equation (7.3.1), this becomes

$$\rho_0 \left(v_x \frac{\partial v_x}{\partial x} + v_y \frac{\partial v_x}{\partial y} \right) = \rho_0 g \beta (T - T_\infty) + \mu \frac{\partial^2 v_x}{\partial y^2}. \qquad (7.5.2)$$

Pohlhausen (1921a) found a similarity solution for the flow due to a uniformly heated flat plate. Since there is no velocity outside of the boundary layer, there is no reference velocity to use in the solution, so the previous general form, dependent as it is on a Reynolds number, does not hold. By choosing stream and temperature functions in the form of a power of x times a function of η, a variable that is of the form y times another power of x, he found that the momentum and energy equations along the plate could be put in similarity form if he chose

$$\psi(x, \eta) = 4 \frac{\mu}{\rho} C x^{3/4} f(\eta), \quad T = T_\infty + (T_{\text{wall}} - T_\infty) \Theta(\eta, \text{Pr}), \quad \text{where } \eta = C y x^{-1/4}.$$

The constant C is given by $C = \left[\frac{g\beta(T_{\text{wall}} - T_\infty)}{4(\mu/\rho)^2} \right]^{1/4}$. Choosing the temperature as being solely a function of the similarity variable is dictated by the requirement that the temperature be constant on the boundary.

Insertion of these into the momentum and energy equations gives

$$\frac{d^3 f}{d\eta^3} + 3f \frac{d^2 f}{d\eta^2} - 2 \left(\frac{df}{d\eta} \right)^2 + \Theta = 0, \qquad (7.5.3)$$

$$\frac{d^2 \Theta}{d\eta^2} + 3 \, \text{Pr} \, f \frac{d\Theta}{d\eta} = 0, \qquad (7.5.4)$$

subject to the boundary conditions

$$f(0) = 0, \quad \frac{df}{d\eta}(0) = 0, \quad \Theta(0) = 1, \quad \frac{df}{d\eta}(\infty) = 0, \quad \Theta(\infty) = 0. \qquad (7.5.5)$$

Pohlhausen (1921a and b), Schmidt and Beckmann (1930), and Ostrach (1953) all contributed to the solution of this problem. Their principle results are shown in Table 7.5.1.

TABLE 7.5.1   Wall values for natural convection on an isothermal vertical wall

Pr	0.01	0.72	0.733	1.0	2.0	10.0	100	1000
$\frac{d\Theta}{d\eta}\Big|_{\text{wall}}$	0.080592	0.50463	0.50789	0.56714	0.716483	1.168	2.1914	3.97
$\frac{d^2 f}{d\eta^2}\Big|_{\text{wall}}$	0.9862	0.6760	0.6741	0.6421	0.5713	0.4192	0.2517	0.1450

The rate of heat transfer from the wall per unit area is given by

$$q''_{\text{wall}}(x) = -k \left.\frac{\partial T}{\partial y}\right|_{\text{wall}} = -\sqrt{2}(T_{\text{wall}} - T_\infty)\frac{k}{x}Gr_x^{1/4}\left.\frac{d\Theta}{d\eta}\right|_0, \qquad (7.5.6)$$

where Gr_x is the local **Grashof number** given by

$$Gr_x = \frac{g\beta|T_{\text{wall}} - T_\infty|x^3}{\nu^2}. \qquad (7.5.7)$$

The Grashof number is a dimensionless quantity that represents the ratio of buoyancy forces to viscous forces in natural convection.

7.6 Natural Convection on a Vertical Plate with Uniform Heat Flux

Sparrow and Gregg (1956) were able to find the flow on a vertical plate in a manner similar to that used in the previous section. This time the issue in determining the similarity variables is that the temperature gradient normal to the wall be constant. Again, choosing stream and thermal functions in the form of a power of x times a function of η, a variable that is of the form y times another power of x, they found the momentum and energy equations along the plate could be put in similarity form if they chose

$$\psi(x, \eta) = 4\mu C x^{3/4} f(\eta)/\rho, \quad T = T_\infty + (T_{\text{wall}} - T_\infty)\Theta(\eta, \text{Pr}), \qquad (7.6.1)$$

where $\eta = Cyx^{-1/4}$. The constants are given by $C_1 = \left(\rho^2 g\beta q''_{\text{wall}}/5k\mu^2\right)^{1/5}$ and $C_2 = \left(5^4 g\beta q''_{\text{wall}}\mu^3/\rho^3 k\right)^{1/5}$. This time the choice of the various powers is dictated by the requirement that the temperature gradient be constant on the boundary,

Insertion of this stream function and temperature distribution into the momentum and energy equations gives

$$\frac{d^3 f}{d\eta^3} + 4f\frac{d^2 f}{d\eta^2} - 3\left(\frac{df}{d\eta}\right)^2 - \Theta = 0, \qquad (7.6.2)$$

$$\frac{d^2\Theta}{d\eta^2} + \text{Pr}\left(4f\frac{d\Theta}{d\eta} - \Theta\frac{df}{d\eta}\right) = 0, \qquad (7.6.3)$$

subject to the boundary conditions

$$f(0) = 0, \quad \frac{df}{d\eta}(0) = 0, \quad \frac{d\Theta}{d\eta}(0) = 1, \quad \frac{df}{d\eta}(\infty) = 0, \quad \Theta(\infty) = 0. \qquad (7.6.4)$$

The principle results are given in Table 7.6.1.

TABLE 7.6.1   Wall values for natural convection on a constant heat flux vertical wall

Pr	0.1	1.0	10.0	100.0
Θ_{wall}	-2.7507	-1.3574	-0.76746	-0.46566
$\left.\frac{d^2 f}{d\eta^2}\right\|_{\text{wall}}$	1.6434	0.72196	0.30639	0.12620

The temperature at the wall is given by

$$T_{\text{wall}} = T_\infty - \frac{x q''_{\text{wall}}}{k} Gr_x^{-1/5} \Theta(0), \qquad (7.6.5)$$

where Gr_x is the local **Grashof number** given by

$$Gr_x = \frac{g\beta q''_{\text{wall}} x^4}{5k(\mu/\rho)^2}. \qquad (7.6.6)$$

7.7   Thermal Boundary Layer on Inclined Flat Plates

The results of the previous two sections can also be used for nonvertical plates by replacing the gravitational acceleration g by $g\cos\theta$, where θ is the angle that the plate makes with the vertical. As the departure angle becomes too large, gravity also tends to induce motion normal to the plate, which acts to thicken the momentum boundary layer and causes separation. Details on some of the investigations can be found in the works of Rich (1953), Vliet (1969), Fujii and Imura (1972), and Pern and Gebhart (1972).

7.8   Integral Method for Natural Convection on an Isothermal Vertical Plate

The integral method used in previous sections can also be adapted to free convection problems. Not having an outer flow makes the definitions of displacement and momentum thicknesses useless, so we return to the integrated momentum and energy equation and take the limit as U goes to zero. The result is

$$\frac{d}{dx}\int_0^\delta \rho_0 v_x^2 dy = \int_0^\delta \rho_0 g\beta(T - T_\infty)dy - \mu \left.\frac{\partial v_x}{\partial y}\right|_{\text{wall}}, \qquad (7.8.1)$$

$$\frac{d}{dx}\left[\rho_0 c_p \int_0^\delta v_x (T - T_\infty)\, dy\right] = -k \left.\frac{\partial T}{\partial y}\right|_{\text{wall}}. \qquad (7.8.2)$$

For the velocity and temperature profile choose

$$v_x = u\frac{y}{\delta}\left(1 - \frac{y}{\delta}\right)^2, \qquad (7.8.3)$$

$$T = T_\infty + (T_{\text{wall}} - T_\infty)\left(1 - \frac{y}{\delta}\right)^2. \qquad (7.8.4)$$

These satisfy the constant temperature and no-slip conditions at the wall. Both are zero and have zero derivatives at the edge of the boundary layer, meaning that the shear stress and heat flux are both zero there.

Substituting these forms into equation (7.8.1) and (7.8.2) yields

$$\frac{1}{105}\frac{d}{dx}\left(u^2\delta\right) = \frac{1}{3}g\beta(T_{\text{wall}} - T_\infty)\delta - \frac{\mu u}{\rho\delta},$$

$$\frac{1}{30}\frac{d}{dx}(u\delta) = \frac{2k}{\rho_0 c_p \delta}. \tag{7.8.5}$$

$$\frac{1}{30}\frac{d}{dx}(u\delta) = \frac{2k}{\rho_0 c_p \delta}. \tag{7.8.6}$$

From previous results it seems reasonable to assume that u and δ both behave as powers of x. Choose $\delta = Cx^m$, and substitute this into equation (7.8.7) to find

$$u_x\delta = \frac{60kx^{-m+1}}{(-m+1)\rho_0 c_p C}, \quad \text{or}$$

$$u_x = \frac{60\mu x^{-2m+1}}{\rho(1-m)\text{Pr}\,C^2}. \tag{7.8.7}$$

Substituting this into equation (7.8.5), and requiring that the power of x in each term is the same, gives $m = 1/4$. With this at hand, the constant C can be found to be

$$C = 240^{1/4}\text{Pr}^{-1/2}\left(\frac{20}{21} + \text{Pr}\right)^{1/4}\left[\frac{\rho^2 g\beta(T_{\text{wall}} - T_\infty)}{\mu^2}\right]^{-1/4}$$

$$= 3.936\text{Pr}^{-1/2}\left(\frac{20}{21} + \text{Pr}\right)^{1/4}\left[\frac{\rho^2 g\beta(T_{\text{wall}} - T_\infty)}{\mu^2}\right]^{-1/4}. \tag{7.8.8}$$

Then

$$\delta = 3.936\text{Pr}^{-1/2}\left(\frac{20}{21} + \text{Pr}\right)^{1/4}\left[\frac{\rho^2 g\beta(T_{\text{wall}} - T_\infty)}{\mu^2}\right]^{-1/4}x^{1/4}, \tag{7.8.9}$$

$$u_x = 5.164\nu\left(\frac{20}{21} + \text{Pr}\right)^{-1/2}\left[\frac{\rho^2 g\beta(T_{\text{wall}} - T_\infty)}{\mu^2}\right]^{1/2}x^{1/2}. \tag{7.8.10}$$

7.9 Temperature Distribution in an Axisymmetric Jet

The solution for a submerged jet given in Section 5.4.5 can have a temperature distribution added to it. Let

$$Q = 2\pi\rho c_p \int_0^\pi v_x TR^2 \sin\vartheta d\vartheta \tag{7.9.1}$$

be the total heat flux in the jet, and also let the temperature be given in the form

$$T(R, \vartheta) = \frac{1}{R}G(\varsigma), \quad \varsigma = \cos\vartheta. \tag{7.9.2}$$

The energy boundary layer equation then becomes

$$\rho c_p \left(\frac{df}{d\varsigma} G + f \frac{dG}{d\varsigma} \right) = k \frac{d}{d\varsigma} \left[\left(1 - \varsigma^2 \right) \frac{dG}{d\varsigma} \right]. \tag{7.9.3}$$

Here, the stream function f is as given in Section 5.4.5. The boundary conditions to be applied to G are

$$\frac{dG}{d\vartheta} = 0 \quad \text{at} \quad \vartheta = 0 \quad \text{and} \quad \vartheta = \pi. \tag{7.9.4}$$

These conditions ensure that there is no heat transfer across the axis of the jet.

Equation (7.9.3) can be integrated once to give

$$\left(1 - \varsigma^2 \right) \frac{dG}{d\varsigma} = \Pr f \ G, \tag{7.9.5}$$

and again to give

$$G(\varsigma) = A \left(\frac{a}{a+1-\varsigma} \right)^{\Pr}, \tag{7.9.6}$$

where equation (5.4.31) has been used for f. The constant of integration in equation (7.9.5) was chosen as zero to satisfy the condition on $\vartheta = 0$.

The second constant of integration, A, is determined from equation (7.9.1) to be $A = Q/2\pi \rho c_p \nu \int_{-1}^{1} \frac{df(\varsigma)}{d\varsigma} G(\varsigma) d\varsigma$. Since f and G are both known and are relatively simple, the integration can be carried out, giving, finally,

$$A = \frac{Q}{4\pi \rho c_p \frac{\mu}{\rho} \left\{ \frac{a+2}{2\Pr+1} \left[1 - \left(\frac{a}{a+2} \right)^{2\Pr+1} \right] - \frac{a}{2\Pr-1} \left[1 - \left(\frac{a}{a+2} \right)^{2\Pr-1} \right] \right\}}. \tag{7.9.7}$$

In the special case where $a \ll 1$, this simplifies to

$$A \simeq \rho Q \frac{2\Pr+1}{8\pi \rho c_p \mu}. \tag{7.9.8}$$

Problems—Chapter 7

7.1 A point source of heat such as a candle produces a rising thermal plume. Using cylindrical polar coordinates and starting with the basic boundary layer equations in the form

$$\frac{\partial v_r}{\partial r} + \frac{v_r}{r} + \frac{\partial v_z}{\partial z} = 0,$$

$$\rho_\infty \left(v_r \frac{\partial v_z}{\partial r} + v_z \frac{\partial v_z}{\partial z} \right) = \mu \left(\frac{\partial^2 v_z}{\partial r^2} + \frac{1}{r} \frac{\partial v_z}{\partial r} \right) + (\rho - \rho_\infty) g,$$

$$\rho_\infty c_p \left(v_r \frac{\partial T}{\partial r} + v_z \frac{\partial T}{\partial z} \right) = k \left(\frac{\partial^2 T}{\partial r^2} + \frac{1}{r} \frac{\partial T}{\partial r} \right),$$

show that a similarity solution can be found in the form

$$\psi = \frac{4\mu}{\rho_\infty} f(\eta), \quad (\rho - \rho_\infty)g = g\frac{d\rho}{dT}(T - T_\infty) = \frac{G\rho_\infty}{z\mu}\Theta(\eta), \quad \eta = \left(\frac{\rho_\infty^2 G}{4\mu^3}\right)^{1/4}\frac{r}{z^{1/4}}.$$

Here, G is a constant related to the strength of the heat source, defined by

$$G = \int_0^\infty 2\pi r v_z(\rho - \rho_\infty)g\,dr.$$

Give the similarity equations and the boundary conditions. Do not solve the equations. (Closed form solutions have been found for Prandtl numbers of 1 and 2.)

7.2 A line source of heat produces a rising thermal plume in the form of a sheet. Using Cartesian coordinates and starting with the basic boundary layer equations in the form

$$\frac{\partial v_x}{\partial x} + \frac{\partial v_y}{\partial y} = 0, \quad \rho_\infty\left(v_x\frac{\partial v_x}{\partial x} + v_y\frac{\partial v_x}{\partial y}\right) = \mu\frac{\partial^2 v_x}{\partial y^2} + (\rho - \rho_\infty)g,$$

$$\rho_\infty c_p\left(v_x\frac{\partial T}{\partial x} + v_y\frac{\partial T}{\partial y}\right) = k\frac{\partial^2 T}{\partial y^2},$$

show that a similarity solution can be found in the form

$$\psi = \left(\frac{Gx^3\mu^2}{\rho_\infty^3}\right)^{1/5} f(\eta), \quad \eta = \frac{1}{5}\left(\frac{\rho_\infty^2 G}{4\mu^3}\right)^{1/5}\frac{y}{x^{2/5}},$$

$$(\rho - \rho_\infty)g = g\frac{d\rho}{dT}(T - T_\infty) = \frac{1}{125}\left(\frac{\rho_\infty^3 G^4}{x^3\mu^2}\right)^{1/5}\Theta(\eta).$$

Here, G is a constant related to the strength of the heat source, defined by $G = \int_{-\infty}^\infty v_x(\rho - \rho_\infty)g\,dy$. Give the similarity equations and the boundary conditions. Do not solve the equations. (Closed form solutions have been found for Prandtl numbers of 5/9 and 2.)

7.3 For a plane Poiseuille flow between parallel planes, the velocity distribution is $v_x = U_0\left(1 - \frac{y^2}{a^2}\right) + U_{\text{lower}} + \left(U_{\text{upper}} - U_{\text{lower}}\right)\frac{y}{a}$. Find the temperature distribution including the effects of dissipation. The upper wall temperature is T_{wall}, and the lower wall heat flux is $q = k\frac{dT}{dy}$.

7.4 A horizontal semi-infinite plate is submerged in a uniform stream of a very low Prandtl number fluid (a liquid metal). The fluid is at a temperature of T_∞ with constant properties, and the wall is maintained at a constant temperature T_{wall}. Using the integral method with a temperature distribution in the form $T = A + B\sin(Cy)$, find A, B, and C; the thermal boundary layer thickness; and the heat transfer at the wall. (Hint: The very low Prandtl number means that the thermal boundary layer is much thicker than the momentum boundary layer.)

7.5 Use the integral method and the following velocity and temperature profiles to find the boundary layer on a vertical flat plate with constant heat flux q.

$$v_x = u\frac{y}{\delta}\left(1 - \frac{y}{\delta}\right)^2, \quad T = T_\infty + \frac{\delta q}{2k}\left(1 - \frac{y}{\delta}\right)^2.$$

Hint: As in the temperature-specified natural convection case, expect that u and δ will be of the form of powers of x.

Low Reynolds Number Flows

Flows that occur at small values of the Reynolds number are important for studying the swimming of microorganisms, motion of small particles, determining the viscosity of suspensions, motion of glaciers, micro- and nanotechnology, and many other applications. The mathematical difficulties involved are somewhat subtler than those found in large Reynolds number flows but are fundamentally much the same.

8.1 Stokes Approximation

For very small values of the Reynolds number Stokes (1851) proposed that the convective acceleration terms could be neglected and the Navier-Stokes equations replaced by

$$\rho \frac{\partial \mathbf{v}}{\partial t} = -\nabla p + \mu \nabla^2 \mathbf{v}, \tag{8.1.1}$$

$$\nabla \cdot \mathbf{v} = 0.$$

Taking the divergence of the first of equation (8.1.1) gives

$$\nabla^2 p = 0, \tag{8.1.2}$$

and taking the curl gives

$$\rho \frac{\partial \boldsymbol{\omega}}{\partial t} = \mu \nabla^2 \boldsymbol{\omega}, \quad \text{with } \boldsymbol{\omega} = \nabla \times \mathbf{v}. \tag{8.1.3}$$

If the time dependence of the velocities is of the form $e^{\sigma t}$, then it follows from equation (8.1.1) that

$$\nabla p = \mu \nabla^2 \mathbf{v} - \rho \sigma \mathbf{v}. \tag{8.1.4}$$

Since the pressure satisfies Laplace's equation, a particular solution of equation (8.1.4) is then

$$\mathbf{v}_{\text{particular}} = -\frac{1}{\rho \sigma} \nabla p. \tag{8.1.5}$$

This is useful in solving for pressure once the velocity is known.

Stokes originally solved the problem of flow about a sphere in terms of the stream function. Lamb (1932) presented a general solution for the pressure in terms of a Taylor series. He also presented a general solution for the velocity in the form

$$\mathbf{v}(\mathbf{R}) = \nabla \phi + \nabla \times \nabla \times \mathbf{A} + \mathbf{R} \times \nabla \chi, \quad \text{where}$$

$$\nabla^2 \phi = 0, \quad \left(\rho \frac{\partial}{\partial t} - \mu \nabla^2 \right) \nabla^2 \mathbf{A} = 0, \quad \text{and} \quad \left(\rho \frac{\partial}{\partial t} - \mu \nabla^2 \right) \chi = 0. \tag{8.1.6}$$

Since attention will be restricted here to simple geometries, these three approaches are too complicated for our purposes. Instead, Stokes original solution supplies three fundamental solutions of equation (8.1.1): a doublet, a *stokeslet*, and a *rotlet*.[1]

1. **Doublet**. This is the same doublet as that found previously for irrotational inviscid flow—namely,

$$\phi(r) = \frac{\mathbf{A} \cdot \mathbf{R}}{R^2}, \quad \mathbf{R} = x\mathbf{i} + y\mathbf{j} + z\mathbf{k}. \tag{8.1.7}$$

Its velocity is given by $\mathbf{v} = \nabla \phi = \dfrac{\mathbf{A} \times \mathbf{R}}{R^3}$. Here, $\mathbf{A}$ is a vector that can either be constant or a function of time.

2a. **Stokeslet for steady flows**. The velocity field

$$\mathbf{v} = \frac{\mathbf{B}}{R} + \frac{\mathbf{B} \cdot \mathbf{R}}{R^3} \mathbf{R}, \tag{8.1.8}$$

where $\mathbf{B}$ is a constant vector, satisfies the time-independent case of (8.1.1). It corresponds to the second term in (8.1.4). It contributes a force on a body, but not a moment.

2b. **Stokeslet for unsteady flows**. The velocity vector

$$\mathbf{v} = e^{-i\nu k^2 t} \left\{ -\mathbf{B} \nabla^2 f + \nabla \left[(\mathbf{B} \cdot \nabla) f \right] \right\}, \quad \text{where}$$

$$f = \frac{1}{k^2} \left(ik + \frac{1 - e^{ikR}}{R} \right) \tag{8.1.9}$$

and where $k = (1 + i)\sqrt{\rho \omega / 2\mu}$ and $\mathbf{B}$ are constants, satisfies equation (8.1.1) for unsteady flows. The form for f has been chosen so that in the limit as k approaches zero, the steady-state form equation (8.1.8) is recovered.

---

[1] The terms *Stokeslet* and *rotlet* were coined in the later half of the twentieth century.

3a. **Rotlet for steady flows**. The velocity vector

$$\mathbf{v} = \mathbf{R} \times \nabla \left(\frac{\mathbf{C} \cdot \mathbf{R}}{R^3} \right), \tag{8.1.10}$$

where $\mathbf{C}$ is a constant vector, has been termed a **rotlet**. There is no pressure field associated with it, nor does it contribute to the force acting on a body. It can, however, exert a turning moment on a body.

3b. **Rotlet for unsteady flows**. The velocity vector

$$\mathbf{v} = \mathbf{R} \times \nabla \left(\frac{\mathbf{C} \cdot \mathbf{R}}{R^3} h \right), \quad \text{where } h = e^{ikR - \mu k^2 t/\rho}(1 - ikR), \tag{8.1.11}$$

satisfies equation (8.1.1) for unsteady flows. Notice that in the limit as k approaches zero, this velocity becomes equal to that given in equation (8.1.10).

8.2   Slow Steady Flow Past a Solid Sphere

For slow flow past a sphere of radius a, Stokes superimposed a uniform stream, doublet, and stokeslet, all oriented in the z direction. Thus,

$$\begin{aligned}
\mathbf{v} &= U\mathbf{k} + B\nabla \left(\frac{1}{R^2} \frac{\partial R}{\partial z} \right) + C\nabla \times \nabla \times (\mathbf{k}R) \\
&= \left(U + \frac{1}{R^3}B - \frac{1}{R}C \right)\mathbf{k} + \left(-\frac{3z}{R^5}B - \frac{z}{R^3}C \right)\mathbf{R}.
\end{aligned} \tag{8.2.1}$$

On $R = a$ the velocity must vanish. Thus,

$$\frac{1}{a^3}B - \frac{1}{a}C = -U, \quad -\frac{3}{a^5}B - \frac{1}{a^3}C = 0,$$

yielding

$$B = -\frac{a^3}{4}U, \quad C = \frac{3a}{4}U. \tag{8.2.2}$$

Thus,

$$\begin{aligned}
\mathbf{v} &= U \left(1 - \frac{a^3}{4R^3} - \frac{3a}{4R} \right)\mathbf{k} + U \left(\frac{3a^3 z}{4R^4} - \frac{3az}{4R^2} \right)\mathbf{e}_R \\
&= U \left(1 - \frac{3a}{2R} + \frac{a^3}{2R^3} \right)\mathbf{e_R}\cos\beta + U \left(-1 + \frac{3a}{4R} + \frac{a^3}{4R^3} \right)\mathbf{e}_\beta\sin\beta.
\end{aligned}$$

From this and equation (8.1.1) the pressure can be found as

$$p = p_0 - \frac{3a\mu}{2R^2}U\cos\beta \tag{8.2.3}$$

and the stress components as

$$\tau_{RR} = -p + 2\mu \frac{\partial v_R}{\partial R} = \mu U \left(\frac{9a}{2R^2} - \frac{3a^3}{R^4} \right)\cos\beta,$$

$$\tau_{R\beta} = \mu \left(\frac{1}{R}\frac{\partial v_R}{\partial \beta} + \frac{\partial v_\beta}{\partial R} - \frac{v_\beta}{R} \right) = -\frac{3a^3\mu}{2R^4}U\sin\beta. \tag{8.2.4}$$

On the surface of the sphere, these stresses become $|\tau_{RR}|_a = \frac{3\mu}{2a} U \cos\beta$, $\tau_{R\beta}|_a = -\frac{3a^3\mu}{2a} U \sin\beta$. The force on the sphere is thus given by

$$F_z = \int_0^\pi \left[(\tau_{RR} \cos\beta - \tau_{R\beta} \sin\beta) 2\pi R \sin\beta \right]_a a\, d\beta$$

$$= 3\mu\pi a U \int_0^\pi (\cos^2\beta + \sin^2\beta) \sin\beta\, d\beta \qquad (8.2.5)$$

$$= 6\pi\mu U a.$$

The preceding solution has had many uses. Perhaps the most noteworthy was determination of the charge on an electron by Robert A. Millikan, who was awarded the 1923 Nobel Prize in physics for his work. First he sprayed a few tiny drops of oil between two parallel plates. He then measured the diameter of the drops and, using the fact that $F_{drag} = Weight - Buoyant\ Force$, determined the diameter of the drops. He then applied an electric charge on the plates in a direction opposite to gravity and again measured velocity. The electric charge induces a force qE/D, where q is the charge on the electron, E is the strength of the electric field, and D is drop diameter. Since the number of electrons attached to a given drop is unknown, a number of measurements had to be taken. The experiment does not give an exact answer, since because of small droplet size Brownian motion causes an appreciable percentage of error. Nevertheless, his results gave a far more accurate estimate than was previously obtainable.

Stokes's result for the force on a sphere is also used in viscometry. By dropping a small sphere of known diameter in a fluid and measuring the terminal velocity, the viscosity can be deduced.

Einstein (1906, 1911) also used Stokes's equations to provide an estimate of the viscosity of dilute suspensions. He found that the equivalent viscosity was given by

$$\mu_{equivalent} = \mu_{fluid}(1 + 2.5\phi), \quad \text{where } \phi = \frac{4}{3}\pi R^3 c = \frac{\text{volume of spheres}}{\text{volume of suspension}}.$$

Here, c is the concentration of the suspended material. (The preceding formula is a corrected version of Einstein's results, as found in the text by Landau and Lifshitz (1959)).

When there are clusters of spheres falling in a fluid—an unusual behavior where some particles pass others and the passed particles speed up and pass the original passers—has been observed and explained in part using Stokes's solution for the drag force (Hocking, 1964).

8.3 Slow Steady Flow Past a Liquid Sphere

A result similar to the preceding for liquid spheres is also useful, as was demonstrated by Millikan's experiments. Expect that equation (8.2.1) can be used outside of the sphere, but because it is infinite at the origin, a different nonsingular solution must be found inside the sphere. Also, the doublet and stokeslet are not applicable inside the sphere for the same reason. The rotlet can be used if **C** is chosen as

$$\mathbf{C} = \frac{1}{6}\mathbf{k}\left[\left(x^2 + y^2\right)^2 + 1.5z^2\left(x^2 + y^2\right) - \frac{13}{3}z^4 \right]. \qquad (8.3.1)$$

This satisfies equation (8.1.1) and has the velocity field $z\mathbf{e}_r - 2r^2\mathbf{k}$ expressed in cylindrical polar coordinates. Then, using this rotlet plus a uniform stream of strength D in the z direction, the result is

$$\mathbf{v} = \begin{cases} \left(U - \dfrac{2B}{R^3} - \dfrac{C}{R}\right)\cos\beta\,\mathbf{e}_R + \left(-U - \dfrac{B}{R^3} + \dfrac{C}{R}\right)\sin\beta\,\mathbf{e}_\beta, & R \geq a, \\[2mm] (D - R^2 E)\cos\beta\,\mathbf{e}_R + \left(-D + 2R^2 E\right)\sin\beta\,\mathbf{e}_\beta, & R \leq a. \end{cases} \tag{8.3.2}$$

Requiring that both of the $\mathbf{v}_\beta$'s be the same at $R = a$ gives $U + \frac{B}{a^3} - \frac{C}{a} = D - 2a^2 E$.

Requiring that both of the $\mathbf{v}_R$'s be zero at $R = a$ (so that the sphere does not change in size) gives

$$U - \frac{2B}{a^3} - \frac{C}{a} = D - a^2 E = 0.$$

Thus,

$$\tag{8.3.3}$$

$$B = a^3 U - Ca^2, \quad D = -\frac{3U}{2} + \frac{C}{4a}, \quad E = -\frac{3U}{2a^2} + \frac{C}{4a^3}.$$

The constant C is determined by requiring that $\tau_{R\beta}$ be continuous on the surface of the sphere. Using $\tau_{R\beta} = \mu(\frac{1}{R}\frac{\partial v_R}{\partial \beta} + \frac{\partial v_\beta}{\partial R} - \frac{v_\beta}{R})$ gives

$$C = -\frac{Ua^3}{4}\frac{\mu_{\text{inner}}}{\mu_{\text{inner}} + \mu_{\text{outer}}}. \tag{8.3.4}$$

Then

$$\mathbf{v} = \begin{cases} U\cos\beta\left(1 - \dfrac{2\mu_{\text{outer}} + 3\mu_{\text{inner}}}{\mu_{\text{outer}} + \mu_{\text{inner}}}\dfrac{a}{2R} + \dfrac{\mu_{\text{inner}}}{\mu_{\text{outer}} + \mu_{\text{inner}}}\dfrac{a^3}{2R^3}\right)\mathbf{e}_R \\[3mm] \quad + U\sin\beta\left(-1 + \dfrac{2\mu_{\text{outer}} + 3\mu_{\text{inner}}}{\mu_{\text{outer}} + \mu_{\text{inner}}}\dfrac{a}{4R} + \dfrac{\mu_{\text{inner}}}{\mu_{\text{outer}} + \mu_{inner}}\dfrac{a^3}{4R^3}\right)\mathbf{e}_\beta \quad \text{for } R \geq a, \\[3mm] \dfrac{\mu_{\text{outer}}}{2(\mu_{\text{outer}} + \mu_{\text{inner}})}U\left[\cos\beta\left(\dfrac{R^2}{a^2} - 1\right)\mathbf{e}_R + \sin\beta\left(1 - \dfrac{2R^2}{a^2}\right)\mathbf{e}_\beta\right] \quad \text{for } R \leq a. \end{cases}$$
$$\tag{8.3.5}$$

The portion of the pressure field is found by substituting equation (8.3.5) into the steady form of equation (8.1.2). A body force must be included to provide the mechanism for the flow. Here, gravity has been chosen as the most common body force. The result is

$$p = \begin{cases} \cos\beta\left(-U\dfrac{\mu_{\text{outer}}}{2}\dfrac{a}{R^3}\dfrac{2\mu_{\text{inner}} + 3\mu_{\text{outer}}}{\mu_{\text{inner}} + \mu_{\text{outer}}} - \rho_{\text{outer}}Rg\right) & \text{for } R \geq a, \\[3mm] \cos\beta\left(U\dfrac{5\mu_{\text{outer}}\mu_{\text{inner}}}{\mu_{\text{inner}} + \mu_{\text{outer}}}\dfrac{R}{a^2} - \rho_{\text{inner}}Rg\right) & \text{for } R \leq a. \end{cases} \tag{8.3.6}$$

Having the velocity field, the stresses can be found as

$$\tau_{R\beta} = \mu\left(\frac{\partial v_\beta}{\partial R} - \frac{v_\beta}{R} + \frac{1}{R}\frac{\partial v_R}{\partial \beta}\right) = \begin{cases} -3U\dfrac{\mu_{\text{outer}}\mu_{\text{inner}}}{2(\mu_{\text{outer}} + \mu_{\text{inner}})}\dfrac{a^3}{R^4}\sin\beta, & R \geq a, \\[3mm] -3U\dfrac{\mu_{\text{outer}}\mu_{\text{inner}}}{2(\mu_{\text{outer}} + \mu_{\text{inner}})}\dfrac{R}{a^2}\sin\beta, & R \leq a, \end{cases}$$

$$\tau_{RR} = -p + 2\mu \frac{\partial v_R}{\partial R}$$

$$= \cos\beta \begin{cases} 3\mu_{\text{outer}}U\left[\dfrac{1}{2}\left(\dfrac{2\mu_{\text{outer}}+3\mu_{\text{inner}}}{\mu_{\text{outer}}+\mu_{\text{inner}}}\right)\dfrac{a}{R^2} - \left(\dfrac{\mu_{\text{inner}}}{\mu_{\text{outer}}+\mu_{\text{inner}}}\right)\dfrac{a^3}{R^4}\right] \\[3mm] +\rho_{\text{outer}}Rg, \quad R \geq a, \\[4mm] -3\mu_{\text{outer}}U\left(\dfrac{\mu_{\text{inner}}}{\mu_{\text{outer}}+\mu_{\text{inner}}}\right)\dfrac{R}{a^2} + \rho_{\text{inner}}Rg, \quad R \leq a, \end{cases}$$

Therefore, on $R = a$,

$$\tau_{RR} = \cos\beta \begin{cases} \left[\dfrac{3U}{2a}\mu_{\text{outer}}\left(\dfrac{2\mu_{\text{outer}}+3\mu_{\text{inner}}}{\mu_{\text{outer}}+\mu_{\text{inner}}}\right) - \dfrac{\mu_{\text{inner}}\mu_{\text{outer}}}{\mu_{\text{outer}}+\mu_{\text{inner}}}\right] - \rho_{\text{outer}}ag, \quad R = a+, \\[4mm] -\dfrac{3U}{a}\dfrac{\mu_{\text{outer}}\mu_{\text{inner}}}{\mu_{\text{outer}}+\mu_{\text{inner}}} - \rho_{\text{inner}}ag, \qquad\qquad\qquad R = a-. \end{cases}$$

Requiring the two forms of the normal stress τ_{RR} be the same at $R = a$ gives

$$\left[\frac{3U}{2a}\mu_{\text{outer}}\left(\frac{2\mu_{\text{outer}}+3\mu_{\text{inner}}}{\mu_{\text{outer}}+\mu_{\text{inner}}}\right) - \frac{\mu_{\text{inner}}\mu_{\text{outer}}}{\mu_{\text{outer}}+\mu_{\text{inner}}}\right] - \rho_{\text{outer}}ag$$

$$= -\frac{3U}{a}\frac{\mu_{\text{outer}}\mu_{\text{inner}}}{\mu_{\text{outer}}+\mu_{\text{inner}}} - \rho_{\text{inner}}ag,$$

yielding for the liquid sphere's terminal velocity

$$U = \frac{2}{3}\frac{(\rho_{\text{outer}}-\rho_{\text{inner}})a^2g}{\mu_{\text{outer}}}\left(\frac{\mu_{\text{outer}}+\mu_{\text{inner}}}{2\mu_{\text{outer}}+3\mu_{\text{inner}}}\right). \tag{8.3.7}$$

Limiting cases of this which are of interest are

$$U \to \frac{2ga^2\left(\rho_{\text{outer}}-\rho_{\text{inner}}\right)}{9\mu_{\text{outer}}} \quad \text{as } \mu_{\text{inner}} \to \infty \text{ (solid sphere)}, \tag{8.3.8}$$

$$U \to \frac{ga^2\left(\rho_{\text{outer}}-\rho_{\text{inner}}\right)}{3\mu_{\text{outer}}} \quad \text{as } \mu_{\text{inner}} \to 0 \text{ (vapor sphere)}.$$

In the preceding analysis, surface tension has been neglected because the combination of a perfectly spherical liquid drop and surface tension is incompatible with gravity. Generally speaking, providing the Reynolds numbers are small, surface tension will play a minor role. However, for very small gas bubbles rising in a liquid, it has been found that their velocity is closer to that predicted for a solid sphere rather than the second of equation (8.3.8). In this case, it is believed that the presence of surface-active impurities in the liquid can form a mesh of large molecules on the interface, causing it to behave like a rigid surface (Levich, 1962).

8.4 Flow Due to a Sphere Undergoing Simple Harmonic Translation

If a solid sphere undergoes simple harmonic motion along the z-axis in a fluid that is otherwise at rest, the flow can be described by a stokeslet-doublet combination. A stream function approach can be used and the development in Lamb (1932, page 643) followed.

The calculations are a bit lengthy, so to simplify things, complex numbers are used for the stream function, velocity components, and stresses, with the understanding that the real portions of the expressions are intended.

Inserting the general form $\psi = \sin^2 \beta f(r) U$ with $U = U_0 e^{i\omega t}$ into equation (8.1.1) gives solutions

$$f(R) = \frac{A}{R} + \frac{B}{R}(1 + kR)e^{-k(R-a)}.$$

Here, $k = \sqrt{i\rho\omega/\mu} = (1+i)\alpha$ with $\alpha = \sqrt{\rho\omega/2\mu}$. In obtaining solutions of equation (8.1.1) those portions that die out as R becomes large have been selected.

Applying the no-slip boundary conditions on the surface of the sphere, the constants A and B are found to be

$$A = \frac{3a}{2k^2}\left(1 + ka + \frac{1}{3}k^2a^2\right), \quad B = -\frac{3a}{2k^2},$$

giving

$$\psi = \frac{3\sin^2\beta}{2k^2}\frac{a}{R}\left[\left(1 + ka + \frac{1}{3}k^2a^2\right) - (1+kR)e^{-k(R-a)}\right]U, \qquad (8.4.1)$$

This yields velocity components

$$v_R = \frac{3\cos\beta}{k^2a^2}\left(\frac{a}{R}\right)^3 U\left[\left(1 + ka + \frac{1}{3}k^2a^2\right) - (1+kR)e^{-k(R-a)}\right],$$

$$v_\beta = \frac{3\sin\beta}{2k^2a^2}\left(\frac{a}{R}\right)^3 U\left[\left(1 + ka + \frac{1}{3}k^2a^2\right) - \left(1+kR+k^2aR\right)e^{-k(R-a)}\right]. \qquad (8.4.2)$$

Having found the stream function and velocity, the pressure can be found from equation (8.1.5). Since the pressure satisfies the Laplace equation, it must be associated with the irrotational portion of the flow: the doublet. Thus, the pressure satisfies

$$\nabla p = -\rho\frac{\partial v_{\text{doublet}}}{\partial t}.$$

The pressure is thus given by

$$p = i\rho\omega U A \cos\beta/R^2. \qquad (8.4.3)$$

The force needed to produce this motion is found as before by integrating the shear stresses over the surface of the sphere as in equation (8.2.5). The result is

$$F_z = -\int_0^\pi \left(\tau_{RR}\cos\vartheta - \tau_{R\vartheta}\sin\vartheta\right)_a 2\pi a^2 \sin\vartheta d\vartheta$$

$$= \frac{4}{3}\pi\rho a^3\left(\frac{1}{2} + \frac{9}{4a\alpha}\right)(i\omega U) + 3\pi\rho a^3\omega\frac{1}{a\alpha}\left(1 + \frac{1}{a\alpha}\right)U \qquad (8.4.4)$$

$$= \frac{4}{3}\pi\rho a^3\left(\frac{1}{2} + \frac{9}{4}\sqrt{\frac{2\mu}{\rho\omega a^2}}\right)\frac{dU}{dt} + 3\pi\rho a^3\omega\sqrt{\frac{2\mu}{\rho\omega a^2}}\left(1 + \sqrt{\frac{2\mu}{\rho\omega a^2}}\right)U.$$

The one-half in the acceleration term represents the added mass found in potential flow theory, while the accompanying term represents a viscous correction to the added mass. We can see that as the frequency becomes small, the result reduces to that found for the stationary sphere.

8.5 General Translational Motion of a Sphere

The results of the previous section can be generalized to arbitrary translational motion by use of the Fourier transform. If the velocity of the sphere is taken to be $U(t)$, then Fourier transform theory tells us that there is a Fourier representation of the velocity in terms of the pair of functions

$$U(t) = \int_{-\infty}^{\infty} F(\omega) e^{-i\omega t} d\omega \quad \text{and} \quad F(\omega) = \frac{1}{2\pi} \int_{-\infty}^{\infty} U(t) e^{i\omega t} dt. \quad (8.5.1)$$

Landau and Lifshitz (1959, page 96) showed that the drag for each harmonic component for the previous problem is given by

$$\pi \rho a^3 e^{i\omega t} \left[\frac{6\mu}{\rho a^2} F(\omega) + \left(\frac{2}{3} + 3(1+i)\sqrt{\frac{2\mu}{\rho \omega}} \right) \dot{F}(\omega) \right], \quad (8.5.2)$$

where $\dot{F}(\omega)$ is the Fourier transform of $\frac{dU}{dt}$. The integration of this to produce the force requires some careful handling of the path of integration (a not uncommon problem with Fourier integrals), the net result being

$$F(t) = 2\pi \rho a^3 \left[\frac{1}{3} \frac{dU}{dt} + 3 \frac{\mu}{\rho a^2} U + \frac{3}{a} \sqrt{\frac{\mu}{\pi \rho}} \int_{-\infty}^{t} \frac{dU(\tau)}{d\tau} \frac{d\tau}{\sqrt{t-\tau}} \right]. \quad (8.5.3)$$

8.6 Oseen's Approximation for Slow Viscous Flow

If the solution of Section 8.2 for slow flow past a sphere were to be investigated further, it would be found that at large distances from the sphere the neglected convective acceleration terms were actually more important than the viscous terms retained in our analysis. The ratio of the inertia to viscous forces is in fact of the order of $\rho U r / \mu$. Thus, any efforts made to try to use the Stokes approximation as a starting point for solutions of the Navier-Stokes equations for somewhat larger values of the Reynolds number would rapidly run into insurmountable difficulties. This has been termed the **Whitehead paradox**, after the mathematician Alfred North Whitehead.

Even more embarrassing, if an attempt were made to solve the Stokes equations for the slow flow past a circular cylinder, it would be found that the solution for the stream function is of the form

$$\psi = \left(A r^3 + B r \ln r + C r + \frac{D}{r} \right) \sin \theta. \quad (8.6.1)$$

The constants A and B have to be rejected because they give velocities that grow faster with r than does a uniform stream. C represents a uniform stream, so what is left is one constant, D, to satisfy two boundary conditions! This is known as **Stokes paradox**. (A dictionary definition of *paradox* is "an argument that apparently derives self-contradictory conclusions by valid deduction from acceptable premises." As we will see in the next section, in this case the "acceptable premises" were not valid.)

Oseen (1910) suggested an alternative approach, approximating the Navier-Stokes equations by

$$\rho\left(\frac{\partial \mathbf{v}}{\partial t} + U\frac{\partial \mathbf{v}}{\partial x}\right) = -\nabla p + \mu \nabla^2 \mathbf{v}, \tag{8.6.2}$$

U being the free stream velocity in the x direction. Oseen obtained a solution for the circular cylinder in terms of

$$v_r = \frac{\partial \phi}{\partial r} + \frac{1}{2k}\frac{\partial \chi}{\partial r} - \chi \cos\theta,$$

$$v_\theta = \frac{1}{r}\frac{\partial \phi}{\partial \theta} + \frac{1}{2kr}\frac{\partial \chi}{\partial \theta} + \chi \sin\theta, \tag{8.6.3}$$

where

$$\phi = Ua\left[\frac{r}{a}\cos\theta + A_0\ln\frac{r}{a} + \sum_{n=1}^{\infty}\frac{1}{n}A_n\left(\frac{a}{r}\right)^n\cos n\theta\right],$$

$$\chi = Ue^{kr\cos\theta}\sum_{n=0}^{\infty}B_nK_n(kr)\cos n\theta. \tag{8.6.4}$$

Here, $k = \rho U/\mu$ and the K_n's are the modified Bessel functions. The A_n and B_n are determined from the no-slip boundary conditions.

Oseen also used his approximation to find the slow flow past a sphere. The solution is more complicated than the Stokes equations and involves much tedious computation. His result for the drag force is

$$F = 6\pi\mu aU\left(1 + \frac{3}{8}\text{Re}\right), \quad \text{where Re} = \rho Ua/\mu. \tag{8.6.5}$$

Stokes's formula is seen to be the limit as Re goes to zero.

Goldstein (1929) carried out a series expansion in the Reynolds number for the sphere using Oseen's equations. His results for the drag coefficient gave

$$C_D = \frac{24}{\text{Re}}\left(1 + \frac{3}{16}\text{Re} - \frac{19}{1,280}\text{Re}^2 + \frac{71}{20,480}\text{Re}^3 - \frac{30,179}{34,406,400}\text{Re}^4\right.$$

$$\left. + \frac{122,519}{550,502,400}\text{Re}^5 \pm \cdots\right). \tag{8.6.6}$$

Here the Reynolds number is based on the diameter. As we will see in the next section, this result is not correct, since an expansion solely in powers of Re is inadequate.

The drag force per unit length of a circular cylinder as determined by the Oseen equation was found by Lamb (1911) to be

$$F_{\text{per unit length}} = \frac{4\pi\mu U}{\frac{1}{2} - \gamma - \log\left(\dfrac{\rho Ua}{4\mu}\right)}, \tag{8.6.7}$$

where γ is Euler's constant (also called Mascheroni's constant)

$$\gamma = \lim_{n\to\infty}\left(\sum_{m=1}^{n}\frac{1}{m} - \log n\right) = 0.577215664901533. \tag{8.6.8}$$

Another version of the drag force on a circular cylinder was given by Tomotika and Aoi (1950) as

$$C_D = \frac{8\pi}{\text{Re}\,S}\left[1 - \frac{1}{32}\left(\frac{5}{16S} - \frac{1}{2} + S\right)\text{Re}^2 + O(\text{Re}^4 \ln^2 \text{Re})\right], \qquad (8.6.9)$$

where $\text{Re} = \frac{\rho U D}{\mu}$ and $S = \frac{1}{7} - \gamma + \ln\frac{8}{\text{Re}}$. Their result was achieved by using matched asymptotic expansions.

For all of the formidable theory and numerous calculations that went into these results, comparison with experiments (Van Dyke, 1965, page 164) seems best for the result given in (8.6.9) up to a Reynolds number based on diameter of about 1.4.

8.7   Resolution of the Stokes/Whitehead Paradoxes

While Lamb (1932) points out many of the questions raised by the Stokes/Oseen equations, it wasn't until the mid-1950s that the questions were finally resolved. Proudman and Pearson (1957) and Kaplun (1957) independently pointed out that the Stokes equation hold only in the vicinity of the boundary and the Oseen equations hold only away from the boundary. Using matched asymptotic expansions, they found that an intermediate region could be found where both Stokes and Oseen were valid. By matching terms in that region, they were able to develop a solution. The result was an equation for the drag coefficient of the sphere given by Proudman and Pearson as

$$C_D = \frac{6\pi}{\text{Re}}\left(1 + \frac{3}{8}\text{Re} + \frac{9}{40}\text{Re}^2 \ln \text{Re} + O(\text{Re}^2)\right). \qquad (8.7.1)$$

The logarithmic term is forced by the matching process and the logarithms of the radius that appear in both the Stokes and Oseen solutions. Any expansions involving the Navier-Stokes equations should be aware that logarithms of the Reynolds numbers must be anticipated. This is presaged in equation (8.6.7).

Thus, both the large Reynolds number and the small Reynolds number solutions have a common nature. For the large Reynolds number case, Prandtl's boundary layer equations must be used near the boundary, and the Euler equations must be used away from the boundary. For the small Reynolds number case, Stokes equations must be used near the boundary, and Oseen's equations must be used away from the boundary. This is true for the lowest-order expansions in each category. As more than one term in the expansion is considered, the necessary equations become modified and, as should be anticipated, more complex.

One final remark must be made about the order of the various terms in these expansions, particularly terms involving $\ln \text{Re}$. As pointed out by Van Dyke (1964, Sections 1.3, 10.5), when the next unknown term in an expansion is of order Re^n, then even though mathematically it is smaller than $\text{Re}^n \ln \text{Re}$, since it numerically can be of equal magnitude, it should be included. That is, even though

$$\lim_{\text{Re}\to 0}\frac{\text{Re}^n}{\text{Re}^n \ln \text{Re}} \to 0, \quad \text{for } \text{Re} = 0.01\text{— say,}$$

$$\frac{\text{Re}^2}{\text{Re}^2 \ln \text{Re}} = \frac{10^{-4}}{10^{-4}\ln 0.01} = -\frac{1}{4.605} = -0.217,$$

which means that for practical purposes $\text{Re}^2 \ln \text{Re}$ and Re^2 can contribute equally to a calculation.

Problems—Chapter 8

8.1 Show that the flow field given by $\mathbf{v} = R\mathbf{e}_R \times \mathbf{e}_z$, $p = 0$ is a solution of the Stokes equations. Use it along with a steady rotlet to find the velocity and pressure for a fluid contained between two concentric spheres of radii a and b, where the outer sphere is rotating with angular velocity Ω_{outer} about the x-axis and the inner sphere is rotating with angular velocity Ω_{inner} about the same axis and in the same direction.

8.2 Millikan's experiment for making the first accurate measurement of the charge on an electron involved spraying a few tiny droplets of oil between two electrically charged plates. He first determined the diameter of the small drops by measuring their terminal velocity without an electric field ($V_{withoutE}$) and using Stokes's formula. He then applied a voltage E (the electric potential in volts ($N - m/C$)) across the plates, resulting in an upward force Eq/d on the drop and a terminal velocity V_{withE}. Here, d is the plate spacing in meters, and q is the electron charge in Coulombs (C). Making repeated measurements, he found that the values of q he determined were always integer multiples of a specific number q_1, the charge due to a single electron. Determine first a formula for the droplet diameter when no electric field is present. Then determine the terminal velocity when the electric field is present. Solve for q in terms of the weight of the droplet minus the buoyancy force, the terminal velocities with and without an electric field, and the quantity E/d.

8.3 In a repeat of Millikan's experiment (Problem 8.2), the voltage was 115 volts, the plate spacing 4.53 mm, the oil density was $920 \, kg/m^3$, and the air viscosity was 1.82×10^{-4} poise. Two horizontal lines were drawn 1.73 mm apart, and the time of traverse of these two lines was measured with a stopwatch. By changing the direction of the electric filed, it was possible to use the same droplet for each measurement, although each set of measurements did have different numbers of electrons attached to the droplet. Times of transit were as follows:

72.2 seconds averaged over 10 measurements, no electric field, direction down
41.1 seconds averaged over 8 measurements, electric field present, direction up
23.5 seconds averaged over 2 measurements, electric field present, direction up
16.5 seconds averaged over 5 measurements, electric field present, direction up
12.5 seconds averaged over 3 measurements, electric field present, direction up
8.06 seconds averaged over 7 measurements, electric field present, direction up

Find the terminal velocities for the six sets of data, and estimate the charge due to a single electron. Be sure to observe the proper sign on the velocities.

Note: Millikan also found that Brownian motion due to the small size of these droplets gives errors in measuring velocity, resulting in charge estimates that are about 27% too high.

Flow Stability

9.1 Linear Stability Theory of Fluid Flows

Chapters 3 and 4 examined several examples that involved the stability of inviscid flows, such as the round jet, the vortex street, and several cases of interfacial waves. In those cases an infinitesimal disturbance was added to a primary flow, and then it was determined whether there was a possibility of the disturbance growing, decaying, or remaining unchanged. The driving mechanisms involved in those examples were gravity and surface tension interacting with the momentum of the flow.

In this chapter the same type of analysis is used and applied to laminar viscous flows as well. The presence of viscosity in some cases will dampen the destabilizing influences and in others act to enhance them and thus destabilize the primary flow. Flows frequently are laminar for sufficiently small values of a dimensionless parameter, such as the Reynolds number, and become turbulent once the parameter becomes sufficiently large. While this transition value of the parameter can be found experimentally, as Reynolds did for pipe flow, in many cases a flow has so many describing parameters that a complete experimental study becomes too costly. An analytic approach then can be desirable, with the added advantage of possibly shedding some light on the physical mechanisms involved in the transition.

The classical approach in studying such transitions, introduced in the nineteenth century in the study of inviscid flows, is to take the solution for the laminar flow and add to it a very small disturbance. There is usually some experimental knowledge of the form of the instability, and the disturbance is tailored to reflect this. This combined flow is then put into the Navier-Stokes equations, and the original laminar flow terms are subtracted. By themselves, the primary flow satisfies the Navier-Stokes equations.

However, some interaction terms between the primary and secondary flows remain behind by this process. The remaining equations are then linearized in the disturbance quantities. This set of partial differential equations is then solved to determine whether the disturbance will grow or decay in time. For viscous flows, the mathematics for finding the solution of the stability equations is generally much more complicated than it is for inviscid flows, due to the higher order of the resulting differential equations and also possible additional physical mechanisms for supporting the instability.

The use of infinitesimal disturbances results in linear equations, which by themselves have proven difficult to solve. In some of the cases where nonlinear formulations have been made and solutions found, the efforts naturally increase the difficulty of finding a solution. Landau and Lifshitz (1964) present some of the procedures and assumptions involved with these nonlinear equations.

Because a linear approach is being used, if our solution predicts an unstable solution of the primary flow, this by itself cannot predict the nature of the secondary flow. In some cases a direct transition to turbulent flow exists, and in others, secondary laminar flows evolve, more complicated in nature than the original primary flow but still not the fully complex character associated with turbulent flow. Some flows also can evolve to have tertiary flows that are still laminar. Such is the complexity of fluid flow behavior!

9.2   *Thermal Instability in a Viscous Fluid— Rayleigh-Bénard Convection*

One of the least mathematically complicated examples of the analysis of viscous flow instability is due to Lord Rayleigh (1916a). Several investigators in the nineteenth century had noted that tesselated structures were frequently seen in the drying of horizontal thin layers of fluids. Bénard (1900, 1901) performed quantitative experiments on this phenomenon, most of the details later explained by Rayleigh's analysis. This instability can be seen in the drying of paints with metallic particles and also in the stratified atmosphere, where it results in interesting cloud patterns.

While there are many versions of the boundary conditions that can be studied, only the case of a fluid confined between two horizontal plates, each with horizontal extent large compared to their vertical spacing, will be considered here. (A thorough description of the extensive literature available on this problem is given in Koschmeider (1993)). The plates are each heated to a constant temperature, with the lower plate being the hottest, as shown in Figure 9.2.1. Thus, there is a basic gravitational instability, with lighter fluid on the bottom. Initially, the fluid between the plates is at rest, with a

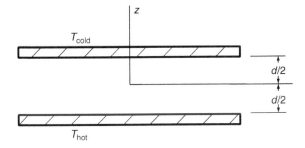

Figure 9.2.1   Physical layout for studying Bénard stability

vertical hydrostatic pressure gradient. The density will be taken as varying linearly with temperature according to

$$\rho = \rho_{\text{mean}} \left[1 - \beta(T - T_{\text{mean}}) \right], \quad \text{with } T_{\text{mean}} = \frac{T_{\text{hot}} + T_{\text{cold}}}{2}. \tag{9.2.1}$$

Here, reference density ρ_{mean} and reference temperature T_{mean} have been taken as the density and temperature midway between the plates. Beta is given by

$$\beta = -\frac{1}{\rho}\frac{d\rho}{dT} = -\frac{1}{\rho}\frac{d\rho}{dz}\frac{dz}{dT} = -\frac{1}{\rho}\frac{\Delta\rho}{\Delta T}. \tag{9.2.2}$$

The plate spacing will be chosen as b, and the origin will be taken halfway between the plates.

Since there is no mean flow for this problem, the Navier-Stokes and energy equations for the primary undisturbed flow are given by

$$T_0 = T_{\text{mean}} - \frac{z}{b}\Delta T, \ \rho_0 = \rho_{\text{mean}}\left(1 - \frac{z}{b}\beta\Delta T\right), \quad \text{with } \Delta T = T_{\text{hot}} - T_{\text{cold}}. \tag{9.2.3}$$

Here, primary flow components $p_0(z), \rho_0$, and $T_0(z)$) are denoted by the subscript 0, and **k** is a unit vector in the direction of gravity (z). From these and the boundary conditions, and taking the origin halfway between the plates, find for the quiescent primary flow that

$$0 = -\nabla p_0 - g\mathbf{k}\rho_0, \tag{9.2.4a}$$

$$0 = k\frac{d^2 T_0}{dy^2}. \tag{9.2.4b}$$

To investigate the stability of this flow, introduce a small disturbance and see whether it grows or decays. For the disturbed flow, velocities, pressure, density and temperatures will be denoted by primes. The combined primary and disturbance flows must satisfy the continuity, Navier-Stokes, and energy equations.

It shall be assumed that even though the basic density does vary with z, the main influence of the density variation on the flow stability is going to be felt in the body force and not in the acceleration term. Thus, the ρ_0 multiplying the acceleration term is taken at its mean value and is considered constant, and the flow will be regarded as being incompressible. This is called the **Boussinesq approximation**.

After linearizing on the small disturbance terms and then subtracting out the primary flow terms, the terms' first order in the primed quantities are

$$\rho_{\text{mean}}\nabla \cdot \mathbf{v}' = 0, \tag{9.2.5}$$

$$\rho_{\text{mean}}\frac{\partial \mathbf{v}'}{\partial t} = -\nabla p' + \rho_{\text{mean}}\beta g T'\mathbf{k} + \mu\nabla^2\mathbf{v}', \tag{9.2.6}$$

$$\rho_{\text{mean}}c_p\left(\frac{\partial T'}{\partial t} - v_z'\frac{\Delta T}{b}\right) = k\nabla^2 T'. \tag{9.2.7}$$

Together with the conditions that $\mathbf{v}'$ and T' vanish at each plate, the mathematical statement of the problem has been completed. Now these equations must be made manageable.

The first step is to eliminate the pressure. Taking the divergence of equation (9.2.6) and using the continuity condition equation (9.2.5) gives

$$0 = -\nabla^2 p' + \rho_{\text{mean}}\beta g\frac{\partial T'}{\partial z}. \tag{9.2.8}$$

Similarly, operating on the z component of equation (9.2.6) with the Laplace operator gives

$$\rho_{\text{mean}} \frac{\partial}{\partial t} \nabla^2 v'_z - \frac{\partial}{\partial z} \left(\nabla^2 p' \right) + \rho_{\text{mean}} \beta g \nabla^2 T' + \mu \nabla^2 \left(\nabla^2 v'_z \right). \tag{9.2.9}$$

Using equation (9.2.8) to eliminate the pressure in equation (9.2.9) gives

$$\rho_{\text{mean}} \frac{\partial}{\partial t} \nabla^2 v'_z = -\frac{\partial}{\partial z} \left(\rho_{\text{mean}} \beta g \frac{\partial T'}{\partial z} \right) + \rho_{\text{mean}} \beta g \nabla^2 T' + \mu \nabla^2 \left(\nabla^2 v'_z \right),$$

which simplifies to

$$\left(\rho_{\text{mean}} \frac{\partial}{\partial t} - \mu \nabla^2 \right) \nabla^2 v'_z = \rho_{\text{mean}} \beta g \nabla_1^2 T',$$

$$\text{where} \quad \nabla_1^2 = \nabla^2 - \frac{\partial^2}{\partial z^2} = \frac{\partial^2}{\partial x^2} + \frac{\partial^2}{\partial y^2}. \tag{9.2.10}$$

Next, rewrite equation (9.2.7) in the form

$$\left(\rho_{\text{mean}} c_p \frac{\partial}{\partial t} - k\nabla^2 \right) T' = \rho_{\text{mean}} c_p \frac{\Delta T}{b} v'_z. \tag{9.2.11}$$

Equations (9.2.10) and (9.2.11), then, are two equations in the two unknowns v'_z and T'.

Since the plates are considered to be large in extent, no boundary conditions will be imposed at the edges of the plates. The boundary conditions imposed are simply

$$v'_x = v'_y = v'_z = T' = 0 \quad \text{on} \quad z = \pm\frac{b}{2}.$$

At this point the velocity components and the temperature can be taken as being of the form

$$\mathbf{v}'(x, y, z, t) = [U(z), V(z), W(z)]f(x, y)e^{st}, \tag{9.2.12}$$

$$T'(x, y, z, t) = \Theta(z)f(x, y)e^{st}, \tag{9.2.13}$$

where

$$\frac{\partial^2 f}{\partial x^2} + \frac{\partial^2 f}{\partial y^2} = -\frac{a^2}{b^2}f. \tag{9.2.14}$$

This form allows separation of variables and is sufficiently general to describe the observed flows.

If you have studied separation of variables in a mathematics course, you many be surprised that this separable form had to be assumed rather than have it follow from the analysis. The reason is that classical presentations of the method of separation of variables all deal with second-order partial differential equations, while here our system is actually of sixth order. Generally, systems of order higher than the second do not separate except under very special boundary conditions. The governing equation introduced for f allows for separating the variables and is compatible with a number of boundary conditions. Further comments on the nature of f will be made a bit later.

Putting these forms into our system of equations, using equation (9.2.9) and cancelling f where it appears in every term of the equation, equations (9.2.5), (9.2.10), and (9.2.11) become

$$U \frac{\partial f}{\partial x} + V \frac{\partial f}{\partial y} + f \frac{dW}{dz} = 0, \tag{9.2.15}$$

$$\left[\rho_{\text{mean}}s - \mu\left(\frac{d^2}{dz^2} - \frac{a^2}{b^2}\right)\right]\left(\frac{d^2}{dz^2} - \frac{a^2}{b^2}\right)W = -\frac{a^2}{b^2}\rho_{\text{mean}}\beta g\Theta, \qquad (9.2.16)$$

$$\left[\rho_{\text{mean}}c_p s - k\left(\frac{d^2}{dz^2} - \frac{a^2}{b^2}\right)\right]\Theta = \frac{\Delta T}{b}W. \qquad (9.2.17)$$

Note from equation (9.2.15) that $U = V = 0$ on the boundaries implies that $\frac{dW}{dz} = 0$ on the boundaries also.

It is possible that the time constant s could be complex. The real portion would be associated with a growth or decay rate, and the imaginary part with an oscillatory behavior. To show that the oscillatory behavior cannot occur, perform the following operations:

1. The momentum equation (9.2.16) is multiplied by the complex conjugate of W, denoted as W^*, and integrated over the gap between the plates. Using integration by parts and the boundary conditions then gives

$$\frac{s\rho_{\text{mean}}}{b^2}J_1 + \frac{\mu}{b^4}J_2 = -\rho_{\text{mean}}\beta g\frac{a^2}{b^2}\int_{-b/2}^{b/2} W^*\Theta dz, \qquad (9.2.18)$$

where

$$J_1 = b^2\int_{-b/2}^{b/2}\left(\left|\frac{dW}{dz}\right|^2 + \frac{a^2}{b^2}|W|^2\right)dz,$$

$$J_2 = b^4\int_{-b/2}^{b/2}\left(\left|\frac{d^2W}{dz^2}\right|^2 + 2\frac{a^2}{b^2}\left|\frac{dW}{dz}\right|^2 + \frac{a^4}{b^4}|W|^2\right)dz$$

are both positive definite.

2. The energy equation (9.2.17) is multiplied by the complex conjugate of Θ, denoted as Θ^*, and integrated over the gap between the plates. Using integration by parts and the boundary conditions gives

$$\left(\frac{\rho_{\text{mean}}bc_p\Delta T}{k}\right)^2\left(s\rho_{\text{mean}}c_p I_0 + \frac{k}{b^2}I_1\right) = -\rho_{\text{mean}}c_p\frac{\Delta T}{b}\int_{-b/2}^{b/2} W\Theta^* dz,$$

$$\text{where } I_0 = \left(\frac{k}{\rho_{\text{mean}}bc_p\Delta T}\right)^2\frac{1}{C^2}\int_{-b/2}^{b/2}|\Theta|^2\,dz,$$

$$(9.2.19)$$

$$I_1 = \left(\frac{bk}{\rho_{\text{mean}}bc_p\Delta T}\right)^2\int_{-b/2}^{b/2}\left(\left|\frac{d\Theta}{dz}\right|^2 + \frac{a^2}{b^2}|\Theta|^2\right)dz$$

are both positive definite. The rather awkward-appearing combination of parameters appearing before the Is are to make the dimensions coincide in the final result.

Notice that, except for multiplying constants, the right-hand sides of equations (9.2.18) and (9.2.19) are the complex conjugates of one another. From this it follows that

$$\left(\frac{s\rho_{\text{mean}}c_p b^2}{k}\frac{1}{\text{Pr}}J_1 + J_2\right) = a^2 Ra\left(\frac{s^*\rho_{\text{mean}}c_p b^2}{k}I_0 + I_1\right), \qquad (9.2.20)$$

where Pr is the Prandtl number and $Ra = \frac{b^3 g\rho_{\text{mean}}^2 c_p\beta\Delta T}{k\mu}$ is called the ***Rayleigh number***.

Taking the imaginary portion of both sides of equation (9.2.20) gives

$$s_i J_1 = -s_i PrRa I_0, \quad \text{giving } s_i(J_1 + PrRa I_0) = 0. \qquad (9.2.21)$$

It follows from this that if Ra is positive (Ra is positive if ΔT is positive, so heavier fluid is on the bottom), s must be real.

If the real portion of both sides of equation (9.2.20) is taken, the result is

$$\frac{s_r \rho_{\text{mean}} c_p b^2}{k} \left(\frac{1}{Pr} J_1 - a^2 Ra I_0 \right) = a^2 Ra I_1 - J_2. \qquad (9.2.22)$$

This states that the real part of s can be either positive or negative if ΔT is positive.

The results of the previous paragraph are often referred to as the **principle of exchange of stabilities**. While the appropriateness of this title is unclear, the usefulness of the principle is without doubt.

Since equations (9.2.16) and (9.2.17) have constant coefficients, their solution is in the form of exponentials and they can be found in a standard manner. That is, letting $W = Ae^{cz/b}$, $\Theta = Be^{cz/b}$, equations (9.2.16) and (9.2.17) become

$$\left[\rho_{\text{mean}} s b^2 - \mu \left(c^2 - a^2 \right) \right] \left(c^2 - a^2 \right) A = -a^2 b^2 \rho_{\text{mean}} \beta g B,$$

$$\left(\rho_{\text{mean}} c_p b^2 s - k \left(c^2 - a^2 \right) \right) B = \rho_{\text{mean}} c_p b \Delta T A.$$

Eliminating A and B between these two equations, find that c must satisfy

$$\left(c^2 - a^2 \right) \left(c^2 - a^2 - \frac{\rho_{\text{mean}} c_p b^2}{k} s \right) \left(c^2 - a^2 - \frac{\rho_{\text{mean}}}{\mu} s \right) + a^2 Ra = 0, \qquad (9.2.23)$$

where Ra is the Rayleigh number as just defined.

Equation (9.2.23) is a cubic equation for c^2, whose solution is somewhat messy. The calculations can be simplified by saying that only the neutrally stable case is desired to be solved; that is, only the results where the real part of s vanishes will be studied. In that case the roots become

$$c_n^2 = a^2 \left[1 - \left(\frac{Ra}{a} \right)^{1/3} \left(\cos \frac{2\pi n}{3} + i \sin \frac{2\pi n}{3} \right) \right], \quad n = 0, 1, 2, \ldots. \qquad (9.2.24)$$

The solution can now be expressed in terms of these roots. Since they occur in complex conjugate pairs, write

$$W = \sum_{n=0}^{2} \left(A_n \cosh \frac{c_n z}{b} + B_n \sinh \frac{c_n z}{b} \right),$$

$$\Theta = \frac{-\mu}{g a^2 b^2} \sum_{n=0}^{2} \left(c_n^2 - a^2 \right)^2 \left(A_n \cosh \frac{c_n z}{b} + B_n \sinh \frac{c_n z}{b} \right).$$

Require that U, V, W, and Θ vanish on $z = \pm b/2$, or equivalently, from continuity require that W, dW/dz, and Θ vanish at these boundaries. The result is

$$0 = \sum_{n=0}^{2} \left(A_n \cosh \frac{c_n}{2} + B_n \sinh \frac{c_n}{2} \right),$$

$$0 = \sum_{n=0}^{2} \left(A_n \cosh \frac{c_n}{2} - B_n \sinh \frac{c_n}{2} \right),$$

$$0 = \sum_{n=0}^{2} c_n \left(A_n \sinh \frac{c_n}{2} + B_n \cosh \frac{c_n}{2} \right),$$

$$0 = \sum_{n=0}^{2} c_n \left(-A_n \sinh \frac{c_n}{2} + B_n \cosh \frac{c_n}{2} \right),$$

$$0 = \sum_{n=0}^{2} \left(c_n^2 - a^2 \right) \left(A_n \cosh \frac{c_n}{2} + B_n \sinh \frac{c_n}{2} \right),$$

$$0 = \sum_{n=0}^{2} \left(c_n^2 - a^2 \right) \left(A_n \cosh \frac{c_n}{2} - B_n \sinh \frac{c_n}{2} \right).$$

(9.2.25)

These are six homogeneous equations in the six sets of unknown constants A_n, B_n. There is only a trivial solution unless the determinant of the coefficients of the A_n and B_n vanishes. In this case the 6 by 6 determinant can be written as the product of two 3 by 3 determinants. The result of the calculation is that either

$$\begin{vmatrix} \sinh \frac{c_0}{2} & \sinh \frac{c_1}{2} & \sinh \frac{c_2}{2} \\ c_0 \cosh \frac{c_0}{2} & c_1 \cosh \frac{c_1}{2} & c_2 \cosh \frac{c_2}{2} \\ \left(c_0^2 - a^2 \right)^2 \sinh \frac{c_0}{2} & \left(c_1^2 - a^2 \right)^2 \sinh \frac{c_1}{2} & \left(c_2^2 - a^2 \right)^2 \sinh \frac{c_2}{2} \end{vmatrix} = 0. \qquad (9.2.26)$$

or

$$\begin{vmatrix} \cosh \frac{c_0}{2} & \cosh \frac{c_1}{2} & \cosh \frac{c_2}{2} \\ c_0 \sinh \frac{c_0}{2} & c_1 \sinh \frac{c_1}{2} & c_2 \sinh \frac{c_2}{2} \\ \left(c_0^2 - a^2 \right)^2 \cosh \frac{c_0}{2} & \left(c_1^2 - a^2 \right)^2 \cosh \frac{c_1}{2} & \left(c_2^2 - a^2 \right)^2 \cosh \frac{c_2}{2} \end{vmatrix} = 0. \qquad (9.2.27)$$

The first 3 by 3 determinant came from the hyperbolic cosine terms in W and is referred to as the **even disturbance** because of the hyberbolic cosine terms in W. The second 3 by 3 determinant came from the hyperbolic sine terms in W and is referred to as the **odd disturbance**.

Expansion and simplification of the determinants in equations (9.2.26) and (9.2.27) give

$$\left(c_0^2 - a^2 \right) c_0 \tanh \frac{c_0}{2} + \left(c_1^2 - a^2 \right) c_1 \tanh \frac{c_1}{2} + \left(c_1^2 - a^2 \right) c_2 \tanh \frac{c_2}{2} = 0 \qquad (9.2.28)$$

for even disturbances, and

$$c_0 \coth \frac{c_0}{2} + \frac{1 + i\sqrt{3}}{2} c_1 \coth \frac{c_1}{2} + \frac{-1 + i\sqrt{3}}{2} c_2 \coth \frac{c_2}{2} = 0 \qquad (9.2.29)$$

for odd disturbances. Further simplification is possible by noting that c_0 is pure imaginary and c_2 is the complex conjugate of c_1. Therefore, both equations (9.2.28) and (9.2.29) are real.

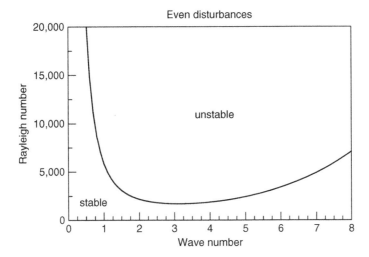

Figure 9.2.2   Bénard stability curve for even disturbances

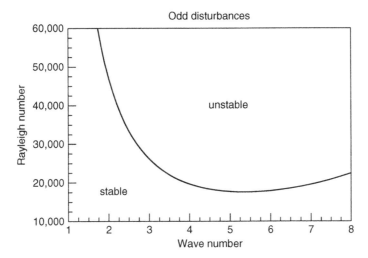

Figure 9.2.3   Bénard stability curve for odd disturbances

Plots of the roots of these equations are shown in Figure 9.2.2 for even disturbances and Figure 9.2.3 for odd disturbances. Even disturbances occur at the lowest value of the Rayleigh number—approximately 1,707.76—occurring at a wave number a of 3.117. For values of the Rayleigh number below this, according to our analysis the disturbances are all stable. For values of the Rayleigh number above this critical value, there will be some range of wave number a for which the disturbance will grow. Thus the flow will be unstable. For odd disturbances, the corresponding values of the critical Rayleigh and wave numbers are 17,610.39 and 5.365.

Other boundary conditions that have been used in the solution include the case where the top surface is free ($Ra_{\text{critical}} = 1,100.7$, $a_{\text{critical}} = 2.682$) and when both surfaces are free ($Ra_{\text{critical}} = 657.5$, $a_{\text{critical}} = 2.221$).

As yet, the shape factor f has not been found. The actual form of f will be determined by the plate geometry. There are, in fact, many geometric possibilities for the cell shape. For instance, f proportional to either $\sin\frac{ax}{b}$ or $\sin\frac{ay}{b}$ will give long cylindrical rollers observed frequently in the atmosphere; f proportional to $\sin \frac{px}{b} \sin \frac{qy}{b}$, where $p^2 + q^2 = a^2$, will give rectangular cells; f proportional to $2\cos \frac{\sqrt{3}ax}{2b} \cos \frac{ay}{2b} + \cos \frac{ay}{b}$ gives hexagonal cells. Bénard observed hexagonal cells in his experiments, but this shape was likely due to surface tension on the free surface at the top of his apparatus. The cell shape is influenced by the boundary shape, and circular rollers can be observed if round plates are used (Koschmeider, 1993).

Research has shown that the secondary flow just described leads to at least one further laminar flow if the temperature distance between the plates is increased beyond the critical limit.

The preceding is an exact solution for the neutrally stable curves for the linear stability problem. While the computations are tedious, the mathematical questions encountered are relatively uncomplicated. This generally is not true when the primary flow is nonquiescent, as shall be seen.

9.3 Stability of Flow Between Rotating Circular Cylinders—Couette-Taylor Instability

Couette (1890) studied the possibility of using flow between rotating cylinders to determine the viscosity of liquids. His general formula for the flow in cylindrical polar coordinates was

$$v_\theta = Ar + \frac{B}{r}, \tag{9.3.1}$$

where

$$A = \frac{r_1^2 \Omega_1 - r_2^2 \Omega_2}{r_1^2 - r_2^2}, \quad B = -\frac{r_1^2 r_2^2 (\Omega_1 - \Omega_2)}{r_1^2 - r_2^2}. \tag{9.3.2}$$

The subscript 1 refers to the inner cylinder and 2 to the outer cylinder. Couette built an apparatus with a rotating outer cylinder and an inner cylinder supported on a fine torsion wire to measure torque. His measurements were not particularly accurate, but he did notice that as the angular speed was increased, the graph of torque versus angular speed departed from the straight line expected for laminar flow. His results prompted Rayleigh (1916b) to examine the flow and find that for the flow to be stable, it is necessary that

$$r_{outer}^2 \Omega_{outer} > r_{inner}^2 \Omega_{inner}. \tag{9.3.3}$$

Since Couette's apparatus did not meet this criterion, departure of the flow from the state predicted by equation (9.3.1) was a possibility.

Taylor (1923) constructed an apparatus similar to that used by Couette but that allowed for visualization of the flow between the cylinders. He also provided a theoretical stability analysis and thus was able to confirm his findings. The analysis proceeded as follows.

In the manner used in Rayleigh's convection problem, a small disturbance is added to the flow of equation (9.3.1) and introduced onto the Navier-Stokes equations. The disturbance, based on Taylor's observations, was taken to be axisymmetric and of the form

$$u_r = u(r) \cos kz \ e^{st},$$

$$u_\theta = v(r) \cos kz \ e^{st}, \tag{9.3.4}$$

$$u_z = w(r) \sin kz \ e^{st}.$$

Insertion of these into the Navier-Stokes equations gives

$$\rho\left(su - \frac{2V}{r}v\right) = -\frac{dp}{dr} + \mu\left(\frac{d^2u}{dr^2} + \frac{1}{r}\frac{du}{dr} - k^2u - \frac{u}{r^2}\right), \tag{9.3.5}$$

$$\rho\left(sv + \left[\frac{dV}{dr} + \frac{V}{r}\right]u\right) = \mu\left(\frac{d^2v}{dr^2} + \frac{1}{r}\frac{dv}{dr} - k^2v - \frac{v}{r^2}\right), \tag{9.3.6}$$

$$\rho sw = kp + \mu\left(\frac{d^2w}{dr^2} + \frac{1}{r}\frac{dw}{dr} - k^2w\right), \tag{9.3.7}$$

along with the continuity condition

$$\frac{du}{dr} + \frac{u}{r} - kw = 0. \tag{9.3.8}$$

These are to be solved subject to the boundary conditions

$$u = v = w = 0 \text{ at } r = r_1 \quad \text{and} \quad \text{at } r = r_2. \tag{9.3.9}$$

Solution of the system of equations (9.3.5) through (9.3.9) is complicated by the order of the system (6th), the cylindrical polar coordinates that introduce nonconstant coefficients and suggest the need for Bessel functions, and also the number of parameters needed to describe the problem. (Remember that in 1923 the most sophisticated computer available was an adding machine.) Taylor restricted the problem to the case of small gap spacing and therefore was able to obtain an approximate solution in terms of trigonometric and hyperbolic functions. His and later results are shown in Figure 9.3.1. The dashed line in the lower right-hand corner is the Rayleigh stability criterion. The top

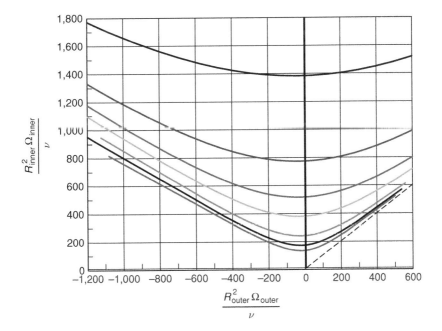

Figure 9.3.1   Stability curves for Taylor/Couette stability

curve is for $r_2/r_1 = 1.1$. In successive curves this ratio increases by 0.05. The curves are from empirical formulae in Coles (1967). Taylor's small-gap theory showed that as the gap size goes to zero, for the case where the outer cylinder is stationary,

$$T = \text{Taylor number} = \frac{4 \left(r_{\text{outer}} - r_{\text{inner}} \right) \Omega^2}{\nu^2 \left(r_{\text{outer}}^2 / r_{\text{inner}}^2 - 1 \right)} \approx 1708. \qquad (9.3.10)$$

The instability that Taylor observed in Couette flow consisted of a series of stacked rings, with the flow occurring along helical paths in each ring. The flow is complicated but still laminar. Subsequent experimenters have found that higher speeds result in the forming of wavy, ropey cells before turbulent flow is achieved and that the pattern observed can depend on the history of how the final pattern is obtained. See Koschmeider (1993) for a summary.

Note that, as in the previous case of convection flow, the cause of the flow is inertial effects. In this case it is the centrifugal and Coriolis acceleration, as contrasted with gravity and buoyancy in the Rayleigh problem.

As an interesting sidelight, Taylor, a descendant of George Boole, who discovered Boolean algebra, on which computer logic is based, was given a lifetime stipend by the British government upon completing his studies with the freedom to do whatever he wished in science. He chose to do his work at Cambridge University, where Rayleigh was located at the time Taylor worked on this problem. Taylor has said that Rayleigh discouraged him from pursuing this work, as it was "unlikely to be fruitful." Fortunately, Taylor did not heed this advice, as it was one of his early successes. Taylor's collected works compile six volumes.

While Couette pursued this work initially for viscometric purposes, these types of flows occur in many other situations. In high-speed situations in instruments such as gyroscopes it is necessary to maintain constant speed, which may require temperature control to ensure constant viscosity. This requires an understanding of the heat transfer occurring in the flow, which differs greatly if the cellular rings form. Other situations where this type of inertial instability forms are in flows between stationary curved parallel plates and curved pipes, where similar cellular patterns occur. The curved plate geometry was investigated by Dean (1928).

9.4 Stability of Plane Flows

The stability of plane flows has a long and interesting history. The equations were first formulated by Orr (1906–1907), and exact solutions were presented by him for the special case where the velocity varied linearly between two plates. The results, however, were indefinite, since even though the solutions were in terms of well-known (but complicated) functions, the calculations that had to be done were still formidable. Later efforts by such famous physicists as Sommerfeld (1908) and Heisenberg (1924) found asymptotic solutions for the stability of two-dimensional flow between parallel plates. Their results were incorrect because of the difficult nature of the solution, involving the proper way to traverse turning points. The first successful asymptotic solution was by C. C. Lin (1945). His theoretical results were confirmed by a series of elaborate experiments performed at the National Bureau of Standards by Schubauer and Skramstad (1943). Since that first successful analysis, with the advent of digital computers and guided by Lin's results, numerical solutions have produced more accurate numerical results without the need for elaborate asymptotic analysis.

The physical nature of the instability of parallel flows is quite different than for the Rayleigh-Bénard instability just studied. Here, the disturbance is of the form of a traveling wave rather than a standing cellular pattern. For plane Poiseuille flow with plates at $y = 0$ and $y = b$, the primary flow is $\mathbf{v} = (U(y), 0, 0)$, with $U(y) = U_m(y/b - y^2/b^2)$ and a pressure $P = -vxU_m/b^2$. Take a disturbance of the form $(u'(x, y, t), v'(x, y, t), 0)$, where

$$u'(x, y, t) = u(y)e^{ik(x-ct)},$$

$$v'(x, y, t) = v(y)e^{ik(x-ct)}, \qquad (9.4.1)$$

and

$$p'(x, y, t) = p(y)e^{ik(x-ct)}.$$

The preceding is a two-dimensional disturbance that is sufficient to study the stability of this flow, since it was shown by Squire (1933) that two-dimensional disturbances always became unstable at lower Reynolds numbers than do three-dimensional disturbances. Here, $c = c_r + \mathbf{i}c_i$, where c_r is the speed at which the wave travels in the x direction, and kc_i is the growth rate of the disturbance. If $kc_i < 0$, the disturbance decreases with time and the wave is **stable**. If $kc_i > 0$, the wave grows with time and there is **instability**. If $kc_i = 0$, the wave neither grows nor decays with time, and there is **neutral stability**.

From substituting this form for the velocities into the incompressible continuity equation in the form

$$\frac{\partial(U + u')}{\partial x} + \frac{\partial v'}{\partial y} = 0,$$

find that

$$iku' = -\frac{\partial v'}{\partial y}. \qquad (9.4.2)$$

The applicable Navier-Stokes equations for this flow are

$$\rho\left[\frac{\partial(U + u')}{\partial t} + (U + u')\frac{\partial(U + u')}{\partial x} + v'\frac{\partial(U + u')}{\partial y}\right]$$

$$= -\frac{\partial(P + p')}{\partial x} + \mu\left(\frac{\partial^2(U + u')}{\partial x^2} + \frac{\partial^2(U + u')}{\partial y^2}\right),$$

$$\rho\left[\frac{\partial v'}{\partial t} + (U + u')\frac{\partial v'}{\partial x} + v'\frac{\partial v'}{\partial y}\right] = -\frac{\partial(P + p')}{\partial y} + \mu\left(\frac{\partial^2 v'}{\partial x^2} + \frac{\partial^2 v'}{\partial y^2}\right).$$

Substituting the forms of equation (9.4.2), subtract out the terms from the primary flow equilibrium

$$0 = -\frac{\partial P}{\partial x} + \mu\frac{\partial^2 U}{\partial y^2}.$$

Linearizing any terms of order greater than the first in u, v, and p, we are left with

$$\rho\left(-ikcu + Uiku + v\frac{dU}{dy}\right) = -ikp + \mu\left(\frac{d^2 u}{dy^2} - k^2 u\right) \qquad (9.4.3)$$

and

$$\rho\left(-ikcv + Uikv\right) = -\frac{dp}{dy} + \mu\left(\frac{d^2 v}{dy^2} - k^2 v\right). \qquad (9.4.4)$$

Solving for the pressure p from equation (9.4.3) gives

$$ik\frac{p}{\rho} = \frac{\mu}{\rho}\left(\frac{d^2u}{dy^2} - k^2u\right) - ik(U-c)u - \nu\frac{dU}{dy}.\tag{9.4.5}$$

Differentiating this expression for p with respect to y and using the result together with equation (9.4.4) to eliminate p and u, after some rearrangement the result is

$$\frac{\mu}{\rho}\left(\frac{d^2}{dy^2} - k^2\right)\left(\frac{d^2}{dy^2} - k^2\right)v = ik(U-c)\left(\frac{d^2}{dy^2} - k^2\right)v - ik\frac{d^2U}{dy^2}v.\tag{9.4.6}$$

This is to be solved subject to the boundary conditions $u = v = 0$ at $y = 0$ and $y = b$, or equivalently,

$$v = \frac{dv}{dy} = 0 \text{ at } y = 0 \text{ and at } y = b.\tag{9.4.7}$$

If equation (9.4.6) is made dimensionless by letting $U' = U/U_m c' = c/U_m$, $k' = kb$, $\eta = y/b$, and $Re = \rho U_m b/\mu$, the result is

$$\left(\frac{d^2}{d\eta^2} - k'^2\right)^2 v = ik'Re\left[(U'-c')\left(\frac{d^2}{d\eta^2} - k'^2\right)v - v\frac{d^2U'}{d\eta^2}\right].\tag{9.4.8}$$

with the boundary conditions

$$v = \frac{dv}{d\eta} = 0 \text{ at } \eta = 0 \text{ and at } \eta = 1.\tag{9.4.9}$$

This dimensionless form of the stability equation for parallel flows is referred to as the **Orr-Sommerfeld equation**. It is also used as an approximation for nearly parallel flows, such as the flow in a boundary layer. (Notice that in the preceding dimensionalization U_m was any convenient velocity—that is, either the mean or maximum of U or any other value could in fact be used.)

The differential equation (9.4.8) and the boundary conditions of equation (9.4.9) are all homogeneous, so there is only the trivial solution $v = 0$, except for special combinations of the parameters R, k', and c'. These combinations are what must be determined to answer the flow stability question.

At first glance, the solution of equation (9.4.8) may not appear to be all that difficult. The equation is fourth order, compared to the sixth-order equations governing Rayleigh-Bénard and Couette stability. There are, however, two important differences in the two problems that far outweigh the difference in order. One is the fact that the coefficients in the equation are no longer constant ($U' = U'(y)$). The second, less obvious one is that the real part of the coefficient $U' - c'$ can, and in fact does, change sign in the region $0 \leq \eta \leq 1$ for the conditions of interest. If you think of the difference between the solutions of the two more familiar equations $y'' + y = 0$ (simple harmonic oscillations) and $y'' - y = 0$ (exponential behavior), it can be imagined that both behaviors are represented in various parts of the y region. The disturbance in fact tends to be fairly rapidly oscillating between the walls and where $U' = c'_r$ (this is, of course, the place where momentum transfer between the primary flow and the disturbance is easiest), and more gradually varying in the central core.

Viscosity in these flows plays a dual role. On the one hand, it has its traditional dissipative role, whereby it dissipates the energy of the disturbance. The second role of viscosity is to set up phase differences between the pressure and the various velocity

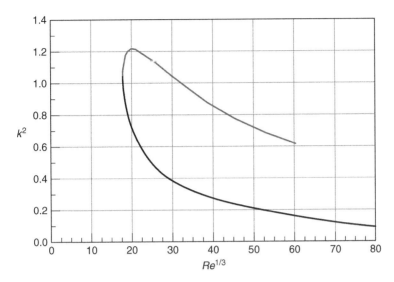

Figure 9.4.1 Stability curve for plane Poiseuille flow

components that enhance the energy transfer from the primary flow. It is this role that is responsible for allowing the growth of the instability. Experimentally it is known that the instability that occurs is a transition from laminar flow to turbulent flow.

The technical details of the asymptotic solutions are more complicated than can be dealt with here. See Lin's works listed in the References for more details. Use of matched asymptotic expansions to compute higher-order terms in the asymptotic expansion are given in Graebel (1966). A plot of the neutral stability curve k'^2 versus $Re^{1/3}$ for plane Poiseuille flow is given in Figure 9.4.1. Inside the curve the flow is unstable, while outside it is stable. The minimum value of Re is approximately 5,772 at a dimensionless wave number of 1.0206.

A similar curve for the Blasius boundary layer profile is given in Figure 9.4.2. The figure is based on data in Shen (1954), which was found to agree well with the classic experiments of Schubauer and Skramstad (1947) at the U.S. Bureau of Standards. Their experiments used a wand that oscillated at a variable rate ω as a disturbance source. R_1 is a Reynolds number based on boundary layer thickness as introduced by Lin (1946). The minimum Reynolds number for stability was found by Shen to be about 416.

Problems—Chapter 9

9.1 If viscous terms on equation (9.4.5) are neglected, the Orr-Sommerfeld equation becomes $(U - c)\left(\frac{d^2v}{dy^2} - k^2v\right) - \frac{d^2U}{dy^2}v = 0$, with $f(a) = f(b) = 0$. (This is sometimes called the Rayleigh equation.) The real part of the complex number c is the wave speed. The imaginary part relates to the rate of growth or decay of the disturbance. Because of the difficulty in solving the full Orr-Sommerfeld equation, much early work was done on the inviscid version. The problem is complicated that $U - c$ likely has a zero somewhere in the interval. To investigate this, first divide the inviscid equation by $U - c$. Next, multiply it by v^*, the complex conjugate of v, and then integrate over the

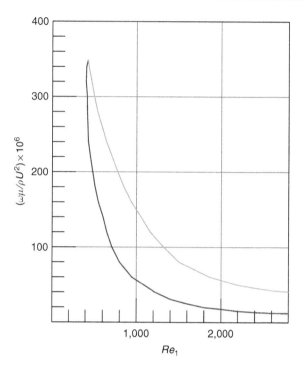

Figure 9.4.2 Stability curve for Blasius flow

range—say, a to b. Form positive-definite integrals, and then, after splitting into real and imaginary parts, make some conclusions concerning c.

9.2 An alternate approach to solving the preceding Rayleigh equation $(U - c)\left(\frac{d^2v}{dy^2} - k^2v\right) - \frac{d^2U}{dy^2}v = 0$ was introduced by Howard (1961). He first made the change of variables $f = (c - U)v$. Then the Rayleigh equation becomes $\frac{d}{dy}\left[(c - U)^2\frac{df}{dy}\right] - k^2(c - U)^2f = 0$. To investigate this, multiply it by f^*, the complex conjugate of f, and then integrate over the range—say, a to b. Form positive-definite integrals, and then, after splitting into real and imaginary parts, make some conclusions concerning c.

9.3 Show that the equations governing stability of flow between rotating cylinders (Couette-Taylor instability, equations 9.3.5, 9.3.6, 9.3.7) reduce to

$$(D^2 - \lambda^2 - sR)v = 2RA'u,$$

$$(D^2 - \lambda^2 - sR)(D^2 - \lambda^2)u = 2\lambda^2 R\omega v,$$

where

$$k = \frac{r_1\lambda}{r_2 - r_1}, \quad R = \frac{\rho\Omega_1(r_2 - r_1)^2}{\mu}, \quad \omega = \frac{A}{\Omega_1} + \frac{B}{r_1^2\Omega_1}\left[1 + \frac{r_2 - r_1}{r_1}\eta\right]^{-2},$$

$$\sigma = \frac{\rho(r_2 - r_1)^2}{\mu}s, \quad \eta = \frac{r - r_1}{r_2 - r_1}, \quad D = \frac{d}{d\eta},$$

and A and B are as in equation (9.3.2). To aid in your analysis, let $\varepsilon = (r_2 - r_1)/r_1$ be a very small parameter.

Turbulent Flows

tur-bu-lence (circa 1598): the quality or state of being turbulent: as a: wild commotion b: irregular atmospheric motion esp. when characterized by up-and-down currents c: departure in a fluid from a smooth flow.

tur-bu-lent (1538) 1: causing unrest, violence, or disturbance 2 a: characterized by agitation or tumult; TEMPESTUOUS b: exhibiting physical turbulence

turbulent flow (circa 1922): a fluid flow in which the velocity at a given point varies erratically in magnitude and direction; compare LAMINAR FLOW

10.1 The Why and How of Turbulence

These dictionary definitions lay out the problems with the study of turbulent flow: *wild, commotion, irregular, unrest, violence, disturbance, agitation, tumult, tempestuous, erratic*. To that list could be added adjectives such as *chaotic, random, confused, unorganized*—well, you see the point. Clearly, while the study of laminar flow is difficult, the study of turbulent flow requires a completely different approach.

Experience tells us that turbulence is often, but not necessarily, associated with high Reynolds numbers and is all too common in the atmosphere, oceans, rivers, lakes—even the plumbing in our buildings and our bodies. Sometimes it serves good purposes. If you want to mix together two liquids, cool off your coffee by stirring, or stir a can of paint, the introduction of turbulence comes in handy. On the other hand, if you want to increase lift on airplane wings or decrease drag on autos, ships, and planes, keeping things laminar as long as possible may be better.

Thus, it is necessary to understand the rudiments of turbulence, both from theory and experiments, and to find ways to predict what turbulence does to fluid flow. This will not be as clear-cut as in the case of laminar flow, but some information can be obtained.

Several approaches have been used to develop equations suited for the study of turbulent flow. Since the Navier-Stokes equations are valid for both laminar and turbulent flows, they usually serve as a general starting point. The complexity and rich detail of what transpires in turbulent flow, however, limits our abilities to fully tackle the Navier-Stokes equations directly. This is because the major effect of turbulence is that of mixing. As the Reynolds number increases, the range of the length scales of the various eddies in the flow also increases. Such a range of detail in the flow cannot be dealt with entirely by analytic means, and it poses a challenge to the largest supercomputers available. Computer simulations of the full Navier-Stokes equations are usually limited to flows where periodicity of the flow can be assumed and the boundaries are of particularly simple shape, usually rectangular. Numerical grids used must be small enough to resolve the smallest significant *eddy scale*[1] present in the flow, and the simulation must be carried out for a sufficiently long time that initial conditions have died out and significant features of the flow have evolved. The necessary memory and speed requirements on computers has in fact been one of the principle driving factors in the evolution of supercomputers and distributed computing architecture.

Covering the complete spectrum of eddy length scales is necessary, since different physical mechanisms are important at various places in the spectrum. Large-scale eddies contain much of the kinetic energy and little of the vorticity. The eddies tend to be anisotropic (i.e., the turbulence behaves differently in different spatial directions), and viscous effects are not important. These large eddies gradually break down into smaller eddies that have little kinetic energy but much vorticity. The small eddies tend to be isotropic, and viscous effects are important for them because it is the mechanism for converting the kinetic energy of the turbulence finally into heat.

10.2 Statistical Approach—One-Point Averaging

The vast amount of detail in turbulent flows requires that many features be looked at in a statistical sense. That is, for any turbulent quantity (e.g., velocity or pressure), split the flow into a time mean quantity and a turbulent fluctuation quantity. For example, letting unmarked quantities denote the total value, superposed bars denote an average, and primes denote the turbulent portion, for velocity, temperature, and pressure, write

$$\mathbf{v} = \bar{\mathbf{v}} + \mathbf{v}', \qquad (10.2.1)$$

$$p = \bar{p} + p', \qquad (10.2.2)$$

where the average of a quantity (pressure for example) is given by

$$\bar{p} = \frac{1}{\tau} \int_0^\tau p \, dt. \qquad (10.2.3)$$

---

[1] An *eddy* is defined as an entity that contains vorticity. See Lighthill, in Rosenhead (1963), Introduction. Boundary Layer Theory, Chapter II.

This is also called the *first moment* of the pressure. The averaging time τ is a time scale sufficiently long compared to the fluctuation times present in the flow. Notice that for any fluctuating quantity, it follows by definition that

$$\overline{p'} = 0. \tag{10.2.4}$$

The *root mean square value* (designated by *rms*, and also called the second moment) gives some idea as to the size of a fluctuation quantity. It is defined in the case of pressure, for example, by

$$rms\ p = \sqrt{\overline{p'p'}} = \sqrt{\frac{1}{T}\int_0^T p'p'dt}. \tag{10.2.5}$$

This term is never zero unless pressure fluctuations are completely absent from the flow. This second-order moment is a correlation of p' with itself and is thus called an *autocorrelation*.

Reynolds first looked at the Navier-Stokes equations in this manner and introduced what are now called the *Reynolds averaged Navier-Stokes equations*. To see how the equations are found, start first with the continuity equation for a constant density flow.[2] Splitting the velocity as in equation (10.2.1) results in

$$\nabla \cdot \mathbf{v} = \nabla \cdot \overline{\mathbf{v}} + \nabla \cdot \mathbf{v'} = 0. \tag{10.2.6}$$

Averaging equation (10.2.6) and using the fact that the average of a fluctuation is zero, this reduces to

$$\nabla \cdot \overline{\mathbf{v}} = 0. \tag{10.2.7}$$

Substituting this into equation (10.2.6) gives also

$$\nabla \cdot \mathbf{v'} = 0. \tag{10.2.8}$$

Consider next the component of the Navier-Stokes equations in the x_i direction. It is convenient to use index notation for much of what follows, so we will alternate between using $\mathbf{v} = (u,v,w)$ and $\mathbf{v} = (v_1,v_2,v_3)$ in what follows. It is also convenient to write the Navier-Stokes equation in the index notation form (with the help of continuity equation (10.2.6)):

$$\rho\left(\frac{\partial v_i}{\partial t} + \frac{\partial(v_i v_j)}{\partial x_j}\right) = -\frac{\partial p}{\partial x_i} + \mu\frac{\partial}{\partial x_j}\left(\frac{\partial v_i}{\partial x_j} + \frac{\partial v_j}{\partial x_i}\right). \tag{10.2.9}$$

The summation convention on repeated subscripts is being used here—that is,

$$\frac{\partial\left(v_i v_j\right)}{\partial x_j} = \sum_{j=1}^3 \frac{\partial\left(v_i v_j\right)}{\partial x_j} = \frac{\partial\left(v_i v_1\right)}{\partial x_1} + \frac{\partial\left(v_i v_2\right)}{\partial x_2} + \frac{\partial\left(v_i v_3\right)}{\partial x_3}, \quad i = 1, 2 \text{ or } 3.$$

---

[2] In this presentation, to avoid a large degree of additional complication temperature and fluid properties such as density and viscosity will be assumed to be constant. Variations in these quantities can be included in a similar fashion with substantial additional effort.

Decomposing the velocities into mean and turbulent parts, inserting these into equation (10.2.9) and then averaging, results in

$$\rho\left(\frac{\partial \overline{v}_i}{\partial t} + \frac{\partial(\overline{v}_i\overline{v}_j)}{\partial x_j} + \frac{\partial(\overline{v_i'v_j'})}{\partial x_j}\right) = -\frac{\partial \overline{p}}{\partial x_i} + \mu\frac{\partial^2 \overline{v}_i}{\partial x_j^2}, \qquad (10.2.10)$$

or upon rearranging,

$$\rho\left(\frac{\partial \overline{v}_i}{\partial t} + \frac{\partial(\overline{v}_i\overline{v}_j)}{\partial x_j}\right) = -\frac{\partial \overline{p}}{\partial x_i} + \frac{\partial}{\partial x_j}\left(\mu\frac{\partial \overline{v}_i}{\partial x_j} - \rho\overline{v_i'v_j'}\right). \qquad (10.2.11)$$

The terms in equation (10.2.11) that involve averages of the products of the turbulent fluctuations

$$-\rho\overline{v_1'^2}, -\rho\overline{v_2'^2}, -\rho\overline{v_3'^2}, -\rho\overline{v_1'v_2'}, -\rho\overline{v_1'v_3'}, -\rho\overline{v_2'v_3'}, \text{ or}$$
$$-\rho\overline{u'^2}, -\rho\overline{v'^2}, -\rho\overline{w'^2}, -\rho\overline{u'v'}, -\rho\overline{u'w'}, -\rho\overline{v'w'}, \qquad (10.2.12)$$

are called the **Reynolds stresses**. They are the terms through which the turbulence fluctuations interact with and change the mean flow. At a solid boundary they must vanish, but usually they have large gradients near a boundary and reach maximum values very close to the wall. They can be measured by means of hot wire anemometers or laser doppler anemometry, with suitable processing of the signals. They represent what are called *second-order correlations* of the velocity at a given point.

Notice that in the averaging process, we have gone from a fully determinate system, one where the number of unknowns is equal to the number of equations, to one that is indeterminate, in that we now have more unknowns than equations. To attempt to correct this, we could first multiply the Navier-Stokes equations by some quantity such as a velocity component, then average the equations so obtained. In the process we would introduce averages of three velocity fluctuation components (*third-order correlations*) and find that the system is even more indeterminate than before. This game is one that *can't* be won, as this statistical averaging process will continue to generate new unknowns faster than it generates new equations. As the order of a correlation increases, the difficulties of both accurate measurement and physical understanding decrease, and seldom is it necessary or useful to go beyond fourth-order correlations—if indeed that far!

It should be noticed that although the statistical approach has many advantages and is one of the few well-developed mathematical procedures available for analyzing turbulent flows, it also has disadvantages. Averaging hides rather than reveals the underlying physical processes that take place in such flows. Mollo-Christiansen (1971) has given the example of a blind person using a single sensor on a single road as an attempt to discover what kinds of vehicles used that road. "Happening to use a road traveled only by limousines and motorcycles, he concluded that the average vehicle is a compact car with 2.4 wheels."[3]

At first it may seem strange that although the Navier-Stokes equations are closed mathematically, the statistical equations obtained from them are not. Batchelor (1971) has given two complementary answers to this. He notes that the Navier-Stokes equations

---

[3] The remark was made to point out the inadequacy of using a single sensor.

are open in time, since any one flow realization must be integrated over the whole of future time before exact values can be found for the various correlations. This is contrasted with the equations for the correlations, which can be made independent of time. Second, additional correlations give less and less additional information so that only a finite number of correlations is required to determine the properties of a turbulent field to sufficient accuracy.

The problem of finding a way to get the same number of unknowns as equations is called the ***closure problem*** of turbulence. In the elementary fluids course you may have taken, you might have seen algebraic attempts at closure such as mixing length and eddy viscosity. There, the Reynolds stresses were approximated in terms of mean quantities and their gradients. While these closure models are useful for providing a correlation of experimental data in simple flows such as pipe flow, they often fail when they are applied to situations outside of the range of the original experiments. They have nevertheless been developed to a high degree of complexity (see, for instance, Section 5.2.3 of Bradshaw (1978)) and can give good results for boundary layer–type flows.

Before going to a discussion of closure formulas, three more correlation equations will be developed. These are the transport equations for the important correlations—namely, Reynolds stresses, kinetic energy, and dissipation. Deriving the equations is an involved process, and it is necessary to use index notation because of the nature of the terms involved in the needed multiplications. Also, it eases the physical interpretation of the various terms.

Derivation of the first correlation equation starts with equation (10.2.9) with the velocity written in decomposed form. This is next subtracted from equation (10.2.11), giving

$$\rho\left(\frac{\partial v_i'}{\partial t} + \overline{v}_k\frac{\partial v_i'}{\partial x_k}\right) \equiv \rho\frac{\overline{D}v_i'}{Dt}$$

$$= -\frac{\partial p'}{\partial x_i} + \frac{\partial}{\partial x_k}\left[\mu\frac{\partial v_i'}{\partial x_k} - \rho\left(\overline{v_i'v_k'} + \overline{v}_iv_k' + v_i'v_k'\right)\right]. \tag{10.2.13}$$

Multiplying equation (10.2.13) by v_j' and adding to this the same quantity but with the i and j interchanged,[4] we get

$$\rho\frac{\overline{D}v_i'v_j'}{Dt} = -\rho\left(v_j'v_k'\frac{\partial\overline{v}_i}{\partial x_k} + v_i'v_k'\frac{\partial\overline{v}_j}{\partial x_k}\right) + p'\left(\frac{\partial v_j'}{\partial x_i} + \frac{\partial v_i'}{\partial x_j}\right) - 2\mu\frac{\partial v_i'}{\partial x_k}\frac{\partial v_j'}{\partial x_k}$$

$$- \frac{\partial}{\partial x_k}\left[p'\left(v_j'\delta_{ik} + v_i'\delta_{jk}\right) - \mu\frac{\partial\left(v_i'v_j'\right)}{\partial x_k} + \rho v_i'v_j'v_k'\right]. \tag{10.2.14}$$

Time averaging equation (10.2.14) then gives

$$\frac{DR_{ij}}{Dt} = P_{ij} + T_{ij} - D_{ij} - \frac{\partial J_{ijk}}{\partial x_k}, \tag{10.2.15}$$

$$\text{I} \qquad \text{II} \quad \text{III} \quad \text{IV} \qquad \text{V}$$

---

[4] For example, for an equation $a_i = b_i$, multiplication gives $v_ja_i = v_jb_i$, so after the interchange and addition, we have $v_ja_i + v_ia_j = v_jb_i + v_ib_j$.

where

$$D_{ij} = 2 \frac{\mu}{\rho} \overline{\frac{\partial v_i'}{\partial x_k} \frac{\partial v_j'}{\partial x_k}} = \text{dissipation of Reynolds stress,}$$

$$J_{ijk} = \frac{1}{\rho} \left(\overline{p' v_j'} \delta_{ik} + \overline{p' v_i'} \delta_{jk} - \mu \frac{\partial R_{ij}}{\partial x_k} \right) + \overline{v_i' v_j' v_k'} = \text{flux of Reynolds stress,}$$

$$P_{ij} = - \left(R_{ik} \frac{\partial \overline{v}_k}{\partial x_j} + R_{jk} \frac{\partial \overline{v}_k}{\partial x_i} \right) = \text{production of Reynolds stress,}$$

$$R_{ij} = \overline{v_i' v_j'} = -\text{Reynolds stress}/\rho,$$

$$T_{ij} = \frac{1}{\rho} \overline{p' \left(\frac{\partial v_i'}{\partial x_j} + \frac{\partial v_j'}{\partial x_i} \right)} = \text{pressure strain rate.}$$

The roman numerals under the various terms indicate the source of that term and the role it plays in affecting the turbulence. The physical interpretation of the terms are as follows:

- *Transport of the correlation*—in this case, the Reynolds stresses. This term represents convection of the correlation with the mean velocity.

- *Generation or production.* The presence of mean velocity and temperature gradients allows energy to be transferred between the mean and fluctuating field. The energy can in principal flow either way—that is, from the mean flow into the turbulence or vice versa. The first is generally more prevailing. These terms arise from the fluctuating portion of the convective term.

- *Redistribution* (also called pressure-strain). In an incompressible flow the interactions of the pressure fluctuations can neither create nor destroy energy. Rather, they cause it to be redistributed among the various components of the correlation.

- *Destruction.* This term comes from the viscous stresses and results in a decrease of the correlation expressed in transport of the correlation.

- *Diffusion.* These terms arise from the fluctuating convection, pressure, and viscous terms. They are characterized by being expressible as the gradient of a tensor.

For the second correlation equation, the disturbance velocity vector is dotted with the Navier-Stokes equations and then time averaged. The result can be obtained from the previous result by contracting on i and j, giving

$$\frac{\overline{D}k}{Dt} = P - D - \frac{\partial J_k^*}{\partial x_k}, \tag{10.2.16a}$$

$$\text{I} \qquad \text{II} \quad \text{IV} \quad \text{V}$$

where

$$D = \frac{1}{2} D_{ii} = \frac{\mu}{\rho} \overline{\frac{\partial v_i'}{\partial x_k} \frac{\partial v_i'}{\partial x_k}} = \text{homogeneous dissipation of Reynolds stress,}$$

$$J_k^* = \frac{1}{2} J_{iik} = \frac{1}{\rho} \overline{p' v_k'} + \frac{1}{2} \overline{v_i' v_i' v_k'} - \frac{\mu}{\rho} \frac{\partial k}{\partial x_k} = \text{flux of Reynolds stress,}$$

$$J_k = \frac{1}{\rho} \overline{p' v_k'} + \frac{1}{2} \overline{v_i' v_i' v_k'} - 2 \frac{\mu}{\rho} \overline{v_i' \left(\frac{\partial v_i'}{\partial x_k} + \frac{\partial v_k'}{\partial x_i} \right)}$$

$$k = \frac{1}{2}q^2 = \frac{1}{2}R_{ii} = \frac{1}{2}\overline{v'_i v'_i} = \text{turbulent kinetic energy,}$$

$$P = \frac{1}{2}P_{ii} = -R_{ij}\frac{\partial \overline{v_i}}{\partial x_j} = \text{rate of kinetic energy production,}$$

$$\varepsilon = \frac{1}{2}\frac{\mu}{\rho}\overline{\left(\frac{\partial v'_i}{\partial x_j} + \frac{\partial v'_j}{\partial x_i}\right)\left(\frac{\partial v'_i}{\partial x_j} + \frac{\partial v'_j}{\partial x_i}\right)} = \text{total rate of energy dissipation.}$$

This equation is also found in the literature written in the alternate form

$$\frac{\overline{D}k}{Dt} = P - \varepsilon - \frac{\partial J_k}{\partial x_k}. \qquad (10.2.16b)$$

$$\text{I} \qquad \text{II} \quad \text{IV} \quad \text{V}$$

The Roman numerals under the various terms are interpreted as for equation (10.2.16).

The third equation is found by first differentiating the Navier-Stokes equations and then multiplying it by $\frac{\mu}{\rho}\frac{\partial v'_i}{\partial x_j}$, adding the interchanged equation, and averaging the result. This gives

$$\frac{\overline{D}D}{Dt} = -W - \frac{\partial H_k}{\partial x_k}, \qquad (10.2.17)$$

where

$$W = 2\frac{\mu}{\rho}\left[\overline{\frac{\partial v'_i}{\partial x_j}\frac{\partial v'_j}{\partial x_k}\frac{\partial v'_k}{\partial x_i}} + \overline{v'_j\frac{\partial v'_i}{\partial x_k}\frac{\partial^2 v_m}{\partial x_j \partial x_k}} + \overline{\frac{\partial v'_i}{\partial x_j}\frac{\partial v'_i}{\partial x_k}\frac{\partial \overline{v_j}}{\partial x_k}} + \overline{\frac{\partial v'_i}{\partial x_k}\frac{\partial v'_j}{\partial x_k}\frac{\partial \overline{v_i}}{\partial x_j}}\right]$$

$$+ 2\left(\frac{\mu}{\rho}\right)^2 \overline{\frac{\partial^2 v'_i}{\partial x_j \partial x_j}\frac{\partial^2 v'_i}{\partial x_k \partial x_k}}, \qquad (10.2.18)$$

$$H_j = \frac{\mu}{\rho}\left(\overline{v'_j\frac{\partial v'_i}{\partial x_k}\frac{\partial v'_j}{\partial x_k}} + \overline{\frac{\partial v'_j}{\partial x_k}\frac{\partial p'}{\partial x_k}} - \frac{\partial D}{\partial x_k}\right).$$

The various correlations introduced so far are called **one-point correlations**. This means that every velocity and/or pressure term in the correlation is measured at the same point. At times **two-point correlations** are also used in turbulence studies, when one is trying to ascertain how what happens at a downstream point is affected by what happens at an upstream point.[5] Similar correlation transport equations can be developed for such correlations, but, of course, even more algebra is involved. Two-point correlations have been used in turbulence calculations but to a much more limited extent than one-point correlations.

At least five different approaches have been used to provide closure formulae for our equations:

1. *Zero-equation models*. This approach uses equation (10.2.11) as the only partial differential equation. The Reynolds stresses are modeled by algebraic

---

[5] One example of such a correlation could be the degree to which the proverbial butterfly flapping his wings in Beijing affects the climate in your location.

expressions, based largely on eddy viscosity and mixing length concepts, and equation (10.2.11) is used to determine mean flow. This is widely used in practical applications, and it also crops up in the one- and two-equation models.

2. *One-equation models*. Another partial differential equation involving a turbulence velocity scale is used along with equation (10.2.11), often equation (10.2.16). The length scale often used is based on q or ε.

3. *Two-equation models*. Two partial differential equations involving two turbulence velocity scales are used along with equation (10.2.11), usually equations (10.2.16) and (10.2.17). Length scales often are based on q and ε.

4. *Stress-equation models*. Partial differential equations are used for all Reynolds stress components and often for a length scale as well.

5. *Large eddy simulations*. Computations are made for both large-scale and small-scale eddies.

In the following sections these methods will be discussed in more detail. A number of reviews are available in the literature and should be consulted for more information. Among them are Bradshaw (1972), Mellor and Herring (1973), Reynolds (1974, 1976), Reynolds in Bradshaw (1978), and Bradshaw, Cebeci, and Whitelaw (1981).

Much of the early work in theoretical turbulence models was done for special models: *homogeneous turbulence*, where the turbulence is the same in every location (e.g., Batchelor 1967); *isotropic turbulence*, where at a given point the turbulence is both homogeneous and independent of direction (e.g., Taylor, 1935–1938); and turbulent shear flows (e.g., Townsend 1956).[6] While the concepts seem simple in hindsight, the results together with Taylor's experiments demonstrated the power of the statistical approach (Kármán 1937b, 1938; Komogoroff 1941; Batchelor 1947).

10.3 Zero-Equation Turbulent Models

The eddy viscosity ν_T and mixing length ℓ formulas

$$-\rho\overline{u'v'} = \rho\ell^2 \left|\frac{d\overline{u}}{dy}\right| \frac{d\overline{u}}{dy} \quad \text{and} \quad -\rho\overline{u'v'} = \rho\nu_T \frac{d\overline{u}}{dy} \tag{10.3.1}$$

form the basis for most of these models. Many variations on them exist. For instance, at large Reynolds numbers (Re > 5,000), it is common to neglect the viscous sublayer and split the boundary layer region into two parts. An example of this is

$$\ell = \begin{cases} 0.4y, & y_0 \le y \le y_c, \\ 0.075\delta, & y_c \le y \le \delta, \end{cases}$$

$$\nu_T = \begin{cases} \ell^2 \left|\frac{d\overline{u}}{dy}\right|, & y_0 \le y \le y_c, \\ 0.0168 \left|\int_0^x (\overline{u}_c - \overline{u})\,dy\right|, & y_c \le y \le \delta. \end{cases} \tag{10.3.2}$$

[6] It should be noted that although Reynolds introduced the concept of what are now called Reynolds stresses, the concept of statistical analysis of turbulent flows languished until Taylor in 1935 revived the idea and brought it to the attention of researchers.

Here, $y_0 = 40\nu/\overline{u}_c$ is the thickness of the viscous sublayer measured from the wall, and y_c is another distance found from making the mixing length continuous. The parameter δ is a boundary layer thickness, defined, for example, where the mean velocity is at 0.5% of the outer flow.

A number of empirical corrections of these formulae have become feasible over the years as computing memory and speed have increased. One due to Cebeci and Smith (1967, 1974) gives improved results for low Reynolds numbers, transition, mass transfer, pressure gradients, and boundary curvature. The eddy viscosity is given by

$$\nu_T = \begin{cases} (\nu_T)_i \\ (\nu_T)_o \end{cases} = \begin{cases} \ell^2 \left| \dfrac{d\overline{u}}{dy} \right|, & y \leq y_{\text{crossover}}, \\ 0.0168 \left| \int_0^x (\overline{u}_c - \overline{u})\, dy \right| F_{\text{Klebanoff}}, & y > y_{\text{crossover}}. \end{cases} \tag{10.3.3}$$

The quantity $y_{\text{crossover}}$ is where the outer eddy viscosity is greater than the inner viscosity. The other quantities are

$$\ell = 0.4y \left\{ 1 - e^{-y^+/A^+} \right\},$$

$$A^+ = 26 \left(1 + \frac{y}{\tau_{\text{wall}}} \frac{dp}{dx} \right)^{-1/2}, \tag{10.3.4}$$

$$F_{\text{Klebanoff}} = \left[1 + 5.5 \left(\frac{y}{\delta} \right)^6 \right]^{-1},$$

$F_{\text{Klebanoff}}$ being a factor introduced by Klebanoff to account for intermittancy in the flow. The quantity y^+ is given by $y^+ = \rho u^* y/\mu$, where $u^* = \sqrt{\tau_{\text{wall}}/\rho}$. The expression for the mixing length is the van Driest fit to velocity measurements near the wall. This model was extended to three dimensions and also accounts for compressibility effects. It was developed by Cebeci (1974), Cebeci and Abbott (1975), and Cebeci et al. (1975).

Because of its robustness, a model by Baldwin and Lomax (1978) is commonly used in quick design iterations. It does not capture all details of the turbulence but still gives useful design information. Its details are given by

$$\nu_T = \begin{cases} (\nu_T)_i \\ (\nu_T)_0 \end{cases} = \begin{cases} \ell^2 |\Omega|, & y \leq y_{\text{crossover}}, \\ K C_{\text{cp}} F_{\text{wake}} F_{\text{Klebanoff}}, & y > y_{\text{crossover}}. \end{cases} \tag{10.3.5}$$

The y^+ is as given in equation (10.3.4). Other quantities are

$$\Omega_{ij} = \frac{1}{2} \left(\frac{\partial \overline{u}_i}{\partial x_j} + \frac{\partial \overline{u}_j}{\partial x_i} \right), \quad |\Omega| = \sqrt{\Omega_{ij} \Omega_{ij}}, \quad \ell = \kappa \left(1 - e^{-y^+/A^+} \right),$$

$$F_{\text{wake}} = \min \left(y_{\text{max}} F_{\text{max}};\ C_{\text{wk}} y_{\text{max}} u_{\text{dif}}^2 / F_{\text{max}} \right),$$

y_{max} and F_{max} are the maximum of the function $F(y) = y |\Omega| \left(1 - e^{-y^+/A^+} \right)$,

$$F_{\text{Klebanoff}} = \left[1 + 5.5 \left(\frac{y C_{\text{Klebanoff}}}{y_{\text{max}}} \right)^6 \right]^{-1},$$

$$u_{\text{dif}} = \max \left(\sqrt{\overline{u}_i \overline{u}_i} \right) - \min \left(\sqrt{\overline{u}_i \overline{u}_i} \right).$$

The various constants are given by

A^+	C_{cp}	$C_{Klebanoff}$	C_{wk}	κ	α
26	1.6	0.3	0.25	0.4	0.0168

Granville (1987) has suggested improvements in these constants. Further comments have been made by Wilcox (1998).

10.4 One-Equation Turbulent Models

An early model due to Prandtl (1945) suggested that

$$\nu_T = \sqrt{k}\ell, \tag{10.4.1}$$

with k being the turbulent kinetic energy and the mixing length coming from an algebraic formula. The turbulent kinetic energy is usually determined from equation (10.2.16b) after suitable modeling of the terms on the right-hand side. A model used at Stanford University in the STAN-5 code let

$$\nu_T = c_2 q\ell,$$

$$R_{ij} = \frac{1}{3}q^2\delta_{ij} - \nu_T\left(\frac{\partial \overline{u}_i}{\partial x_j} + \frac{\partial \overline{u}_j}{\partial x_i}\right),$$

$$D = c_3 q^2/\ell, \tag{10.4.2}$$

$$J_j^* = -(c_4\nu_T + \nu)\frac{\partial q^2}{\partial x_j},$$

where the length scale is modeled by means described in the zero-equation model. Typically the zero-equation model is used for $y^+ \leq 2A^+$, and the preceding is used for greater values of y^+. STAN-5 uses $c_2 = 0.38$, $c_3 = 0.055$, $c_4 = 0.59$, all determined from experimental results. A similar model was developed at Imperial College (Wolfshtein, 1969). Further forms for one-equation models can be found in the review by Mellor and Herring (1973).

10.5 Two-Equation Turbulent Models

The previous models required specification of a turbulent-length scale for their realization. To avoid this, equation (10.2.17) is frequently used to compute the dissipation. For the case of isotropic turbulence,

$$D = -\frac{1}{2}q^2 \quad \text{and} \quad \frac{\overline{DD}}{Dt} = -W. \tag{10.5.1}$$

For large values of the Reynolds number, after making the assumption that W is of the form $W = c_7 D^2/q^2$, the solution of equation (10.5.1) is

$$q^2 = q_0^2 (1 + t/a)^{-n}, \quad D = D_0 (1 + t/a)^{-n-1}, \tag{10.5.2}$$

where $a = nq_0^2/2D_0$, $n = 2/(c_7 - 2)$. Early experiments suggest that $c_7 = 4$ and n is in the range 1.1 to 1.3. Later arguments by Reynolds (1976) suggest that this model works best for short wavelength disturbances. If the entire spectrum is included, Reynolds suggests that an improvement on equation (10.5.2) is

$$q^2 = \alpha \left(\frac{1}{m+1} + \frac{3}{2} \right) \left(\frac{aD^{2/3}}{A} \right)^{-2/(3m+5)}, \quad D = C \left(q^2 \right)^{(3m+5)/(2m+3)}, \qquad (10.5.3)$$

where $m = 2$ seems the most reasonable choice. This corresponds to $c_7 = 11/3$ and $n = 6/5$ in the previous result.

A number of other models have been introduced, each with pluses and minuses. A heuristic model by Saffman and Wilcox (1974) introduces a pseudovorticity Ω and uses the transport equation

$$\frac{\overline{D}\Omega}{Dt} = \left[\alpha \sqrt{\frac{\partial \overline{u}_i}{\partial x_j} \frac{\partial \overline{u}_i}{\partial x_j}} - \beta\Omega \right] \Omega^2 + \frac{\partial}{\partial x_j} \left[(\nu + \nu_T) \frac{\partial \Omega^2}{\partial x_j} \right] \qquad (10.5.4)$$

in place of equation (10.2.17). For equation (10.2.15) they use

$$P = \alpha^* \left[\frac{1}{2} \left(\frac{\partial \overline{u}_i}{\partial x_j} + \frac{\partial \overline{u}_j}{\partial x_i} \right) \left(\frac{\partial \overline{u}_i}{\partial x_j} + \frac{\partial \overline{u}_j}{\partial x_i} \right) \right]^{1/2}, \quad D = \beta^* q^2 \Omega/2, \qquad (10.5.5)$$

with recommended values

$$\alpha = 0.1638, \quad \alpha^* = 0.3, \quad \beta = 0.15, \quad \beta^* = 0.09, \quad c_4 = 0.5, \quad \text{and} \quad c_4^* = 0.5.$$

This model has been tested against a number of examples and has given reasonable results—in some cases better than the more theoretical models proposed by others. For updates, see Wilcox (1998).

Since the two-equation models frequently use equations (10.2.15) and (10.2.17), which involve the kinetic energy and dissipation, they are frequently referred to in the literature as $k - \varepsilon$ models. The Saffman-Wilcox model is referred to as a $k - \Omega$ model.

10.6 Stress-Equation Models

The one- and two-equation models still require modeling of the various terms in the equations, and the results are often at best mixed. In view of these difficulties, attention has also gone back to the original equation for the Reynolds stresses, equation (10.2.15). For a given mean velocity field, the production term does not need to be modeled, but the remaining ones do.

The type of modeling needed depends on the flow being studied. For instance, at high Reynolds numbers the small-scale dissipative structures are practically isotropic, and it has become customary to use

$$D_{ij} = \frac{2}{3}\delta_{ij}D \qquad (10.6.1)$$

The pressure strain rate term for incompressible flow has the property that

$$T_{ii} = 0, \qquad (10.6.2)$$

and for flows with no mean strain rate it is responsible for the return to isotropic conditions. It is known, however, that for deforming flows the situation is much more complex, and a number of diverse methods have been proposed to model it. Examples are found in Tennekes and Lumley (1972); Rotta (1951); Hanjalic and Launder (1972); Launder, Reece, and Rodi (1975); Norris and Reynolds (1975); Kwak, Reynolds, and Ferziger (1975); and many others. A similar situation exists for the J term.

To establish the validity of a model, we compare the results of well-established experiments. For incompressible flows, one of the gold standards is the rearward-facing step, where reattachment occurs behind the eddy formed at the base of the step. This was selected as a predictive case at a conferences at Stanford University in 1980 and 1981 (Kline et al., 1981). The algebraic stress model of Baldwin and Lomax (1978) appeared to produce the most satisfactory results at that time.

Several classes of flows have provided the data used for determining the empirical constants. These provide an indication of the flow situations that might be computed from these closure models with the highest degree of confidence. It is, of course, hoped that an even broader class of flows is covered by the models. Confident realization of that hope is only possible by calculation and subsequent comparison with experiment

1. *Decay of grid turbulence*: Batchelor and Townsend (1948), Lin and Lin (1973).
2. *Return to isotropy of distorted turbulence*: Rotta (1962), Tucker and Reynolds (1968), Uberoi (1957).
3. *Free shear flows (jets, wakes, mixing layers)*: Bradbury (1965), Bradshaw et al. (1964), Champagne et al. (1977), Chevray and Kovaszney (1969), Tailland and Mathield (1967), Webster (1964), Wygnanski and Fiedler (1970).
4. *Wall turbulence*: Arya and Plate (1969), Hanjalic and Launder (1972), Johnson (1959), Klebanoff (1955).

Much research is still being done today to find improvements on these models. Results so far indicate the following:

Zero-equation models: These generally are simple, efficient, and numerically stable. The concept of a length scale, however, is difficult to define. Also, history effects are not accounted for.

One-equation models: These are naturally more difficult to code than the zero-equation models. Again, a length scale is not well defined. Often it is only marginally better than zero-equation models.

Two-equation models: The length scale here is easily defined, but the equations that result are numerically stiff ("stiff" implies extreme sensitivity to coefficients in equations—see Chapter 11 and the Appendix). They do, however, provide better results than the other two cases.

10.7 Equations of Motion in Fourier Space

Knowing the distribution of length scales in turbulence is important because the various length scales behave in different ways. For example, if you want to reduce the level of turbulence in a wind tunnel, the air is passed through arrays of fine screens or closely spaced bars. After passing through these filters, the air turbulence is of fine length scale, whose energy is dissipated rapidly by viscosity. Fourier analysis is used to determine the distribution of length scales. The formal definition of the Fourier transform in three dimensions is

$$F(k_x, k_y, k_z) = \frac{1}{(2\pi)^3} \iiint_V e^{-i(xk_x + yk_y + zk_z)} f(x, y, z) dx dy dz,$$

$$f(x, y, z) = \iiint_V e^{i(xk_x + yk_y + zk_z)} F(k_x, k_y, k_z) dk_x dk_y dk_z,$$

(10.7.1)

where F is the Fourier transform of f, and f is "retrieved" or inverted by the second of equation (10.7.1).

Differential equations for transformed quantities can be found by taking the Fourier transform of equations such as (10.2.15), (10.2.16), and (10.2.17). Fourier transforms are used with two-point correlations such as

$$R_{ij}(\mathbf{r}) = \overline{v_i(\mathbf{x}, t) v_j(\mathbf{x} + \mathbf{r}, t)},$$

(10.7.2)

the integration performed on $\mathbf{r}$ while $\mathbf{x}$ is held fixed, or with two-time correlations such as

$$R_{ij}(\tau) = \overline{v_i(\mathbf{x}, t) v_j(\mathbf{x}, t + \tau)}.$$

(10.7.3)

The purpose of the two-point correlations is to investigate the effects of either different spatial scales or different time scales in the turbulence and to find the region of coherence of the correlations. When dealing with the equations for **two-point correlations**, it is customary to make three basic assumptions: (1) mean velocity and temperature gradients are constant, (2) the turbulence is homogeneous, and (3), the presence of walls can be ignored in the equations. Under these assumptions Fourier transforms are useful.

The advantage of using Fourier transforms is that there are no space derivatives involved in the equations. Certainly, a disadvantage is that we are now "transformed" to a space where we have even less physical understanding of the new quantities.

Signals from hot wire or hot film anemometers can be transformed by signal processors (dedicated computers called **spectrum analyzers**) capable of time averaging, multiplication, time delay, and fast Fourier transforming (FFT). FFT is an approximation to the integration process, which is well suited for use in a computer.

One of the triumphs of isotropic turbulence studies has been the explanation of the energy content as a function of wavelength. When the log (base 10) of the **energy spectrum** is plotted versus $\log_{10} k/k_{\text{Kolmogoroff}}$, where $k_{\text{Kolmogoroff}} = (\varepsilon/\nu^3)^{1/4}$, the plot has a linear region with downward slope $k - 3/5$ in the range $-4 < \log_{10} k/k_{\text{Kolmogoroff}} < -1$. This is known as the **inertial subrange**. Below this range, from $-1 < \log_{10} k/k_{\text{Kolmogoroff}} < 0$, is the **dissipation range**. The combination of the two is called the **universal equilibrium range**.

It is easiest to describe what is happening by denoting wave number k by its reciprocal, the wave length λ, so the more familiar dimension of length is being dealt with. Then $\log_{10} k/k_{\text{Kolmogoroff}} = -\log_{10} \lambda/\lambda_{\text{Kolmogoroff}}$. An eddy introduced into the flow tends to cascade down this energy-length curve in the following manner. This process is called, naturally, the **energy cascade**.

For large wavelengths—say, for argument, wavelengths of the order greater than $10^4 \lambda_{\text{Kolmogoroff}}$—the disturbance is highly direction-dependent and nonisotropic. This is the range where energy production takes place. Eddies of a given size are stretched by large eddies, and mixing and stretching takes place. Eddy size tends to be reduced in the process, and bigger eddies are converted into smaller eddies. In the process, they also tend more toward the isotropic turbulence state.

When an eddy gets into the inertial range, dissipation is still minor, but straining of eddies continues (the T_{ij} in our earlier discussion), and the eddy size continues to diminish until it reaches the Kolmogoroff length scale, at which point viscous dissipation takes over and eventually destroys the eddy. This process seems to have the universality of being applicable at all length scales, whether consideration is of a hurricane or a "tempest in a teapot."

10.8 Quantum Theory Models

Quantum theory approaches have interested a number of theoretical researchers and have resulted in several models. It has a particular appeal in that it strives to provide a model without empirical constants and thus endeavors to be universal and absolute. Most work, however, is mainly intent on investigating general consequences of a given model rather than on predicting the characteristics of a particular turbulent flow. Few if any actual flow calculations appear to have been made using these models, and the theory does not appear to be at a point where such is feasible except in the simplest possible case.

In the following, the basic equations are presented along with brief statements about the approaches used in their derivation so readers may provide their own details. No one reference exists that deals with all the equations simultaneously, but partial versions of the equations can be found in the standard references dealing with turbulence (e.g., Bradshaw (1978), Craya (1958), Favre et al. (1976), Hinze (1975), Leslie (1973), and Lesieur (1978)).

The quantum theory starts with the Fourier-transformed one-point averaged version of the Navier-Stokes equations, with the assumption that the turbulence is isotropic. In that case they reduce to

$$\left(\frac{\partial}{\partial t} + \nu k^2\right) v_j(\mathbf{k}) = M_{jmn}(\mathbf{k}) \sum^{\Delta} v_m(\mathbf{p}) v_n(\mathbf{r}) + f_j(\mathbf{k}), \qquad (10.8.1)$$

where

$$\sum^{\Delta} v_i(\mathbf{p}) v_n(\mathbf{r}) = \iiint_{\mathbf{r}} \iiint_{\mathbf{p}} v_i(\mathbf{k}) \nu u_n(\mathbf{r}) \delta(\mathbf{k} - \mathbf{p} - \mathbf{r}) d^3\mathbf{p} d^3\mathbf{r} \text{ and}$$

$$M_{jmn}(\mathbf{k}) = -\frac{1}{2} i \left[k_n \Delta_{jm}(\mathbf{k}) + k_m \Delta_{jn}(\mathbf{k}) \right], \qquad (10.8.2)$$

$$\Delta_{jm}(\mathbf{k}) = \delta_{jm} - k_j k_m / k^2.$$

The f function is a hypothetical "stirring force," which models the interaction of the mean flow with the Reynolds stress, and δ is the Dirac delta function, infinite when the argument is zero, zero everywhere else. The $\sum^{\Delta}$ term arises from the convective acceleration and pressure terms and illustrates the triad interaction that is characteristic of this model. In other words, two wave numbers, $\mathbf{p}$ and $\mathbf{r}$, combine algebraically to form a third wave number, $\mathbf{k} = \mathbf{p} + \mathbf{r}$.

The first attempts at closure of this model (Millionshchtikov, 1941; Proudman and Reid, 1954) termed it the ***quasi-normal approximation*** (QN). An equation is formed from the ensemble average of the triple correlation $\overline{v_j(-\mathbf{k}) v_m(\mathbf{p}) v_n(\mathbf{r})}$, which involves fourth-order correlations. The fourth-order correlation is assumed to be expressible as products of the second-order correlations, which would be exactly true if the correlations were Gaussian in nature. The equations representing time derivatives of the second- and third-order correlations thus represent a closed set. While the assumptions made in developing the theory appear reasonable, calculations show that the ensemble average of $v_i(\mathbf{k}) v_i(-\mathbf{k})$ becomes negative, which is physically impossible.

To correct this defect in the theory, the QN model was modified (Orszag, 1970) to the ***eddy-damped quasi-normal approximation*** (EDQN). This theory adds linear combinations of the third-order correlations to the QN forms for the fourth-order correlation. This necessarily introduces vector parameters (eddy viscosities) that multiply the third-order correlations. To assure that the second-order correlation remains positive, the additional assumption is made that the relaxation time of the triple correlations is much weaker than the evolution time of the energy spectrum. The model has now become the EDQNM, or EDQN-"Markovianized," model (Leith, 1971). A further model is needed for specification of the eddy viscosity vectors (Kraichnan, 1971; Sulen et al., 1975).

Another quantum theory model that has received attention and that is at a high level of development is the ***direct interaction*** (DI) model (Kraichnan, 1958, 1959; Leslie, 1973). A velocity field in the form $v_i + \delta v_i$, where δv_i is an infinitesimally small velocity perturbation, is introduced into equation (10.8.1). This equation is then linearized and a Green's function is introduced. The resulting equation is called the response equation, and the Green's function is called the response function. The DI model is then determined by proceeding in the following manner:

1. The nonlinear terms in equation (10.8.1) are taken to be multiplied by a small parameter.

2. The velocity v_i and the response function are expanded as power series in the small parameter.

3. The lowest-order term of the velocity expansion is assumed to be Gaussian in nature.

4. The expansion is stopped at the lowest-order terms that do not vanish when averaged statistically. At that stage, the small parameter is set to unity!

5. In the equations that result from this process, the lowest-order terms in the expansion for the velocity and the response function are replaced by their exact values.

The justification for this process is necessarily indirect because of the complexity of the original equations. Kraichnan (1961) introduced a model related to, but much simpler than, the response equation. This model can be solved exactly. He shows that DI works reasonably well for this model. This equation is then modified by a process that does not depend on the nonlinear term being small into another equation for which DI gives the exact statistical average. The same process can be used to generate a similar set of model equations from the Navier-Stokes equations. The DI approximations to the Navier-Stokes equations are then the exact statistical average of the second set of model equations.

This process is unorthodox and does not provide any assurance that the DI approximation lies close to the exact solution. In fact, Kraichnan put forth the statement that existing mathematical techniques are incapable of providing such assurance. However, to quote Leslie (1973, page 60), one of the proponents of the method, "DI, and other methods which are broadly like it, do give answers where the earlier methods (e.g., QN) gave nothing. These answers are certainly interesting and may well be relevant, but at the moment it cannot be proven that the whole structure rests on a secure basis." Perhaps more confidence can be gained from the fact that DI has been shown to give reasonable agreement with some computer simulations and that the number of investigators of these ideas, and interest in them, has grown.

The DI approximation as first presented was Eulerian in concept (it is sometimes labeled EDI to emphasize this) and suffered from the flaw that it was not invariant to a Galilean transformation. As a result, one of the response integrals diverges at one of its limits of integration. The correction of this (Kraichnan, 1964, 1965, 1966; Edwards, 1968) is most easily carried out in a Lagrangian reference frame and has been termed the ***Lagrangian history direct interaction*** (LHDI) model.

A direct Lagrangian approach is difficult, and closure approximations become virtually impossible in such a frame. To simplify these matters, Kraichnan (1966) has introduced a further model, the ***abridged Lagrangian interaction*** (ALI) model, which introduces an approach somewhere between the Eulerian and Lagrangian models. To illustrate this, consider the notation $f(\mathbf{x}, t_0 | t)$, which reads as "the property f at time t of a fluid particle that was at $\mathbf{x}$ at time t_0." Thus, if $\mathbf{z}$ is the current position of a fluid particle and $\mathbf{v}$ its velocity,

$$\mathbf{v}(\mathbf{x}, t_0 | t) = \frac{\partial \mathbf{z}(\mathbf{x}, t_0 | t)}{\partial t}. \tag{10.8.3}$$

As a simple example, consider an irrotational vortex and flow in a square corner. Then $z = \sqrt{x_1^2 + x_2^2}(\cos(\theta + \phi), \sin(\theta + \phi), 0)$ for the vortex, with $\theta = \frac{(t-t_0)\Gamma}{x_1^2 + x_2^2}, \phi = \tan^{-1}\frac{x_1}{x_2}$, Γ as the circulation, and $\mathbf{z} = (x_1 e^{-k(t-t_0)}, x_2 e^{-k(t-t_0)}, 0)$ for the square corner.

Eulerian mechanics integrates along a line where $t = t_0$, while Lagrangian mechanics integrates along the line $t_0 = $ constant. ALI is an approach that integrates the equations first along the line $t = t_0$ and then along a line of constant t. This results in a model that is approximately Lagrangian in character but yet is simple enough to allow a DI closure approximation.

ALI is extremely complicated and difficult to work with. To get around this, Kraichnan (1970, 1971) modified EDI in a purely Eulerian framework to arrive at an almost-Markovlan model that is invariant to a Galilean transformation. He called it the ***test field model*** (TFM). TFM is much more efficient in computing time than ALI, in that TFM implies an exponential time dependency, whereas ALI must compute the time dependency. It is thought also that TFM may represent the diffusion of a scalar passive better than does ALI (Leslie, 1973, Chapter 12). In a series of computations (Herring and Kraichnan, 1972) where the predictions of these various models were compared with computer solutions (Orszag and Patterson, 1972) and experiments (Grant, Stewart, and Moillet, 1962), the following conclusions were reached:

1. At low turbulent Reynolds numbers, all quantum theory models are usable. The accuracy depends on which quantity is being compared, and no one model stands out as being obviously better. The skewness factor, which is particularly sensitive to differences in the models, turns out to be best predicted by a version of TFM.

2. At high turbulent Reynolds numbers, the models that are not Galilean-invariant fail completely, and only TFM and ALI are usable. These seem to agree quite well with each other. Because the TFM model is much simpler than ALI, the comparison favors TFM.

10.9 Large Eddy Models

In 1973 a group at Stanford University begin a long-term project to study the large-scale end of the turbulence spectrum (Kwak, Reynolds, and Ferziger 1975) by simulation on a supercomputer. The original calculations were done on a grid with 16^3 points, but this was later increased to 32^3 points and more. Reasonable agreement was made with previous experimental results, as long as strong filtering was done on the numerical results. Experiments such as this are useful for testing turbulent flow models and also to illustrate local behavior that might not be simply visualized. The memory requirements, however, make this a research tool rather than a practical tool for applications.

10.10   Phenomenological Observations

Since O. Reynolds's original dye experiments on the stability of pipe flow, it has been known that for a given point in a flow at a given (sufficiently large) Reynolds number, the flow would alternate between the laminar and turbulent states. Investigators defined an ***intermittency factor*** γ as the percent of time at which a flow is turbulent at a fixed point. Experimentally, this is done when measuring a flow by circuitry that produces an ***intermittency signal*** $I(t)$, where

$$I(t) = \begin{cases} 0 & \text{if the flow is nonturbulent,} \\ 1 & \text{if the flow is turbulent.} \end{cases} \qquad (10.10.1)$$

Then the intermittency is found as

$$\gamma = \overline{I(t)} = \text{probability of occurrence of turbulence at that point.} \qquad (10.10.2)$$

Measurements made by Townsend (1951) in the wake behind a thin wire[7] boundary layer show that near the center line of the wake γ is approximately unity. Starting around 18%, or the averaged wake half-diameter, the intermittency appears to drop in an exponential manner, so at 60%, or the averaged wake half-diameter, the intermittency has reduced to less than 0.1. Generally, it is necessary to correct measurements to account for the local degree of intermittency.

In many ways turbulent flow can be thought of as a tangle of vortex filaments, stretching and moving as they interact. (Think of a container filled with squirming earthworms!) Vorticity must start at a wall and then be convected to the interior flow. It has been observed that vortex filaments are created at the wall and oriented perpendicular to the flow. These filaments are generated in bursts in the sublayer, the bursting phenomenon appearing to be cyclical. It has been suggested that the bursting process is part of a cycle in which interaction between the wall region and the outer region of the boundary layer is important. New bursts are created as a result of a disruption of the primary flow by the previous burst (Willmarth, 1979). Once created, the filament can then develop kinks and distort into a ***hairpin vortex***, a U with long arms. The base of the U then lifts from the wall and is carried off into the interior flow.

The interactions between the filaments are governed by the ***Biot-Savart law***, first developed to explain the magnetic field around an electric current-carrying wire. If Γ is the circulation for a given vortex filament of length δs, then at a point a distance R from the filament a speed δv is induced, where

$$\delta v = \frac{\Gamma}{4\pi} \frac{\sin \beta}{R^2} \delta s. \qquad (10.10.3)$$

The angle β is the angle between the vortex filament and the line connecting the filament and the point. The proof of the law comes from the use of the velocity potential of the vortex and is purely kinematic in nature.

---

[7] The wire diameter was approximately 1/16 inch, and the Reynolds number based on wire diameter was 1,360.

Vortex filaments can be both desirable and undesirable in a flow. Recently, high altitude research by NCAR (National Center for Atmospheric Research) indicates that clear air turbulence, the bane of air travelers, is due to rapidly spinning horizontal vortex tubes, presumably on the scale of modern aircraft. On the other hand, Boeing airplanes in the 7X7 family have long used vortex generators on the upper wing to create artificial turbulence that acts to stabilize the boundary layer, reducing drag. Winglets are used to change the vortex pattern in a plane's wake, reducing induced drag. And airplane engine designers are turning to jet engines with chevron rims on the exit nozzles. The serrated, or scalloped, edges on these rims introduce a sheet of vorticity that allows the outer and inner flows to interact more rapidly than if a straight rim were used. The result is a reduction in perceived engine noise with little loss in engine efficiency.

10.11 Conclusions

In the previous material a number of different models were used. The zero-, one-, and two-equation models are used in many commercial CFD packages, although often the details of the parameters are not made known to the user. To be optimistic, there may come a day where one turbulence model fits all—every flow situation, every value of the Reynolds number, every geometry, every Mach number—but that day is still far away.

In a lecture given before a turbulence class some years ago, Dave Wilcox of DCW Industries, an active turbulence modeler, stated that the following is required in turbulence modeling:

Tools of the trade for a turbulence modeler:

1. Ingenuity

2. Computer literacy

3. Knowledge of numerical methods

4. Knowledge of perturbation methods

5. Lots of self esteem

Fundamental premises of a turbulence modeler:

- Try to model the physics, not the partial differential equation.

- Avoid adjusting coefficients from one class of flow to the next.

- Strive for elegance and simplicity.

- Know what other workers in the field are doing, but keep at arm's length.

Good advice!

In summary, turbulence is a difficult subject. No one *really* understands turbulence!

Chapter 11

Computational Methods—Ordinary Differential Equations

11.1 Introduction

Many of the problems that were solved in the previous chapters required exacting calculations that were originally done either by hand computation or at best with the aid of an adding machine. Numerical methods were invented as far back as Newton, and techniques suited to the computer, such as the use of Green's functions, were known in the nineteenth century, but it wasn't until the mid-twentieth century that mainframe computers were available to researchers. Even then, the need to use storage devices such as punched cards required significant time and effort for all but the smallest calculation.

Today, the personal computer has made the power of the most powerful computer of mid-nineteenth century available to all. Many languages have been invented to make calculating the problems previously covered relatively easy. The first of these was FORTRAN (Formula Translator), developed at IBM in the 1940s. This was followed by similar languages such as BASIC, QUICKBASIC, PASCAL, VISUAL BASIC, C++, and many others. BASIC originally was included as a chip in the original IBM PC, and QUICKBASIC was provided as part of the disk operating system (DOS) until version 4.5 of DOS. These languages required a fair degree of programming skills on the part of the user. Today, MATHCAD, MATLAB, and symbolic manipulators such as MAPLE and MATHEMATICA have greatly increased the availability of mathematical computations.[1]

---

[1] Newer languages are not necessarily more productive than the newer versions of FORTRAN, such as versions FORTRAN 90 and 95. See www.nr.com/CiP97.pdf for comments on this matter.

A few of the concepts and techniques used in computational fluid mechanics are presented in this Chapter and Chapter 12, along with sample programs that illustrate the methodology. The programs are written in FORTRAN, the most persistent of these languages, and can be easily transformed to most of the BASIC class of language, as well as others. In all of these programs, lines preceded by either a capital C or an exclamation mark are comments to the reader and are not used in computations. (The C is used for FORTRAN 90, and the ! for FORTRAN 95 programs.) The ampersand (&) is used to denote line continuation.

There are a number of versions of the FORTRAN compiler available from several sources. Programs presented in this book use the extensions .FOR (FORTRAN 90 fixed source form); .F90 (FORTRAN 90 free source form); or .F95 (FORTRAN 95). Most present-day compilers are capable of handling these files.

As the first example, Program 11.1.1 is a program for determination of the separation point on a circular cylinder using both Thwaites's and Stratford's criteria. The velocity

```
      PROGRAM    THWAITES
C     Thwaite - Stratford separation prediction of separation on a circular cylinder
C     Schmitt-Wenner velocity profile
C.............................................................
      DIMENSION SX(10)
      A=-0.451
      B=-.00578
      SMAX=1.1976
      UMAX=1.606299
      U=0.
      UP=2.
      DS=.001
      G=0.
      G1=0.
      FINT=0.
      FOLD=0.
      GOLD=0.
      DO I=1,1600
          S=I*DS
          S2=S*S
          U=S*(2.+S2*(A+B*S2))
          UP=2.+S2*(3.*A+5.*B*S2)
          FO=-2.*(U**6)/UP
          G2=U**5
          FINT=FINT+0.5*(G1+G2)*DS
          F=FO-FINT
          DEG=S*180./3.14159
          G1=((S-SMAX)**2)*(1.-(U/UMAX)**2)*((-2.*U*UP/(UMAX*UMAX))**2)
     &           -.0104
          IF (S.GT.SMAX)  G=G1
          WRITE(6,100) S,U,UP,F,G,DEG
           G1=G2
           IF ((F*FOLD).LT.0.) THEN
                  SX(1)=S
                  SX(2)=F
                  SX(3)=FOLD
                  SX(4)=DEG
                  WRITE(*,*) "THWAITES----------------------------------"
           END IF
           IF ((G*GOLD).LT.0.) THEN
                  SX(5)=S
                  SX(6)=G
                  SX(7)=GOLD
                  SX(8)=DEG
                  WRITE(*,*) "STRATFORD---------------------------------"
           END IF
           FOLD=F
           GOLD=G
      END DO
```

Program 11.1.1—Separation criteria for a circular cylinder—program by author

```
                    WRITE(*,*)
                    WRITE(*,*) "    S          F          FOLD      DEGREES"
                    WRITE(*,*) "Thwaites Criterion"
                    WRITE(*,102) SX(1),SX(2),SX(3),SX(4)
                    WRITE(*,*) "Stratford Criterion"
                    WRITE(*,102) SX(5),SX(6),SX(7),SX(8)
          100       FORMAT(6F10.4)
          102       FORMAT(4F10.4)
                    END
```

Program 11.1.1—(Continued)

profile is a fifth-order polynomial used by Schmitt and Wenner (1941) and derived from their experiments. The quantity S is the arc distance along the surface of the cylinder, measured from the stagnation point. The criteria are put in the form $F(S) = 0$, and $F(S)$ is calculated at progressively increasing values of S until F changes sign, indicating that the zero point has been passed. Values of S and F are printed to the screen, and zero-crossing is pointed out for each criteria.

A more complicated programming example is Program 11.1.2, the calculation of the various regions for the behavior of the Mathieu functions. When encountering such sophisticated problems, it is worthwhile (and the better part of valor) to first determine if prepared programs are available. Good starting points for this are the IMSL Math/Library (1987) and Press, Flannery, Teukolsky, and Vetterling (1986). In this case, however,

```
           PROGRAM Mathieu
C     ================================================================
C     Purpose: This program computes a sequence of characteristic
C              values of Mathieu functions using subroutine CVA1
C
C                          y'' + (m^2 - 2q cos 2t) y = 0
C
C     Input :  m   --- Order of Mathieu functions
C                 q  --- Parameter of Mathieu functions
C                 KD --- Case code
C                        KD=1 for cem(x,q)  ( m = 0,2,4,....)
C                        KD=2 for cem(x,q)  ( m = 1,3,5,....)
C                        KD=3 for sem(x,q)  ( m = 1,3,5,....)
C                        KD=4 for sem(x,q)  ( m = 2,4,6,....)
C
C     It is adapted from programs given in Computation of Special Functions
C        by S. Zhang and J. Jin, Wiley, 1996.
C     ================================================================
           IMPLICIT DOUBLE PRECISION (A-H,O-Z)
           DIMENSION CV1(200),CV2(200),CVE(30,30),CVS(30,30),A(30,30),
      &         IC(30),IB(65)
           OPEN(7,FILE='A:MATHIEU.DAT',STATUS='UNKNOWN')
           DO 19 MMAX=1,12
             DO 11 K=1,28
                    Q=K
                    CALL CVA1(1,MMAX,Q,CV1)
                    CALL CVA1(2,MMAX,Q,CV2)
                    DO 10 J=1,MMAX/2+1
                      CVE(2*J-1,K)=CV1(J)
                      CVE(2*J,K)=CV2(J)
          10        END DO
          11     END DO
```

Program 11.1.2—Mathieu function stability regions adapted from program given in *Computation of Special Instruction* by S. Zhang and J. Jin, Wiley, 1996. (The first adaptation was by Ben Barrowes (barrowes@alum.mit.edu) and can be found on the internet.)

```
            DO 16 K=1,28
                Q=K
                CALL CVA1(3,MMAX,Q,CV1)
                CALL CVA1(4,MMAX,Q,CV2)
                DO 15 J=1,MMAX/2+1
                        CVS(2*J,K)=CV1(J)
                        CVS(2*J+1,K)=CV2(J)
15                  END DO
16          END DO
            J=MMAX
            WRITE(*,*)'-------------------------------------------'
            WRITE(*,*) J
            WRITE(*,*) "  ALPHA  Q       a(J)      b(J)"
            WRITE(*,*)'-------------------------------------------'
            DO 17 K=1,28
                Q=K
                A(2*J-1,K)=CVE(J+1,K)
                A(2*J,K)=CVS(J+1,K)
                WRITE(*,50)Q,CVE(J+1,K),CVS(J+1,K)
17          END DO
19      END DO
            DO 22 L=1,28
              IC(L)=L
                IB(2*L-1)=L*L
                IB(2*L)=L*L
22      END DO
            J=0
            IXB=0
C               WRITE ROWS
            WRITE(*,46)J,IXB,IB(1),IB(2),IB(3),IB(4),IB(5),IB(6),IB(7),
     &      IB(8),IB(9),IB(10),IB(11),IB(12),IB(13),IB(14),IB(15)
            WRITE(7,46)J,IXB,IB(1),IB(2),IB(3),IB(4),IB(5),IB(6),IB(7),
     &      IB(8),IB(9),IB(10),IB(11),IB(12),IB(13),IB(14),IB(15)
            WRITE(*,46)J,IXB,IC(1),IC(2),IC(3),IC(4),IC(5),IC(6),IC(7),
     &      IC(8),IC(9),IC(10),IC(11),IC(12),IC(13),IC(14),IC(15)
            WRITE(7,46)J,IXB,IC(1),IC(2),IC(3),IC(4),IC(5),IC(6),IC(7),
     &      IC(8),IC(9),IC(10),IC(11),IC(12),IC(13),IC(14),IC(15)
            DO 24 J=1,14
                WRITE(*,45)J,IB(J),A(J,1),A(J,2),A(J,3),A(J,4),A(J,5),A(J,6),
     &          A(J,7),A(J,8),A(J,9),A(J,10),A(J,11),A(J,12),A(J,13),A(J,14),
     &          A(J,15),A(J,16),A(J,17),A(J,18),A(J,19),A(J,20),A(J,21),
     &          A(J,22),A(J,23),A(J,24),A(J,25),A(J,26),A(J,27),A(J,28)
                WRITE(7,45)J,IB(J),A(J,1),A(J,2),A(J,3),A(J,4),A(J,5),A(J,6),
     &          A(J,7),A(J,8),A(J,9),A(J,10),A(J,11),A(J,12),A(J,13),A(J,14),
     &          A(J,15),A(J,16),A(J,17),A(J,18),A(J,19),A(J,20),A(J,21),
     &          A(J,22),A(J,23),A(J,24),A(J,25),A(J,26),A(J,27),A(J,28)
24      END DO
C               WRITE COLUMNS
            DO 25 J=1,28
                WRITE(*,30)J,IB(J),A(1,J),A(2,J),A(3,J),A(4,J),A(5,J),A(6,J),
     &          A(7,J),A(8,J),A(9,J),A(10,J),A(11,J),A(12,J)
                WRITE(7,30)J,IB(J),A(1,J),A(2,J),A(3,J),A(4,J),A(5,J),A(6,J),
     &          A(7,J),A(8,J),A(9,J),A(10,J),A(11,J),A(12,J)
25      END DO
        CLOSE(7,STATUS='KEEP')
30       FORMAT(2(I10,","),11(F10.5,","),F10.5)
45       FORMAT(2(I10,","),27(F10.5,","),F10.5)
46       FORMAT(16(I10,","),I10)
50       FORMAT((F6.1,","),F10.4,",",F10.4)
        END
```

Program 11.1.2—(Continued)

```
C=================================================================
C
        SUBROUTINE CVA1(KD,M,Q,CV)
C
C       =================================================================
C       Purpose: Compute a sequence of characteristic values of
C                Mathieu functions
C       Input :  M  --- Maximum order of Mathieu functions
C                q  --- Parameter of Mathieu functions
C                KD --- Case code
C                       KD=1 for cem(x,q)  ( m = 0,2,4,úúú )
C                       KD=2 for cem(x,q)  ( m = 1,3,5,úúú )
C                       KD=3 for sem(x,q)  ( m = 1,3,5,úúú )
C                       KD=4 for sem(x,q)  ( m = 2,4,6,úúú )
C       Output:  CV(I) --- Characteristic values; I = 1,2,3,...
C                For KD=1,CV(1), CV(2), CV(3),..., correspond to
C                the characteristic values of cem for m = 0,2,4,...
C                For KD=2,CV(1), CV(2), CV(3),..., correspond to
C                the characteristic values of cem for m = 1,3,5,...
C                For KD=3,CV(1), CV(2), CV(3),..., correspond to
C                the characteristic values of sem for m = 1,3,5,...
C                For KD=4,CV(1), CV(2), CV(3),..., correspond to
C                the characteristic values of sem for m = 0,2,4,...
C       =================================================================
        IMPLICIT DOUBLE PRECISION (A-H,O-Z)
        DIMENSION G(200),H(200),D(500),E(500),F(500),CV(200)
        EPS=1.0D-14
        ICM=INT(M/2)+1
        IF (KD.EQ.4) ICM=M/2
        IF (Q.EQ.0.0D0) THEN
           IF (KD.EQ.1) THEN
              DO 10 IC=1,ICM
10               CV(IC)=4.0D0*(IC-1.0D0)**2
           ELSE IF (KD.NE.4) THEN
              DO 15 IC=1,ICM
15               CV(IC)=(2.0D0*IC-1.0D0)**2
           ELSE
              DO 20 IC=1,ICM
20               CV(IC)=4.0D0*IC*IC
           ENDIF
        ELSE
           NM=INT(10+1.5*M+0.5*Q)
           E(1)=0.0D0
           F(1)=0.0D0
           IF (KD.EQ.1) THEN
              D(1)=0.0D0
              DO 25 I=2,NM
                 D(I)=4.0D0*(I-1.0D0)**2
                 E(I)=Q
25               F(I)=Q*Q
              E(2)=DSQRT(2.0D0)*Q
              F(2)=2.0D0*Q*Q
           ELSE IF (KD.NE.4) THEN
              D(1)=1.0D0+(-1)**KD*Q
              DO 30 I=2,NM
                 D(I)=(2.0D0*I-1.0D0)**2
                 E(I)=Q
30               F(I)=Q*Q
```

Program 11.1.2—(Continued)

```
                         ELSE
                            D(1)=4.0D0
                            DO 35 I=2,NM
                               D(I)=4.0D0*I*I
                               E(I)=Q
         35                    F(I)=Q*Q
                         ENDIF
                         XA=D(NM)+DABS(E(NM))
                         XB=D(NM)-DABS(E(NM))
                         NM1=NM-1
                         DO 40 I=1,NM1
                            T=DABS(E(I))+DABS(E(I+1))
                            T1=D(I)+T
                            IF (XA.LT.T1) XA=T1
                            T1=D(I)-T
                            IF (T1.LT.XB) XB=T1
         40              CONTINUE
                         DO 45 I=1,ICM
                            G(I)=XA
         45                 H(I)=XB
                         DO 75 K=1,ICM
                            DO 50 K1=K,ICM
                               IF (G(K1).LT.G(K)) THEN
                                  G(K)=G(K1)
                                  GO TO 55
                               ENDIF
         50                 CONTINUE
         55                 IF (K.NE.1.AND.H(K).LT.H(K-1)) H(K)=H(K-1)
         60                 X1=(G(K)+H(K))/2.0D0
                            CV(K)=X1
                            IF (DABS((G(K)-H(K))/X1).LT.EPS) GO TO 70
                            J=0
                            S=1.0D0
                            DO 65 I=1,NM
                               IF (S.EQ.0.0D0) S=S+1.0D-30
                               T=F(I)/S
                               S=D(I)-T-X1
                               IF (S.LT.0.0) J=J+1
         65                 CONTINUE
                            IF (J.LT.K) THEN
                               H(K)=X1
                            ELSE
                               G(K)=X1
                               IF (J.GE.ICM) THEN
                                  G(ICM)=X1
                               ELSE
                                  IF (H(J+1).LT.X1) H(J+1)=X1
                                  IF (X1.LT.G(J)) G(J)=X1
                               ENDIF
                            ENDIF
                            GO TO 60
         70                 CV(K)=X1
         75              CONTINUE
                      ENDIF
                      RETURN
                      END
```

Program 11.1.2—(Continued)

a search of the World Wide Web provided the needed information in the form of a downloadable subroutine.

The next program (11.1.3) uses FORTRAN's capability to deal with complex numbers, avoiding tedious computations to arrive at the real and imaginary portions of

```
      PROGRAM Joukowsky
C===============================================================
C               GENERATES A JOUKOWSKY AIRFOIL
C===============================================================
C      w = zU* + Ua^2/z + i*GAMMA/2*pi ln z/a
C
C      B = GAMMA/4*pi*a*|U| = sin(alpha + sin-1 YCP/a)
C
C      dw/dz = U* - U*a^2/z^2 + iGAMMA/2*pi*z
C
C      ZP = Z'                  ZPP = Z''
C
C      Z' =Zc' + Z             Z'' = Z' + b^2/Z'
C===============================================================
      COMPLEX AI,DWDZ,U,USTAR,W,Z,ZCP,ZP,ZPP,ZSP,ZPSP,ZPPSP,ZTE,ZPTE,ZPPTE
      PI=3.1415927
      A=1.
      NPOINT=180
      NP=360/NPOINT
      AI=CMPLX(0,1.)
      DTHETA=2.*PI/NPOINT
      OPEN(1, FILE="A:JOUKOWSKY.DAT", STATUS='REPLACE')
      WRITE(*,*)
      WRITE(*,*) "In this program a and Uinfinity are taken as unity."
      WRITE(*,*)
1     WRITE(*,*) "Input Zc':"
      WRITE(*,*)
      WRITE(*,*) "  Thickness increases with Xc'."
      WRITE(*,*) "  Camber increases with Yc'."
      WRITE(*,*) "  Xc' = Yc' = 0 stops the program.'"
      WRITE(*,*)
      WRITE(*,*) "    Xc' first. It must lie in -1< Xc' <0.  "
      READ(*,*) XCP
      WRITE(*,*) "  Now input Yc'.  It must lie in -1< Yc' <1. "
      READ (*,*) YCP
      WRITE(*,*)
      WRITE(*,*) "  Enter the angle of attack <degrees>.  "
      READ(*,*) ALPHA
      B=XCP+SQRT(1.-YCP*YCP)
      B2=B**2
      BETA=ALPHA+ASIN(YCP/A)
      GAMMA=4.0*PI*A*SIN(BETA)
      R=XCP*XCP+YCP*YCP
      SING=(B+XCP)**2+YCP**2-A**2
      IF (R.EQ.0.) THEN
           STOP
        ELSE IF (R.GT.1.0) THEN
          WRITE(*,*) " The point Zc' must lie within the unit circle.
     &Try again."
          GOTO 1
        ELSE IF (SING.GE.0) THEN
          WRITE(*,*) " The point Z'=-b must lie within the unit circle.
     &     You will have a singularity in the flow field. Try again."
          GOTO 1
        ELSE IF (B.LE.0.) THEN
          WRITE(*,*) "The parameter b <= 0, try again."
          GOTO 1
      END IF
```

Program 11.1.3—Joukowsky Airfoil (program by the author)

```
           ALP=ALPHA*PI/180.
           U=CMPLX(COS(ALP),SIN(ALP))
           USTAR=CMPLX(COS(ALP),-SIN(ALP))
           ZCP=CMPLX(XCP,YCP)
           WRITE(1,5)
           DO I=1, NPOINT+1
              THETA=(I-1)*DTHETA
              Z=A*CMPLX(COS(THETA),SIN(THETA))
              ZP=Z+ZCP
              ZPP=ZP+B2/ZP
              CALL SPEED(B2,GAMMA,PI,Z,ZP,U,USTAR,UM)
              CP=1.-0.5*UM*UM
              XPP=REAL(ZPP)
              YPP=AIMAG(ZPP)
              J=(I-1)*NP
              WRITE(*,*) J,XPP,YPP,UM,CP
              WRITE(1,100)  J,XPP,YPP,UM,CP
           END DO
           WRITE(*,*)
           WRITE(*,*) "Note: The speed calculations are not accurate in "
           WRITE(*,*) "  the vicinity of the trailing edge. Use"
           WRITE(*,*) "  l'Hospital's rule here."
           WRITE(*,*)
           WRITE(*,*) "Press <ENTER> to continue."
           READ(*,*)
           ZTE=CMPLX(A*COS(ALPHA-BETA),A*SIN(ALPHA-BETA))
           ZPTE=ZTE+ZCP
           ZPPTE=ZPTE+B2/ZPTE
           XTE=REAL(ZPPTE)
           YTE=AIMAG(ZPPTE)
           ZSP=CMPLX(-A*COS(ALPHA+BETA),-A*SIN(ALPHA+BETA))
           ZPSP=ZSP+ZCP
           ZPPSP=ZPSP+B2/ZPSP
           XSP=REAL(ZPPSP)
           YSP=AIMAG(ZPPSP)
           WRITE(*,*)
           WRITE(*,*) " Trailing edge:     ",XTE,YTE
           WRITE(*,*) " Stagnation point: ",XSP,YSP
           WRITE(*,*) " Xc'  , Yc':        ",XCP,YCP
           WRITE(*,*) " b, beta :          ",B,BETA
           WRITE(*,*) " Gamma:             ",GAMMA
           WRITE(1,10) XCP,YCP
           WRITE(1,20) B,BETA
           WRITE(1,101) XTE,YTE
           WRITE(1,102) XSP,YSP
           WRITE(1,103) GAMMA
           WRITE(*,*)
           CLOSE(1)
      5    FORMAT("Angle",',',"x'",',',"y'",',',"Speed",',',"Cp")
     10    FORMAT(" Xc' = ",F10.7,',',", Yc' = ",F10.7)
     20    FORMAT(" b = ",F10.7,',',", beta = ",F10.7," radians")
    100    FORMAT (I4,2(',',F10.7),',',F15.7,',',F15.7)
    101    FORMAT(" Trailing edge   :    ",2F10.7)
    102    FORMAT(" Stagnation point:    ",2F10.7)
    103    FORMAT(" Gamma:               ",F15.7)
           END PROGRAM
```

Program 11.1.3—(Continued)

```
C=======================================================
C                    SUBROUTINE SPEED
C=======================================================
          SUBROUTINE SPEED(B2,GAMMA,PI,Z,ZP,U,USTAR,UM)
          COMPLEX AI,DWDZ,Z,ZP,U,USTAR
          AI=CMPLX(0,1.)
          DWDZ=USTAR-U/(Z**2)+AI*GAMMA/(2.*PI*Z)
          DWDZ=DWDZ/(1.-B2/(ZP*ZP))
          UM=ABS(DWDZ)
          RETURN
          END SUBROUTINE
```

Program 11.1.3—(Continued)

complicated complex expressions. The user inputs the three parameters involved in the Joukowsky transformations described in Chapter 3, as well as the angle of attack of the airfoil. The program then generates the shape of the airfoil, as well as the velocity and pressure distributions on its surface.

Program 11.1.4 is an example of the generation of a potential flow about a half-body consisting of a source in a uniform stream. After specifying source strength and

```
        PROGRAM   RANKINE      ! RANKINE.FOR
!----------------------------------------------------------------------
!              Generates a Rankine half-body in two dimensions
!              Potential consists of a source in a uniform stream
!                 This program is easily generated to the axisymmetric
!                    half-body with a change in the potential
!              from a 2-dimensional source to a three-dimensional source.
!           More general shapes can be generated by the addition of more sources
!                              and/or sinks
!              Remember that to generate a closed body the sum of the source
!                     and sink strengths must be zero.
!----------------------------------------------------------------------
!              Data generated here was used in generating data for Figure 2.2.9
!       Values used:
!       A=m/2*PI*U      XSP = Stag point @ -A     Half-width = WH = pi*A
!              The data in File 1 is easily inserted in any spreadsheet.
!         This was made possible by including the commas in FORMAT statement 200.
!----------------------------------------------------------------------
!       Configure maximum data size for storage.
        DIMENSION S(8),XO(26),Y
                       O(26),XXS(26),YYS(26),XX(26),YY(11,71)
!       Open file for data storage
        OPEN(1, FILE='A:RANKINE.DAT', STATUS='UNKNOWN')
        A=.125
        PI=3.14159
!         Generates PSI = 0 streamline on x axis, infinity to source
        DO I=1,25
           XO(I)=-6.+(12./24.)^(I-1)
           YO(I)=0.
        END DO
!         Generates PSI = 0 streamline off of x axis, starts at stagnation point
        CALL BODYSHAPE(A,PI,XXS,YYS)
!         Generates other streamlines
        XSTART=-6.
        YSTART=0.
```

Program 11.1.4—Potential flow past a half-body (program by the author)

```
        YS=0.
        DX=0.5
        DO M=1,25                                      ! X LOOP
            XX(M)=XSTART+(M-1)*DX
            X=XX(M)
            DO N=1,8                                   ! PSI LOOP, PSI incremented by 0.2
                PSI=0.2*N
                CALL SEARCH(PSI,A,X,Y,PSIS,ATD,IS,PI)
                YY(N,M)=Y
                FER=PSIS-PSI
                WRITE(*,100) IS,PSI,FER,X,Y,ATD
            END DO
            WRITE(*,101)
        END DO
        WRITE(*,*) "  IS      PSI        PSIS        X         Y"
        DO I=1,25
            SM=-YYS(I)
            DO J=1,8
                S(J)=-YY(J,I)
            END DO
            WRITE(1,200) X0(I),Y0(I),XXS(I),YYS(I),SM,XX(I),YY(1,I),
     &      YY(2,I),YY(3,I),YY(4,I),YY(5,I),YY(6,I),YY(7,I),YY(8,I),
     &      S(1),S(2),S(3),S(4),S(5),S(6),S(7),S(8)
        END DO
100     FORMAT(I6,F10.4,F10.6,3F10.4)
101     FORMAT("=================================================")
200     FORMAT(21(F8.4,','),F8.4)
        CLOSE (1)
        END PROGRAM
!=====================================================================
        SUBROUTINE BODYSHAPE(A,PI,XXS,YYS)
!---------------------------------------------------------------------
!       Generates body shape for a 2-D Rankine half-body
!       Generates PSI = 0 streamline off of x axis, starts at stagnation point
!          Fixes Y, searches for X
!---------------------------------------------------------------------
        DIMENSION XXS(26),YYS(26)
        WH=A*PI
        XSP=-A
        DY=WH/25
        XXS(1)=XSP
        YYS(1)=0.
        II=1
        XOLD=XSP
        Y=DY
        X=XOLD-.01
        FOLD=Y+A*(ATAN(-X/Y)-.5*PI)
        WRITE(*,*) A,WH
!       Y loop covers a range of Y's, staring with 0.04*halfwidth
!          and continuing to 0.96* half-width.
        DO I=1,24
                Y=I*DY
                X=XOLD
                IG=0
                WRITE(*,*) "============================="
                WRITE(*,*) X,Y,FOLD
!       X loop covers a range of X. Since Y is fixed, a starting point is used (XOLD)
!          and then incremented a small amount. The search procedure is much as used
!          in the subroutine SEARCH, except the roles of X and Y are reversed.
!       Comments made in the SEARCH subroutine are true here also regarding ATAN.
```

Program 11.1.4—(Continued)

```
                DO J=1,1500
                    DX=.0001
                    IF(X.GT.0.234) DX=.0009
                    IF(I.EQ.24) DX=0.0015
                    X=DX*(J-1)+XOLD
                    AT=ATAN(-X/Y)
                    F=Y+A*(AT-.5*PI)
!       Test for convergence.
                    FTEST=FOLD*F
                    ATD=AT*180./PI
                    IF((FTEST.LE.0.).AND.(IG.EQ.0)) THEN
                        WRITE(*,200)I,J,X,Y,F,ATD
                        II=I+1
                        XXS(II)=X
                        YYS(II)=Y
                        XS=X
                        YS=Y
                        IG=1
                    ENDIF
                    FOLD=F
                END DO
                XOLD=XS-.01
                FOLD=YS+A*(ATAN(-(XS-.1)/YS)-.5*PI)
        END DO
        XXS(25)=6.
        YYS(25)=WH
        WRITE(*,*) "  X     Y     FOLD"
        WRITE(*,*) "  I    J     X    Y     F    ATD"
  100   FORMAT(F10.5,','F10.5)
  200   FORMAT(2I5,3F10.5,F12.5)
        RETURN
        END SUBROUTINE
!-------------------------------------------------------------------
        SUBROUTINE SEARCH(PSI,A,X,YS,PSIS,ATD,IS,PI)
!        Generates streamlines
!         Fixed X, searches for Y
!        The search procedure used is not very elegant, but it works (a lot to be
!            said for that).
!        A desired value of PSI is inputted, along with a value of x.
!        A value of y is guessed at, then a value for PSI based on that is computed.
!        The first guess for y is always taken above the desired streamline so it
!            gives a value of PSI which is too large.
!        The y value is decremented a certain amount dy and the computation is
!            repeated, until the error changes sign.
!        Then x is incremented and the process repeated.
!        Two considerations arise in the computation:
!            1.  At and near the stagnation point the slope of the streamline is
!                steep, thus small dy (0.0001) must be used. Later on larger values
!                of dy (0.001) are used.
!            2.  Arctangents are multivalued, and FORTRAN assumes that you want
!                values between - pi/2 and + pi/2.  Thus in using the ATAN function
!                care has to be taken.
!-------------------------------------------------------------------
        WH=A*PI
        XSP=-A
        DY=0.0001
        IF(X.GE.-.6) DY=.001
```

Program 11.1.4—(Continued)

```
            FOLD=Y+A*(ATAN(-X/(Y-DY))-.5*PI)
            IG=0
            Y=PSI-.015
!              This is the Y loop, where Y is continually incremented.
            DO I=1,1500
               Y=Y+DY
               PSIOLD=PSI-.05
               AT=ATAN(-X/Y)
               F=Y+A*(AT-.5*PI)
               FTEST=F-PSI
!              Test to see if the desired streamline has been crossed.
!              If it has, linear interpolation is used to improve result.
               IF((FTEST.GE.0.).AND.(IG.EQ.0)) THEN
                  IS=I
                  FF=(F-PSIOLD)/(PSI-PSIOLD)
                  PSIS=F
                  YS=Y
                  IG=1
                  ATD=AT*180./PI
               ENDIF
               PSIOLD=F
            END DO
            RETURN
            END SUBROUTINE
```

Program 11.1.4—(Continued)

uniform stream velocity, the body shape and surrounding streamlines are found. The computation is complicated by the fact that on a streamline it is not possible to solve directly for y as a function of x. An iteration procedure must be used. The appropriate source strength on the panel is computed so as to make the velocity tangent to the body.

11.2 Numerical Calculus

Many schemes are available for computing derivatives and integrals of functions, depending on the degree of accuracy needed. Many methods are based on the Taylor series expansion of a function—that is, for $y(x) = f(x)$, on

$$y(x) = \sum_{n=0}^{\infty} \frac{1}{n!} f^{(n)}(x_0)(x - x_0)^n. \tag{11.2.1}$$

Here $f^{(n)}(x_0)$ stands for the nth order derivative of f evaluated at x_0. If y is known at neighboring points and knowledge of its derivative is required, then to lowest order in the point spacing, write $y(x + \Delta x) = y(x) + f'(x)\Delta x + O(\Delta x^2)$, giving

$$f'(x) = \frac{y(x + \Delta x) - y(x)}{\Delta x} + O(\Delta x). \tag{11.2.2}$$

If greater accuracy is needed, start with

$$y(x + \Delta x) = y(x) + f^{(1)}(x)\Delta x + \frac{1}{2}f^{(2)}(x)\Delta x^2 + O(\Delta x^3), \tag{11.2.3a}$$

$$y(x - \Delta x) = y(x) - f^{(1)}(x)\Delta x + \frac{1}{2}f^{(2)}(x)\Delta x^2 + O(\Delta x^3). \tag{11.2.3b}$$

Subtracting one of the two equations from the other gives

$$f^{(1)}(x) = \frac{y(x+\Delta x) - y(x-\Delta x)}{2\Delta x} + O(\Delta x^2). \qquad (11.2.4)$$

Further accuracy can be found by including more terms and more points in the Taylor expansion. Some examples follow:

$$\text{Central differences:} \quad \frac{df(x)}{dx} \approx \frac{1}{z\Delta x} [f(x+\Delta x) - f(x-\Delta x)] + O(\Delta x^2).$$

$$\text{Forward differences:} \quad \frac{df(x)}{dx} \approx \frac{1}{\Delta x} [f(x+\Delta x) - f(x)] + O(\Delta x). \qquad (11.2.5)$$

$$\text{Backward differences:} \quad \frac{df(x)}{dx} \approx \frac{1}{\Delta x} [f(x) - f(x-\Delta x)] + O(\Delta x).$$

Newton's method is frequently used for finding the roots of an equation of the form $f(x) = 0$. (Other methods can be found in Press, Flannery, Teukolsky, and Vetterling (1986), and in Carnahan, Luther, and Wilkes (1969).) Newton suggested making an initial guess and then repeatedly improving it according to the formula

$$x_{n+1} = x_n - \frac{f(x_n)}{f'(x_n)}, \qquad (11.2.6)$$

the prime denoting the derivative of f. This is merely a rearrangement of equation (11.2.2). Providing that the first guess is reasonably close to the root sought, the iteration proceeds quickly to the desired root.

Newton's rule can be used for the computation of the stability curves for Couette-Rayleigh instability. First, to simplify the calculations, introduce $\tau = -(c_0^2 - a^2)/a^2$. Then

$$c_n^2 = a^2 \left[1 - \tau \left(\cos \frac{2\pi n}{3} + i \sin \frac{2\pi n}{3} \right) \right], \qquad n = 0,\ 1,\ 2,\ldots.$$

Since we expect that $\tau > 1$,

$$c_0 = ai\sqrt{\tau - 1},$$

$$c_1 = a\sqrt{1 + \tau \left(\frac{1}{2} - i\frac{\sqrt{3}}{2} \right)} = a\sqrt{1 + \tau + \tau^2}\ e^{-i\theta/2},$$

$$c_2 = a\sqrt{1 + \tau \left(\frac{1}{2} + i\frac{\sqrt{3}}{2} \right)} = a\sqrt{1 + \tau + \tau^2}\ e^{i\theta/2},$$

$$\text{with } \theta = \tan^{-1} \frac{\sqrt{3}\tau}{2 + \tau}.$$

To start the computation, use the known minimum critical values for even disturbances to compute the Rayleigh number, and then move first to slightly smaller values of the wave number, understanding that the corresponding Rayleigh number is always slightly larger than the previous value. Use Newton's method repeatedly to guess at the Raleigh number for this new wave number. (The derivative is computed using equation (11.2.2).) Repeat the procedure for wave numbers larger than critical, and finally the process is repeated for odd disturbances. Successive iteration improves the accuracy by narrowing the interval at which the derivative is computed.

In Program 11.2.1 a subroutine is used to calculate the stability criteria to simplify understanding of the program.

```
C.....................BENARD.FOR.......................
C.............Rayleigh-Benard Stability.............
C...........R = Rayleigh NUMBER, a = Wave Number........
C.....................................................
        DIMENSION AE(200),AO(200),RE(200),RO(200)
        OPEN(1,FILE='A:BENARD.DAT', STATUS='UNKNOWN')
C.....................................................
C.............Even Disturbances.......................
C.....................................................
        WRITE(6,*) "Even Disturbances"
        AE(28)=3.117
        RE(28)=1707.762
        CALL CALC(AE(28),RE(28),FEVEN,FODD)
        WRITE(6,*) " a              Ra        FEVEN"
        WRITE(6,*) AE(28),RE(28),FEVEN
        WRITE(6,*) "              J     a       RSAVE      FEVEN"
        FEMIN=FEVEN
        RMIN=RE(28)
        DR=0.1
C.............Even - Less Than Minimun.................
        FEVEN=FEVEN-0.1
        ROLD=1700.-DR
        DO I=1,27
            J=0
            a=3.1-(I-1)*0.1
            DO WHILE (FEVEN<0.)
                J=J+1
                R=ROLD+DR
                CALL CALC(a,R,FEVEN,FODD)
                rat=-FEOLD/(-FEOLD+FEVEN)
                RSAVE=R+rat*DR
                IF(rat<0.) WRITE(6,*)FEVEN,FEOLD,rat
                FEOLD=FEVEN
                ROLD=R
            END DO
            WRITE(6,100) J,a,RSAVE,FEVEN
            FEVEN=FEVEN-1.
                AE(28-I)=a
                RE(28-I)=RSAVE
        END DO
C.............Even - Greater Than Minimun.................
        FEVEN=FEVEN-0.1
        ROLD=1700.-DR
        DO I=1,59
            J=0
            a=3.2+(I-1)*0.1
            DO WHILE (FEVEN<0.)
                J=J+1
                R=ROLD+DR
                CALL CALC(a,R,FEVEN,FODD)
                rat=-FEOLD/(-FEOLD+FEVEN)
                RSAVE=R+rat*DR
                FEOLD=FEVEN
                ROLD=R
            END DO
```

Program 11.2.1—Computation of the neutral stability curves for Rayleigh-Bénard stability (program by the author)

```
                     WRITE(6,101) J,a,RSAVE,FEVEN
                     FEVEN=FEVEN-1.
                     AE(28+I)=a
                     RE(28+I)=RSAVE
           END DO
           WRITE(1,104)
           WRITE(1,105)
           DO I=1,87
                     WRITE(1,106) I,AE(I),RE(I)
           END DO
C......................................................
C..............Odd Disturbances.........................
C......................................................
           WRITE(6,*) "Odd Disturbances"
           AO(38)=5.365
           RO(38)=17610.39
           CALL CALC(AO(38),RO(38),FEVEN,FODD)
           WRITE(6,*) " a              Ra     FODD"
           WRITE(6,*) AO(38),RO(38),FODD
           WRITE(6,*) "            J       a       RSAVE     FODD"
           FOMIN=FODD
C.............Odd - Less Than Minimun.................
           FODD=FOMIN-0.1
           ROLD=17600.-DR
           DO I=1,37
              J=0
              a=5.3-(I-1)*0.1
              DO WHILE (FODD<0.)
                     J=J+1
                     R=ROLD+DR
                     CALL CALC(a,R,FEVEN,FODD)
                     rat=-FOOLD/(-FOOLD+FODD)
                     RSAVE=R+rat*DR
                     FOOLD=FODD
                     ROLD=R
              END DO
              WRITE(6,102) J,a,RSAVE,FODD
              FODD=FODD-1.
              AO(38-I)=a
              RO(38-I)=RSAVE
           END DO
C.............Odd - Greater Than Minimun.................
           FODD=FOMIN-0.1
           ROLD=17600.-DR
           DO I=1,38
              J=0
              a=5.3+(I-1)*0.1
           DO WHILE (FODD<0.)
                     J=J+1
                     R=ROLD+DR
                     CALL CALC(a,R,FEVEN,FODD)
                     rat=-FOOLD/(-FOOLD+FODD)
                     RSAVE=R+rat*DR
                     FOOLD=FODD
                     ROLD=R
              END DO
```

Program 11.2.1—(Continued)

```
                    WRITE(6,103) J,a,RSAVE,FODD
                    FODD=FODD-1.
                    A0(38+I)=a
                    R0(38+I)=RSAVE
                END DO
                WRITE(1,107)
                WRITE(1,105)
                DO I=1,75
                    WRITE(1,106) I,A0(I),R0(I)
                END DO
                CLOSE(1, STATUS='KEEP')
100  FORMAT("  Even 1  ",I5,F8.4,F12.4,F10.6)
101  FORMAT("  Even 2  ",I5,F8.4,F12.4,F10.6)
102  FORMAT("   Odd 1  ",I5,F8.4,F12.4,F10.6)
103  FORMAT("   Odd 2  ",I5,F8.4,F12.4,F10.6)
104  FORMAT("                       EVEN DISTURBANCE")
105  FORMAT("    WAVE NUMBER                    RAYLEIGH NUMBER")
106  FORMAT(I3,F12.4,F10.6)
107  FORMAT("                        ODD DISTURBANCE")
     END
C...............Subroutine.................................
     SUBROUTINE CALC(a,R,FEVEN,FODD)
     TAU=((R/a)**(1./3.))/a
     S3=SQRT(3.)
     Q0=a*SQRT(TAU-1.)
     Q1=a*SQRT(.5*SQRT(1.+TAU*(1.+TAU))+.5*(1.+.5*TAU))
     Q2=a*SQRT(.5*SQRT(1.+TAU*(1.+TAU))-.5*(1.+.5*TAU))
     P1=Q1+Q2*S3
     P2=Q1*S3-Q2
     FEVEN=Q0*TAN(.5*Q0)+(P1*SINH(Q1)+P2*SIN(Q2))/(COSH(Q1)+COS(Q2))
     FODD=-Q0/TAN(.5*Q0)+(P1*SINH(Q1)-P2*SIN(Q2))/(COSH(Q1)-COS(Q2))
     RETURN
     END SUBROUTINE
```

Program 11.2.1—(Continued)

Numerical integration is performed in much the same manner as differentiation, again using the Taylor series expansion and integrating it term by term, stopping when the needed order of accuracy is obtained. The lowest order of accuracy is termed the ***trapezoidal rule***, in the form

$$\int_{x}^{x+\Delta x} f(x)dx \approx \frac{1}{2}\left[f(x+\Delta x) + f(x)\right]\Delta x. \tag{11.2.7}$$

This approximates the true area of the integral by the area of a trapezoid with corners $(x, 0)$, $(x, f(x))$, $(x+\Delta x, f(x+\Delta x))$, $(x+\Delta x, 0)$. Integration over a finite interval amounts to adding up the contributions of the small intervals.

An example of the use of the trapezoidal rule is given in Program 11.2.2, where the integration of the Elliptic Integrals of the first and second kind are carried out. Tables available traditionally use degrees instead of k (degrees $= \sin^{-1} k$). The program uses step sizes of 0.01 degrees and prints results at every degree. Agreement with published tables (given to three significant figures) is very good.

Another often used method for integration is ***Simpson's rule***. It essentially fits a polynomial to the integrand and deals with points at x, Δx, and $2\Delta x$ in one step. Letting $f_0 = f(x)$, $f_1 = f(x+\Delta x)$ and $f_2 = f(x+2\Delta x)$, the integral of f from x to $x+2\Delta x$ is

$$I = \int_{x}^{x+2\Delta x} f(x)dx \approx \frac{1}{3}\Delta x \left(f_0 + 4f_1 + f_2\right) + O(\Delta x^5). \tag{11.2.8}$$

```
C       PROGRAM EllfUN
C          Calculates incomlete and complete Elliptic Integrals of the first & second kind.
C=====================================================================================
        DOUBLE PRECISION, DIMENSION (18,100) :: AE,AF
        DOUBLE PRECISION AFT,AFT,AKDI,DI,G,GI,GOLD,GIOLD,PHI,PI
        PI=3.1415927
        DI=0.01*PI/180.
        OPEN(1, FILE='a:EllFun.dat', STATUS='REPLACE')
        WRITE(1,90)
C              VARIABLE INTEGRATION
        DO I=1,18
             IDEGR=5*I
             AK=SIN(IDEGR*PI/180.)
             GOLD=1.
             GIOLD=1.
             AE(I,1)=0.
             AF(I,1)=0.
             AET=0.
             AFT=0.
             ICOUNT=0
             JPRINT=0
C            INTEGRATION ITERATION
             DO   J=1,9000
                    ICOUNT=ICOUNT+1
                    PHI=J*DI
                    G=SQRT(1.-AK*AK*SIN(PHI)*SIN(PHI))
                    GI=1./G
                    AET=AET+0.5*(G+GOLD)*DI
                    AFT=AFT+0.5*(GI+GIOLD)*DI
                    GOLD=G
                    GIOLD=GI
                    IF(ICOUNT.EQ.100) THEN
                           JPRINT=JPRINT+1
                           JDEGR=JPRINT
                           AF(I,JPRINT)=AFT
                           AE(I,JPRINT)=AET
                           WRITE(*,*) AK,IDEGR,JDEGR,AF(I,JPRINT),AE(I,JPRINT)
                           WRITE(1,100) AK,IDEGR,JDEGR,AF(I,JPRINT),AE(I,JPRINT)
                           ICOUNT=0
                    ENDIF
             END DO
        END DO
        CLOSE(1)
   90   FORMAT("  K        ARCSIN(K)       PHI  F(K,PHI)  E(K,PHI)")
  100   FORMAT(F10.7,',',I10,',',F10.7,',',F10.7)
        END
```

Program 11.2.2—Elliptic Integrals of the first and second kind (program by the author)

The formula

$$I - \int_x^{x+3\Delta x} f(x)dx \approx \frac{3}{8}\Delta x \left(f_0 + 3f_1 + 3f_2 + f_3\right) + O(\Delta x^5), \tag{11.2.9}$$

referred to as Simpson's second rule, is also useful.

11.3   Numerical Integration of Ordinary Differential Equations

Frequently a first-order ordinary differential equation (or set of equations if y and f are vectors) of the type

$$\frac{dy}{dx} = f(x, y) \tag{11.3.1}$$

cannot be integrated in closed form, and numerical methods must be resorted to if the solution is to be found. One of the simplest methods of doing this is ***Euler's method***, where the solution is found by repeating the process

$$y(x + \Delta x) = y(x) + f[y(x), x]\Delta x. \tag{11.3.2}$$

This technique is easy to implement in a computer code and is readily extended to the case where f is an array. It is not, however, particularly accurate, the error being of the order of Δx squared.

An example of more accurate methods is the family of **Runge-Kutta** methods. They require evaluation at intermediate points in the interval to achieve their accuracy. The method is used for systems of first-order differential equations, which can be linear or nonlinear. Two versions of Runge-Kutta methods are listed here.

Second-order Runge-Kutta (the error is of order Δx cubed)

$$
\begin{aligned}
&y(x + \Delta x) = y(x) + [(1 - b)k_1 + bk_2]\Delta x, \text{ with} \\
&k_1 = f[y(x), x], \\
&k_2 = f[y(x) + pk_1, x + p], \\
&p = \Delta x/2b.
\end{aligned}
\tag{11.3.3}
$$

The constant b can be chosen arbitrarily in the range 0 to 1. Common choices are 1/2 and 1. When b is chosen as 1 (modified Euler-Cauchy method), then

$$
\begin{aligned}
&y(x + \Delta x) = y(x) + k_2\Delta x, \text{ with} \\
&k_1 = f[y(x), \ x], \\
&k_2 = f[y(x) + k_1\Delta x/2, \ x + \Delta x/2].
\end{aligned}
\tag{11.3.4}
$$

When b is chosen as 1/2 (Heun's method), equation (11.3.3) becomes

$$
\begin{aligned}
&y(x + \Delta x) = y(x) + 0.5(k_1 + k_2)\Delta x, \\
&\text{with} \\
&k_1 = f[y(x), x], \\
&k_2 = f[y(x) + k_1\Delta x, x + \Delta x].
\end{aligned}
\tag{11.3.5}
$$

Both choices of b (1/2 and 1) give results with the same order of accuracy.

Fourth-order Runge-Kutta (the error is of order Δx to the fifth power.)

The general form for a fourth-order accurate Runge-Kutta scheme is

$$y(x + \Delta x) = y(x) + \sum_{j=1}^{4} w_i k_i. \tag{11.3.6}$$

There are many different combinations of the w and k that all give the same accuracy. All are easy to program. Here are three sets of commonly used formulas:

1. (Credited to Kutta.)

$$
\begin{aligned}
&y(x + \Delta x) = y(x) + \frac{1}{6}(k_1 + 2k_2 + 2k_3 + k_4), \text{ with} \\
&k_1 = f[y(x), \ x]\Delta x, \\
&k_2 = f[y(x) + k_1\Delta x/2, \ x + \Delta x/2]\Delta x, \\
&k_3 = f[y(x) + k_2\Delta x/2, \ x + \Delta x/2]\Delta x, \\
&k_4 = f[y(x) + k_3\Delta x, \ x + \Delta x]\Delta x.
\end{aligned}
\tag{11.3.7a}
$$

2. (Credited also to Kutta.)

$$y(x+\Delta x) = y(x) + \frac{1}{8}(k_1 + 3k_2 + 3k_3 + k_4), \text{ with}$$

$$k_1 = f[y(x), \ x]\Delta x,$$

$$k_2 = f[y(x) + k_1\Delta x/3, \ x + \Delta x/3]\Delta x, \qquad (11.3.7b)$$

$$k_3 = f[y(x) - k_1\Delta x/3 + k_2\Delta x, \ x + 2\Delta x/3]\Delta x,$$

$$k_4 = f[y(x) + k_1\Delta x - k_2\Delta x + k_3\Delta x, \ x + \Delta x]\Delta x.$$

3. (Credited to Gill. Minimizes storage space needed.)

$$y(x+\Delta x) = y(x) + \frac{1}{6}\left[k_1 + 2\left(1 - \frac{1}{\sqrt{2}}\right)k_2 + 2\left(1 + \frac{1}{\sqrt{2}}\right)k_3 + k_4\right], \text{ with}$$

$$k_1 = f[y(x), x]\Delta x,$$

$$k_2 = f[y(x) + k_1\Delta x/2, x + \Delta x/2]\Delta x, \qquad (11.3.8)$$

$$k_3 = f[y(x) + (-1 + \sqrt{2})k_1\Delta x/2 + (1 - \sqrt{2})k_2\Delta x/2, x + \Delta x/2]\Delta x,$$

$$k_4 = f[y(x) - k_2\Delta x/\sqrt{2} + (1 + 1/\sqrt{2})k_3\Delta x, x + \Delta x]\Delta x.$$

Again, all three of these choices give results with the same order of accuracy.

Not many of the ordinary differential equations encountered so far have been couched as a set of first order equations, but they all can be easily converted to such. For instance, an equation such as

$$f''' + af'' + bf' + cf = 0 \qquad (11.3.9)$$

can be converted by letting

$$y_1 = f, \quad y_2 = f', \quad y_3 = f''.$$

Then

$$\frac{d}{dx}\begin{Bmatrix} y_1 \\ y_2 \\ y_3 \end{Bmatrix} = \begin{Bmatrix} y_2 \\ y_3 \\ -ay_2 - by_1 - cy \end{Bmatrix}. \qquad (11.3.10)$$

Note that it is not necessary that the a, b, c be constants, or the right-hand side of equation (11.3.1), be linear in y. It is, however, necessary that all y be known at the starting point of the integration.

The similarity solutions encountered in investigating the boundary layer are all good candidates for a Runge-Kutta scheme. Since some of the conditions are known at the wall, it is necessary to guess them and then see if they do converge to the proper values away from the boundary. Also, it is necessary to make a choice for "infinity." As we saw in the previous solutions, good choices can be 5, 7, or even 10. After reasonably satisfactory values for the unknown boundary conditions are found, moving the infinite boundary further out can be used to improve the solution.

Program 11.3.1 was used to compute the values for the Falkner-Skan class of boundary layer solutions. Suggestions are made for the unknown condition ($f''(0)$ in this case) based on published values, and interpolations from these can be easily made for other values of m. Screen prints of f, f', and f'' are made. A good test of whether

```
      PROGRAM FALKNER
!========================================
!    Infinity is set at 7
!========================================
      DIMENSION F0(3),Fk1(3)
      H=.001
      WRITE(*,*)
      WRITE(*,*) " Solves the Falkner-Skan equation f''' + 0.5(m + 1)ff'' + m[1 - (f')^2] = 0"
      WRITE(*,*) "    using a 4th order Runge-Kutta scheme."
      WRITE(*,*)
      WRITE(*,*) "Suggested value pairs"
      WRITE(*,*) "Beta:   054    1872   3690    108    144    180    216 "
      WRITE(*,*) "m:      0    0.0526 0.1111 0.1765 0.25   0.3333 0.4287 0.6667 1    1.5 "
      WRITE(*,*) "f''(0): 0.332 0.4259 0.5113 0.5936 0.6759 0.7569 0.8418 1.022  1.232 1.497"
      WRITE(*,*)
      WRITE(*,*) "    Enter m.    "
      READ (*,*) AM
      WRITE(*,*)
      DO 3 JJ=1,50
          WRITE(*,*) "  Enter f''(0).  "
          WRITE(*,*) "   A ZERO stops the program."
          READ(*,*) FG
          IF (FG.EQ.0.)  GOTO 4
          F0(3)=FG
          F0(1)=0.
          F0(2)=0.
          X=0.
          A=0.-.5
          WRITE(*,*) "    eta        f          f'          f'''"
          DO 1 J=1,10000
                 IF (X.GE.A) THEN
                        WRITE(*,100) X,F0(1),F0(2),F0(3)
                        A=A+.1
                 ENDIF
                 CALL RK4(F0,H,AM)
                 X=X+H
                 IF (X.GT.7.05)  GOTO 2
  1       END DO
  2       WRITE(*,*)" m             f''(0)     f'(large)"
          WRITE(*,*) AM,FG,F0(2)
  3   END DO
100   FORMAT(4F12.7)
  4   END PROGRAM
!========================================
!   Runge_Kutta solver - Press, Flannery, et al
!========================================
      Subroutine RK4(F0,H,AM)
      DIMENSION F0(3),F1(3),F2(3),F3(3),Fk1(3),Fk2(3),Fk3(3),Fk4(3)
      CALL DERIVS(F0,Fk1,AM)
      DO 11 I=1,3
          F1(I)=F0(I)+0.5*H*Fk1(I)
 11   END DO
      CALL DERIVS(F1,Fk2,AM)
      DO 12 I=1,3
          F2(I)=F0(I)+0.5*H*Fk2(I)
 12   END DO
      CALL DERIVS(F2,Fk3,AM)
      DO 13 I=1,3
          F3(I)=F0(I)+H*Fk3(I)
 13   END DO
      CALL DERIVS(F3,Fk4,AM)
      DO 14 I=1,3
          F0(I)=F0(I)+H*(Fk1(I)+2.*Fk2(I)+2.*Fk3(I)+Fk4(I))/6.
 14   END DO
      RETURN
      END SUBROUTINE
```

Program 11.3.1—Falkner-Skan boundary layers (main program by the author, subroutine from Press, Flannery, et al. (1986))

```
!====================================================
!   Defines equation
!====================================================
        SUBROUTINE DERIVS(FF,DFF,AM)
        DIMENSION FF(3),DFF(3)
        DFF(1)=FF(2)
        DFF(2)=FF(3)
        DFF(3)=-0.5*(AM+1.)*FF(1)*FF(3)+AM*(FF(2)*FF(2)-1.)
        RETURN
        END SUBROUTINE
```

Program 11.3.1—(Continued)

the choice of $f''(0)$ was suitable is how f' behaves as the independent variable becomes large. The program regards $\eta = 7$ to be infinity.

As long as there is a suitable first guess for the wall shear term, the procedure is simple to use and can be made to converge rather quickly by using Newton's method or, even simpler, just repeating the calculation over a range of points. In problems such as von Kármán's rotating disk or the thermal boundary layers where there are more than one unknown at the origin, the search is over a two-dimensional set of parameters and becomes more difficult. A starting point would be to use the Kármán-Pohlhausen method to calculate a first guess. Generally, if one starts close to the desired pair of values, improving the guess, no matter how crude the method, is doable.

Another possible approach would be a version of the steepest descent method, whereby one sets one of the parameters and then considers the results of a series of computations performed by varying the second parameter. Finding which value of the second parameter gives the "best result," the first parameter could then be incremented and then the process repeated. Defining "best result" might not, however, be obvious, and the convergence and programming just might be challenging.

A method that has been proposed in IMSL (1987, pages 660–671, programs BVPFD/DBVPFD) is to multiply the nonlinear terms by a parameter p and then starting with $p = 0$, and slowly increase p until it reaches unity—in which case the full original equation is reached. This method is designed for two-point boundary value problems so initial guessing of derivatives at the wall is avoided.

A method similar to this would be to insert time derivatives into each of the first-order equations, thus making an initial value problem. The hope would be, of course, that a steady state would eventually be reached.

Still another method for integrating a system of first-order ordinary differential equations is based on the method of **cubic splines**. Cubic splines became useful and popular with the advent of computer graphics. The idea is to take a function defined over an interval and break the interval into a number of subintervals. Within each subinterval the function is defined by a cubic polynomial. The polynomials are joined together in such a manner that the function and its first and second derivatives are continuous throughout the larger interval. To accomplish this, within each subinterval the function is defined by

$$y(x) = y''(x_i)\frac{(x_{i+1} - x)^3}{6\Delta_i} + y''(x_{i+1})\frac{(x - x_i)^3}{6\Delta_i} + \left[\frac{y(x_i)}{\Delta_i} - \frac{1}{6}y''(x_i)\Delta_i\right](x_{i+1} - x)$$

$$+ \left[\frac{y(x_{i+1})}{\Delta_i} - \frac{1}{6}y''(x_{i+1})\Delta_i\right](x - x_i), \tag{11.3.11}$$

where the subinterval extends from x_i to x_{i+1} and is of length $\Delta_i = x_{i+1} - x_i$.

To use this in integrating the system of equations

$$\frac{dy}{dx} = f(x, y) \text{ in } a \leq x \leq b, \qquad (11.3.12)$$

first divide the interval into n subintervals and let

$$y(x) = \sum_{i=1}^{n} \delta_i g_i(x), \text{ where } \delta_i = \begin{cases} 1 & \text{for } x_i \leq x \leq x_{i+1} \\ 0 & \text{otherwise.} \end{cases} \qquad (11.3.13)$$

The g_i are the cubic splines as given in equation (11.3.11). Now insert equation (11.3.13) into equation (11.3.12), multiply each side of the resulting equation by γ_j and integrate over the interval, where

$$\gamma_j = \begin{cases} 1 & \text{for } x_{j-1} \leq x \leq x_{j+1} \\ 0 & \text{otherwise.} \end{cases}$$

The result is

$$y_{j+1} - y_{j-1} = \int_{x_{j-1}}^{x_{j+1}} f\left(x, \sum_{i=1}^{n} \delta_i g_i(x)\right) dx. \qquad (11.3.14)$$

The integrals on the right-hand side can be carried out by any of the previously discussed integration schemes.

There are a number of other methods for numerically integrating a system of first-order ordinary differential equations, such as predictor-corrector schemes like Adams-Bashforth-Moulton. Discussions of these can be found in Press, Flannery, Teukolsky, and Vetterling (1986) or Carnahan, Luther, and Wilkes (1969).

To improve accuracy and speed, some integration approaches use variable step size. For instance, one technique used for the Runge-Kutta fourth-order scheme is to first perform the integration over the interval $\Delta x = x_2 - x_1$, and then the integration is repeated by returning to x_1 and repeating the integration by doing it in two steps and using a step size of one-half Δx. The integration proceeds now from x_2 but this time using a step size $\Delta x_{\text{new}} = \left|\frac{y_2 - y_1}{DMC}\right|^{1/n} \Delta x_{\text{previous}}$, where $y_2 - y_1$ is the computed change in the solution, DMC is the desired maximum change in the solution, and n is the order of the error of the method—5 in this case. The choice of DMC is at the discretion of the user.

11.4 The Finite Element Method

A method that is capable of extension to higher dimensional space is the ***finite element method***. To illustrate this method, consider the equation

$$\frac{d}{dx}\left(p(x)\frac{dy}{dx}\right) - q(x)y = f(x), \text{ where } y(0) = y(1) = 0. \qquad (11.4.1)$$

This equation is referred to as the ***Sturm-Liouville equation***, and many of the special functions of mathematics are generated from it. It has the following important property. Multiplication of it by $y(x)$ (if the coefficients are complex, use the complex conjugate of y) and subsequent integration of the result over the interval yields

$$py\frac{dy}{dx}\bigg|_0^1 - \int_0^1 \left[p\left(\frac{dy}{dx}\right)^2 + qy^2\right] dx = \int_0^1 yf\,dx.$$

By virtue of the boundary conditions, the first term vanishes. Providing that p and q are both positive and nonzero, it also tells us that the integral on the right-hand side must be negative. Rearranging, find that

$$\int_0^1 \left[p \left(\frac{dy}{dx} \right)^2 + qy^2 + yf \right] dx = 0. \tag{11.4.2}$$

Considering now the function $F(y)$ defined by

$$F(y) = \int_0^1 \left[p \left(\frac{dy}{dx} \right)^2 + qy^2 + 2yf \right] dx, \tag{11.4.3}$$

it can be demonstrated that the exact solution of equation (11.4.1) is one that minimizes equation (11.4.3). To do this, let z be an approximation to the exact solution y such that $z(x) = y(x) + \delta(x)$ and $z(0) = z(1) = 0$. Introduce this into equation (11.4.3), and find after integration by parts and use of the boundary conditions that

$$F(z) = \int_0^1 \left[p \left(\frac{dz}{dx} \right)^2 + qz^2 + 2zf \right] dx + 2 \int_0^1 \delta \left[-p \frac{dy}{dx} + qy + f \right] dx$$

$$+ \int_0^1 \left[p \left(\frac{d\delta}{dx} \right)^2 + q\delta^2 \right] dx.$$

The second integral is zero by virtue of equation (11.4.2), and the third is equal or greater than zero. Thus,

$$F(z) \geq F(y).$$

For the finite element method (FEM), break the interval into n regions (elements) and let

$$y(x) = \sum_{j=1}^n y_j N_j(x),$$

where

$$N_j(x) = \begin{cases} (x - x_{j-1})/(x_j - x_{j-1}) & \text{for } x_{j-1} \leq x \leq x_j \\ (x - x_{j+1})/(x_j - x_{j+1}) & \text{for } x_j \leq x \leq x_{j+1}, \\ 0 & \text{elsewhere.} \end{cases} \tag{11.4.4}$$

These chosen functions are triangles of a maximum height of one and are defined as nonzero over just two intervals. Inserting equation (11.4.4) into equation (11.4.3) and differentiating with respect to the y_j to minimize F, the result is a set of n by n linear algebraic equations in the y_j in the form $Ay = b$, where

$$a_{ij} = \int_0^1 \left(p \frac{dN_i}{dx} \frac{dN_j}{dx} + qN_i N_j \right) dx, \quad b_i = - \int_0^1 f N_i dx. \tag{11.4.5}$$

If n is not too large, the set of linear equations can be solved by elementary methods such as Gaussian elimination, whereby line by line, the terms to the left of the diagonal are eliminated, thereby giving a set of equations solvable by back substitution. However, when this is applied to more than one dimension, or when n is very large, methods dedicated to the solution of a large set of algebraic equations must be used.

In this example, the approximating functions have been chosen to be linear. This may be adequate in many cases. If not, higher-order polynomials or other functions may be used that cover the interval in a similar fashion.

11.5 Linear Stability Problems—Invariant Imbedding and Riccati Methods

In Chapter 9 several flow stability problems were considered—namely, Rayleigh-Bénard, Couette and Poiseuille viscous flows, and several interfacial-wave-type problems for inviscid flows. With the exception of the Rayleigh-Bénard convection flows, all of the viscous flow problems involved rather serious and sophisticated computations, even though the equations were linear and therefore traditional series expansion methods and superposition could be used. When numerical methods are considered, however (sometimes surprisingly), there is no advantage to having linear equations, and exchanging them for nonlinear equations may have some advantage.

To introduce the concept of invariant imbedding, consider a very simple equation that often appears when discussing mechanical problems (the Euler column theory is one example):

$$\frac{d^2y}{dx^2} + \lambda^2 y = 0, \quad y(0) = y(1) = 0. \tag{11.5.1}$$

This is a classic eigenvalue (characteristic value) problem with eigenvalues λ, where the solution is zero unless λ takes on the values $\lambda = n\pi$, $n = 1, 2, 3, \ldots$. In that case the solution is $y(x) = A\sin(\lambda x)$, where A is indeterminate.

If equation (11.5.1) is changed to

$$\frac{d^2y}{dx^2} + \lambda^2 y = 1, \quad y(0) = y(1) = 0. \tag{11.5.2}$$

the solution becomes determinate, with the solution

$$y(x) = \frac{1}{\lambda^2}\left[1 - \cos\lambda x + (-1 + \cos\lambda)\frac{\sin\lambda x}{\sin\lambda}\right]. \tag{11.5.3}$$

This is a well-behaved solution as long as λ does not equal one of the eigenvalues associated with the homogeneous problem (11.5.)!

Similarly, if equation (11.5.1) had been changed to

$$\frac{d^2y}{dx^2} + \lambda^2 y = 0, \quad y(0) = 0, \quad y(1) = 1, \tag{11.5.4}$$

the solution would be

$$y(x) = \frac{\sin\lambda x}{\sin\lambda}. \tag{11.5.5}$$

Thus, either change from the homogeneous problem, whether it be in the differential equation or the boundary conditions, results in a solution that is well defined except when λ is one of the eigenvalues.

Of the two techniques of the section title, the Riccati method is the easiest to describe. Start with equation (11.5.1) and let $\frac{dy}{dx} = R(x)y$. Then $\frac{d^2y}{dx^2} = \frac{dR}{dx}y + R\frac{dy}{dx} = \left(\frac{dR}{dx} + R^2\right)y$.

The second derivative of y is known from equation (11.5.1). Thus, $\frac{d^2y}{dx^2} = -\lambda^2 y = \left(\frac{dR}{dx} + R^2\right) y$, or, upon dividing by y, the result is

$$\frac{dR}{dx} = -\lambda^2 - R^2. \tag{11.5.6}$$

For later reference, note that in this simple case, equation (11.5.6) can be integrated to give

$$R(x) = \lambda \frac{\cos \lambda x}{\sin \lambda x}. \tag{11.5.7}$$

Equation (11.5.6) falls in the general form of what is termed a Riccati equation. (The general form of a Riccati equation is $R' = A + BR + RC + RDR$, where R can be a vector and A, B, C, D square matrices.) It is nonlinear and ideally suited to a numerical method of integration such as Runge-Kutta.

To start the integration, the value of R at $x = 0$ is needed. From the boundary conditions, since y is to be set to zero and its derivative must be nonzero, it is seen that $R(0)$ is singular. This can be gotten around in at least two ways. First, looking at equation (11.5.6), notice that when R and its derivative are large, then $\frac{dR}{dx} \cong -R^2$, so with a little imagination it can be concluded that R behaves like $1/x$ at the origin.

Alternately, we can work with the reciprocal (more generally, the inverse) of R, letting $S = 1/R$, so that $\frac{dS}{dx} = -\frac{1}{R^2}\frac{dR}{dx} = -\frac{1}{R^2}(-\lambda^2 - R^2) = 1 + \lambda^2 S^2$. If R is infinite at the origin, S will be zero there, so the integration can proceed.

To do so, however, a value for λ must be set first. Choose an arbitrary value that seems in a reasonable range, and then carry out the integration until R becomes infinity (defined here as being outside the bounds of numbers your computer is capable of handling). Then this value where R becomes infinite is the length at which the chosen value of λ is the correct eigenvalue. Thus, the problem has been turned around, finding the value of the spacing associated with an eigenvalue rather than the usual reverse of this.

One thing to note is that while S is finite throughout most of the region of integration, it probably will be infinite at an intermediate point. (From equation (11.5.7), we can see that this is true at $x = \pi/2\lambda$.) Thus, programming requires that we start with the S equation, then shift to the R equation, then (to improve accuracy) shift back to the S equation as the singularity in R is approached. To accomplish this, fortunately, does not require particularly skilled programming abilities.

The invariant imbedding technique follows very closely the Riccati approach. It recognizes that the value of the eigenvalue (λ in this case) depends on the length and along the lines of equation (11.5.4) changes equation (11.5.1) to

$$\frac{\partial^2 y(x,z)}{\partial x^2} + \lambda^2 y(x,z) = 0, \quad y(0,z) = 0, \quad y(z,z) = 1. \tag{11.5.8}$$

R is chosen as $R(z) = \partial y(z,z)/\partial z$, differentiation of R with respect to z and using equation (11.5.6) to eliminate the derivative, leads to an equation practically identical to equation (11.5.6).

In the problem considered here, there is not much flexibility in the choice of R. In higher-order differential systems, depending on the boundary conditions, it may be possible to avoid having to deal with singularities and the need to invert R.

These methods have been successfully used in computing the neutral stability curves for Couette-Taylor stability (Curl and Graebel, 1972; Wilks and Sloan, 1976), for Poiseuille flow between parallel plates (Curl and Graebel, 1972; Davey, 1977; Sloan, 1977) and for the Blasius flat plate solution (Wilks and Bramley, 1977). They are at least as accurate as previous methods (Orszag, 1971) and (perhaps even better) easier to use.

To illustrate the method for more complex flows, consider the Orr-Sommerfeld equation for plane Poiseuille flow. Taking the origin at $y = 0$, the primary flow is given by $U = U_{max}\left(1 - y^2\right)$, where U_{max} is the centerline velocity and y has been made dimensionless by the half-spacing of the plates. First, recall equation (9.4.8) in the expanded form

$$\frac{d^4v}{dy^4} - 2k^2\frac{d^2v}{dy^2} + k^4v = ikR\left[(U - c)\left(\frac{d^2v}{dy^2} - k^2v\right) - v\frac{d^2U}{dy^2}\right]. \qquad (9.4.8)$$

To simplify the notation, the primes have been dropped and η replaced by y.

Only disturbances of u even in y will be considered, as they are known to be less stable. By the continuity relation, disturbances even in u are odd in v—thus,

$$\frac{dv}{dy} = \frac{d^3v}{dy^3} = 0 \text{ at } y = 0 \quad \text{and} \quad u = v = \frac{dv}{dy} = 0 \text{ at } y = 1.$$

Following Davey (1977), let

$$\begin{aligned} Z_1 &= v, \quad Z_3 = v'' - k^2v, \\ Z_2 &= v', \quad Z_4 = v''' - k^2v'. \end{aligned} \qquad (11.5.9)$$

and

$$\begin{pmatrix} Z_2 \\ Z_4 \end{pmatrix} = \begin{pmatrix} R_1 & R_2 \\ R_3 & R_4 \end{pmatrix} \begin{pmatrix} Z_1 \\ Z_3 \end{pmatrix}. \qquad (11.5.10)$$

From the boundary conditions at $y = 0$, Z_2 and Z_4 are both zero at $y = 0$. Since Z_1 and Z_3 are not necessarily zero there, all of the Rs must be set to zero at $y = 0$. Since Z_1 and Z_2 must both be zero at $y = 1$, $R_2 = 0$ at $y = 1$.

Next, differentiate the first of equation (11.5.10),

$$Z_2 = R_1 Z_1 + R_2 Z_3, \qquad (11.5.11)$$

obtaining

$$Z_2' = R_1' Z_1 + R_1 Z_1' + R_2' Z_3 + R_2 Z_3', \qquad (11.5.12)$$

where here primes denote derivatives with respect to y. From equation (11.5.9) note that

$$Z_1' = v' = Z_2, \quad Z_2' = v'' = Z_3 - k^2Z_1, \quad Z_3' = Z_4, \qquad (11.5.13)$$

so equation (11.5.12) can be rewritten as

$$Z_2' = Z_3 + k^2 Z_1 = R_1' Z_1 + R_1 Z_2 + R_2' Z_3 + R_2 Z_4. \qquad (11.5.14)$$

Z_2 and Z_4 can be eliminated from equation (11.5.14) by using equation (11.5.10), which, upon collection of terms, gives

$$Z_1 \left(R_1' + R_1^2 + R_2 R_3 - k^2 \right) + Z_3 \left(R_2' + R_1 R_2 + R_2 R_4 - 1 \right) = 0 \qquad (11.5.15)$$

From equation (9.4.8), the fourth derivative of v is given by

$$v'''' = 2k^2 v'' - k^4 v + ik \ \text{Re} \left[(U - c) \left(v'' - k^2 v \right) - \frac{d^2 U}{dy^2} v \right], \ \text{or equivalently,}$$

$$v'''' = 2k^2 Z_3 + k^4 Z_1 + ik \ \text{Re} \left[(U - c) Z_3 - \frac{d^2 U}{dy^2} Z_1 \right]. \qquad (11.5.16)$$

Repeating the previous procedure with the second equation of (11.5.2) and using equation (11.5.16) gives

$$Z_1 \left[R_3' + R_1 R_3 + R_3 R_4 - 2ik \ \text{Re} \right]$$
$$+ Z_3 \left[R_4' + R_2 R_3 + R_4^2 - k^2 - ik \ \text{Re} \left(1 - y^2 - c \right) \right] = 0 \qquad (11.5.17)$$

Equations (11.5.15) and (11.5.17) are two homogeneous equations in the two unknowns Z_1 and Z_3 which must hold over a range of y. The conclusion is then that the coefficients of Z_1 and Z_3 must vanish, leading to the system of four differential equations

$$\begin{aligned}
R_1' &= -R_1^2 - R_2 R_3 + k^2, \\
R_2' &= -R_1 R_2 - R_2 R_4 + 1, \\
R_3' &= -R_1 R_3 - R_3 R_4 + 2ik \ \text{Re}, \\
R_4' &= -R_2 R_3 - R_4^2 + k^2 + ik \ \text{Re} \left(1 - y^2 - c \right).
\end{aligned} \qquad (11.5.18)$$

These are the Riccati equations associated with the problem.

Notice that the second and third equations are the same except for the magnitude of the constant. Thus,

$$R_3 = 2ik \ \text{Re} \ R_2. \qquad (11.5.19)$$

The procedure requires that values of the wave number, Reynolds number, and the wave speed be set a priori. Integration then precedes until $R_2 = 0$, or at least nearly zero. If the initial guess is close to a true eigenvalue, then this will occur near to 1. Using the dimensions of the various quantities, once the location of the eigenvalue is known, it is possible to correct the starting parameters to make the zero occur at 1. Since integration is in the complex plane, it is entirely possible that a zero of R_2 will not be found for a particular choice of the three input parameters. It is best to start near known results and make small steps away from them. A discussion of dealing with more general problems using the Riccati method is given in the Appendix.

Program 11.5.1 illustrates one realization of a program for this problem. Table 11.5.1 suggests several combinations of input data near the minimum Reynolds number.

```
             PROGRAM Orr
             COMPLEX :: R(3)
             DIMENSION  AMR(3)
             DY=.001
             Y=0.
             DO I=1,3
               R(I)=0.
             END DO
             R2R=0.
             R2I=0.
             R2ROLD=.00001
             WRITE(*,*)
             WRITE(*,*) " Solves the Orr-Sommerfeld equation using a Riccati method and also"
             WRITE(*,*) "    using a 4th order Runge-Kutta scheme."
             WRITE(*,*)
             WRITE(*,*) "   Enter the real part of the wave speed c.     "
             READ (*,*) C
             WRITE(*,*) "   Enter the wave number k.     "
             READ (*,*) AK
             WRITE(*,*) "   Enter the Reynolds number Re.     "
             READ (*,*) Re
             WRITE(*,*)
             WRITE(*,*) "   y          R(1)        R(2)        R(3)      R(2)real    R(2)imag"
             WRITE(*,100) Y,AMR(1),AMR(2),AMR(3),R2R,R2I
             DO I=1,1050
                 Y=Y+DY
                 CALL RK4(R,AK,C,DY,Re,Y)
                 DO  J=1,3
                     AMR(J)=ABS(R(J))
                 END DO
                 R2R=REAL(R(2))
                 R2I=AIMAG(R(2))
                 WRITE(*,100) Y,AMR(1),AMR(2),AMR(3),R2R,R2I
                 IF((R2R*R2ROLD)<0.)  THEN
                     RN=RE*(Y**3)
                     CN=(C-1.+Y**2)/Y**3
                     YN=Y+R2R*DY/(R2ROLD-R2R)
                     R2RS=R2R
                     R2IS=R2I
                 END IF
                 R2ROLD=R2R
             END DO
             WRITE(*,*)
             WRITE(*,*) "         Input c, k, Re: ",C,AK,Re
             WRITE(*,*) " Computed y, c, k, Re: ",YN,CN,AK,RN
             WRITE(*,*) "  R(2) real, imaginary: ",R2RS,R2IS
             WRITE(*,*)
100   FORMAT(6(F12.7,','))
             END PROGRAM
!=======================================================
!   Runge_Kutta solver - Press, Flannery, et al
!=======================================================
      SUBROUTINE RK4(R,AK,C,DY,RE,Y)
      COMPLEX :: R(3),RT(3),DR(3),DRM(3),DRT(3)
      DYH=0.5*DY
      YH=Y+DYH
      YPH=Y+DY
      CALL DERIVS(R,DR,AK,C,RE,Y)
      DO  I=1,3
         RT(I)=R(I)+DYH*DR(I)
      END DO                              ! 11
      CALL DERIVS(RT,DRT,AK,C,RE,YH)
      DO  I=1,3
         RT(I)=R(I)+DYH*DRT(I)
      END DO                              ! 12
```

Program 11.5.1—Neutral stability curves for the Orr-Sommerfeld equation (program by the author, subroutine from Press, Flannery, et al. (1986))

```
                  CALL DERIVS(RT,DRM,AK,C,RE,YH)
                  DO  I=1,3
                      RT(I)=R(I)+DY*DRM(I)
                      DRM(I)=DRT(I)+DRM(I)
                  END DO                              ! 13
                  CALL DERIVS(RT,DRT,AK,C,RE,YPH)
                  DO  I=1,3
                      R(I)=R(I)+DY*(DR(I)+DRT(I)+2.*DRM(I))/6.
                  END DO                              ! 14
                  RETURN
                  END SUBROUTINE
  !----------------------------------------------------------------
  !   Defines equations
  !----------------------------------------------------------------
                  SUBROUTINE DERIVS(R,DR,AK,C,RE,Y)
                  COMPLEX :: AI,R(3),DR(3)
                  AI=CMPLX(0.,1.)
  !               R(3)=2*I*K*Re*R(2)
                  DR(1)=-R(1)*R(1)-R(2)*(2.*AI*AK*RE*R(2))+AK**2
                  DR(2)=-R(1)*R(2)-R(2)*R(3)+1.
                  DR(3)=-R(2)*(2.*AI*AK*RE*R(2))-R(3)*R(3)+AK**2+AI*AK*RE*(1.-Y**2-C)
                  RETURN
                  END SUBROUTINE
```

Program 11.5.1—(Continued)

TABLE 11.5.1   Values of Re, k, and c near the minimum Reynolds number

Re	k	c	Re	k	c
5773.9761	1.0163	0.2634788	5772.0947	1.0203	0.2639588
5773.728	1.0168	0.2635388	5771.9966	1.0205564	0.2639906
5773.2251	1.0173	0.2635988	5772.0869	1.0205617	0.2639905
5772.9077	1.0178	0.2636588	5772.0879	1.0205617	0.2639905
5772.6406	1.0183	0.2637188	5772.0908	1.0208	0.2640188
5772.4238	1.0188	0.2637788	5772.4077	1.0209	0.2640288
5772.2617	1.0193	0.2638388	5772.0591	1.02056	0.2639905
5772.1499	1.0198	0.2638988			

11.6   Errors, Accuracy, and Stiff Systems

There are many issues that affect computations made with computers. Computers can deal only with a finite number of digits when making computations, usually some multiple of 8, depending on the computer and the program. This means that ***truncation errors*** naturally occur during an operation, as well as ***round-off errors***. These errors can sometimes be minimized by avoiding the operation of subtraction and also by minimizing the numbers of operation. If the error introduced by one round-off is ε, after N operations you might be lucky and find that the accumulated error is of the order of $\sqrt{N}\varepsilon$, providing that the errors accumulated randomly. (On the other hand, if the errors do not accumulate randomly, the error could be of the order $N\varepsilon$.) Subtraction between two nearly equal numbers can be disastrous. That is why use of the formula $y = \frac{-b \pm \sqrt{b^2 - 4ac}}{2a}$ should be avoided if $4ac \approx b^2$. A better choice is $y_1 = -\frac{b + \text{sgn}(b)\sqrt{b^2 - 4ac}}{2a}$, $y_2 = -\frac{2c}{b + \text{sgn}(b)\sqrt{b^2 - 4ac}}$, which avoids the subtraction.

The numerical method used in approximating a given operation introduces further errors. In the case of finite difference formulations of integral and differential calculus operations, as we have seen, these errors are determined by the order of approximation made in deriving the finite difference form. It is often suggested that after performing a procedure at one step size, the procedure should be repeated using half the original step size. This, of course, may not be practical in large calculations, and reducing the step size too far may increase the errors discussed in the previous paragraph.

Sometimes calculations can become unstable. An example of this would be the solution of an ordinary differential equation where a parameter in the equation means that there are a family of solutions generated by solving the differential equation for various values of the parameter. If at one point in the solution space the members of the family lie close together, round-off errors can cause the procedure to leave the original solution and drift to a neighboring solution. A simple model of this is to compute integer powers of the Golden Mean $\phi = \frac{\sqrt{5}-1}{2}$. While powers can be computed by successive multiplications, it is easy to verify that $\phi^{n+1} = \phi^{n-1} - \phi^n$. Thus, knowing $\phi^0 = 1$ and $\phi^1 = 0.61803398$, successive powers can be computed by simple subtraction.

Unfortunately, while higher powers of ϕ rapidly decrease, the function $-\frac{\sqrt{5}+1}{2}$ satisfies the same recursion relation, and this function in absolute value is greater than unity. Unavoidable computational errors indicate that a little round-off error means that use of the recurrence relationship will soon give completely wrong errors, usually around the power 16 (Press et al., 1986).

Another situation where instability occurs is in dealing with **stiff systems**. These systems are those where there is more than one length scale to the problem, and at least two of the scales differ greatly in magnitude. An example is the ordinary differential equation $\frac{d^2y}{dx^2} - 100y = 0$. This has the general solution $y(x) = Ae^{10x} + Be^{-10x}$, with length scales differing by a factor of one hundred. So, even if the boundary conditions are such that A is zero—for example, if $y(0) = 1$, $y(\infty) = 0$—round-off errors would result in a small nonzero value for A, and the solution would grow rapidly.

The general form of this example equation is $\frac{d^2y}{dx^2} - \lambda^2y = 0$. Here, the values of λ are the eigenvalues of the problem, and the equation is stiff because in the example the two values of λ differ substantially. A similar example occurs in the solution of algebraic equations. For a set of algebraic equations $Ay = b$, where A is an n by n matrix and y and b are column vectors, the eigenvalues of the matrix A are the solution of the determinant $|A - \lambda I| = 0$, I being the identity matrix. Again, if the values of the eigenvalue λ differ too greatly, the usual solution methods can run into trouble. Since finite difference methods frequently reduce the computation to a set of algebraic equations, it should not be surprising that stiffness can occur in both algebra and calculus.

Should the Runge-Kutta scheme not work for a particular set of equations, another method that has been successful in some cases[2] is the **Bulirsch-Stoer method**, introduced in 1966. The method is based on three different concepts, and so programming it is reasonably complicated. Programs can be found in Press et al. (1986) or on the Web. The concepts are Richardson's deferred approach to the limit (an extrapolation procedure), choice of a proper fitting functions (the original choice was rational polynomials, but ordinary polynomials have been used), and the use of a method whose error function is even in the step size. Further details can be found in the preceding sources.

---

[2] One may *hope*—but never *expect*—to find one method that works for all problems. See www.nr.com/CiP97.pdf for some comments on this.

Problems—Chapter 11

If available, use one of the programing languages mentioned in the book. If no such languages are available to you, the problems can be done (if less conveniently) through the use of spreadsheets such as Excel or Quattro Pro.

11.1 Use the trapezoidal rule to integrate the function $\cos(\pi x)$ over the range $0 \leq x \leq 1$. Perform the integration three times, using successively the intervals 0.2, 0.1, and 0.5. Compare your computed answer with the exact result at each step of the integration.

11.2 A technique known either as **Richardson's deferred approach to the limit** or as **Richardson's extrapolation** enables improvement of the accuracy of the integration carried out in the previous problem. For half-stepping, as was done there, if I_1 is the result for a step size of h and I_2, the result for a step size $h/2$, then the improved result is given by $I_{\text{improved}} = \frac{4I_2 - I_1}{3}$. Use this method to improve your results of problem 11.1.

11.3 Compare the accuracy of the finite difference approximation to the derivative of the sine of x for the following three formulas:

$$\textit{Forward difference}: \frac{dy}{dx} \simeq \frac{y(x+h) - y(x)}{h},$$

$$\frac{dy}{dx} \simeq \frac{-y(x+2h) + 4y(x+h) - 3y(x)}{2h}.$$

$$\textit{Centered difference}: \frac{dy}{dx} \simeq \frac{y(x+h) - y(x-h)}{2h}.$$

$$\textit{Backward difference}: \frac{dy}{dx} \simeq \frac{y(x) - y(x-h)}{h},$$

$$\frac{dy}{dx} \simeq \frac{3y(x) - 4y(x-h) + y(x-2h)}{2h}.$$

Do the computation at $x = \pi/4$ with $h = 0.1\pi$.

11.4 Use Euler's method with $\Delta x = 0.1$ to solve $\frac{dy}{dx} = y + x$, $y(0) = 0$. Carry out the integration to $x = 10$, and compare your result with the exact solution.

11.5 Repeat the above problem, this time using a fourth-order Runge-Kutta scheme for integration.

11.6 The Runge-Kutta scheme of fifth-order accuracy is given by

$$y(x+h) = y(x) + \left[\frac{1}{24}k_1 + \frac{5}{48}k_4 + \frac{27}{56}k_5 + \frac{125}{336}k_6 \right] h, \text{ where}$$

$$k_1 = f(x, y),$$

$$k_2 = f\left(x + \frac{1}{2}h, y + \frac{1}{2}k_1\right),$$

$$k_3 = f\left(x + \frac{1}{2}h, y + \frac{1}{4}k_1 + \frac{1}{4}k_2\right),$$

$$k_4 = f(x + h, y - k_2 + 2k_3),$$

$$k_5 = f\left(x + \frac{2}{3}h, y + \frac{7}{27}k_1 + \frac{10}{27}k_2 + \frac{1}{27}k_3\right),$$

$$k_6 = f\left(x + \frac{1}{5}h, y + \frac{28}{625}k_1 - \frac{1}{5}k_2 + \frac{546}{625}k_3 + \frac{54}{625}k_4 - \frac{378}{625}k_5\right).$$

Modify the procedure used in the previous problem and solve the same equation set.

11.7 Use the finite element method to solve the equation $\frac{d^2y}{dx^2} - \lambda^2 y = f(x)$, $y(0) = y(2) = 0$. Use the triangular elemental functions of equation (11.4.4) to find that $y_{j+1} - 2y_j + y_{j-1} - \frac{h^2\lambda^2}{6}(y_{j+1} + 4y_j + y_{j-1}) = \frac{h^2}{6}(f_{j+1} + 4f_j + f_{j-1})$, $j = 1, 2, \cdots N - 1$. The parameter h is the constant spacing, with $h = 2/N$ for this problem. Solve with $\lambda = 1$, $f(x) = x$, $N = 10$.

11.8 Find exact solutions to the following equations, and consider how the small parameter $\varepsilon > 0$ might affect a numerical solution.

a. $\dfrac{d^2y}{dx^2} + 2\dfrac{dy}{dx} - \varepsilon y = 0$, $y(0) = 1$, $y(\alpha) = 0$,

b. $\dfrac{d^2y}{dx^2} + 2\varepsilon\dfrac{dy}{dx} + y = 0$, $y(0) = 1$, $y(\alpha) = 0$,

c. $\varepsilon\dfrac{d^2y}{dx^2} + 2\dfrac{dy}{dx} + y = 0$, $y(0) = 1$, $y(\alpha) = 0$.

11.9 Find the eigenvalues of the system $\frac{d^4y}{dx^4} + \lambda^4 y = 0$, $y(0) = \frac{dy}{dx}(0) = y(1) = \frac{dy}{dx}(1) = 0$ using the invariant imbedding method. This equation is associated with the vibration of a clamped-clamped beam.

Hint: Let $Z_1 = y$, $Z_2 = y'$, $Z_3 = y''$, $Z_4 = y'''$, and let $Z_1 = R_1 Z_3 + R_2 Z_4$, $Z_2 = R_3 Z_3 + R_4 Z_4$. The motivation for this choice is that y and y' are known at the origin.

Chapter 12

Multidimensional Computational Methods

12.1 Introduction

When dealing with differential equations involving more than one dimension, it is important to know the classification of the equation, as the method of numerical computation must be tied to the behavior of the equation's solution. The three archetypical equations are the following:

$$\text{Elliptic:} \qquad \nabla^2 V = 0$$

$$\text{Parabolic:} \qquad \nabla^2 V = \alpha \frac{\partial V}{\partial t} \qquad\qquad (12.1.1)$$

$$Hyperbolic: \quad \nabla^2 V = \frac{1}{c^2} \frac{\partial^2 V}{\partial t^2}$$

The wave equation is the most familiar form of hyperbolic equation. A propagation velocity c is associated with the time derivative. Thus, a disturbance of an existing condition at any point takes a finite time before its effect is noted at distant points. Whereas solutions of the other two classes tend to be smooth, hyperbolic equations can have abrupt discontinuities (shocks) in the solutions.

Parabolic equations are associated with diffusion processes such as heat, mass, and concentration diffusion. There is no wave speed associated with this, so mathematically a point an infinite distance from a place of change of conditions knows of the change instantly. The Prandtl boundary layer equations, wherein the x second derivative term is neglected, is frequently referred to as the **parabolized Navier-Stokes equations,** since the highest order of the stream-wise derivative is one.

The elliptic type of equation, of which Laplace's equation is the prime example, has each location communicating at all times with all other locations in the domain, as indicated by the mean value theorem that states that the value of the function at the center of a circle or sphere is the average of the values on the surface. Thus, a change at one point in the domain instantly affects every other point. Formally, the steady-state Navier-Stokes equations belong in this class because of the order of the viscous terms. At large Reynolds numbers, however, the magnitude of the convective acceleration essentially overcomes this, and the behavior becomes either more like the parabolic or hyperbolic classes.

Even for inviscid flows with free surfaces, the fact that surface and interfacial waves can propagate at finite speeds indicates behavior more of a hyperbolic than elliptic nature. This is brought about by the boundary conditions. So while classifications are useful, one should keep in mind that there are other influences on the nature of the solution.

Elliptic Partial Differential Equations

12.2   Relaxation Methods

Since the Laplace equation is perhaps the oldest and most used of the equations of engineering physics, there are quite naturally the greatest number of methods for numerical calculation—many existing long before the advent of the computer. In the case of rectangular boundaries, a rectangular grid can be superimposed on the boundary containing rectangles of size Δx by Δy. Using either the numerical differentiation procedures from Chapter 11 or the mean value theorem, Laplace's equation can be approximated by

$$0 = \nabla^2 V \approx \frac{V_{j+1,k} - 2V_{j,k} + V_{j-1,k}}{\Delta x^2} + \frac{V_{j,k+1} - 2V_{j,k} + V_{j,k-1}}{\Delta y^2} \qquad (12.2.1)$$

to second order accuracy in the grid spacing, or

$$0 = \nabla^2 V \approx \frac{V_{j+2,k} - 16V_{j+1,k} + 30V_{j,k} - 16V_{j-1,k} + V_{j-2,k}}{\Delta x^2}$$
$$+ \frac{V_{j,k+2} - 16V_{j,k+1} + 30V_{j,k} - 16V_{j,k-1} + V_{j,k-2}}{\Delta y^2} \qquad (12.2.2)$$

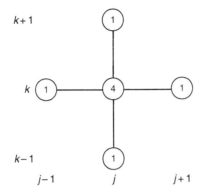

Figure 12.2.1a   Computational molecule for the relaxation method

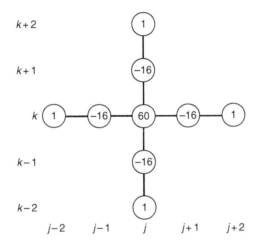

Figure 12.2.1b   Computational molecule for the SOR method

to fourth-order accuracy. The computational molecules for the two cases are shown in Figures 12.2.1a and b. When $\Delta x = \Delta y$, these equations reduce to

$$V_{j,k} \approx \frac{V_{j+1,k} + V_{j-1,k} + V_{j,k+1} + V_{j,k-1}}{4}, \tag{12.2.1a}$$

$$V_{j,k} \approx \frac{\left(-V_{j+2,k} + 16V_{j+1,k} + 16V_{j-1,k} - V_{j-2,k}\right) + \left(-V_{j,k+2} + 16V_{j,k+1} + 16V_{j,k-1} - V_{j,k-2}\right)}{60}. \tag{12.2.2a}$$

If the values of the function V are known on the boundary (***Dirichlet problem***), such as when posing a flow situation in terms of a stream function, the boundary conditions are easily handled. If the normal derivatives of the function V are known on the boundary (***Neumann problem***), such as when solving for the velocity potential, the boundary conditions require additional equations to accommodate the derivatives. In analyzing a given flow using the velocity potential, the body shape and the normal derivatives are known on the surface, so this is a Neumann problem and the pressure coefficient can be found. In the case of a design problem, the body shape is unknown except for the fact that it is a stream surface, and dealing with it as a Dirichlet problem has advantages. Often the pressure coefficient is also known in a design situation.

The resulting algebraic equations can be handled in a number of ways.

1. Use traditional methods such as Gaussian elimination. Unfortunately this will require $(N+1)!$ multiplications (N being the number of nodes), and possible round-off errors can be introduced in the solution process.

2. Use the **relaxation method**. For the case with Dirichlet conditions and a square grid, for second-order accuracy use equation (12.2.1a) in the form

$$V_{j,k}^{i+1} \approx \frac{V_{j+1,k}^i + V_{j-1,k}^i + V_{j,k+1}^i + V_{j,k-1}^i}{4}. \tag{12.2.1b}$$

The superscript denotes the number of the sweep through all of the nodes. Start off by assigning arbitrary values to all nodes. This is sweep number 0. Use your best judgement in assigning these first values, but it is not necessary to be perfectly accurate. Next, go from point to point, changing the value of the point you are at to the average of its neighbors (sweep number 2). After you have swept through every point, repeat the process again and again until the change is negligible. In practice, rather than using old values of V in computing the right-hand side of equation (12.2.1b), the most recently computed values are used, thus speeding up the process slightly.

While such a boring procedure is perfectly designed for a computer, it is sobering to reflect that in days gone by this was done with an adding machine (or maybe not), pencil, paper, eraser, and a human!

3. Use the **successive over-relaxation method (SOR)**. In this case equation (12.2.1a) is replaced by

$$V_{j,k}^{i+1} \approx V_{j,k}^i + \frac{\varpi}{4} \left(V_{j+1,k}^i + V_{j-1,k}^i + V_{j,k+1}^i + V_{j,k-1}^i - 4V_{j,k}^i \right). \tag{12.2.1c}$$

Again, the i superscript denotes the number of the integration, and ϖ is a parameter between 1 and 2 used to speed up the calculations. The "best" value of ϖ to use for a particular problem is determined by making a few trial runs for various values of ϖ, which can be time-consuming. Choosing a value somewhere around 1.7 or so does a good job.

4. Use the **successive line over-relaxation method (SLOR)**. This is the same as the SOR method, but rather than going around from point to point, a row of points (line) is solved using previous values and a method such as Gauss elimination.

5. If information is sought only in a particular region, **random walk** techniques can be useful. If you want to find the value at a point (j, k), for example, start at that point and randomly choose one of the numbers 1 to 4 (1 to 6 for three-dimensional problems). Do this until a boundary point is reached, where the value of V is B_i, the i standing for the ith iteration. Repeat this process N times. Then $V_{j,k} \approx \frac{1}{N_T} \sum_{i=1}^N N_i B_i$, where N_i is the

number of steps needed to get to the boundary point with value B_i and $N_T = \sum_{i=1}^{N} N_i$ is the total number of steps. The accuracy increases as N_T^{-4}, but clearly many, many steps must be taken.

6. Write the N algebraic equations in N unknowns, and use a traditional algebraic solver. The algebraic equations are sparse, which helps, but the fact that the matrix of the coefficients is not narrow-banded means that special methods tailored to such a problem must be used. These and other procedures can be found in much more detail in Smith (1978), for example.

In the preceding, attention has been paid only to the case where boundaries are rectangular, a fairly restricted case. For irregular boundaries, one could rephrase equation (12.2.1) in a mesh of unequal sides, but then a good part of the computational problem is to determine which boundary point you are near and which variation of equation (12.2.1) is needed. Also, near corners, where changes in the solution can be rapid, accuracy can be lost unless the grid mesh is shrunk. Two (at least) methods have been introduced to overcome this problem.

The ideas of conformal mapping introduced in Chapter 3 are ideally suited to generate grids to fit boundaries of any shape. Thomson, Warsi, and Mastin (1985) present techniques useful in both two and three dimensions for computer generation of grids for all three classes of partial differential equations. Basically, they use the conformality of analytic functions to map the flow space into a rectangle. Control functions can be used to adapt the grid spacing so that spacing is small and cell count denser where the gradient of the function can be expected to be large. After the space is transformed to the rectangle, the equations of interest are also transformed to the new coordinate system and then solved.

The finite element method (FEM, also sometimes FEA for finite element analysis), introduced for one-dimensional problems in Chapter 11, can also be adapted to two- and three-dimensional problems. Programs usually come as a package, including grid generation and solvers. Grid generation is to some degree usually automatic, with provisions for intervention by the user where refinements in the grid are needed. Elements used can be rectangular, triangular, semi-infinite, and a variety of others. The polynomials used on the sides of the elements vary in complexity, depending on the accuracy and order of the derivatives needed.

FEM was originally developed for solution of problems in the linear theory of elasticity, where the equations are strongly elliptic. In elasticity theory energy is conserved, and only the "laminar" state exists. Thus, one would expect that, unless special provisions are made for fluid flow problems, there would be a Reynolds number limitation on computational accuracy. Upwind differencing, described later, has made it possible to extend this limitation somewhat, and great claims have been made for the commercial programs. Many even claim to handle turbulent flows. However, since many of the companies are secretive as to how Reynolds number limitations and turbulence are treated, it is difficult to assess their claims.

FEM programs can be used for irrotational flows with cavities. A simple approach is to first estimate the shape of the cavity, then correct the shape to make it tangent to the computed velocity. In the process, all nodes on the cavity are moved by the process. It is to be repeated until some error norm such as $\sum_{\text{all cavity nodes}} (y_{\text{new}} - y_{\text{old}})^2$ is less than some value. Other methods for cavity flows have been suggested (e.g., Brennen, 1969).

12.3 Surface Singularities

To illustrate the use of surface singularities in two-dimensional flows, the basic starting point is Cauchy's integral formula, which states that for any analytic function (i.e., one that satisfies the Cauchy-Riemann conditions)

$$f(z) = \frac{1}{2\pi i} \oint \frac{f(\varsigma)}{\varsigma - z} d\varsigma, \tag{12.3.1}$$

where the integration is about a closed path traversed in the positive sense. That is, as the path of integration is traversed, the direction taken is such that the interior is always to the left. If z is within the closed path, the integral is zero.

Since the complex velocity $\frac{dw}{dz}$ is an analytic function, we can write

$$u - iv = \frac{1}{2\pi i} \oint \frac{u(\varsigma) - iv(\varsigma)}{\varsigma - z} d\varsigma. \tag{12.3.2}$$

Letting $d\varsigma = e^{i\theta} ds$, where ds is real and θ represents the slope of the integration path, equation (12.3.2) becomes

$$[u(\varsigma) - iv(\varsigma)] d\varsigma = [u(\varsigma) - iv(\varsigma)] e^{i\theta} ds = \left(q_{\text{tangent}} - iq_{\text{normal}} \right) ds,$$

where q_{tangent} and q_{normal} are the tangent and normal velocity components with respect to the path. Thus, equation (12.3.2) becomes

$$u - iv = \frac{1}{2\pi i} \oint \frac{q_{\text{tangent}} - iq_{\text{normal}}}{\varsigma - z} d\varsigma = \oint \frac{q_{\text{normal}}}{2\pi} \frac{ds}{z - \varsigma} + i \oint \frac{q_{\text{tangent}}}{2\pi} \frac{ds}{z - \varsigma}. \tag{12.3.3}$$

Recalling that

$$(u - iv)_{\text{source}} = \frac{m}{2\pi} \frac{1}{z - \varsigma} \quad \text{and} \quad (u - iv)_{\text{vortex}} = \frac{i\Gamma}{2\pi} \frac{1}{z - \varsigma},$$

it is seen that the first integral represents a source distribution, while the second integral can be interpreted as a vortex distribution. Notice as shown in Chapter 2, that a vortex distribution and a doublet distribution are equivalent.

In Chapter 2 the panel method was discussed for finding the flow about a submerged body. We next show how for two-dimensional inviscid flows this can be implemented numerically for lifting bodies.

First, consider a closed two-dimensional body made up of a series of N flat panels. An example is shown in Figure 12.3.1, where 12 panels are inscribed within the body. Each panel has a source and a vortex on it. The strength of each source and vortex is constant on a panel but can differ from panel to panel. Two such panels are shown in Figure 12.3.2 to illustrate the geometry. The velocity potential then contains the uniform stream plus the contributions from each of the N panels—that is,

$$\phi(x, y) = Ux + \sum_{j=1}^{N} \frac{m_j}{2\pi} \int_{s_j} \ln \sqrt{(x - x_j')^2 + (y - y_j')^2} ds_j + \sum_{j=1}^{N} \frac{\gamma_j}{2\pi} \int_{s_j} \tan^{-1} \frac{y - y_j'}{x - x_j'} ds_j. \tag{12.3.4}$$

The control points where the boundary will be taken are at the center of each panel and designated by (x_i, y_i). (Note: Panels can be constructed so either their endpoints lie on the surface of the body or the control points lie on the body. Some evidence suggests that the latter is more accurate.)

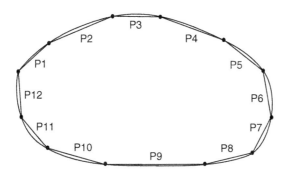

Figure 12.3.1   Panel method—numbering of panels

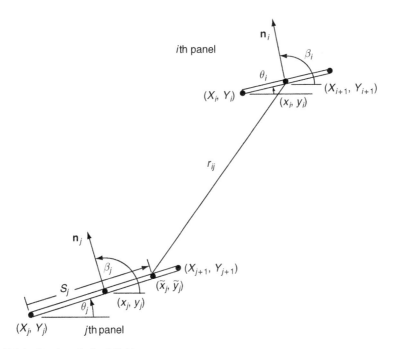

Figure 12.3.2   Panel method—definitions

Applying the boundary conditions, find that

$$\left.\frac{\partial \phi}{\partial n}\right|_{x_i,y_i} = U \cos \beta_i + \frac{m_i}{2} + \sum_{\substack{j=1 \\ j \neq i}}^{N} \frac{m_j}{2\pi} \int_{s_j} \frac{(x - x_j')\cos \beta_i + (y - y_j')\sin \beta_i}{(x - x_j')^2 + (y - y_j')^2} ds_j$$

$$+ \sum_{j=1}^{N} \frac{\gamma_j}{2\pi} \int_{s_j} \frac{(x - x_j')\sin \beta_i - (y - y_j')\cos \beta_i}{(x - x_j')^2 + (y - y_j')^2} ds_j. \tag{12.3.5}$$

Let

$$I_{ij} = \int_{s_j} \frac{(x - x_j')\cos \beta_i + (y - y_j')\sin \beta_i}{(x - x_j')^2 + (y - y_j')^2} ds_j \tag{12.3.6}$$

for the sources and

$$I'_{ij} = \int_{s_j} \frac{(x - x'_j)\sin\beta_i - (y - y'_j)\cos\beta_i}{(x - x'_j)^2 + (y - y'_j)^2} ds_j \qquad (12.3.7)$$

for the vortices. Then

$$\frac{m_i}{2} + \frac{\gamma_i}{2} + \sum_{\substack{j=1 \\ j \neq i}}^{N} \left(\frac{m_j}{2\pi} I_{ij} + \frac{\gamma_j}{2\pi} I'_{ij} \right) = -U\cos\beta_i. \qquad (12.3.8)$$

The first thing to notice is that by applying the boundary conditions, there are N equations in $2N$ unknowns. The Kutta condition also has not yet been applied, which is necessary for a lifting body. Notice also that in using either source or vortex panels, the tangency condition is satisfied only at the control points, and the velocity is infinite at every panel edge.

There are a number of approaches that can be used to model a lifting surface and balance the number of equations and unknowns in the process.

1. Use a unique source strength and the same vortex strength on every panel, giving $N+1$ unknowns. Impose the tangency condition at N control points, and also impose one Kutta condition at the trailing edge. This a fully determinate system.

2. Use a unique source and a parabolic vorticity distribution on the top and bottom panels. Let top and bottom maximum vortex strengths have the same magnitude so that there are $N+1$ unknowns. Tangency conditions at N control points and one Kutta condition make this a fully determinate system.

3. Use a unique source and vortex strength on each panel, giving $2N$ unknowns. Use two control points on each panel and one Kutta condition. This an indeterminate system, requiring a least squares procedure or something similar to resolve the inconsistency.

4. Use a unique source and vortex on each panel, giving $2N$ unknowns. Use two control points on each panel and one Kutta condition. This an indeterminate system, requiring a least squares procedure or something similar.

5. Use a unique source and the same vorticity on each panel, giving $N+1$ unknowns. Use two control points on each panel and one Kutta condition. This an indeterminate system, requiring a least squares procedure or a similar technique.

6. Use a unique vortex on each panel, giving N unknowns. Satisfy tangency at one control point on each panel and one Kutta condition. This an indeterminate system, requiring a least squares procedure or something similar.

7. Use a unique vortex on each panel, giving N unknowns. Use two control points on each panel and one Kutta condition. This an indeterminate system, requiring a least squares procedure or something similar.

8. Use curved panels, perhaps parabolic or cubic in shape, along with singularity distributions that vary on each panel. This would be particularly advantageous

near the rounded nose of an airfoil, where otherwise the number of panels must be increased to fit the geometry. The number of variations on this is unlimited.

Indeterminate systems can be handled as follows:
If the given system is $\sum_{j=1}^{N} A_{ij}x_j = b_i$, $i = 1, 2, \ldots N, N+1, \ldots N+M$, let

$$E^2 = \sum_{i=1}^{N+M} \left(\sum_{j=1}^{N} A_{ij}x_j - b_i \right)^2. \tag{12.3.9}$$

Then, to minimize E^2, let

$$\frac{\partial E^2}{\partial x_k} = 2 \sum_{i=1}^{N+M} A_{ik} \left(\sum_{j=1}^{N} A_{ij}x_j - b_i \right) = 0. \tag{12.3.10}$$

Then

$$\sum_{j=1}^{N} \left(\sum_{i=1}^{N+M} A_{ij}A_{ik} \right) x_j = \sum_{i=1}^{N+M} b_i A_{ik}, \quad k = 1, 2, \ldots N. \tag{12.3.11}$$

This system is determinate.

There are several approaches possible for satisfying the Kutta condition. For example, if a body shape with a sharp trailing edge is to be modeled, the Kutta condition could be imposed at the trailing edge. Consider the two panels surrounding the trailing edge, as shown in Figure 12.3.3. Locally the complex potential will look like

$$w = Az^{n-1}, \tag{12.3.12}$$

where $n = \frac{2\pi}{2\pi - \kappa}$ and κ is the wedge angle. Denoting the tangent velocities on the top panel (panel #1, length S_1) and the bottom panel (panel #N, length S_N), the Kutta condition requires that the velocities be the same on these two panels. The velocities at the control points are

$$v_{t1} = nA \left(\frac{1}{2}S_1 \right)^{n-1}, \qquad v_{tN} = -nA \left(\frac{1}{2}S_N \right)^{n-1},$$
$$\tag{12.3.13}$$
$$\text{therefore} \quad v_{t1} \left(\frac{1}{2}S_1 \right)^{1-n} + v_{tN} \left(\frac{1}{2}S_N \right)^{1-n} - 0.$$

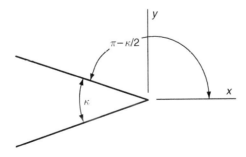

Figure 12.3.3   Kutta condition—first method

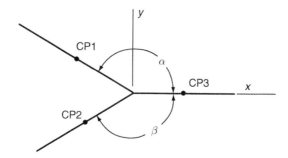

Figure 12.3.4 Kutta condition—second method

An alternate way of meeting the Kutta condition is shown in Figure 12.3.4. Here an extra panel of length S_3 has been added at the trailing edge. Take the lengths of the three panels be given as Δs_1, Δs_2, Δs_3. The result is then

$$w(z) = \begin{cases} Az^{\pi/\alpha} = Ar^{\pi/\alpha}\left(\cos\dfrac{\pi\theta}{\alpha} + i\sin\dfrac{\pi\theta}{\alpha}\right), & 0 \le \theta \le \alpha, \\[2ex] Bz^{\pi/\beta-1} = Br^{\pi/\beta-1}\left(\cos\dfrac{\pi\theta}{\beta} + i\sin\dfrac{\pi\theta}{\beta}\right), & -\beta \le \theta \le 0, \end{cases} \tag{12.3.14}$$

giving

$$\frac{dw}{dz} = \begin{cases} \dfrac{\pi}{\alpha}Az^{\pi/\alpha-1}, & 0 \le \theta \le \alpha, \\[2ex] \dfrac{\pi}{\beta}Bz^{\pi/\beta-2}, & -\beta \le \theta \le 0. \end{cases} \tag{12.3.15}$$

Then at control point 3 there is a discontinuity in ϕ of magnitude

$$\Delta\phi = A\left(\frac{1}{2\Delta s_3}\right)^{\pi/\alpha} - B\left(\frac{1}{2\Delta s_3}\right)^{\pi/\beta}. \tag{12.3.16}$$

The velocities at control points 1 and 2 are

$$\left.\left|\frac{dw}{dz}\right|\right|_{cp1} = \frac{\pi}{\alpha}A\left(\frac{1}{2}\Delta s_3\right)^{(\pi/\alpha)-1} = V_1$$

and

$$\left.\left|\frac{dw}{dz}\right|\right|_{cp2} = \frac{\pi}{\beta}A\left(\frac{1}{2}\Delta s_3\right)^{\pi/\beta-1} = V_2. \tag{12.3.17}$$

Thus $A = \dfrac{\alpha}{\pi}V_1\left(\dfrac{1}{2}\Delta s_1\right)^{1-\pi/\alpha}$, $B = \dfrac{\beta}{\pi}V_2\left(\dfrac{1}{2}\Delta s_2\right)^{1-\pi/\beta}$, so

$$\pi\Delta\phi = \alpha V_1\frac{\Delta s_1}{2}\left(\frac{\Delta s_3}{\Delta s_1}\right)^{\pi/\alpha} + \beta V_2\frac{\Delta s_2}{2}\left(\frac{\Delta s_3}{\Delta s_2}\right)^{\pi/\beta}. \tag{12.3.18}$$

It is next necessary to carry out the integrations and put our equations in final form. With the help of Figure 12.3.2, letting $\tilde{x}_j = X_j + s_j \cos\theta_j$, $\tilde{y}_j = Y_j + s_j \sin\theta_j$, then

$$A_j = (X_j - x_j)\cos\theta_j + (Y_j - y_j)\sin\theta_j,$$
$$B_j = (X_j - x_j)^2 + (Y_j - y_j)^2, \qquad\qquad (12.3.19)$$
$$C_j = -(X_j - x_j)\sin\theta_j + (Y_j - y_j)\cos\theta_j,$$

and

$$I_{ij} = \frac{1}{2}\sin(\theta_i - \theta_j)\ln\left(1 + \frac{S_j(S_j + 2A_j)}{B_j}\right) - \cos(\theta_i - \theta_j)\left[\tan^{-1}\frac{S_j + A_j}{C_j} - \tan^{-1}\frac{A_j}{C_j}\right].$$

$$(12.3.20)$$

Similarly

$$I'_{ij} = -\frac{1}{2}\cos(\theta_i - \theta_j)\ln\left(1 + \frac{S_j(S_j + 2A_j)}{B_j}\right) - \sin(\theta_i - \theta_j)\left[\tan^{-1}\frac{S_j + A_j}{C_j} - \tan^{-1}\frac{A_j}{C_j}\right].$$

$$(12.3.21)$$

With these at hand, a choice can be made as to the number of sources and vortices to retain.

Note that from the point of view of the calculations, just having a determinate system may not be sufficient for obtaining good results. Some of the combinations of sources and vortices result in **stiff systems**. This means that when a system of equations such as (in matrix form) $\mathbf{Ax} = \mathbf{B}$ is being solved, such a system has a set of eigenfunctions and eigenvalues given by $\mathbf{Ay}_j = \lambda_j \mathbf{y}_j$. If the eigenvalues are all of the same order of magnitude, the problem is not stiff, and the equation can be solved by standard methods. If, however, there is a large disparity in the magnitude of the eigenvalues, then the problem is stiff and is said to be ill-conditioned—that is, it is very sensitive to small changes in the magnitudes of terms in either the matrix $\mathbf{A}$ or the known vector $\mathbf{B}$.

Generally, it is not practical to first compute all of the eigenvalues to see whether the system is stiff. It is better to solve the problem and then use back substitution or some other check to see whether the solution is accurate. If it turns out that the method you selected results in a stiff system, there are specialized methods that are designed for dealing with stiff systems. It may, however, be much easier to select a different combination of sources and vortices.

Notice that having a wide range in magnitude of the eigenvalues is reminiscent of the boundary layer, where there are large gradients within the boundary layer and much smaller ones outside it. It should come as no surprise that stiffness questions arise there as well.

Here are some tips for maintaining accuracy. Rubbert and Saaris (1972) recommend using double sheets of vortices inside the body. The following should be observed:

1. It is not necessary to place singularities on the boundary; they may be placed within the boundary. If a single vortex is inserted in a thin body, the sources above and below the vortex, the sources tend to be of opposite signs and act like a doublet, with the strength inversely proportional to the distance between them. This gives a strong gradient in source strength, which is to be avoided.

2. A series of vortices that are inserted with a strength approximating the load distribution generally gives good results. However, if the angle of attack is to be

varied, and the airfoil is highly cambered, two such rows may be needed, one to take care of each problem individually.

A program using the preceding procedure with source panels only to develop the body shape is given as Program 12.3.1. Once the panel strengths are known, there is enough information to find the velocity and pressure at any point in the exterior flow.

```
        PROGRAM PANELBOOK
        REAL*8 LAMBDA
        DIMENSION ACS(40),ASN(40),BETA(40),RHS(40),THETA(40),XCP(40),
     &    XLEN(40),XP(40),YCP(40),YP(40),VEL(40)
        DIMENSION AA(40,40),AA1(40,40),LAMBDA(40),SUM1(40,40),SUM2(40,40)
        PI=3.14159265
        WRITE(*,*)"This is a panel program designed to calculate the inviscid flow about"
        WRITE(*,*)"  an arbitrary body. The body shape is determined by the user. "
        WRITE(*,*)"The program first asks for the number of panels to be used."
        WRITE(*,*)"Next it asks for the X,Y coordinates of the end points of the panels."
        WRITE(*,*)"  These are called the nodes of the panel. For the program to run"
        WRITE(*,*)"  properly the nodal points should be numbered in a clockwise manner"
        WRITE(*,*)"  as one goes around the body."
        WRITE(*,*)"'Finally the free stream velocity is requested."
        WRITE(*,*)
        WRITE(*,*)"The output will consist ofthe velocities calculated at the control points."
        WRITE(*,*)"The control points were chosen so as to be located at the center of the panels."
        WRITE(*,*)
        WRITE(*,*)"Enter the number of panels you wish to use. Maximum allowable is 40. "
        READ(*,*) N
        WRITE(*,*)"The coordinates of the end points of the panels must now be entered."
        WRITE(*,*)"his program is two dimensional, so only x and y coordinates should be entered."
        WRITE(*,*)
        WRITE(*,*)"You will be told which node you are entering. Panel number"
        WRITE(*,*)"one has nodes one and two, panel two has nodes two and three,"
        WRITE(*,*)"and the last panel has nodes n and one."
        WRITE(*,*)
        WRITE(*,*)"Results are in file A:PANELBOOK.DAT"
        WRITE(*,*)
!=========================================================================
!                 ENTER NODES
!=========================================================================
        DO I=1,N
            WRITE(*,*) "Enter node number ",I
            READ(*,*) XP(I),YP(I)
        END DO
        OPEN(1, FILE='A:PANELBOOK.DAT', STATUS='UNKNOWN')
        WRITE(*,*) "Your entered points are listed below:"
        DO  I=1,N
          WRITE(*,100)I,XP(I),I,YP(I)
          WRITE(1,100)I,XP(I),I,YP(I)
        END DO
        WRITE(*,*) " Enter the free steam velocity. "
        READ(*,*) U
!=========================================================================
!       CALCULATE CONTROL POINTS
!=========================================================================
        L=N-1
        WRITE(1,*) "LOCATION OF CONTROL POINTS:"
        DO I=1,L
            XCP(I)=(XP(I)+XP(I+1))/2.
            YCP(I)=(YP(I)+YP(I+1))/2.
            WRITE(1,100)I,XCP(I),I,YCP(I)
        END DO
        XCP(N)=(XP(N)+XP(1))/2.
        YCP(N)=(YP(N)+YP(1))/2.
        WRITE(1,110)N,XCP(N),I,YCP(N)
```

Program 12.3.1—Panel method for flow past a nonlifting body (program by the author)

```
!=============================================================
!     CALCULATE ANGLES
!=============================================================
      DO I=1,L
         IF(XP(I+1).EQ. XP(I).AND. YP(I+1).LT.YP(I)) THEN
                     THETA(I)=-PI/2.
            ELSE IF(XP(I+1).EQ.XP(I).AND.YP(I+1).GT.YP(I)) THEN
                     THETA(I)=PI/2.
               ELSE
                        THETA(I)=ATAN((YP(I+1)-YP(I))/(XP(I+1)-XP(I)))
         END IF
         IF (XP(I+1).LT.XP(I)) THETA(I)=THETA(I)-PI
      END DO
      IF(XP(1).EQ.XP(N).AND.YP(1).LT.YP(N)) THEN
            THETA(N)=-PI/2.
            ELSE IF(XP(1).EQ.XP(N).AND.YP(1).GT.YP(N)) THEN
            THETA(N)=PI/2.
         ELSE
            THETA(N)=ATAN((YP(1)-YP(N))/ (XP(1)-XP(N)))
      END IF
      IF( XP(1) .LT. XP(N)) THETA(N)=THETA(N)-PI
!=============================================================
!     CALCULATE THE LENGTH AND ANGLES FOR ALL PANELS
!=============================================================
      DO  I=1,L
         XLEN(I)=SQRT( (XP(I+1)-XP(I))**2 + ( YP(I+1)-YP(I))**2 )
         ASN(I)= SIN( THETA(I) )
         ACS(I)=COS( THETA(I))
      END DO
      XLEN(N)=SQRT( (XP(1)-XP(N))**2 + (YP(1)-YP(N))**2 )
      ASN(N)=SIN( THETA(N) )
      ACS(N)=COS( THETA(N) )
!=============================================================
!     CALCULATE I(I,J)=INTEGRAL ( D/DN LN(R) DR(J) ) OF THE Ith PANEL
!=============================================================
      DO I=1,N
            DO J=1,N
               IF(J.EQ.I) THEN
                     SUM1(I,J)=PI
                     SUM2(I,J)=0.
               ELSE
                     A=-(XCP(I)-XP(J))*ACS(J)-( YCP(I)-YP(J))*ASN(J)
                     B=( XCP(I)-XP(J))**2 + ( YCP(I)-YP(J) )**2
                     C=SIN( THETA(I)-THETA(J) )
                     D=COS( THETA(I)-THETA(J) )
                     E=( XCP(I)-XP(J))*ASN(J) - ( YCP(I)-YP(J) )*ACS(J)
                     F=LOG( 1.0 +XLEN(J)*( XLEN(J)+2.*A)/ B )
                     G=ATAN( (XLEN(J)+A)/E )-ATAN(A/E)
                     SUM1(I,J)=C*F/2.-D*G
                     SUM2(I,J)=-D*F/2.-C*G
               FND IF
            END DO
      END DO
!=============================================================
!     ASSIGN THE MEMBERS OF THE MATRIX EQUATION TO SOLVE FOR
!     THE INDIVIDUAL PANEL STRENGTH'S.
!=============================================================
      DO  I=1,N
         DO  J=1,N
            AA(I,J)=SUM1(I,J)
         END DO
      END DO
      DO  I=1,N
         DO  J=1,N
            AA1(I,J)=AA(I,J)/AA(1,1)
         END DO
      END DO
      DO  I=1,N
         BETA(I)=THETA(I)+PI/2.
         RHS(I)=-U*COS( BETA(I) )/AA(1,1)
      END DO
```

Program 12.3.1—(Continued)

```
! ===================================================================
!      SOLVE THE MATRIX SYSTEM FOR THE PANEL STRENGTHS
!      USING A GAUSS-SIEDEL ROUTINE
! ===================================================================
      DO  I=1,N
          LAMBDA(I)=0.
      END DO
      IMAX=20
      DO K=1,IMAX
          ICOUNT=1
          DO I=1,N
              XSTAR=LAMBDA(I)
              LAMBDA(I)=RHS(I)
              DO J=1,N
                  IF (J.GT.I) LAMBDA(I)=LAMBDA(I)-AA1(I,J)*LAMBDA(J)
              END DO
! ===================================================================
!      TEST FOR CONVERGENCE
! ===================================================================
              IF( DABS(XSTAR-LAMBDA(I)).GT.0.00001) THEN ICOUNT=0
          END DO
          IF (ICOUNT.EQ.1) GOTO 10
      END DO
! ===================================================================
!      CALCULATE PANEL STRENGTHS
! ===================================================================
   10 DO I=1,N
          VI=U*ACS(I)
          DO  J=1,N
              VI=VI+LAMBDA(J)*SUM2(I,J)
          END DO
          VEL(I)=VI
      END DO
      WRITE(1,120)
      DO  I=1,N
          WRITE(1,130) I,LAMBDA(I)
      END DO
      WRITE(1,140)
      DO I=1,N
          WRITE(6,150) I,VEL(I)
      END DO
  100 FORMAT(1X,'XP(',I2,')=',F10.5,1X,'YP(',I2,')=',F10.5,/)
  110 FORMAT(1X,'XCP(',I2,')=',1X,F10.5,2X,'YCP(',I2,')=',F10.5)
  120 FORMAT("CALCULATED STRENGTHS OF THE PANELS:")
  130 FORMAT(1X,'PANEL',1X,I2,'=',1X,F10.5)
  140 FORMAT(//1X,'THE SOLUTION IS GIVEN BELOW'//)
  150 FORMAT(1X,'THE VELOCITY AT CONTROL POINT',1X,I3,1X,'IS',F10.5)
      CLOSE(1)
      END PROGRAM
```

Program 12.3.1—(Continued)

The use of surface singularity distributions can be continued into three dimensions with, however, the loss of the use of complex functions. Green's second identity in three dimensions is

$$\iint_c \left(h\frac{\partial f}{\partial n} - f\frac{\partial h}{\partial n} \right) dS = \iiint_V (h\nabla^2 f - f\nabla^2 h) dV. \qquad (12.3.22)$$

Let $h = \phi$, where $\nabla^2\phi = 0$, and $f = 1/R$, where $R = |(x - x_0)\mathbf{i} + (y - y_0)\mathbf{j} + (z - z_0)\mathbf{k}|$. Inserting this into equation (12.3.22) gives

$$\phi = \frac{1}{4\pi} \iint_S \left[-\frac{1}{R}\frac{\partial\phi}{\partial n} + \phi\frac{\partial}{\partial n}\left(\frac{1}{R}\right) \right] dS. \qquad (12.3.23)$$

The $1/R$ term represents a surface distribution of sources of strength $\partial\phi/\partial n$, and the second term represents a surface doublet distribution of strength φ. Thus, in principal, the two- and three-dimensional methodologies are the same.

Singularity distribution techniques have been widely used by naval architects and aircraft designers. The approach was pioneered by Hess and Smith (1962, 1967) and widely adapted and used by others. In those industries the primary emphasis is the external flow past bodies. The methodology chosen can either suppress the "flow" within the body or allow for that flow. For these applications the internal flow is ignored as being without physical interest. It is also possible to use these methods for wave problems where there are free surfaces with or without solid boundaries (Schwartz (1981), Forbes (1989), Zhou and Graebel (1990)).

A combination of the surface integral method and the FEM is the ***boundary integral method*** (BIM). It uses the idea of surface singularities to solve for the interior flow and uses paneling as in FEM to solve for the magnitude of the singularities.

Parabolic Partial Differential Equations

12.4   One-Step Methods

One-step methods take the results from a previous computation in the space dimension(s) and advance it in the time dimension.

12.4.1   Forward Time, Centered Space—Explicit

In one space dimension, the diffusion equation becomes

$$\frac{\partial T}{\partial t} = \alpha \frac{\partial^2 T}{\partial x^2}. \tag{12.4.1}$$

Approximating the time derivative by

$$\frac{\partial T}{\partial t} \approx \frac{T_i^{n+1} - T_i^n}{\Delta t} \tag{12.4.2}$$

and the space derivative by

$$\frac{\partial^2 T}{\partial x^2} \approx \frac{T_{i+1}^n - 2T_i^n + T_{i-1}^n}{\Delta x^2}, \tag{12.4.3}$$

equation (12.4.1) can be approximated by

$$\frac{T_i^{n+1} - T_i^n}{\Delta t} \approx \alpha \frac{T_{i+1}^n - 2T_i^n + T_{i-1}^n}{\Delta x^2}. \tag{12.4.4}$$

Here, the superscript refers to the time dimension, and the subscript refers to the space dimension (Figure 12.4.1). Solving for T at the most recent time gives

$$T_i^{n+1} \approx T_i^n + \frac{\alpha \Delta t}{\Delta x^2}\left(T_{i+1}^n - 2T_i^n + T_{i-1}^n\right). \tag{12.4.5}$$

This is an ***explicit*** (i.e., it's not necessary to solve a system of equations, just solve one equation at a time) equation for T in terms of its value at three points at the previous time. It is a stable method as long as $\alpha\Delta t/\Delta x^2 < 1/2$. This stability criterion is called the

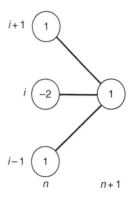

Figure 12.4.1 Computational molecule for the forward time centered space explicit method

Courant condition. It guarantees that the solution does not "blow up," but it does not tell us anything about the accuracy of the method. Since equations (12.4.2) and (12.4.3) come about from Taylor series expansions, it can be shown that accuracy is of first order in Δt and second order in Δx. If more terms in the Taylor expansions are taken, providing we choose $\alpha \Delta t / \Delta x^2 < 1/6$, the accuracy improves to second order in Δt and fourth order in Δx.

12.4.2   Dufort-Frankel Method—Explicit

Changing the derivatives somewhat in the previous method leads to another explicit method. Instead of equation (12.4.2), use

$$\frac{\partial T}{\partial t} \approx \frac{T_i^{n+1} - T_i^{n-1}}{2\Delta t}, \tag{12.4.6}$$

and instead of equation (12.4.3), use

$$\frac{\partial^2 T}{\partial x^2} \approx \frac{T_{i+1}^n - \left(T_i^{n+1} + T_i^{n-1}\right) + T_{i-1}^n}{\Delta x^2}. \tag{12.4.7}$$

Then, with $\beta = \alpha \Delta t / \Delta x^2$, equation (12.4.5) is replaced by

$$T_i^{n+1} \left(\frac{1}{2} + \beta\right) \approx \beta \left(T_{i+1}^n + T_{i-1}^n\right) + \left(\frac{1}{2} - \beta\right) T_i^{n-1}. \tag{12.4.8}$$

The computational molecule is shown in Figure 12.4.2. This method is explicit and unconditionally stable (i.e., stable for all values of β). It does, however, require another method to start up and generate the first line, and β must be small, or the method is not even first-order accurate.

12.4.3   Crank-Nicholson Method—Implicit

If we stay with equation (12.4.2) for the time derivative but replace equation (12.4.3) by

$$\frac{\partial^2 T}{\partial x^2} \approx \frac{T_{i+1}^n - 2T_i^n + T_{i-1}^n}{2\Delta x^2} + \frac{T_{i+1}^{n+1} - 2T_i^{n+1} + T_{i-1}^{n+1}}{2\Delta x^2}, \tag{12.4.9}$$

instead of equation (12.4.5), we have

$$-T_{i+1}^{n+1} + 2\frac{1+\beta}{\beta} T_i^{n+1} - T_{i-1}^{n+1} \approx T_{i+1}^n + 2\frac{1-\beta}{\beta} T_i^n + T_{i-1}^n, \tag{12.4.10}$$

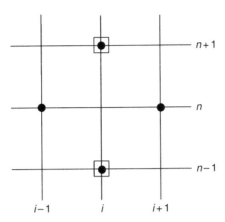

Figure 12.4.2 Computational molecule for the Dufort-Frankel method

where $\beta = \alpha\Delta t/\Delta x^2$. This is an ***implicit method***, meaning a system of equations must be solved simultaneously, but since the system is tridiagonal, the solution method is easy. Depending on boundary conditions, it is usually unconditionally stable (i.e., stable for all values of β), but it can be unstable if the boundary conditions include derivatives of the independent variable. Also, if there are two or more space dimensions, it becomes block tridiagonal, increasing the difficulty of solution.

12.4.4   Boundary Layer Equations—Crank-Nicholson

We saw earlier that similarity solutions of the boundary layer equations could be obtained, reducing a partial differential equation to an ordinary equation. This, however, depended on having a minimum number of dimensions and also special outer-velocity profiles. With the present-day easy access to personal computers, a numerical program can easily be written to solve the two-dimensional steady equations with nonsimilar solutions with little difficulty.

An easy procedure for the two-dimensional boundary layer equations uses the Crank-Nicolson method. Start with the steady form of the boundary layer equations—namely,

$$u\frac{\partial u}{\partial x} + v\frac{\partial u}{\partial y} = \frac{\partial U}{\partial t} + U\frac{\partial U}{\partial x} + \frac{\mu}{\rho}\frac{\partial^2 u}{\partial y^2} \tag{12.4.11}$$

The second-order derivative in y and the first in x tell us that this is a parabolic partial differential equation. It is convenient to first suppress as much as is easily possible the growth of the boundary layer thickness in the x direction. Introduce variables suggested by our Falkner-Skan solution according to

$$\xi = x/L, \quad \eta = y\sqrt{\frac{\rho U x}{\mu}}, \quad \psi = \nu\sqrt{\frac{\rho U x}{\mu}}f(\xi, \eta). \tag{12.4.12}$$

Then, since

$$\frac{\partial}{\partial x} = \frac{1}{L}\frac{\partial}{\partial\xi} + \frac{\eta}{2}\left(-\frac{1}{x} + \frac{1}{U}\frac{dU}{dx}\right)\frac{\partial}{\partial\eta} \quad \text{and} \quad \frac{\partial}{\partial y} = \sqrt{\frac{\rho U}{\mu x}}\frac{\partial}{\partial\eta},$$

we have

$$u = \frac{\partial \psi}{\partial y} = \sqrt{\frac{\rho U}{x \mu}} \frac{\partial}{\partial \eta} \left(\frac{\mu}{\rho} \sqrt{\frac{\rho U x}{\mu}} f \right) = U \frac{\partial f}{\partial \eta}, \tag{12.4.13}$$

$$v = -\frac{\partial \psi}{\partial x} = -\frac{\mu}{\rho} \left[\frac{1}{2} \left(\sqrt{\frac{\rho x}{U \mu}} \frac{dU}{dx} + \sqrt{\frac{\rho U}{x \mu}} \right) f + \frac{1}{L} \sqrt{\frac{\rho U x}{\mu}} \frac{\partial f}{\partial \xi} + \frac{\eta}{2} \sqrt{\frac{\rho U x}{\mu}} \left(-\frac{1}{x} + \frac{1}{U} \frac{dU}{dx} \right) \frac{\partial f}{\partial \eta} \right].$$

Substituting these into equation (12.4.1), find, after some arranging, that

$$U \frac{\partial f}{\partial \eta} \left[\frac{dU}{dx} \frac{\partial f}{\partial \eta} + \frac{U}{L} \frac{\partial^2 f}{\partial \eta \partial \xi} + \frac{\eta U}{2} \left(-\frac{1}{x} + \frac{1}{U} \frac{dU}{dx} \right) \frac{\partial^2 f}{\partial \eta^2} \right]$$

$$- U^2 \frac{\partial^2 f}{\partial \eta^2} \left[\frac{1}{2} \left(\frac{1}{U} \frac{dU}{dx} + \frac{1}{x} \right) f + \frac{1}{L} \frac{\partial f}{\partial \xi} + \frac{\eta}{2} \left(-\frac{1}{x} + \frac{1}{U} \frac{dU}{dx} \right) \frac{\partial f}{\partial \eta} \right] = \frac{U^2}{x} \frac{\partial^3 f}{\partial \eta^3} + U \frac{dU}{dx}.$$

For convenience in writing introduce the parameter $\beta = (x/U) dU/dx$. Multiply the preceding equation by x/U^2 and simplify. The result is

$$\frac{\partial f}{\partial \eta} \left(\beta \frac{\partial f}{\partial \eta} + \xi \frac{\partial^2 f}{\partial \eta \partial \xi} \right) - \frac{\partial^2 f}{\partial \eta^2} \left(\frac{\beta+1}{2} f + \xi \frac{\partial f}{\partial \xi} \right) = \frac{\partial^3 f}{\partial \eta^3} + \beta. \tag{12.4.14}$$

It is seen that equation (12.4.14) is a third-order partial differential equation. In numerical solutions, it is somewhat easier to deal with first-order partial differential equations. To get to that point here, make the further change of variables

$$F = \frac{\partial f}{\partial \eta}, \quad V = v \sqrt{\frac{\rho x}{\mu U}},$$

where F is then the dimensionless x velocity component and V is a scaled dimensionless version of the y velocity component. Then, from the second of equation (12.4.13) find that

$$V = - \left(\frac{\beta+1}{2} f + \xi \frac{\partial f}{\partial \xi} + \frac{\eta(\beta-1)}{2} \frac{\partial f}{\partial \eta} \right). \tag{12.4.15}$$

Differentiating this with respect to η gives

$$\frac{\partial V}{\partial \eta} = - \left(\beta F + \xi \frac{\partial F}{\partial \xi} + \frac{\eta(\beta-1)}{2} \frac{\partial F}{\partial \eta} \right). \tag{12.4.16}$$

Our momentum equation (12.4.4) can now be written as

$$F \left(\beta F + \xi \frac{\partial F}{\partial \xi} \right) + \frac{\partial F}{\partial \eta} \left(V + \frac{\eta}{2} (\beta-1) F \right) = \frac{\partial^2 F}{\partial \eta^2} + \beta. \tag{12.4.17}$$

To put these equations in numerical form, set up a grid with spacing $\Delta \eta$ by $\Delta \xi$, with grid points at

$$\eta = (n-1) \Delta \eta, \quad n = 1, 2, \ldots N,$$

$$\xi = (m-1) \Delta \xi, \quad m = 1, 2, \ldots M.$$

Using centered differences and keeping in mind that we want to finish up with a set of linear algebraic equations, the various derivatives can be replaced by

$$\frac{\partial V}{\partial \eta} = \frac{1}{2\Delta \eta}\left(V_{m+1,n} - V_{m+1,n-1} + V_{m,n} - V_{m,n-1}\right),$$

$$\beta F = \frac{\beta_{m+1}}{4}\left(F_{m+1,n} + F_{m+1,n-1} + F_{m,n} + F_{m,n-1}\right),$$

$$\xi\frac{\partial F}{\partial \xi} = \frac{\xi_{m+0.5}}{2\Delta \xi}\left(F_{m+1,n} - F_{m+1,n-1} + F_{m,n} - F_{m,n-1}\right),$$

$$\frac{\eta(\beta - 1)}{2}\frac{\partial F}{\partial \eta} = \frac{\eta_{n-0.5}(\beta_{m+0.5} - 1)}{4\Delta \eta}\left(F_{m+1,n} - F_{m+1,n-1} + F_{m,n} - F_{m,n-1}\right),$$

$$\beta(1 - F^2) = \beta_{m+1}\left(1 - F_{m+1,n}F_{m,n}\right), \tag{12.4.18}$$

$$\xi F\frac{\partial F}{\partial \xi} = \frac{\xi_{m+0.5}}{\Delta \xi}F_{m,n}\left(F_{m+1,n} - F_{m,n}\right),$$

$$\left(V + \frac{\eta(\beta - 1)}{2}F\right)\frac{\partial F}{\partial \eta} = \frac{1}{4\Delta \eta}\left(V_{m,n} + \frac{\eta_n(\beta_m - 1)}{2}F_{m,n}\right)\left(F_{m+1,n} - F_{m+1,n-1} + F_{m,n} - F_{m,n-1}\right),$$

$$\frac{\partial^2 F}{\partial \eta^2} = \frac{1}{2\Delta \eta^2}\left(F_{m+1,n+1} - 2F_{m+1,n} + F_{m+1,n-1} + F_{m,n+1} - 2F_{m,n} + F_{m,n-1}\right).$$

Putting these into equation (6.6.5) find that

$$V_{m+1,n} = V_{m+1,n-1} - V_{m,n} + V_{m,n-1} + 2\Delta \eta(P_m F_{m+1,n} + Q_m F_{m+1,n-1}$$
$$+ R_m F_{m,n} + S_m F_{m,n-1}), \tag{12.4.19}$$

with

$$P_m = -\frac{\beta_{m+1}}{4} - \frac{\xi_{m+0.5}}{2\xi} - \frac{\eta_{n-0.5}}{4\Delta \eta}(\beta_{m+0.5} - 1),$$

$$Q_m = -\frac{\beta_{m+1}}{4} - \frac{\xi_{m+0.5}}{2\xi} + \frac{\eta_{n-0.5}}{4\Delta \eta}(\beta_{m+0.5} - 1),$$

$$\tag{12.4.20}$$

$$R_m = -\frac{\beta_{m+1}}{4} + \frac{\xi_{m+0.5}}{2\zeta} - \frac{\eta_{n-0.5}}{4\Delta \eta}(\beta_{m+0.5} - 1),$$

$$S_m = -\frac{\beta_{m+1}}{4} + \frac{\xi_{m+0.5}}{2\xi} + \frac{\eta_{n-0.5}}{4\Delta \eta}(\beta_{m+0.5} - 1).$$

For the momentum equation (12.4.14), similar substitutions give

$$-\beta_{m+1}\left(1 - F_{m+1,n}F_{m,n}\right) + \frac{\xi_{m+0.5}}{\Delta \xi}F_{m,n}\left(F_{m+1,n} - F_{m,n}\right)$$

$$+ \frac{1}{4\Delta \eta}\left(V_{m,n} + \frac{\eta_n(\beta_m - 1)}{2}F_{m,n}\right)(F_{m+1,n} - F_{m+1,n-1} + F_{m,n} - F_{m,n-1})$$

$$- \frac{1}{2\Delta \eta^2}\left(F_{m+1,n+1} - 2F_{m+1,n} + F_{m+1,n-1} + F_{m,n+1} - 2F_{m,n} + F_{m,n-1}\right) \tag{12.4.21}$$

$$= 0.$$

These two equations can be rearranged in a form suitable for solution as follows:

$$A_{m,n}F_{m+1,n-1} + B_{m,n}F_{m+1,n} + C_{m,n}F_{m+1,n+1} = D_{m,n}, \qquad (12.4.22)$$

$$V_{m+1,n+1} = V_{m+1,n} + E_{m,n}, \qquad (12.4.23)$$

with $1 < n < N$, and the parameters defined by

$$A_{m,n} = -\frac{V_{m,n} + 0.5\eta_n(\beta_m - 1)F_{m,n}}{4\Delta\eta} - \frac{1}{2\Delta\eta^2},$$

$$B_{m,n} = \beta_{m+1}F_{m,n} + \frac{\xi_{m+0.5}F_{m,n}}{\Delta\xi} + \frac{1}{\Delta\eta^2},$$

$$C_{m,n} = \frac{V_{m,n} + 0.5\eta_n(\beta_m - 1)F_{m,n}}{4\Delta\eta} - \frac{1}{2\Delta\eta^2}, \qquad (12.4.24)$$

$$D_{m,n} = \beta_{m+1} + \frac{\xi_{m+0.5}F_{m,n}^2}{\Delta\xi} - \frac{[V_{m,n} + 0.5\eta_n(\beta_m - 1)F_{m,n}](F_{m,n+1} - F_{m,n-1})}{4\Delta\eta}$$
$$+ \frac{F_{m,n+1} - 2F_{m,n} + F_{m,n-1}}{2\Delta\eta^2},$$

$$E_{m,n} = -V_{m,n} + V_{m,n-1} + 2\Delta\eta(P_m F_{m+1,n} + Q_m F_{m+1,n-1} + R_m F_{m,n} + S_m F_{m,n-1}).$$

This numerical integration procedure requires knowledge of F and V at the initial line $m = 1$. These starting conditions usually would come from a Falkner-Skan or other of the similarity solutions previously discussed. We also must know β as input data at the various values of ξ.

For a given value of m, the $A_{m,n}$, $B_{m,n}$, and $C_{m,n}$ are all known. Equation (12.4.22) is then a tridiagonal set of algebraic equations, which is an algebraic system that is particularly amenable to solution by a number of methods. From the boundary conditions we know that $F_{m,1} = 0$, $V_{m,1} = 0$, and $F_{m,N} = 1$, so equation (12.4.13) represents $N - 2$ equations in terms of the $N - 2$ unknowns $F_{m,n}$. Solve for the $F_{m,n}$ first, and then go to equation (12.4.24) and solve for the $V_{m,n}$ step by step. Then increment m by one, and repeat the procedure. As the solution proceeds downstream, it is a good idea to check the wall velocity gradient as the integration goes forward in ξ. When it becomes zero, flow reversal will occur downstream from this point, and the boundary layer approximation is no longer valid. This would normally determine M.

Numerical solution of these boundary layer equations would usually proceed by taking a relatively coarse grid spacing (small M and N), solving the algebraic system, and then repeating the solution after halving the grid spacing (doubling M and N). This procedure would repeat until the changes are acceptably small.

The numerical procedure just outlined is not the only numerical scheme that could have been used. It has, however, the virtue of leading to a tridiagonal matrix, which is about the simplest algebraic system to solve. It also is a method well within the capabilities of a personal computer and provides results that appear to be at least as good as those found by other numerical methods.

12.4.5   Boundary Layer Equation—Hybrid Method

The previous methods can be combined in various ways into a hybrid method for the boundary layer equations. Write

$$\frac{u_{n+1,j} - u_{n,j} + u_{n+1,j-1} - u_{n,j-1}}{2\Delta x} + \frac{v_{n+1,j-v} - v_{n+1,j-1}}{\Delta y} = 0, \qquad (12.4.25)$$

$$\left[ru_{n+1,j} + (1-r)u_{n,j} \right] \frac{u_{n+1,j} - u_{n,j}}{\Delta x}$$

$$+ \left[rv_{n+1,j} \frac{u_{n+1,j+1} - u_{n+1,j-1}}{2\Delta y} + (1-r)v_{n,j} \frac{u_{n,j+1} - u_{n,j-1}}{2\Delta y} \right]$$

$$= \left[rU_{n+1} + (1-r)U_n \right] \frac{U_{n+1} - U_n}{\Delta x}$$

$$+ \frac{\mu}{\rho \Delta y^2} \left\{ r \left[(u_{n+1,j+1} - u_{n+1,j}) - (u_{n+1,j} - u_{n+1,j-1}) \right] \right. \qquad (12.4.26)$$

$$\left. + (1-r) \left[(u_{n,j+1} - u_{n,j}) - (u_{n+1,j} - u_{n+1,j-1}) \right] \right\}.$$

Notice the following:

1. $r = 0$: The method is explicit, and the error is first order in x and second in y. For stability,

$$\frac{2\mu \Delta x}{\rho u_{n,j} \Delta y^2} \le 1, \qquad \frac{\rho v_{n,j}^2 \Delta x}{\mu u_{n,j}} \le 2.$$

2. $r = 1/2$: This becomes the Crank-Nicholson method. The error is second order in both Δx and Δy. There is no stability constraint.

3. $r = 1$: The method now is fully implicit. The error now is first order in Δx and second order in Δy. There is no stability constraint.

12.4.6   Richardson Extrapolation

From the preceding it is seen that there is always a trade-off in stability, accuracy, and ease of solution. For the case of nonlinear differential equations, such as the boundary layer equations, the nonlinearity adds additional complexity. Sometimes, however, it is possible to make some gains in accuracy without necessarily sacrificing computational time and complexity.

Suppose that we have a nonlinear first order system of equations

$$\frac{dy}{dx} = f(x, y), \quad y(a) = y_0. \qquad (12.4.27)$$

Let u_j be a finite difference solution at point x_j; for example, let $u_{j+1} = u_j + hf(x_j, u_j)$. Further, let e_j be the error at point x_j; $e_j = u_j - y(x_j)$. If $e_0 = h\varepsilon(0)$, where $\varepsilon(0)$ does not depend on h, then $e_j = h\varepsilon(x_j) + O(h^2)$ where $\varepsilon(x_j)$ solves $\frac{d\varepsilon}{dx} = \varepsilon \frac{\partial f}{\partial y}(x, y(x)) - \frac{1}{2}\frac{d^2 y}{dx^2}$, $\varepsilon(a) = \varepsilon(0)$. Then $u_j(h) = y(x_j) + h\varepsilon(x_j) + O(h^2)$. If the problem is solved again with half the step size, then $u_j\left(\frac{1}{2}h\right) = y(x_j) + \frac{1}{2}h\varepsilon(x_j) + O(h^2)$. Subtracting these two results to eliminate the first-order term in h gives

$$\overline{u_j} \equiv 2u_j\left(\frac{1}{2}h\right) - u_j(h) = y(x_j) + O(h^2). \qquad (12.4.28)$$

Thus, by doing the numerical solution twice with different step sizes, the solution represented by equation (12.4.28) has an extra order of accuracy.

12.4.7   Further Choices for Dealing with Nonlinearities

What makes solution techniques in computational fluid mechanics unique is how they deal with the convective acceleration terms $u\frac{\partial u}{\partial x} + v\frac{\partial u}{\partial y}$. There are a number of different ways to do this.

1. *Lagging computation.* Evaluate u, v at their old position. Then

$$u\frac{\partial u}{\partial x} + v\frac{\partial u}{\partial y} \approx u_{n,j}\frac{u_{n+1,j} - u_{n,j}}{\Delta x} + v_{n,j}\frac{u_{n,j+1} - u_{n,j}}{\Delta y}. \tag{12.4.29}$$

   This is first order in the "marching coordinate" x and will cause trouble when x derivatives become large.

2. *Simple iterative update.* Use the lagging computation, and then repeat the computation using equation (12.4.29) but with the newly computed $u_{n+1,j}$. The process can be repeated several times.

3. *Newton linearization.* The convergence of a method is improved if the continuity and momentum equations are solved in a coupled manner.

4. *Extrapolation of coefficients.* When using $u_{n+1,j}$ compute it using

$$u_{n+1,j} = u_{n,j} + \frac{\partial u}{\partial x}(x_n, y_j)\Delta x, \text{ where}$$

$$\frac{\partial u}{\partial x}(x_n, y_j) = \frac{u_{n,j} - u_{n-1,j}}{\Delta x}. \tag{12.4.30}$$

12.4.8   Upwind Differencing for Convective Acceleration Terms

Perhaps the most popular method for improving calculation of convective terms is that of upwind differencing. Traditional methods for calculating $Lf = \frac{\partial f}{\partial t} + u\frac{\partial f}{\partial x}$ use

$$Lf \approx \frac{f_{n+1,j} - f_{n,j}}{\Delta t} + u\frac{f_{n,j+1} - f_{n,j}}{\Delta x}. \tag{12.4.31}$$

In upwind differencing, provided u is positive, instead use

$$Lf \approx \frac{f_{n+1,j} - f_{n,j}}{\Delta t} + u\frac{f_{n,j} - f_{n,j-1}}{\Delta x}. \tag{12.4.32}$$

Comparing these two equations shows that the only change has been in the x differencing, which is now in the upwind direction rather than the current (downwind) direction. The effect of this change is to introduce an artificial viscosity into the problem so that if we were trying to solve $Lf = 0$, using upwind differencing we would actually be solving

$$\frac{\partial f}{\partial t} = \frac{\partial(uf)}{\partial x} + \mu_{\text{artificial}}\frac{\partial^2 f}{\partial x_2} + \text{higher-order terms and differences.}$$

The artificial viscosity is given by $\mu_{\text{artificial}} = \frac{1}{2}u\left(\Delta x - u\Delta t\right)$. Providing this artificial viscosity is positive, the use of upwind differencing is necessary for stability of the computations.

12.5   Multistep, or Alternating Direction, Methods

12.5.1   Alternating Direction Explicit (ADE) Method

Consider the differential equation

$$\frac{\partial f}{\partial t} = \alpha \frac{\partial^2 f}{\partial x^2} = \alpha \frac{\partial}{\partial x}\left(\frac{\partial f}{\partial x}\right). \tag{12.5.1}$$

Since in the previous section we saw that it was advantageous at times to take derivatives in different ways, one suggestion for approximating equation (12.5.1) is

$$\frac{f_{n+1,j} - f_{n,j}}{\Delta t} = \alpha \frac{\left(f_{n,j+1} - f_{n,j}\right) - \left(f_{n+1,j} - f_{n+1,j-1}\right)}{\Delta x^2}. \tag{12.5.2}$$

Solving for updated values gives

$$f_{n+1,j} = \frac{1-\beta}{1+\beta} f_{n,j} + \frac{\beta}{1+\beta} f_{n,j+1} + \frac{\beta}{1+\beta} f_{n+1,j-1}. \tag{12.5.3}$$

If we are clever enough to sweep through our equations in the direction of increasing j, the last term on the right-hand side is known, and we have a simple explicit method.

The good news is that, for equation (12.5.1), this method is unconditionally stable, with errors of second order in both Δt and Δx. The bad news is that, if convection terms are added to equation (12.5.1), the method may be unconditionally unstable.

12.5.2   Alternating Direction Implicit (ADI) Method

As our starting point we will be more ambitious than in the previous section, dealing with one time dimension and two space directions. For our model equation let

$$\frac{\partial f}{\partial t} = -u \frac{\partial f}{\partial x} - v \frac{\partial f}{\partial y} + \alpha\left(\frac{\partial^2 f}{\partial x^2} + \frac{\partial^2 f}{\partial y^2}\right). \tag{12.5.4}$$

To do our time differencing proceed in two steps:

$$\text{Step1.}\ \frac{f_{n+1/2} - f_n}{\Delta t/2} = -u \frac{\partial f_{n+1/2}}{\partial x} - v \frac{\partial f_n}{\partial y} + \alpha\left(\frac{\partial^2 f_{n+1/2}}{\partial x^2} + \frac{\partial^2 f_n}{\partial y^2}\right), \tag{12.5.5}$$

$$\text{Step2:}\ \frac{f_{n+1} - f_n}{\Delta t/2} = -u \frac{\partial f_{n+1/2}}{\partial x} - v \frac{\partial f_{n+1}}{\partial y} + \alpha\left(\frac{\partial^2 f_{n+1/2}}{\partial x^2} + \frac{\partial^2 f_{n+1}}{\partial y^2}\right). \tag{12.5.6}$$

Notice that although each of the two steps are implicit, the resulting equations are tridiagonal and thus easy to solve. The error is second order in each of the three incremental deltas.

Formally, this method has unconditional stability. However, it can become unstable if too big a time step is used. This is especially true if the boundary conditions are allowed to lag in time. See Roache (1976, page 92).

In equations (12.5.5) and (12.5.6) details on the x and y differencing have been omitted. One might use

$$\frac{\partial^2 f_{n,i,j}}{\partial x^2} \approx \frac{f_{n,i+1,j} - 2f_{n,i,j} + f_{n,i-1,j}}{\Delta x^2}\ \text{and}$$

$$\frac{\partial^2 f_{n,i,j}}{\partial y^2} \approx \frac{f_{n,i,j+1} - 2f_{n,i,j} + f_{n,i,j-1}}{\Delta y^2}.$$

Further, to keep second-order accuracy, use $u_{n+1/2}$, v_n in equation (12.5.5) and $u_{n+1/2}$, v_{n+1} in equation (12.5.6).

To use this method to solve the full Navier-Stokes equations would require an implicit coupled solution of an equation like (12.5.4) in the stream function and another one in vorticity, both solutions to be carried out at the two times $n+1/2$ and $n+1$. This would be a formidable task! These two methods, wherein the differencing of space derivatives is split into two time steps, are the first of many splitting methods which have been introduced.

Hyperbolic Partial Differential Equations

12.6 Method of Characteristics

Early in this chapter, we briefly examined the three basic classes of partial differential equations and saw examples. We will now examine them more generally and derive a method suitable for solving hyperbolic differential equations.

The forms of the three equations given in equation (12.1.1) can in two dimensions be generalized by the form

$$A\frac{\partial^2 f}{\partial x^2}+2B\frac{\partial^2 f}{\partial x\partial y}+C\frac{\partial^2 f}{\partial y^2}=E. \tag{12.6.1}$$

Here A, B, C, and E are real and may depend on f and its first derivatives. Coordinates x and y are not necessarily space coordinates; one could also be a time dimension. Next, ask the question, When and where are infinitesimal variations of the first derivatives of f allowed, or alternatively, Where are first derivatives of f continuous? This question can be phrased in terms of the solution of three equations in three unknowns—that is,

$$A\frac{\partial^2 f}{\partial x^2}+2B\frac{\partial^2 f}{\partial x\partial y}+C\frac{\partial^2 f}{\partial y^2}=E,$$

$$dx\frac{\partial^2 f}{\partial x^2}+dy\frac{\partial^2 f}{\partial x\partial y}=d\left(\frac{\partial f}{\partial x}\right), \tag{12.6.2}$$

$$dx\frac{\partial^2 f}{\partial x\partial y}+dy\frac{\partial^2 f}{\partial y^2}=d\left(\frac{\partial f}{\partial y}\right),$$

where the second and third lines represent the differentials of the first derivatives.

Using Cramer's rule, this system is easily solved, giving

$$\frac{\partial^2 f}{\partial x^2}=\frac{N_1}{D},\quad \frac{\partial^2 f}{\partial x\partial y}=\frac{N_2}{D},\quad \frac{\partial^2 f}{\partial y^2}=\frac{N_3}{D}, \tag{12.6.3}$$

with

$$D=\begin{vmatrix} A & 2B & C \\ dx & dy & 0 \\ 0 & dx & dy \end{vmatrix},\quad N_1=\begin{vmatrix} E & 2B & C \\ d\left(\frac{\partial f}{\partial x}\right) & dy & 0 \\ d\left(\frac{\partial f}{\partial y}\right) & dx & dy \end{vmatrix},$$

$$N_2=\begin{vmatrix} A & E & C \\ dx & d\left(\frac{\partial f}{\partial x}\right) & 0 \\ 0 & d\left(\frac{\partial f}{\partial y}\right) & dy \end{vmatrix},\quad N_3=\begin{vmatrix} A & 2B & E \\ dx & dy & d\left(\frac{\partial f}{\partial x}\right) \\ 0 & dx & d\left(\frac{\partial f}{\partial y}\right) \end{vmatrix}. \tag{12.6.4}$$

By the usual restriction of Cramer's rule, this is an acceptable solution except where the denominator D vanishes. Expanding this determinant, we find it vanishes where $Ady^2 + Cdx^2 - 2Bdxdy = 0$; that is, along the lines of slope

$$\frac{dy}{dx} = \frac{B \pm \sqrt{B^2 - AC}}{A}. \tag{12.6.5}$$

This pair of lines are called the ***characteristic lines***, or just ***characteristics***, of equation (12.6.1). Notice the following:

1. If $B^2 - AC < 0$, both of the characteristics are imaginary and play no decisive role in the solution. This is the general definition of ***elliptic partial differential equations***.

2. If $B^2 - AC = 0$, there is only one real characteristic. This is the general definition of ***parabolic partial differential equations***.

3. If $B^2 - AC > 0$, both of the characteristics are real. This is the general definition of ***hyperbolic partial differential equations***. We will concentrate further attention on this latter case.

If a region covered by characteristics is to have a solution, it is necessary that the numerators in equation (12.6.3) also vanish, so l'Hospital's rule can be used. Expanding all three of the determinants, we find with the help of equation (12.6.5) that

$$N_1 = \left(\frac{dy}{dx}\right)^2 N_3 \quad \text{and} \quad N_2 = -\frac{dy}{dx} N_3, \tag{12.6.6}$$

so all three numerators are zero on the characteristics provided $N_3 = 0$, which occurs when

$$d\left(\frac{\partial f}{\partial y}\right) = -\frac{A}{C}\frac{dy}{dx}d\left(\frac{\partial f}{\partial x}\right) + \frac{E}{C}dy. \tag{12.6.7}$$

With the aid of equations (12.6.5) and (12.6.7), a solution can now be constructed.

Suppose in Figure 12.6.1 that the derivatives of the function f is known on curve 1-2. Using equation (12.6.5), construct a series of characteristics on a closely spaced number of points on this curve. Label the characteristics $+$ if the plus sign in front of the square root was used, and $-$ otherwise. Notice that in the roughly triangular region 1-2-3 the $+$ and $-$ characteristics intersect, whereas outside this region there are no intersections.

Next, magnify the small region a-b-c in Figure 12.6.1 as in Figure 12.6.2. Point c is at the position (x_c, y_c), determined by solving

$$y_c - y_a = \left(\frac{B + \sqrt{B^2 - AC}}{A}\right)_a (x_c - x_a), \quad y_c - y_b = \left(\frac{B - \sqrt{B^2 - AC}}{A}\right)_b (x_c - x_b).$$
$$\tag{12.6.8}$$

Similarly, from equation (12.6.7) write

$$\frac{\partial f_c}{\partial y} - \frac{\partial f_a}{\partial y} = -\left(\frac{B + \sqrt{B^2 - AC}}{C}\right)_a \left(\frac{\partial f_c}{\partial x} - \frac{\partial f_a}{\partial x}\right) + \left(\frac{E}{C}\right)_a (y_c - y_a),$$
$$\tag{12.6.9}$$
$$\frac{\partial f_c}{\partial y} - \frac{\partial f_b}{\partial y} = -\left(\frac{B - \sqrt{B^2 - AC}}{C}\right)_b \left(\frac{\partial f_c}{\partial x} - \frac{\partial f_b}{\partial x}\right) + \left(\frac{E}{C}\right)_b (y_c - y_b),$$

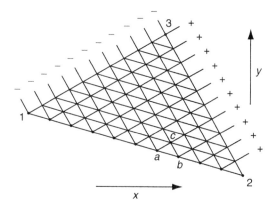

Figure 12.6.1 Method of characteristics—region of solubility

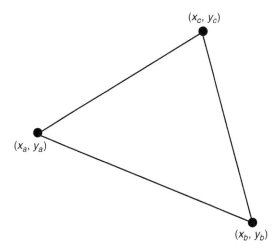

Figure 12.6.2 Method of characteristics—detail of two intersecting characteristic lines

thereby solving for the derivatives of f at c. This process can be continued repeatedly throughout the region 1-2-3, obtaining all derivatives of the function within this region. Outside this region, no solution is possible due to insufficient data.

Note that if there is a rigid boundary (wall) or a line of symmetry, then a point c can land on a wall and there is no $-$ characteristic to meet it. The wall boundary condition replaces this, however, so a solution is still possible.

In inviscid compressible isentropic flow, an equation similar to (12.6.1) governs the flow with f the velocity potential, and

$$A = a^2 - u^2, \quad B = -uv, \quad C = a^2 - v^2, \quad a^2 = a_0^2 - \frac{k-1}{2}\left(u^2 + v^2\right), \quad (12.6.10)$$

u and v being the x and y velocity components (derivatives of "f") and a the speed of sound. Where the flow is supersonic, the equations become hyperbolic.

The method of characteristics predates computers by almost a century. If one is analyzing or designing flow in a nozzle, it is a problem that can easily be solved with a pencil, eraser, and adding machine. With modern computers, however, programs using this methodology become somewhat complicated, since even if the original data is given on a straight line, the subsequent line will more than likely be curved. Often interpolation is used to keep a straight grid in the region of solution. Characteristics can be lines of discontinuity of derivatives of f, but not of f itself.

Even first-order differential equations can have characteristics. If the same procedure is pursued on the equation

$$\frac{\partial u}{\partial t} + u \frac{\partial u}{\partial x} = f(u, x, t), \qquad (12.6.11)$$

for example, the result is

$$\frac{\partial u}{\partial t} = \frac{f\,dx - u\,du}{dx - u\,dt}, \quad \frac{\partial u}{\partial x} = \frac{du - f\,dt}{dx - u\,dt}. \qquad (12.6.12)$$

The equation of the characteristic is $dx = u\,dt$, and along the characteristic $u\,du = f\,dx$.

As an example of the use of the method of characteristics, the equations for analysis of a water hammer are

$$g \frac{\partial H}{\partial x} + V \frac{\partial V}{\partial x} + \frac{\partial V}{\partial t} + \frac{f}{2D} V\,|V| = 0,$$

$$\frac{\partial}{\partial x}(\rho A V) + \frac{\partial}{\partial t}(\rho A) = 0, \quad \text{with} \qquad (12.6.13)$$

$$gH = \frac{p}{\rho} + gz, \quad \frac{\partial \rho}{\partial t} = \frac{\rho}{K} \frac{\partial p}{\partial t}, \quad K = \text{bulk modulus}.$$

The method of characteristics has been used for this problem with much success.

12.7 Leapfrog Method—Explicit

For the one-dimensional wave equation

$$\frac{\partial^2 f}{\partial t^2} = a^2 \frac{\partial^2 f}{\partial x^2}, \qquad (12.7.1)$$

a method that gives interesting results is the leapfrog method. If centered differences are used for the two second derivatives, equation (12.7.1) becomes

$$\frac{f_{n+1,j} - 2f_{n,j} + f_{n-1,j}}{\Delta t} = a^2 \frac{f_{n,j+1} - 2f_{n,j} + f_{n,j-1}}{\Delta x}. \qquad (12.7.2)$$

Solving for f at the latest time step gives

$$f_{n+1,j} = 2\left(1 - A^2\right) f_{n,j} - f_{n-1,j} + A^2\left(f_{n,j+1} + f_{n,j-1}\right), \qquad (12.7.3)$$

where $A = a\Delta t/\Delta x$ is the Courant number. Notice that when $A = 1$, the value at the same position but preceding time drops out of the calculation.

The exact solution to equation (12.7.1) is known to be

$$f(x, t) = F(x - at) + G(x + at), \qquad (12.7.4)$$

where F and G depend on initial conditions according to

$$f(x, 0) = F(x) + G(x), \quad \frac{\partial f}{\partial t}(x, 0) = a\left[-F'(0) + G'(0)\right]. \qquad (12.7.5)$$

Taking (x_1, t_1) as a starting point and letting $x_i = x_1 + (i-1)\Delta x$, $t_n = t_1 + (n-1)\Delta t$, then

$$x - at = \alpha + i\Delta x - na\Delta t, \quad \alpha = (x_1 - \Delta x) - a(t_1 - \Delta t), \text{ and}$$
$$x + at = \beta + i\Delta x + na\Delta t, \quad \beta = (x_1 - \Delta x) + a(t_1 - \Delta t). \tag{12.7.6}$$

Inserting these into equation (12.7.3) with $A = 1$ find that

$$
\begin{aligned}
-f_{n-1,j} + f_{n,j+1} + f_{n,j-1} &= F[\alpha + (i - j - 1)\Delta x] + G[\beta + (i + j - 1)\Delta x] \\
&\quad + F[\alpha + (i - j + 1)\Delta x] + G[\beta + (i + j + 1)\Delta x] \\
&\quad - F[\alpha + (i - j - 1)\Delta x] - G[\beta + (i + j - 1)\Delta x] \quad (12.7.7) \\
&= F[\alpha + (i - j - 1)\Delta x] + G[\beta + (i + j + 1)\Delta x] \\
&= f_{n+1,j}.
\end{aligned}
$$

Thus, for a Courant number of unity, the method gives the exact solution.

The reason for this happy result can be found by comparing the calculation procedure with the characteristics of equation (12.7.1), which are given by $\frac{dx}{dt} = \pm a$. The slopes of the lines connecting $x_{n+1,j}$ with $x_{n,j-1}$ and $x_{n,j+1}$ are $\frac{\Delta x}{\Delta t} = \pm a/A$, so when the Courant number is unity, the leapfrog method corresponds to the method of characteristics. If we were to choose $A < 1$, we are within the triangle of the characteristic lines, and our computation is using valid information and the result is stable. If, however, we were to choose $A > 1$, we would be using information that is known not to be pertinent to the calculation, and the calculation becomes unstable.

One drawback of the leapfrog method is that to start the calculations, there is a need to know the solution at t_0. To handle this, place a fictitious row at t_0 and let

$$f_{1,j} = F(x_1 - at_1) + G(x_1 + at_1) \text{ and}$$
$$f_{0,j} = f_{2,j} - 2a\left[-F'(x_1 - at_1) + G'(x_1 + at_1)\right]\Delta t. \tag{12.7.8}$$

Then

$$
\begin{aligned}
f_{2,j} &= f_{1,j-1} + f_{1,j+1} - f_{0,j} = f_{1,j-1} + f_{1,j+1} - f_{2,j} \\
&\quad + 2a\left[-F'(x_1 - at_1) + G'(x_1 + at_1)\right]\Delta t \text{ or} \tag{12.7.9} \\
f_{2,j} &= \frac{1}{2}\left(f_{1,j-1} + f_{1,j+1}\right) + a\left[-F'(x_1 - at_1) + G'(x_1 + at_1)\right]\Delta t.
\end{aligned}
$$

This is the starting formula for the method.

12.8 Lax-Wendroff Method—Explicit

Many of the equations of fluid mechanics are of the type

$$\frac{\partial \mathbf{U}}{\partial t} + \frac{\partial \mathbf{F}}{\partial x} = 0, \tag{12.8.1}$$

for instance, the one-dimensional compressible flow equation where

$$\mathbf{U} = \begin{pmatrix} \rho \\ u \end{pmatrix}, \quad \mathbf{F} = \begin{pmatrix} \rho u \\ \dfrac{u^2}{2} + \dfrac{a^2}{k-1} \end{pmatrix}. \tag{12.8.2}$$

The first of these equations is the continuity equation, the second the momentum equation. The form of equation (12.8.1) is often referred to as a *conservative form*. The Navier-Stokes equations and the inviscid compressible flow equations are examples that fit this general form.

For the Lax-Wendroff method, let

$$\frac{\partial \mathbf{F}}{\partial t} = \frac{\partial \mathbf{F}}{\partial \mathbf{U}} \frac{\partial \mathbf{U}}{\partial t} = \mathbf{A} \frac{\partial \mathbf{F}}{\partial x}. \tag{12.8.3}$$

Differentiating equation (12.8.1) with respect to time yields

$$\frac{\partial^2 \mathbf{U}}{\partial t^2} = -\frac{\partial}{\partial t}\left(\frac{\partial \mathbf{F}}{\partial x}\right) = -\frac{\partial}{\partial x}\left(\frac{\partial \mathbf{F}}{\partial t}\right) = -\frac{\partial}{\partial x}\left(\mathbf{A}\frac{\partial \mathbf{F}}{\partial x}\right). \tag{12.8.4}$$

From equation (12.8.2) find that

$$\mathbf{A} = \begin{pmatrix} \dfrac{\partial F_1}{\partial U_1} & \dfrac{\partial F_1}{\partial U_2} \\ \dfrac{\partial F_2}{\partial U_1} & \dfrac{\partial F_2}{\partial U_2} \end{pmatrix} = \begin{pmatrix} u & \rho \\ \dfrac{2a}{k-1}\dfrac{da}{d\rho} & u \end{pmatrix} \tag{12.8.5}$$

Then

$$\mathbf{U}_{n+1,j} \approx \mathbf{U}_{n,j} + \Delta t \left(\frac{\partial \mathbf{U}}{\partial t}\right)_{n,j} + \frac{\Delta t^2}{2}\left(\frac{\partial^2 \mathbf{U}}{\partial t^2}\right)_{n,j}$$

$$\approx \mathbf{U}_{n,j} + \Delta t \left(-\frac{\partial \mathbf{F}}{\partial x}\right)_{n,j} + \frac{\Delta t^2}{2}\frac{\partial}{\partial x}\left(\mathbf{A}\frac{\partial \mathbf{F}}{\partial x}\right)_{n,j}$$

$$\approx \mathbf{U}_{n,j} - \frac{\Delta t}{2\Delta x}\left(\mathbf{F}_{n,j+1} - \mathbf{F}_{n,j-1}\right) + \frac{\Delta t^2}{2\Delta x}\frac{\partial}{\partial x}\left(\mathbf{A}_{n,j+1/2}\frac{\partial \mathbf{F}_{n,j+1/2}}{\partial x} - \mathbf{A}_{n,j-1/2}\frac{\partial \mathbf{F}_{n,j-1/2}}{\partial x}\right)$$

$$\tag{12.8.6}$$

$$\approx \mathbf{U}_{n,j} - \frac{\Delta t}{2\Delta x}\left(\mathbf{F}_{n,j+1} - \mathbf{F}_{n,j-1}\right)$$

$$+ \frac{\Delta t^2}{4\Delta x^2}\left[\left(\mathbf{A}_{n,j+1} + \mathbf{A}_{n,j}\right)\left(\mathbf{F}_{n,j+1} - \mathbf{F}_{n,j}\right) - \left(\mathbf{A}_{n,j} + \mathbf{A}_{n,j-1}\right)\left(\mathbf{F}_{n,j} - \mathbf{F}_{n,j-1}\right)\right]$$

This method is explicit and stable providing $0 \le \frac{\Delta t}{\Delta x} \le 1$ with an error of order $\frac{1}{6}\Delta t^2 \frac{\partial^3 \mathbf{U}}{\partial t^3} + \frac{1}{6}\Delta x^2 \frac{\partial^3 \mathbf{U}}{\partial x^3}$. It is usually easy to program, although nonlinear problems require some adjustment to the method. More details can be found in Lax (1954), Smith (1978), and Ferziger (1981).

12.9   MacCormack's Methods

Over the years, R. W. MacCormack has investigated a series of approaches to the conservative form (equation (12.8.1)), which have proven to be useful for computations. Some of these methods are outlined here.

12.9.1   MacCormack's Explicit Method

This method uses a ***predictor-corrector*** approach, where an estimate of one of the variables is first made and then updated in a later calculation. For equation (12.8.1) MacCormack used

$$\text{Predictors: } \overline{U}_{n+1,j} = U_{n,j} - \left(F_{n,j+1} - F_{n,j}\right)\Delta t/\Delta x, \quad \overline{F}_{n+1,j} = F\left(\overline{U}_{n+1,j}\right),$$

$$\text{Corrector: } U_{n+1,j} = \frac{1}{2}\left[U_{n,j} + \overline{U}_{n+1,j} - \left(\overline{F}_{n+1,j} - \overline{F}_{n+1,j-1}\right)\Delta t/\Delta x\right], \tag{12.9.1}$$

the superposed bars denoting the predicted quantities.

Examples of the application of these include the following:

1. The one-dimensional wave equation

For $\frac{\partial f}{\partial t} + a\frac{\partial f}{\partial x} = 0$, application of MacCormack's explicit method (1969) gives

$$\text{Predictor: } \overline{f}_{n+1,j} = f_{n,j} - a\left(f_{n,j+1} - f_{n,j}\right)\Delta t/\Delta x,$$

$$\text{Corrector: } f_{n+1,j} = \frac{1}{2}\left[f_{n,j} + \overline{f}_{n+1,j} - a\left(\overline{f}_{n+1,j} - \overline{f}_{n+1,j-1}\right)\right]. \tag{12.9.2}$$

2. Burger's equation

For $\frac{\partial f}{\partial t} + \frac{\partial F}{\partial x} = \mu\frac{\partial^2 f}{\partial x^2}$, $F = af + \frac{1}{2}bf^2$, application of MacCormack's explicit method gives

$$\text{Predictors: } \overline{f}_{n+1,j} = f_{n,j} - \left(F_{n,j+1} - F_{n,j}\right)\Delta t/\Delta x + r\left(f_{n,j+1} - 2f_{n,j} + f_{n,j-1}\right),$$

$$\overline{F}_{n+1,j} = a\overline{f}_{n+1,j} + \frac{1}{2}b\overline{f}_{n+1,j}^2,$$

$$\text{Corrector: } f_{n+1,j} = \frac{1}{2}\left[f_{n,j} + \overline{f}_{n+1,j} - \left(\overline{F}_{n+1,j} - \overline{F}_{n+1,j-1}\right)\Delta t/\Delta x \right. \tag{12.9.3}$$

$$\left. + r\left(\overline{f}_{n+1,j+1} - 2\overline{f}_{n+1,j} + \overline{f}_{n+1,j-1}\right)\right].$$

where $r = \mu\Delta t/(\Delta x)^2$.

12.9.2   MacCormack's Implicit Method

In 1981 MacCormack suggested that for Burger's equation the predictor-corrector procedure could be varied according to

$$\text{Predictor: } \left(1 + \lambda\Delta t/\Delta x\right)\Delta\overline{f}_{n+1,j} = \left(\Delta f_{n,j}\right)_{\text{explicit}} + \frac{\lambda\Delta t}{\Delta x}\Delta\overline{f}_{n+1,j+1}, \text{ where}$$

$$\Delta\overline{f}_{n+1,j} = \overline{f}_{n+1,j} - f_{n,j},$$

$$\left(\Delta f_{n,j}\right)_{\text{explicit}} = -a\left(f_{n,j+1} - f_{n,j}\right)\Delta t/\Delta x + r\left(f_{n,j+1} - 2f_{n,j} + f_{n,j-1}\right).$$

$$\text{Corrector: } \left(1 + \lambda\Delta t/\Delta x\right)\Delta f_{n+1,j} = \left(\Delta\overline{f}_{n,j}\right)_{\text{explicit}} + \frac{\lambda\Delta t}{\Delta x}\Delta f_{n+1,i-1}, \text{ where}$$

$$f_{n+1,j} = \frac{1}{2}\left(f_{n,j} + \overline{f}_{n+1,j} + \Delta f_{n+1,j}\right), \tag{12.9.4}$$

$$\left(\Delta\overline{f}_{n,j}\right)_{\text{explicit}} = -a\left(\overline{f}_{n+1,j} - \overline{f}_{n+1,j-1}\right)\Delta t/\Delta x + r\left(\overline{f}_{n+1,j+1} - 2\overline{f}_{n+1,j} + \overline{f}_{n+1,j-1}\right).$$

Here, λ is a constant such that $\lambda \geq \max[a + 2\mu/\Delta x - \Delta x/\Delta t, 0]$. This method is unconditionally stable and second-order accurate provided $\mu \Delta t/(\Delta x)^2$ is bounded as Δt and Δx approach zero.

12.10   Discrete Vortex Methods (DVM)

The methods previously discussed all start with an equation and then use finite differences or a similar procedure to transfer the mathematics from calculus to algebra. They deal with flows that are well defined both physically and mathematically throughout a region that is generally fixed. In flows that are separated, however, such as in the wake of a bluff body, a wing, or a propellor, the portion of the flow that is of greatest interest is in the vortices that are shed and find their way downstream. The Kármán vortex street is an example of this.

Discrete vortex methods try to model, or simulate, these effects much in the manner of the Kármán vortex street. Vorticity is generated at a boundary, often in the boundary layer. At the separation point the vorticity tends to leave the boundary and move into a flow region that is more or less inviscid. There the vorticity will move according to $\frac{D\boldsymbol{\omega}}{Dt} \approx 0$. The position of this bit of vorticity will change according to $\frac{d\mathbf{r}}{dt} = \mathbf{v}$, where the velocity is a combination of the inviscid flow plus the induced velocity from all previous bits of shed vorticity.

To make this description more specific, consider a two-dimensional flow in the wake of a cylinder. The boundary layer flow is solved and the separation point determined by some separation criteria such as Stratford's. To get the vortex out of the boundary layer, a random walk procedure can be used. Chorin (1978) suggested that, to avoid the singularities in the vortexes such as discussed in the chapters on inviscid flows, "**vortex blobs**" could be introduced, with stream functions of the form

$$\psi = \begin{cases} \dfrac{\Gamma}{2\pi} \dfrac{|\mathbf{r}|}{\delta} \text{for } |\mathbf{r}| < \delta, \\[2mm] \dfrac{\Gamma}{2\pi} \ln|\mathbf{r}| \text{ for } |\mathbf{r}| \geq \delta. \end{cases} \tag{12.10.1}$$

This gives constant velocity inside the vortex. The choice of the parameter δ is left up to the user. As the blobs are released from the boundary layer, they are moved according to the induced velocities.

There are some matters that remain, such as dealing with combining of vortices if or when they collide, what to do when they leave the computational region, when they reenter the boundary layer, and so forth. These can be dealt with in many ways that fit the specific situation at hand and that deal with practical matters, such as the capacity of the computer and time available. For instance, instead of the description of the vortex blob just given, the formulation $v = \frac{\Gamma}{2\pi(z-z_0)}F$, where F is a vorticity modification function such as $F = \frac{99(r/\delta)^n}{1+99(r/\delta)^n}$, n being an integer greater than or equal to 2. This makes the self-induced velocity zero at the center of the blob and still has the velocity behave as a line vortex away from the flow. More possibilities concerning the implementation of this method can be found in Alexandrou (1986) and Hong (1988).

12.11 Cloud in Cell Method (CIC)

This method is a variation on the discrete vortex method and was first introduced by Christiansen (1973). Instead of tracking individual vortices as in DVM, a grid is placed on the region, and as a vortex moves within a sector of the grid, its vorticity is distributed to the corners of the sector. (See Figure 12.11.1.) For a sector Δx by Δy, if the vortex is at a location $(\delta x, \delta y)$ with respect to the corner of the sector, the vorticity ω might be distributed to the corners according to

$$\omega_{i,j} = \left(1 - \frac{\delta x}{\Delta x}\right)\left(1 - \frac{\delta y}{\Delta y}\right)\omega = A_1\omega,$$

$$\omega_{i,j+1} = \left(1 - \frac{\delta x}{\Delta x}\right)\frac{\delta y}{\Delta y}\omega = A_2\omega,$$

$$\omega_{i+1,j} = \frac{\delta x}{\Delta x}\left(1 - \frac{\delta y}{\Delta y}\right)\omega = A_3\omega, \tag{12.11.1}$$

$$\omega_{i+1,j+1} = \frac{\delta x}{\Delta x}\frac{\delta y}{\Delta y}\omega = A_4\omega.$$

After this distribution, the stream function can be solved from

$$\nabla^2\psi = \omega. \tag{12.11.2}$$

The induced velocities and positions for finding how fast the vortex travels are then found from

$$\frac{d}{dt}\begin{pmatrix}x_n\\y_n\end{pmatrix} = \begin{pmatrix}u_n\\v_n\end{pmatrix} = A_1\begin{pmatrix}u\\v\end{pmatrix}_{i,j} + A_2\begin{pmatrix}u\\v\end{pmatrix}_{i+1,j} + A_3\begin{pmatrix}u\\v\end{pmatrix}_{i,j+1} + A_4\begin{pmatrix}u\\v\end{pmatrix}_{i+1,j+1}. \tag{12.11.3}$$

For 1,000 vortices the CIC method appears to be about 20 times faster than the DVM method. More references to this method and its implementation can be found in Roberts and Christiansen (1972), Milinazzo and Saffman (1977), Chorin (1978), Alexandrou (1986), and Hong (1988).

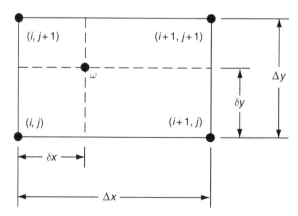

Figure 12.11.1 Computational molecule for the cloud-in-cell method

Problems—Chapter 12

In the following problems use either the program languages specified in the text or a spreadsheet.

12.1 Use simple relaxation (equation (12.2.1a)) to find the values on the center line of the elbow. On the inner boundary of the el (filled dots) the value is 0, on the outer boundary it is 1, and the end points 1 and 10 average the two boundary values. Do at least 10 iterations.

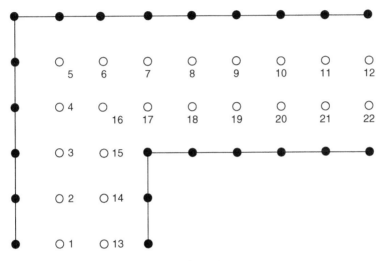

P12.1 Flow in an elbow

12.2 Repeat problem 12.1, this time using successive overrelaxation (SOR). Take the relaxation parameter to be 1.7.

12.3 Use the leapfrog method to solve the one-dimensional wave equation $\frac{\partial^2 f}{\partial x^2} = \frac{1}{c^2}\frac{\partial^2 f}{\partial t^2}$ subject to the conditions

$$f(x, 0) = \sin \pi x, \quad \frac{\partial f}{\partial t}(x, 0) = 0, \quad 0 \le x \le 1; \quad f(0, t) = f(1, t) = 0.$$

To start the solution, introduce a row of fictitious grid points at $t = -\Delta t$, where

$$f(x, -\Delta t) = f(x, 0) - 2\Delta t \frac{\partial f}{\partial t}(x, 0).$$

Let $\Delta x = 0.1$ and $\Delta t = \Delta x/c$. The wave speed can be taken as unity. Compute for at least 20 time steps.

12.4 An explicit method for the diffusion equation $\partial f/\partial t = \partial^2 f/\partial x^2$ is given by $f(i, j + 1) = f(i, j) + \lambda (f(i - 1, j) - 2f(i, j) + f(i + 1, j))$, where $\lambda = \Delta t/\Delta x^2 \le 0.5$. Solve this for the conditions

$$f(0, t) = 0, \quad f(1, t) = 1, \quad f(x, 0) = \begin{cases} 4x, & 0 \le x \le 0.4, \\ -x + 2, & 0.4 \le x \le 1. \end{cases}$$

Take $\lambda = 0.25$ and $\Delta x = 0.1$.

12.5 An implicit formula for solving the unsteady flow in a channel, given by the parabolic equation $\rho\frac{\partial u}{\partial t} = -\frac{\partial p}{\partial x} + \mu\frac{\partial^2 u}{\partial y^2}$, is $u(i,j) - u(i,j-1) = -\Delta t\frac{\partial p}{\partial x} + \lambda\left(u(i-1,j) - 2u(i,j) + u(i+1,j)\right)$, where λ is defined by $\lambda = \mu\Delta t/\rho\Delta x^2$. This scheme is apparently stable for all positive values of λ. Solve for flow starting from rest with

$$u(0) = u(1) = 0, \quad \frac{\partial p}{\partial x} = \begin{cases} 0, & t \leq 0, \\ -1, & t > 0 \end{cases}.$$

Take $\lambda = 0.5$, $\Delta x = 0.1$, $\Delta t = 0.1$, and do 200 time steps. Compare with the exact solution (parabolic profile).

12.6 Repeat the previous problem, this time using the Crank-Nicholson method, with the more accurate finite difference equation in the form

$$u(i,j) - u(i,j-1) = -\frac{\Delta t}{\rho}\frac{\partial p}{\partial x} + \frac{1}{2}\lambda\left(u(i-1,j) - 2u(i,j) + u(i+1,j)\right)$$

$$+ \frac{1}{2}\lambda\left(u(i-1,j-1) - 2u(i,j-1) + u(i+1,j-1)\right).$$

12.7 Repeat problem 12.5, this time using the DuFort-Frankel method for solving the problem. This method is explicit and uses $\frac{u(i-1,j)-u(i,j-1)-u(i,j+1)+u(i+1,j)}{\Delta x^2}$ for the second derivative in x and $\frac{u(i,j+1)-u(i,j-1)}{2\Delta t}$ for the time derivative.

    a. Put these approximations into the Navier-Stokes equation and rearrange to obtain a form suited to solving for the velocity at the grid points.

    b. Draw the computational molecule for this method. Indicate round dots where space derivatives are taken and squares where time derivatives are taken.

12.8 For the diffusion equation in two space derivatives and one time derivative the alternating direction implicit method (ADI) is unconditionally stable. Starting with $\frac{\partial f}{\partial t} = \frac{\partial^2 f}{\partial x^2} + \frac{\partial^2 f}{\partial y^2}$, the method uses the form

$$\frac{f^*(i,j) - f(i,j,k)}{\Delta t/2} = \frac{f^*(i+1,j) - 2f^*(i,j) + f^*(i-1,j)}{\Delta x^2}$$

$$+ \frac{f(i,j+1,k) - 2f(i,j,k) + f(i,j-1,k)}{\Delta y^2}$$

and follows it with

$$\frac{f(i,j,k+1) - f^*(i,j)}{\Delta t/2} = \frac{f^*(i+1,j) - 2f^*(i,j) + f^*(i-1,j)}{\Delta x^2}$$

$$+ \frac{f(i,j+1,k+1) - 2f(i,j,k+1) + f(i,j-1,k+1)}{\Delta y^2}.$$

The first solves for the intermediate values f^*, the second completing the solution for f at the next time step. Rearrange the equations to put them into a form suitable for programming.

12.9 One way of dealing with the nonlinearities of the Navier-Stokes equations is to treat steady-state flows as transient flows starting from a quiescent state. For natural convection on a vertical semi-infinite plate, start with the boundary layer equations in the nondimensional form

$$\frac{\partial u}{\partial t} + u\frac{\partial u}{\partial x} + v\frac{\partial u}{\partial y} = T + \frac{\partial^2 u}{\partial y^2}, \qquad \frac{\partial T}{\partial t} + u\frac{\partial T}{\partial x} + v\frac{\partial T}{\partial y} = \text{Pr}\,\frac{\partial^2 T}{\partial y^2},$$

$$\frac{\partial u}{\partial x} + \frac{\partial v}{\partial y} = 0,$$

and write them in an explicit finite difference form. Propose a scheme for solution.

12.10 For the linear wave equation $\frac{\partial^2 u}{\partial x^2} = \frac{1}{a^2}\frac{\partial^2 u}{\partial t^2}$ use the method of characteristics to solve for a disturbance moving from left to right and striking a wall at $x = 0$. The x interval is in the range $-0.20 \le x \le 0$. The disturbance is initially zero except in the interval $-0.09 \le x \le -0.05$ and $a = 343$ meters/sec. Continue the calculation until the disturbance moves out of the domain.

Appendix

> *I am very well acquainted, too, with matters mathematical,*
> *I understand equations, both the simple and quadratical:*
> *About binomial theorem I'm teeming with a lot of news,*
> *With many cheerful facts about the square of the hypotenuse.*
>
> William S. Gilbert

A.1 Vector Differential Calculus

Derivatives of a vector function of the space coordinates can be performed in any combination of the coordinate directions. A useful differentiation operator is the del operator, denoted by the symbol ∇, or del, defined in Cartesian coordinates as

$$\nabla = \mathrm{del} = \mathbf{i}\frac{\partial}{\partial x} + \mathbf{j}\frac{\partial}{\partial y} + \mathbf{k}\frac{\partial}{\partial z}. \tag{A.1.1}$$

The operator often appears in partial differential equations in its "squared" form

$$\nabla^2 = \nabla \cdot \nabla = \mathrm{Laplacian} = \mathrm{harmonic\ operator} = \frac{\partial^2}{\partial x^2} + \frac{\partial^2}{\partial y^2} + \frac{\partial^2}{\partial z^2}. \tag{A.1.2}$$

To illustrate some of the uses of the del operator, consider a unit vector **a** with direction cosines a_x, a_y, a_z. Then, the operation $\mathbf{a} \cdot \nabla$ acting on a vector **F** gives the derivative of **F** in the direction of **a**, or

$$(\mathbf{a} \cdot \nabla)\mathbf{F} = a_x \frac{\partial \mathbf{F}}{\partial x} + a_y \frac{\partial \mathbf{F}}{\partial y} + a_z \frac{\partial \mathbf{F}}{\partial z}.$$

The operator $\mathbf{v} \cdot \nabla$ is encountered often in fluid mechanics, particularly where **v** is the velocity vector. The operator itself is not a vector, since $\mathbf{v} \cdot \nabla \neq \nabla \cdot \mathbf{v}$, and is referred to as a ***pseudo-vector***.

Frequent uses of the ∇ operator include its operations on a scalar and also in vector multiplications. For instance, if ϕ is a scalar function of the coordinates, then

$$\nabla\phi = \text{grad}\phi = \mathbf{i}\frac{\partial\phi}{\partial x} + \mathbf{j}\frac{\partial\phi}{\partial y} + \mathbf{k}\frac{\partial\phi}{\partial z} \tag{A.1.3}$$

is called the ***gradient*** of ϕ. Its magnitude tells us relative information about the spacing of lines of constant ϕ. Where the magnitude of grad ϕ is large, the constant ϕ lines are closer together than where it is small. The direction of grad ϕ is locally normal to the surface $\phi = $ constant.

The scalar quantity

$$\nabla \cdot \mathbf{F} = \text{div } \mathbf{F} = \frac{\partial F_x}{\partial x} + \frac{\partial F_y}{\partial y} + \frac{\partial F_z}{\partial z} \tag{A.1.4}$$

is called the ***divergence*** of **F**. It tells us the quantity of **F** that passes through a surface.

The vector quantity

$$\nabla \times \mathbf{F} = \text{curl } \mathbf{F} = \mathbf{i}\left(\frac{\partial F_z}{\partial y} - \frac{\partial F_y}{\partial z}\right) + \mathbf{j}\left(\frac{\partial F_x}{\partial z} - \frac{\partial F_z}{\partial x}\right) + \mathbf{k}\left(\frac{\partial F_y}{\partial x} - \frac{\partial F_x}{\partial y}\right) \tag{A.1.5}$$

is called the ***curl*** of **F** and gives information on how much **F** is twisting or rotating. It also tells the direction of this rotation.

Some useful formulas that involve the del operator follow. They can easily be verified by expanding left- and right-hand sides and comparing the results.

$$\nabla(f\mathbf{F}) = f\nabla \cdot \mathbf{F} + (\mathbf{F} \cdot \nabla)f$$

$$\nabla \times (f\mathbf{F}) = f\nabla \times \mathbf{F} + (\nabla f) \times \mathbf{F}$$

$$\nabla \cdot (\mathbf{A} \times \mathbf{B}) = \mathbf{B} \cdot (\nabla \times \mathbf{A}) - \mathbf{A} \cdot (\nabla \times \mathbf{B})$$

$$\nabla \times (\mathbf{A} \times \mathbf{B}) = (\mathbf{B} \cdot \nabla)\mathbf{A} - (\mathbf{A} \cdot \nabla)\mathbf{B} + \mathbf{A}(\nabla \cdot \mathbf{B}) - \mathbf{B}(\nabla \cdot \mathbf{A})$$

$$\nabla(\mathbf{A} \cdot \mathbf{B}) = (\mathbf{A} \cdot \nabla)\mathbf{B} + (\mathbf{B} \cdot \nabla)\mathbf{A} + \mathbf{A} \times (\nabla \times \mathbf{B}) + \mathbf{B} \times (\nabla \times \mathbf{A})$$

$$\nabla \times \nabla f = 0 \text{ for any scalar } f$$

$$\nabla \cdot (\nabla \times \mathbf{F}) = 0 \text{ for any vector } \mathbf{F}$$

$$\nabla \times (\nabla \times \mathbf{F}) = \nabla(\nabla \cdot \mathbf{F}) - \nabla^2\mathbf{F}$$

$$(\mathbf{B} \cdot \nabla)\mathbf{A} = \frac{1}{2}[\nabla \times (\mathbf{A} \times \mathbf{B}) - \mathbf{A} \times (\nabla \times \mathbf{B}) - \mathbf{B} \times (\nabla \times \mathbf{A}) + \nabla(\mathbf{A} \cdot \mathbf{B})$$

$$-\mathbf{A}(\nabla \cdot \mathbf{B}) + \mathbf{B}(\nabla \cdot \mathbf{A})]$$

A theorem due to Helmholtz states that any vector **F** can be expressed in the form

$$\mathbf{F} = \text{grad } \phi + \text{curl } \mathbf{A}, \quad \text{where div } \mathbf{A} = 0. \tag{A.1.6}$$

Here, ϕ is called the **scalar potential** of **F**, and **A** is the **vector potential** of **F**. Since

$$\nabla \cdot \mathbf{F} = \nabla^2 \phi \qquad (A.1.7)$$

and

$$\nabla \times \mathbf{F} = -\nabla^2 \mathbf{A}, \qquad (A.1.8)$$

the scalar potential represents the **irrotational** part of **F** (the "curl-less" portion of **F**), and the vector potential represents the **rotational** portion of **F**.

The Helmholtz decomposition (equation (A.1.6)) results in the two Poisson equations (A.1.7) and (A.1.8) (a Poisson equation is a Laplace equation with a nonhomogeneous "right-hand side"). These can in principle be solved, giving

$$\phi(\mathbf{x}) = \iiint_{V'} \nabla \cdot \mathbf{F}(\mathbf{x}') \, g(\mathbf{x}, \mathbf{x}') \, dV',$$
$$\mathbf{A}(\mathbf{x}) = -\iiint_{V'} \nabla \times \mathbf{F}(\mathbf{x}') \, g(\mathbf{x}, \mathbf{x}') \, dV', \qquad (A.1.9)$$

where

$$g(x, x') = \text{Green's function} = \left(\begin{array}{cc} \dfrac{1}{2} \ln |\mathbf{x} - \mathbf{x}'| & \text{in two dimensions} \\ \dfrac{-1}{4 \, |\mathbf{x} - \mathbf{x}'|} & \text{in three dimensions} \end{array} \right). \qquad (A.1.10)$$

Notice that the **Green's function** is our potential for a source. The solution equation (A.1.9), however, in general does not satisfy the constraint $\nabla \cdot \mathbf{A} = 0$. To take care of this, replace equation (A.1.6) by

$$\mathbf{F} = \text{grad } \phi + \text{curl } (\mathbf{A}' + \text{grad } a). \qquad (A.1.11)$$

Since curl grad $a = 0$ for any a, **F** has not been affected in any manner. Thus, **A**' is given by equation (A.1.9). Since **A**' + grad a now replaces **A**, the requirement div **A** = 0 is replaced by div(**A**' + grad a) = 0. Thus, if a is defined by

$$\nabla^2 a = -\text{div } \mathbf{A}', \qquad (A.1.12)$$

giving

$$a = -\iiint_{V'} \nabla \cdot \mathbf{A}'(\mathbf{x}') \, g(\mathbf{x}, \mathbf{x}') \, dV', \qquad (A.1.13)$$

the constraint has been satisfied. (Note: In two dimensions the indicated volume integrals become surface integrals.)

A.2 Vector Integral Calculus

There are several interesting and useful theorems regarding the ∇ operator and integration that are useful in fluid mechanics, both in deriving the basic equations and in putting them in a form suitable for numerical calculation. They will be listed without proof.

These theorems are all closely related, and the names Gauss, Green, and Stokes are intimately connected with them. In their use, they are closely related to the concept of integration by parts of elementary calculus and can be thought of as a multidimensional extension of that concept.

Gauss's theorem states the following:

In three dimensions, with S a closed surface,

$$\iint_S \mathbf{n} \cdot \frac{(\mathbf{R} - \mathbf{R}_0)}{|\mathbf{R} - \mathbf{R}_0|^3} dS = \begin{pmatrix} 0 & \text{if } \mathbf{R}_0 \text{ is outside of } S \\ 4\pi & \text{if } \mathbf{R}_0 \text{ is inside of } S \end{pmatrix}. \qquad \text{(A.2.1a)}$$

In two dimensions, with C a closed curve,

$$\oint_C \mathbf{n} \cdot \frac{(\mathbf{r} - \mathbf{r}_0)}{|\mathbf{r} - \mathbf{r}_0|^2} ds = \begin{pmatrix} 0 & \text{if } \mathbf{r}_0 \text{ is outside of } S \\ 2\pi & \text{if } \mathbf{r}_0 \text{ is inside of } S \end{pmatrix}. \qquad \text{(A.2.1b)}$$

In both dimensions the integrand will be recognized as the radial velocity component of a source located at $\mathbf{R}_0$ in three dimensions and $\mathbf{r}_0$ in two dimensions.

Stokes's theorem is useful for changing line integrals to surface integrals and vice versa. In its simplest form it is

$$\oint_C \mathbf{t} \cdot \mathbf{F} \, ds = \iint_S \mathbf{n} \cdot (\nabla \times \mathbf{F}) \, dS = \iint_S (\mathbf{n} \times \nabla) \cdot \mathbf{F}) dS, \qquad \text{(A.2.2)}$$

where C is a closed curve bounding the surface S, $\mathbf{n}$ is a unit normal to the surface S, and $\mathbf{t}$ is a unit tangent to the curve C. Note that C and S do not have to lie in a plane and that C could bound an infinite number of different surfaces S.

Variations of Stokes's theorem include the following:

$$\oint_C \mathbf{t} f \, ds = \iint_S \mathbf{n} \times \nabla f \, dS,$$

$$\oint_C \mathbf{t} \times \mathbf{F} \, ds = \iint_S [\mathbf{n}(\nabla \cdot \mathbf{F}) - \nabla(\mathbf{n} \times \mathbf{F})] \, dS = \iint_S (\mathbf{n} \times \nabla) \times \mathbf{F} \, dS,$$

$$\oint_C \mathbf{t} \cdot \mathbf{F} \, ds = \iint_S \mathbf{n} \cdot (\nabla \times \mathbf{F}) \, dS = \iint_S (\mathbf{n} \times \nabla) \cdot \mathbf{F} \, dS,$$

$$\oint_C \mathbf{t} \times \mathbf{n} f \, ds = \iint_S [-\nabla f + \mathbf{n}(\mathbf{n} \cdot \nabla f + f \nabla \cdot \mathbf{n})] \, dS,$$

with

$$\nabla \cdot \mathbf{n} = -\left(\frac{1}{R_1} + \frac{1}{R_2} \right),$$

R_1 and R_2 being the principal radii of curvature of the surface S.

The **divergence theorem**, also called **Green's theorem**, is used for transforming surface integrals to volume integrals and is stated as

$$\iint_S \mathbf{n} \cdot \mathbf{F} \, dS = \iiint_V \nabla \cdot \mathbf{F} \, dV, \qquad \text{(A.2.3)}$$

where S is a surface enclosing the volume V, and $\mathbf{n}$ is again the unit outward normal. The theorem states that the net outflow of $\mathbf{F}$ through S is made up of the sum of the outflows from all of the regions inside of S.

Variations of this theorem include the following:

$$\iint_S \mathbf{n} \times \mathbf{F}\, dS = \iiint_V \nabla \times \mathbf{F}\, dV,$$

$$\iint_S (\mathbf{n} \cdot \nabla)\, \mathbf{F}\, dS = \iiint_V \nabla^2 \mathbf{F}\, dV,$$

$$\iint_S (\mathbf{n} \cdot \nabla) f\, dS = \iiint_V \nabla^2 f\, dV,$$

$$\iint_S \mathbf{n} f\, dS = \iiint_V \nabla f\, dV.$$

On a surface,

$$\iint_S (\mathbf{n} \times \nabla) \times \mathbf{F}\, dS = -\oint_C \mathbf{t} \times \mathbf{F}\, ds,$$

$$\iint_S \mathbf{n} \cdot (\nabla \times \mathbf{F})\, dS = \iint_S (\mathbf{n} \times \nabla) \cdot \mathbf{F}\, dS = \oint_C \mathbf{t} \cdot \mathbf{F}\, ds,$$

$$\iint_S \nabla \cdot \mathbf{F}\, dS = \oint_C (\mathbf{t} \times \mathbf{n}) \cdot \mathbf{F}\, ds,$$

$$\iint_S (\mathbf{n} \times \nabla) f\, dS = \oint_C f\, ds.$$

From these theorems several results follow that are useful in inviscid flow theory.

Green's first identity

$$\iint_S f \frac{\partial h}{\partial n}\, dS = \iiint_V (\nabla f \cdot \nabla h + f \nabla^2 h)\, dV. \tag{A.2.4}$$

Here f and h are any two scalar functions of the coordinates, and S is the surface enclosing V. This identity follows from Gauss's theorem with $\mathbf{F} = f \nabla h$.

Green's second identity

$$\iint_C \left(h \frac{\partial f}{\partial n} - f \frac{\partial h}{\partial n} \right) dS = \iiint_V (h \nabla^2 f - f \nabla^2 h)\, dV. \tag{A.2.5}$$

Here f and h are any two scalar functions of the coordinates. This identity follows from Green's first identity, written first as in equation (A.2.3) and again with f and h interchanged. The result follows by taking the difference of the two.

Green's third identity

$$\phi(\mathbf{P}) = -\iiint_V g(\mathbf{P}, \mathbf{Q}) \nabla^2 \phi(\mathbf{Q})\, dV$$

$$+ \iint_S [g(\mathbf{P}, \mathbf{Q}) \mathbf{n} \cdot \nabla \phi(\mathbf{Q}) - \phi(\mathbf{Q}) \mathbf{n} \cdot \nabla g(\mathbf{P}, \mathbf{Q})]\, dS \tag{A.2.6}$$

where in equation (A.2.5)

$$g(\mathbf{P}, \mathbf{Q}) = \text{Green's function} = \left(\begin{array}{l} \dfrac{1}{2} \ln |\mathbf{P} - \mathbf{Q}| \quad \text{for two dimensional problems} \\[2mm] \dfrac{-1}{4\, |\mathbf{P} - \mathbf{Q}|} \quad \text{for three dimensional problems} \end{array} \right).$$

$$\tag{A.2.7}$$

Here $\mathbf{P}$ is the position vector of a point in the interior of V and $\mathbf{Q}$ is the position vector of a dummy point of integration, on the surface for the surface integral, and in the interior for the volume integral. The function g is called the ***Green's function*** for the Laplace operator and is seen to be the velocity potential for a source.

A.3   Fourier Series and Integrals

Many times it is useful to represent a function in terms of an infinite series of simple functions. A classic example of this is the Fourier series, where if $f(t)$ is defined over an interval $0 \le t \le T$, the Fourier series representation of $f(t)$ is

$$f(t) = \frac{1}{2}a_0 + \sum_{n=1}^{\infty} \left(a_n \cos \frac{nt}{T} + b_n \sin \frac{nt}{T} \right). \tag{A.3.1}$$

This representation makes $f(t)$ periodic with period T; that is, $f(t + nT) = f(t)$ for any integer n.

The coefficients a_n and b_n are determined by the property of the trigonometric functions that, for integers m and n,

$$\int_0^T \sin \frac{nt}{T} \cos \frac{mt}{T} \, dt = 0,$$

$$\int_0^T \sin \frac{nt}{T} \sin \frac{mt}{T} \, dt = \int_0^T \cos \frac{nt}{T} \cos \frac{mt}{T} \, dt = 0 \quad \text{if } n \neq m, \; \frac{T}{2} \text{ if } n = m. \tag{A.3.2}$$

The trigonometric functions are said to be ***orthogonal*** to one another, as the operation of taking a product of two of them and then integrating is analogous to taking a dot product of two vectors.

To use the orthogonality property, multiply both sides of equation (A.3.1) by either the sine or cosine of $m\pi t/T$, and then integrate over the period T. Interchanging integration and summation, the result is

$$a_n = \frac{2}{T} \int_0^T f(t) \, \cos \frac{nt}{T} dt,$$

$$b_n = \frac{2}{T} \int_0^T f(t) \, \sin \frac{nt}{T} dt, \quad \text{for } n = 0, 1, 2, \ldots. \tag{A.3.3}$$

If the function $f(t)$ is discontinuous at either a point interior to, or at either end of, the interval, the series will converge to the average value at that point. Near the discontinuity, the series sum will show oscillations and overshoots, the ***Gibbs*** phenomenon.

The Fourier series representation of a function as given by equations (A.3.1) and (A.3.3) is sometimes also called the ***finite Fourier transform***. By DeMoivre's theorem, the sines and cosines can be replaced by exponentials so that the Fourier series (equation (A.3.1)) becomes

$$f(t) = \frac{1}{2} \left[a_0 + \sum_{n=1}^{\infty} (a_n - ib_n)e^{int/T} + \sum_{n=1}^{\infty} (a_n + ib_n)e^{-int/T} \right]$$

$$= \sum_{n=-\infty}^{\infty} c_n e^{int/T},$$

where

$$c_n = \begin{pmatrix} 0.5(a_n - ib_n) & \text{for } n > 0 \\ 0.5a_0 & \text{for } n = 0 \\ 0.5(a_n + ib_n) & \text{for } n < 0 \end{pmatrix}.$$

There are a number of means of reducing the work needed to determine the coefficients c_n. These are termed *fast Fourier Transform* (FFT) methods.

If the interval T becomes infinite, the Fourier series becomes a *Fourier integral*, with

$$f(t) = \int_{-\infty}^{\infty} F(p) \, e^{-2ipt} \, dp,$$

with F given by

(A.3.4)

$$F(p) = \int_{-\infty}^{\infty} f(p) \, e^{2ipt} \, dp.$$

The equations given in (A.3.4) are analogs of equations (A.3.1) and (A.3.2). The Fourier transform becomes the *Laplace transform* if $f(t) = 0$ for $t < 0$, and if $2\pi p$ is replaced by the notation *is*, i being the root of -1.

The Fourier series expansion is a special case of expansions in terms of orthogonal functions. These expansions are called *spectral representations* of the functions, since the frequencies involved ($2\pi n/T$ in the case of the Fourier series) make up a spectrum, with the c_n's being the amplitudes of the harmonics at these frequencies. For the general case of expansion in terms of orthogonal functions, the expansion is

$$f(x) = \sum_{n=-\infty}^{\infty} c_n \phi_n(x), \quad a \le x \le b, \tag{A.3.5}$$

where the ϕ_n can be shown to be orthogonal easiest if they are solutions of the equation

$$\frac{d[p(x)d\phi]}{dx} + [\lambda^2 r(x) - q(x)]\phi = 0 \tag{A.3.6}$$

with p and q both positive throughout the interval $a \le x \le b$. This is referred to as the *Sturm-Liouville equation*. The boundary conditions which accompany equation (A.3.6) are

$$\alpha\phi(a) + \beta \left.\frac{d\phi}{dx}\right|_a = 0, \quad |\alpha| + |\beta| > 0,$$

(A.3.7)

$$\gamma\phi(b) + \delta \left.\frac{d\phi}{dx}\right|_b = 0, \quad |\gamma| + |\delta| > 0.$$

Providing $pd\phi/dx = 0$ at $x = a$ and $x = b$, it can be shown that

$$\int_a^b r\phi_m\phi_n \, dx = 0 \tag{A.3.8}$$

if m is different from n. The ϕ's are said to be orthogonal to one another with respect to the weight function r. Special cases of orthogonal functions, along with the equations they satisfy, are given in Table A.1.

TABLE A.1 Some well-studied orthogonal functions

Function name	Equation		Interval
	Simple form	Sturm-Liouville form	
Trigonometric	$y'' + n^2 y = 0$	$y'' + n^2 y = 0$	$0 \le x \le T$
Chebeyshev	$(1 - x^2)y'' - xy' + n^2 y = 0$	$(\sqrt{1 - x^2}\, y')' + \frac{n^2}{\sqrt{1-x^2}} y = 0$	$-1 \le x \le 1$
Legendre	$(1 - x^2)y'' - 2xy' + ny = 0$	$(1 - x^2)y'' - 2xy' + ny = 0$	$-1 \le x \le 1$
Laguerre	$xy'' + (1 - x)y' + ny = 0$	$(xe^{-x}y')' + nxe^{x}y = 0$	$0 \le x \le \infty$
Hermite	$y'' - 2xy' + ny = 0$	$\left(e^{-x^2}y'\right)' + ne^{-x^2}y = 0$	$-\infty \le x \le \infty$
Bessel	$x^2 y'' + xy' + (x^2 - n^2)y = 0$	$(xy')' + \left(x - \frac{n^2}{x}\right)y = 0$	$0 \le x \le T$
Mathieu	$y'' + (a - b\cos 2x)y = 0$	$y'' + (a - b\cos 2x)y = 0$	$0 \le x \le \pi$

A.4 Solution of Ordinary Differential Equations

A.4.1 Method of Frobenius

Solutions for many of the equations listed above can be found in the form of power series. The procedure is to assume a solution in the form $y = \sum_{n=0}^{\infty} a_n x^{s+n}$; substitute this into the differential equation, and collect on terms. The lowest power of x is used to determine s, and higher powers are used to give recursion relations to determine the coefficients a_n. This is best demonstrated by considering an example.

Consider Bessel's equation $x^2 y'' + xy' + (x^2 - n^2)y = 0$. Substituting the series form for y into the equation gives

$$\sum_{j=0}^{\infty} a_j \left[\left(s^2 + 2js + j^2 - n^2\right)x^{j+s} + x^{j+s+2}\right] = 0, \text{ or}$$

$$x^s \left\{ a_0(s^2 - n^2) + a_1 \left[(s+1)^2 - n^2\right]x + \sum_{j=2}^{\infty} x^j \left[a_j \left(s^2 + 2js + j^2 - n^2\right) + a_{j-2}\right]\right\} = 0.$$

Since the s power has been inserted to make a_0 the starting term, it cannot be zero and hence $s = \pm n$. The recursion relationship is $a_j = -\frac{1}{(s+j)^2 - n^2} a_{j-2}$, $j \ge 2$, as long as n is not an integer, in which case it would eventually involve division by zero. For the case where n is not an integer, the solution is

$$y = J_n(x) \text{ and } J_{-n}(x), \text{ where } J_n(x) = \sum_{j=0}^{\infty} \frac{(-1)^j x^{n+2j}}{2^{n+2j} j! \Gamma(n+j+1)}. \tag{A.4.1}$$

Thus, two distinct solutions have been found with one effort.

Should n be an integer, we still have one solution $J_n(x)$ but are missing the second. The loss of this second solution can be easiest seen for the case $n = 0$, since $J_0(x) = J_{-0}(x)$. For this case, let the second solution be of the form $J_0(x) \ln x + y_2(x)$. Substituting this into Bessel's equation gives $\ln x \left(x^2 J_0'' + x J_0' + (x^2 - n^2)J_0\right) + 2xJ_0' + x^2 y_2'' + xy_2' + (x^2 - n^2)y_2 = 0$. The J_0 terms multiplied by $\ln x$ vanish, and we are left with

$$x^2 y_2'' + x y_2' + (x^2 - n^2) y_2 = -2x J_0' = -2 \sum_{j=0}^{\infty} \frac{2j(-1)^j x^{2j}}{2^{2j} j! \Gamma(j+1)} = \sum_{j=1}^{\infty} \frac{4j(-1)^{j+1} x^{2j}}{2^{2j} (j!)^2}.$$

We can now use a Taylor series to find the particular integral of this equation.

A.4.2 Mathieu Equations

The solution of Mathieu's equation $\frac{d^2 y}{dx^2} + p^2 (1 - 2\alpha \cos 2x) y = 0$ for general values of p and α is not periodic and is therefore not of much interest. Numerical computation is possibly the best option for solution. However, on each of the boundaries separating the stable and unstable regions, there is one and only one periodic solution. Referring to the Strutt diagram presented in Chapter 3, the solution on these boundaries are given by the following:

p an even integer, period π:

$$\mathrm{ce}_{2p}(x; p, \alpha) = \sum_{n=0}^{\infty} A_{2n}(p, \alpha) \cos 2nx \qquad : (a_{2n})$$

$$p^2 A_0 - \alpha p^2 A_2 = 0, \quad (p^2 - 4) A_2 - \alpha p^2 (2A_0 + A_4) = 0, \qquad (\text{A.4.2})$$

$$(p^2 - 4n^2) A_{2n} - \alpha p^2 (A_{2n-2} + A_{2n+2}) = 0, \quad n = 2, 3, 4, \ldots.$$

$$\mathrm{se}_{2p+2}(x; p, \alpha) = \sum_{n=0}^{\infty} B_{2n+2}(p, \alpha) \sin(2n+2)x \qquad : (b_{2n})$$

$$(p^2 - 4) B_2 - \alpha p^2 B_4 = 0,$$

$$(p^2 - 4n^2) B_{2n} - \alpha p^2 (B_{2n-2} + B_{2n+2}) = 0, \quad n = 2, 3, 4, \ldots.$$

p an odd integer, period 2π:

$$\mathrm{ce}_{2p+1}(x; p, \alpha) = \sum_{n=0}^{\infty} A_{2n+1}(p, \alpha) \cos(2n+1)x \qquad : (a_{2n+1})$$

$$(p^2 - 4) A_2 - \alpha p^2 (2A_0 + A_4) = 0,$$

$$\left[p^2 - (2n+1)^2 \right] A_{2n+1} - \alpha p^2 (A_{2n-1} + A_{2n+3}) = 0, \quad n = 1, 2, 3, \ldots. \qquad (\text{A.4.3})$$

$$\mathrm{se}_{2p+1}(x; p, \alpha) = \sum_{n=0}^{\infty} B_{2n+1}(p, \alpha) \sin(2n+1)x \qquad : (b_{2n+1})$$

$$(p^2 - 1) B_1 - \alpha p^2 (B_3 - B_1) = 0,$$

$$\left[p^2 - (2n+1)^2 \right] B_{2n+1} - \alpha p^2 (B_{2n-1} + B_{2n+3}) = 0, \quad n = 1, 2, 3, \ldots.$$

There are several notations and conventions that are used in the literature and differences in the value for the starting coefficient in each series. The notation chosen here is to emphasize that in the Strutt diagram the curves denoted by a_{2n} and b_{2n} intersect for $\alpha = 0$, as do the curves denoted by a_{2n+1} and b_{2n+1}.

To determine the boundary that separates stable from unstable regions, we are interested in the shape of the dividing curve rather than the solution as a function of the independent variable x. To determine this boundary curve, realize that if we use the recursion relations to set up the infinite set of equations needed to find the A_n's and

B_n's, the equations can be solved unless the determinant of the coefficients vanishes. Setting this determinant to zero is in fact what is needed to determine this curve!

This is illustrated for the function ce_{2p}. From the recursion relations,

$$p^2 A_0 - \alpha p^2 A_2 = 0,$$

$$-2\alpha p^2 A_0 + (p^2 - 4) A_2 - \alpha p^2 A_4 = 0,$$

$$-\alpha p^2 A_2 + (p^2 - 16) A_4 - \alpha p^2 A_6 = 0,$$

$$-\alpha p^2 A_4 + (p^2 - 36) A_6 - \alpha p^2 A_8 = 0, \ldots.$$

Setting the determinant of the coefficients for this function to zero thus gives

$$\begin{vmatrix} p^2 & -\alpha p^2 & 0 & 0 & 0 & 0 & \cdots \\ -2\alpha p^2 & p^2 - 4 & -\alpha p^2 & 0 & 0 & 0 & \cdots \\ 0 & -\alpha p^2 & p^2 - 16 & -\alpha p^2 & 0 & 0 & \cdots \\ 0 & 0 & -\alpha p^2 & p^2 - 36 & -\alpha p^2 & 0 & \cdots \\ \cdots & \cdots & \cdots & \cdots & \cdots & \cdots & \cdots \end{vmatrix} = 0.$$

Similar operations can clearly be carried out on the other three functions.

Notice that the terms on the diagonal of the determinant are growing at a rapid rate! This could have been avoided by dividing each equation by a suitable constant (perhaps the terms which appear on the diagonal) before forming the determinant.

A.4.3   Finding Eigenvalues—The Riccati Method

When using techniques such as separation of variables and when dealing with equations such as Sturm-Liouville systems, determinations of the eigenvalues and eigenfunctions are a key part of the solution. Indeed, in many engineering situations the determination of frequencies involved in flow instabilities, natural frequencies of vibrating structures, acoustics, and many other areas, the determination of frequency is the sole object.

As discussed in Chapter 11, the Riccati method is a powerful tool that is available for determining the characteristic frequency[1] of a system of differential equations. The technique exhibited by examples in Chapter 11 will be explained more generally here. For further discussion, see Scott (1973a and b) and Davey (1977).

Start with the n by n system of linear ordinary differential equations

$$\mathbf{y}' = \mathbf{F}\mathbf{y}, \tag{A.4.4}$$

where $\mathbf{y}$ and $\mathbf{y}'$ are column vectors of order n, and $\mathbf{F}$ is an n by n matrix. Suppose that m less than n of the boundary conditions are known at $x = 0$, and $p = n - m$ are known at $x = L$. Being an eigenvalue problem, these are necessarily homogeneous boundary conditions.

Since the system will have n independent solutions for $\mathbf{y}$, represent the family of the n solutions y, which are involved in the boundary conditions at $x = 0$ by $\mathbf{u}$ (of size m) and the remaining of the y by $\mathbf{v}$ (of size p). In general form, these boundary conditions are expressed as

$$\mathbf{Gu} = 0 \text{ at } x = 0, \quad \mathbf{Hu} = \mathbf{Jv} \text{ at } x = L. \tag{A.4.5}$$

---

[1] The terms *characteristic frequency, natural frequency* and *eigenfrequency* are interchangeable. *Eigen* is the German word for "characteristic."

G is a matrix of order m by n, H of order p by m, and J of order p by p.

The method starts by rewriting the original system in the form

$$\mathbf{u'} = \mathbf{Au} + \mathbf{Bv}, \quad \mathbf{v'} = \mathbf{Cu} + \mathbf{Dv}, \tag{A.4.6}$$

where A, B, C, D are matrices of order m by m, m by p, p by m and p by p, respectively. Also, there will be a matrix $\mathbf{R}$ of order m by p such that

$$\mathbf{u} = \mathbf{Rv}. \tag{A.4.7}$$

It follows from this that there exists the inverse relationship

$$\mathbf{v} = \mathbf{Su}, \quad \text{where } \mathbf{S} = \mathbf{R}^{-1} \tag{A.4.8}$$

is the matrix inverse of $\mathbf{R}$. Differentiation of equation (A.4.6) and use of equation (A.4.7) gives

$$\begin{aligned}
\mathbf{u'} = \mathbf{Au} + \mathbf{Bv} &= (\mathbf{AR} + \mathbf{B})\,\mathbf{v} \\
&= (\mathbf{Rv})' \\
&= \mathbf{R'v} + \mathbf{Rv'} = \mathbf{R'v} + \mathbf{R}(\mathbf{Cu} + \mathbf{Dv}) \\
&= (\mathbf{R'} + \mathbf{RCR} + \mathbf{RD})\,\mathbf{v}.
\end{aligned}$$

Since the equation is true for any $\mathbf{v}$, it follows that

$$\mathbf{R'} = \mathbf{B} + \mathbf{AR} - \mathbf{RD} - \mathbf{RCR}. \tag{A.4.9}$$

This must satisfy the boundary condition $\mathbf{R}(0) = \mathbf{0}$. At the second boundary, $(\mathbf{HR} - \mathbf{J})\,\mathbf{v} = \mathbf{0}$ at $x = L$. This is satisfied providing

$$\det(\mathbf{HR} - \mathbf{J}) = 0. \tag{A.4.10}$$

Generally, it is easiest to set all parameters in F and integrate equation (A.4.9) until equation (A.4.10) is satisfied. This gives the length at which that set of parameters gives the correct eigenvalues. Frequently scaling can be performed so that the situation can be changed to the more usual one of starting with a given L.

In the course of integrating equation (A.4.9), it is not unusual to find that $\mathbf{R}$ becomes singular before the condition of equation (A.4.10) is met. Although it is possible to go to a complex path of integration, it has often been found easier to switch to the inverse problem as $\mathbf{R}$ becomes large. Repeating the process used in developing equation (A.4.9), but this time starting with equation (A.4.8) instead of equation (A.4.7), the result is

$$\mathbf{S'} = \mathbf{C} + \mathbf{BS} - \mathbf{SA} - \mathbf{SBS}. \tag{A.4.11}$$

Thus, at some point the switch is made from the $\mathbf{R}$ to $\mathbf{S}$ equations, and then after the singularity in $\mathbf{R}$ (zero in $\mathbf{S}$) is passed, the switch back to the $\mathbf{R}$ equations is made. The expense in programming and computation time is small, requiring some logical decisions and the inversion of the matrix. Runge-Kutta methods tend to perform well on these problems. There have been reports in the literature that finding eigenvalues higher than the first has produced difficulties and that backward integration to determine eigenfunctions has proven to be unstable.

A.5   Index Notation

Considerable simplicity in writing physical equations is obtained by use of vector notation. Vector notation has the ability to express briefly very powerful ideas, the brevity of the notation in many cases being of great help in understanding terms in the equation. Nevertheless, there are times when vector notation is not useful, either when the quantities of interest are of a more general nature or when you want to deal with an actual problem, when it is always necessary to deal with the components of the vector rather than the vector itself. Index notation has been invented to accommodate both of these cases. It is an adaptation of matrix notation and for physical problems where the indexed quantities are tensors is the same as the notation of Cartesian tensors. Students familiar with matrix algebra and/or computer programming languages such as FORTRAN will already be familiar with many of the concepts, as it is used in most programming languages.

To motivate the notation, consider as an example the vector equation $\mathbf{F} = m\mathbf{a}$. In component form, this becomes, in x, y, z Cartesian coordinates,

$$F_x = ma_x, \quad F_y = ma_y, \quad F_z = ma_z. \tag{A.5.1}$$

Now the three equations in component form just written are similar except for the subscript. The fundamental information is displayed in the first equation, with the remaining equations just giving further detail. In this example, the equations have few terms, and there is no objection to the extra work involved in writing three equations when the first one gives the sense of the rest. Of course, one could write

$$F_x = ma_x \tag{A.5.2}$$

and add "and similarly for the y and z direction," but it can become confusing if this phrase must be written after each equation. Also, if the equation is complicated, perhaps this brief description will not convey all the necessary information, and it will be necessary to give further verbal description. And, certainly, computer programs would prefer to deal with numbers rather than alphabets.

In looking at the component equations again, we can see that the need to write the three equations was due to the original labeling of the axes through the use of different letters. Instead of using alphabetical names, it is more convenient to adopt number names for the coordinate axes, such as x_1, x_2, x_3. Then, $F_x = F_{x1}, F_y = F_{x2}, F_z = F_{x3}$, or, more briefly, since the "x" is now common to all three axes, $F_x = F_1, F_y = F_2, F_z = F_3$. Then the component equations become

$$F_1 = ma_1, \quad F_2 = ma_2, \quad F_3 = ma_3, \tag{A.5.3}$$

or, taking full advantage of the new notation,

$$F_i = ma_i, \quad i = 1, 2, \text{ or } 3. \tag{A.5.4}$$

Thus, one equation stands for three equations, as in vector language. Since the phrase "$i = 1, 2$, or 3" will usually be there, it is the custom to omit it and add a comment only when this phrase is not true.

To tie this concept more closely to vector analysis, let us adopt the notation $\mathbf{g}_1, \mathbf{g}_2, \mathbf{g}_3$, for the base vectors usually denoted by $\mathbf{i}, \mathbf{j}, \mathbf{k}$. Then

$$\mathbf{F} = \sum_{i=1}^{3} \mathbf{g}_i F_i \tag{A.5.5}$$

Now, in (A.5.4), i appeared exactly once in every term of the equation. It is a *free index*—that is, it can take on three different values at the user's choice. In equation (A.5.5), i appears as a *repeated*, or *dummy, index* and takes on the values denoted by the summation sign. Now, in many applications where index notation is used, whenever an index is a dummy index, it appears twice and only twice in that term and always with the summation sign. To make full use of this, henceforth the *summation convention* (attributed to A. Einstein) will be adopted. That is, whenever an index appears twice in a term, summation on that index is implied. Thus, equation (A.5.1) could be written as

$$\mathbf{F} = \mathbf{g}_i F_i \tag{A.5.6}$$

A repeated (dummy) index will always imply summation unless otherwise explicitly stated.

Exercise A.5.1   Index notation
Write out the following in full component form:

$$A = B_i C_i,$$
$$A_i = B_i C_j D_j,$$
$$A_i = B_i C_{jj},$$
$$A_i = B_{ij} C_j,$$
$$A_i = B_{ijj} + C_{ik} D_k, \quad A^2 = A_i A_i.$$

Solution.

$$A = B_i C_i \Rightarrow A = B_1 C_1 + B_2 C_2 + B_3 C_3.$$
$$A_i = B_i C_j D_j \Rightarrow A_1 = B_1 (C_1 D_1 + C_2 D_2 + C_3 D_3),$$
$$A_2 = B_2 (C_1 D_1 + C_2 D_2 + C_3 D_3),$$
$$A_3 = B_3 (C_1 D_1 + C_2 D_2 + C_3 D_3),$$
$$A_i = B_i C_{jj} \Rightarrow A_1 = B_1 (C_{11} + C_{22} + C_{33}),$$
$$A_2 = B_2 (C_{11} + C_{22} + C_{33}),$$
$$A_3 = B_3 (C_{11} + C_{22} + C_{33}),$$
$$A_i = B_{ij} C_j \Rightarrow A_1 = B_{11} C_1 + B_{12} C_2 + B_{13} C_3,$$
$$A_2 = B_{21} C_1 + B_{22} C_2 + B_{23} C_3,$$
$$A_3 = B_{31} C_1 + B_{32} C_2 + B_{33} C_3.$$
$$A_i = B_{ijj} + C_{ik} D_k \Rightarrow A_1 = B_{111} + B_{122} + B_{133} + C_{11} D_1 + C_{12} D_2 + C_{13} D_3,$$
$$A_2 = B_{211} + B_{222} + B_{233} + C_{21} D_1 + C_{22} D_2 + C_{23} D_3,$$
$$A_3 = B_{311} + B_{322} + B_{333} + C_{31} D_1 + C_{32} D_2 + C_{33} D_3,$$
$$|A|^2 = A^2 = A_i A_i \Rightarrow A^2 = A_1 A_1 + A_2 A_2 + A_3 A_3.$$

Since many of the quantities used in mechanics are vectors, it is helpful to see how the new notation can be used to represent familiar quantities and operations. By inspection,

$$\text{grad } \phi = \nabla\phi = \mathbf{g}_i \frac{\partial\phi}{\partial x_i},$$

$$\text{div } \mathbf{A} = \nabla \cdot \mathbf{A} = \frac{\partial A_i}{\partial x_i}, \quad (A.5.7)$$

and

$$\mathbf{A} \cdot \mathbf{B} = A_i B_i.$$

In working with the cross product and curl, we can see that the application of index notation is not as obvious. In fact, it is necessary to introduce a new symbol, the *e symbol* or *alternating tensor*, to accomplish this operation. The components of the alternating tensor are defined by

$$e_{ijk} = \begin{pmatrix} +1 & \text{if } i, j, k \text{ is an even permutation of } 1,2,3, \\ -1 & \text{if } i, j, k \text{ is an odd permutation of } 1,2,3, \\ 0 & \text{otherwise.} \end{pmatrix} \quad (A.5.8)$$

Thus,

$$e_{123} = e_{231} = e_{312} = 1, \quad e_{132} = e_{213} = e_{321} = -1, \quad e_{111} = e_{112} = e_{113} = e_{121} = \cdots = 0.$$

Comparison of components then indicates

$$\mathbf{A} \times \mathbf{B} = \mathbf{g}_i e_{ijk} A_j B_k. \quad (A.5.9)$$

In particular,

$$\nabla \times \mathbf{B} = \text{curl } \mathbf{B} = \mathbf{g}_i e_{ijk} \frac{\partial B_k}{\partial x_j}. \quad (A.5.10)$$

Another symbol that is of use is the *Kronecker delta*, with components δ_{ij}, defined by

$$\delta_{ij} = \begin{pmatrix} 1 & \text{if } i = j \\ 0 & \text{otherwise} \end{pmatrix}. \quad (A.5.11)$$

Thus, $\delta_{11} = \delta_{22} = \delta_{33} = 1$, $\delta_{12} = \delta_{21} = \delta_{23} = \delta_{32} = \delta_{13} = \delta_{31} = 0$. In matrix terminology, the Kronecker delta forms the components of the *identity matrix*.

Notice that since we are dealing with base vectors that are orthogonal to one another and are each of unit length, then

$$\mathbf{g}_i \cdot \mathbf{g}_j = \delta_{ij}.$$

Contraction is the process of multiplying a quantity by the Kronecker delta and then summing on one of the indices. For instance,

$$B_{ij}\delta_{jk} = B_{ik}, \quad B_{ij}\delta_{ji} = B_{ii}. \quad (A.5.12)$$

All of the operations of vector and matrix analysis (and many more as well) can now be written directly in index notation.

Notice that for the Kronecker delta, the value of the component is independent of the order in which the indices are written—that is, $\delta_{ij} = \delta_{ji}$. Such a quantity is said to be *symmetric* in those indices. Similarly, the components of the alternating tensor change sign if two indices are interchanged. That is, $e_{ijk} = -e_{jik}$. Such quantities are said to be *skew-symmetric* in those indices. Note that any quantity with two or more indices can be written as the sum as a symmetric and skew-symmetric form by the decomposition

$$A_{ij} = 0.5[(A_{ij} + A_{ji}) + (A_{ij} - A_{ji})].$$

Exercise A.5.2 Verify the following expressions by expanding them:

$$\delta_{ii} = 3, \quad \delta_{ij}A_j = A_i, \quad e_{ijk}e_{1jk} = 2\delta_{i1}.$$

Solution.

$$\delta_{ii} = \delta_{11} + \delta_{22} + \delta_{33} = 1 + 1 + 1 = 3.$$
$$\delta_{ij}A_j = \delta_{i1}A_1 + \delta_{i2}A_2 + \delta_{i3}A_3. \text{ Thus, since}$$
$$\delta_{1j}A_j = A_1, \quad \delta_{2j}A_j = A_2, \quad \delta_{3j}A_j = A_3,$$

the expression can be summarized as $\delta_{ij}A_j = A_i$.

$$e_{ijk}e_{1jk} = e_{i12}e_{112} + e_{i13}e_{113} + e_{i23}e_{123} + e_{i21}e_{121} + e_{i31}e_{131} + e_{i32}e_{132}$$
$$= 0 + 0 + e_{i23}e_{123} + 0 + 0 + e_{i32}e_{132}$$
$$= 2e_{i23}e_{123} = 2\delta_{i1}.$$

As a last bit of notation, since the typing of a partial differential is awkward on a typewriter or even on a word processor, a comma is frequently used to denote partial differentiation, the subscript after the comma denoting that coordinate is involved in the differentiation. That is,

$$\frac{\partial A—}{\partial x_k} = A—,_k. \tag{A.5.13}$$

As a final example of the use of index notation, the divergence and curl theorems will be shown in index notation. With V as an arbitrary volume, and S the surface of V, the divergence theorem becomes

$$\iiint_V \frac{\partial A_i}{\partial x_i} \, dV = \iint_S A_i n_i \, dS. \tag{A.5.14}$$

Similarly, take surfaces drawn in space with the boundary curve on the periphery of S' and $\mathbf{n}$ as the unit normal to S'. Further, take $\mathbf{t}$ as the unit tangent to C, with integration around C performed in the direction of $\mathbf{t}$ and $\mathbf{n}$ chosen so that on C $\mathbf{n} \times \mathbf{t}$ points inward toward S. Then, if the derivatives exist in the regions chosen, and if A_i are single-valued, the curl theorem is

$$\iint_{S'} n_i e_{ijk} \frac{\partial A_k}{\partial x_j} \, dS' = \iint_C A_i t_i \, ds. \tag{A.5.15}$$

A.6 Tensors in Cartesian Coordinates

In fluid mechanics, as indeed in most branches of physics, we are dealing with fundamental quantities in space that have properties such as magnitude and direction. Familiar examples are scalars (e.g., density that has magnitude but no direction), vectors (e.g., velocity that has magnitude and direction) and more general quantities (e.g., stress that has magnitude and two directions, one for the force and the second for orientation of the area). The generic term that contains all of these categories is a ***tensor***. Thus, a scalar is a tensor of order zero, a vector is a tensor of order one, and stress is a tensor of order two. The ***order*** of a tensor refers to the number of directions that we associate with the physical quantity. While any order is possible, tensors of order greater than two are rare and usually arise as the derivatives of lower-order tensors.

Tensors are written in the following fashion:

$$\rho = \rho \qquad \text{(zero order tensor)} \tag{A.6.1}$$

$$\mathbf{v} = v_i \mathbf{g}_i \qquad \text{(first order tensor)} \tag{A.6.2}$$

$$\tau = \tau_{ij} \mathbf{g}_i \mathbf{g}_j \quad \text{(second order tensor)} \tag{A.6.3}$$

For the second-order tensor, we have two base vectors, one for each direction, and they appear naturally as a product in the right-hand side. (Notice that the appearance of two dummy indices means that there are two summations present, one on i and one on j.) Since we have not put a dot or a cross between the two base vectors, the product is neither a dot product nor a cross product but rather an ***indefinite product***. (It's not really indefinite because we know what the two directions mean. The terminology here is just to identify this new type of product as having a name different from the other two.)

In dealing with tensors of any order, it is convenient to deal with just the components. A test to determine whether a set of quantities are in fact the components of a tensor is to observe how these quantities transform as we rotate axes. (Note: A fundamental point that is all too easily forgotten in learning tensor analysis is that the tensor itself does not change as we rotate the coordinate axes, only the components of the tensor change, since they, and not the tensor itself, are axis-related.) To see this, at a point in space introduce two Cartesian coordinate systems, $\mathbf{x}$ and $\mathbf{y}$, one being obtained from the other by a rigid rotation of axes. Then $y_i = a_{ij} x_j$, and

$$\frac{\partial x_j}{\partial y_i} = \frac{\partial y_i}{\partial y_i} = a_{ij} = a_{ji} = \mathbf{g}_i(\mathbf{y}) \cdot \mathbf{g}_j(\mathbf{x}), \tag{A.6.4}$$

where the a_{ij} are the direction cosines of one set of axes with respect to the other, and the letters in parentheses following the unit base vectors tell the reference frame they refer to. By virtue of the orthogonality of the axes,

$$\mathbf{g}_i(\mathbf{y}) \cdot \mathbf{g}_j(\mathbf{y}) = \mathbf{g}_i(\mathbf{x}) \cdot \mathbf{g}_j(\mathbf{x}) = \delta_{ij}. \tag{A.6.5}$$

By the Pythagorean theorem, and the fact that the angle between the axes x_i and y_k are the same,

$$a_{ik} a_{jk} = a_{ki} a_{kj} = \delta_{ij}. \tag{A.6.6}$$

Since any tensor T must be independent of the choice of axes, then $\mathbf{T}(\mathbf{y}) = \mathbf{T}(\mathbf{x})$, where $\mathbf{T}(\mathbf{y})$ means the tensor quantity $\mathbf{T}$ referred to the $\mathbf{y}$ axes and similarly for $\mathbf{T}(\mathbf{x})$. By equation (A.6.4) the base vectors transform according to

$$\mathbf{g}_i(y) = a_{ij} \mathbf{g}_j(x), \tag{A.6.7}$$

and so for a tensor of any order

$$\mathbf{T}(\mathbf{y}) = T_{ij\ldots}(\mathbf{y})\mathbf{g}_i(\mathbf{y})\mathbf{g}_j(\mathbf{y})\ldots = T_{ij\ldots}(\mathbf{x})a_{im}\mathbf{g}_m(\mathbf{x})a_{jn}\mathbf{g}_n(\mathbf{x})\ldots.$$

But

$$\mathbf{T}(\mathbf{x}) = T_{ij\ldots}(\mathbf{x})\mathbf{g}_i(\mathbf{x})\mathbf{g}_j(\mathbf{x})\ldots.$$

Thus, to have $\mathbf{T}(\mathbf{x}) = \mathbf{T}(\mathbf{y})$—that is, to have the tensor independent of the coordinate system—it is necessary then that the components transform according to

$$T_{mn\text{—}}(\mathbf{x}) = a_{im}a_{jn}\text{—}T_{ij\text{—}},$$

or, multiplying both sides by direction cosines, contracting, and using equation (A.6.5) several times,

$$T_{ij\text{—}}(\mathbf{y}) = a_{im}a_{jn}\text{—}T_{mn\text{—}}(\mathbf{x}). \tag{A.6.8}$$

When there are no subscripts (that is, the tensor is a scalar), then the direction cosines disappear from equation (A.6.8).

To emphasize this transformation law for the orders used the most, we write the following transformation laws:

Zero order tensor $T(y)$: $\qquad\qquad\qquad\qquad T(x) = T(y).$ $\qquad$ (A.6.9)

First order tensor components $T_j(y)$: $\qquad\quad T_i(x) = a_{im}T_m(y).$ $\qquad$ (A.6.10)

Second order tensor components $T_{ij}(y)$: $\qquad T_{ij}(x) = a_{im}a_{jn}T_{mn}(y).$ $\qquad$ (A.6.11)

Third order tensor components $T_{ijk}(y)$: $\qquad T_{ijk}(x) = a_{im}a_{jn}a_{ko}T_{mno}(y).$ $\quad$ (A.6.12)

For a second-order tensor, if the third axis is not rotated, then

$$a_{i3} = a_{3i} = \delta_{i3}.$$

Then the equation

$$T_{ij}(\mathbf{x}) = a_{im}a_{jn}T_{mn}(\mathbf{y}), \quad i = 1, 2, \quad j = 1, 2,$$

is the equation of a circle, called **Mohr's circle**. It affords an easy graphical interpretation of the transformation law in two dimensions.

Tensor components can have symmetry properties. If, for instance, for a second-order tensor the components obey the rule

$$T_{ij} = T_{ji}, \tag{A.6.13}$$

it is said that the second order tensor is **symmetric**. The Kronecker delta is an example of a symmetric tensor. If, on the other hand, the components obey the rule

$$T_{ij} = -T_{ji}, \tag{A.6.14}$$

it is said that the second order tensor is **antisymmetric**, or **skew symmetric**. Any second-order tensor can be written as the sum of a symmetric part and a skew symmetric part—that is,

$$T_{ij} = 0.5(T_{ij} + T_{ji}) + 0.5(T_{ij} - T_{ji}). \tag{A.6.15}$$

The first two terms are clearly symmetric, and the last two are clearly skew symmetric. Components of tensors of order higher than two can be symmetric or skew symmetric in any pair of their indices.

Example A.6.1   Unit vectors
Show that $g_i(x) = a_{ji}g_j(y)$, and thus verify equations (A.6.4) and (A.6.6).

Solution. This result is simply the familiar decomposition of a vector into its components. In the y-axes, the three unit vectors $e_i(x)$ make angles with the axes y_1, y_2, and y_3, whose direction cosines are a_{1i}, a_{2i}, and a_{3i}. Applying the decomposition law, the result follows.

Example A.6.2   Direction cosines
Show by a suitable dot product that the a_{ij} are direction cosines of the vectors $g_i(y)$ referred to the $g_i(x)$ vectors.

Solution. Taking the dot product $g_i(x) \cdot g_j(y)$ and using the results of Example A.6.1, we have $g_i(x) \cdot g_j(y) = a_{ki}g_k(y) \cdot g_j(y) = a_{ki}\delta_{kj} = a_{ji}$ by the orthogonality of the y-axes. QED.

Example A.6.3   Tensor properties of the Kronecker delta and alternating tensors
Show that the alternating tensor and the Kronecker delta are tensors in a Cartesian coordinate system.

Solution. Assume that the Kronecker delta is a second-order tensor. From the transformation law for its components equation (A.6.11),

$$\delta_{ij}(x) = a_{im}a_{jn}\delta_{mn}(y).$$

From the summation property of the Kronecker delta,

$$\delta_{ij}(x) = a_{im}a_{jn}\delta_{mn}(y) = a_{im}a_{jm} = \delta_{ij}(x).$$

Thus, the transformation law is valid.

Assume that the alternating tensor is a third-order tensor. From the transformation law for its components equation (A.6.12),

$$e_{ijk}(x) = a_{im}a_{jn}a_{kp}e_{mnp}(y).$$

By expansion of both sides, it is seen that

$$a_{im}a_{jn}a_{kp}e_{mnp}(y) = e_{ijk}(y) \quad \text{times the determinant of a.}$$

By the orthogonality of the axes, the determinant of the direction cosines is one. Thus,

$$e_{ijk}(x) = e_{ijk}(y),$$

and the transformation law is valid.

Besides the inherent invariant property of a tensor, tensor components also possess invariant properties. For instance, A_iA_i is the same in any coordinate system, as can be easily seen in equation (A.6.6) and in fact is the square of the magnitude of the vector **A**. This is the only independent invariant that can be formed for a first-order tensor. For a second-order tensor, there are at most three independent *invariants*:

$$T_{ii} = I_T, \tag{A.6.16}$$

$$0.5(I_T^2 - T_{ij}T_{ji}) = II_T, \tag{A.6.17}$$

$$\text{determinate of } T_{ij} = III_T. \tag{A.6.18}$$

That these are invariants can be shown through use of the transformation laws.

Example A.6.4 Invariants
Verify that equation (A.6.9) is invariant under rotation of axes.

Solution. To do this, we use the transformation of components relation (equation (A.6.11))—that is,

$T_{ij}(x) = a_{im}a_{jn}T_{mn}(y)$. Then

$$T_{ii}(x) = T_{11}(x) + T_{22}(x) + T_{33}(x) = a_{1m}a_{1n}T_{mn}(y) + a_{2m}a_{2n}T_{mn}(y) + a_{3m}a_{3n}T_{mn}(y)$$

$$= (a_{1m}a_{1n} + a_{2m}a_{2n} + a_{3m}a_{3n})(a_{1p}a_{1q} + a_{2p}a_{2q} + a_{3p}a_{3q})T_{pq}(y)$$

$$= a_{ip}a_{iq}T_{pq}(y).$$

$T_{ij}(x) = a_{im}a_{jn}T_{mn}(y)$. Then

$$T_{ii}(x) = T_{11}(x) + T_{22}(x) + T_{33}(x) = a_{1m}a_{1n}T_{mn}(y) + a_{2m}a_{2m}T_{mn}(y) + a_{3m}a_{3n}T_{mn}(y)$$

$$= (a_{1m}a_{1n} + a_{2m}a_{2n} + a_{3m}a_{3n})T_{mn}(y) = a_{im}a_{in}T_{mn}(y).$$

$T_{ij}(x) = a_{im}a_{jn}T_{mn}(y)$. Then

$$T_{ii}(x) = T_{11}(x) + T_{22}(x) + T_{33}(x) = a_{1m}a_{1n}T_{mn}(y) + a_{2m}a_{2n}T_{mn}(y) + a_{3m}a_{3n}T_{mn}(y)$$

$$= (a_{1m}a_{1n} + a_{2m}a_{2n} + a_{3m}a_{3n})T_{mn}(y) = a_{im}a_{in}T_{mn}(y).$$

But from equation (A.6.6),

$$a_{im}a_{in} = \delta_{mn}, \quad T_{ii}(x) = \delta_{mn}T_{mn}(y) = T_{mm}(y). \quad \text{Thus,} \quad T_{ii}(x) = T_{ii}(y) = I_T.$$

From the preceding result, to prove $0.5(I_T^2 - T_{ij}T_{ji}) = II_T$ is invariant, we need only show that

$$T_{ij}(x)T_{ji}(x) = T_{ij}(y)T_{ji}(y).$$

Using equation (A.6.11), we have

$$T_{ij}(x)T_{ji}(x) = a_{im}a_{jn}T_{mn}(y)a_{jp}a_{iq}T_{pq}(y)$$

$$= a_{im}a_{iq}a_{jn}a_{jp}a_{iq}T_{mn}(y)T_{pq}(y) \text{ (after rearranging)}$$

$$= \delta_{mq}\delta_{np}T_{mn}(y)T_{pq}(y) \text{ (using equation (A.6.6))}$$

$$= T_{qn}(y)T_{nq}(y).$$

This is the desired result.
For the third result, we note from writing out the result in detail that

$$\text{determinate of } T_{ij}(x) = \begin{vmatrix} T_{11} & T_{12} & T_{13} \\ T_{21} & T_{22} & T_{23} \\ T_{31} & T_{32} & T_{33} \end{vmatrix} = \frac{e_{ijk}e_{pqr}T_{ip}T_{jq}T_{kr}}{6}.$$

Using the transformation law (equation (A.6.11)),

$$e_{ijk}e_{pqr}T_{ip}(x)T_{jq}(x)T_{kr}(x)$$
$$= e_{ijk}e_{pqr}a_{is}a_{pt}T_{st}(y)a_{ju}a_{qv}T_{uv}(y)a_{kw}a_{rz}T_{wz}(y)$$
$$- e_{ijk}e_{pqr}a_{is}a_{pt}a_{ju}a_{qv}a_{kw}a_{rz}T_{st}(y)T_{uv}(y)T_{wz}(y)$$

As can be seen by writing out both sides,

$$e_{ijk}a_{is}a_{ju}a_{kw} = e_{suw} \quad \text{determinant of } a, \text{ and}$$
$$e_{pqr}a_{pt}a_q \text{ var } z = e_{tvz} \quad \text{determinant of } a.$$

But the determinant of a is one, since the coordinates are orthogonal. Then

$$e_{ijk}e_{pqr}T_{ip}(x)T_{jq}(x)T_{kr}(x) = e_{suw}e_{tvz}T_{st}(y)T_{uv}(y)T_{wz}(y)$$
$$= e_{suw}e_{suw} \text{ times the determinant of } T(y)$$
$$= 6 \text{ times the determinant of } T(y),$$

that is the desired result.

That there are at most three invariants of a second-order tensor follows from the **Cayley-Hamilton theorem**. This theorem states that the products of components of a tensor are related by

$$T_{im}T_{mn}T_{nj} - I_T T_{im}T_{mj} + II_T T_{ij} - III_T T_{ij} = 0. \tag{A.6.19}$$

Proof of the Cayley-Hamilton theorem is as follows. By contraction (letting i equal j and summing),

$$T_{im}T_{mn}T_{ni} = I_T II_T = I_T(I_T^2 - 2II_T) - II_T I_T + 3III_T = I_T^3 - 3I_T II_T + 3III_T. \tag{A.6.20}$$

Premultiplication of equation (A.6.10) by T_{jk} or higher powers of **T** with the implied contraction shows the desired result. When there is only one subscript, equation (A.6.8) is the familiar form for the decomposition of a vector. When there are two subscripts, equation (A.6.8) is the three-dimensional generalization of Mohr's circle.

A.7 Tensors in Orthogonal Curvilinear Coordinates

The mathematics of tensors in orthogonal curvilinear coordinates is only a bit more complicated than that in Cartesian coordinates, although the application of the general results to a particular coordinate system can be tedious. In the following, we use orthogonal coordinate systems with coordinate axes y_1, y_2, y_3 and with orthogonal base vectors

$$\mathbf{g}_i = \mathbf{ds}/dy_i, \tag{A.7.1}$$

where **ds** is an elemental distance. Define

$$h_1 = \sqrt{\mathbf{g}_1 \cdot \mathbf{g}_1}, \quad h_2 = \sqrt{\mathbf{g}_2 \cdot \mathbf{g}_2}, \quad h_3 = \sqrt{\mathbf{g}_3 \cdot \mathbf{g}_3} \tag{A.7.2}$$

so that the distance between two infinitesimally distant points is

$$ds^2 = (h_1 dy_1)^2 + (h_2 dy_2)_2 + (h_3 dy_3)^2. \tag{A.7.3}$$

It is usually easier to use equation (A.5.4) to find the *h*s than equation (A.5.3) for a particular coordinate system.

The gradient and Laplacian of a vector are given by

$$\text{grad } \phi = \nabla \phi = \left(\frac{1}{h_1} \frac{\partial \phi}{\partial x_1}, \frac{1}{h_2} \frac{\partial \phi}{\partial x_2}, \frac{1}{h_3} \frac{\partial \phi}{\partial x_3} \right), \tag{A.7.4}$$

$$\nabla^2 \phi = \frac{1}{h_1 h_2 h_3} \left[\frac{\partial}{\partial x_1} \left(\frac{h_2 h_3}{h_1} \frac{\partial \phi}{\partial x_1} \right) + \frac{\partial}{\partial x_2} \left(\frac{h_3 h_1}{h_2} \frac{\partial \phi}{\partial x_2} \right) + \frac{\partial}{\partial x_3} \left(\frac{h_1 h_2}{h_3} \frac{\partial \phi}{\partial x_3} \right) \right]. \tag{A.7.5}$$

The divergence of a vector is given by

$$\text{div } \mathbf{V} = \nabla \cdot \mathbf{V} = \frac{1}{h_1 h_2 h_3} \left[\frac{\partial (h_2 h_3 V_1)}{\partial x_1} + \frac{\partial (h_3 h_1 V_2)}{\partial x_2} + \frac{\partial (h_1 h_2 V_3)}{\partial x_3} \right]. \tag{A.7.6}$$

The curl of a vector is given by

$$\text{curl } \mathbf{V} = \nabla \times \mathbf{V} = \frac{1}{h_2 h_3} \left[\frac{\partial (h_3 V_3)}{\partial x_2} - \frac{\partial (h_2 V_2)}{\partial x_3} \right] \mathbf{i} + \frac{1}{h_3 h_1} \left[\frac{\partial (h_1 V_1)}{\partial x_3} - \frac{\partial (h_3 V_3)}{\partial x_1} \right] \mathbf{j} \tag{A.7.7}$$

$$+ \frac{1}{h_1 h_2} \left[\frac{\partial (h_2 V_2)}{\partial x_1} - \frac{\partial (h_1 V_1)}{\partial x_2} \right] \mathbf{k}.$$

Equations (A.7.6) and (A.7.7) are useful in expressing the continuity equation and the vorticity vector in orthogonal curvilinear coordinates, and equation (A.7.5) for the continuity equation for an irrotational flow. For the rate of deformation the expressions for the tensor components are

$$d_{11} = \frac{1}{h_1} \left(\frac{\partial v_1}{\partial x_1} + \frac{v_2}{h_2} \frac{\partial h_1}{\partial h_2} + \frac{v_3}{h_3} \frac{\partial h_1}{\partial h_3} \right),$$

$$d_{22} = \frac{1}{h_2} \left(\frac{\partial v_2}{\partial x_2} + \frac{v_3}{h_3} \frac{\partial h_2}{\partial h_3} + \frac{v_1}{h_1} \frac{\partial h_2}{\partial h_1} \right),$$

$$d_{33} = \frac{1}{h_3} \left(\frac{\partial v_3}{\partial x_3} + \frac{v_1}{h_1} \frac{\partial h_3}{\partial h_1} + \frac{v_2}{h_2} \frac{\partial h_3}{\partial h_2} \right),$$

$$2d_{12} = 2d_{21} = \frac{h_3}{h_2} \frac{\partial}{\partial x_2} \left(\frac{v_3}{h_3} \right) + \frac{h_2}{h_3} \frac{\partial}{\partial x_3} \left(\frac{v_2}{h_2} \right), \tag{A.7.8}$$

$$2d_{23} = 2d_{32} = \frac{h_1}{h_3} \frac{\partial}{\partial x_3} \left(\frac{v_1}{h_1} \right) + \frac{h_3}{h_1} \frac{\partial}{\partial x_1} \left(\frac{v_3}{h_3} \right),$$

$$2d_{31} = 2d_{13} = \frac{h_2}{h_1} \frac{\partial}{\partial x_1} \left(\frac{v_2}{h_2} \right) + \frac{h_1}{h_2} \frac{\partial}{\partial x_2} \left(\frac{v_1}{h_1} \right).$$

The stress components are given by

$$\tau_{11} = (-p + \mu' \nabla \cdot \mathbf{v}) + 2\mu d_{11}, \quad \tau_{22} = (-p + \mu' \nabla \cdot \mathbf{v}) + 2\mu d_{22},$$

$$\tau_{33} = (-p + \mu' \nabla \cdot \mathbf{v}) + 2\mu d_{33}, \tag{A.7.9}$$

$$\tau_{12} = \tau_{21} = 2\mu d_{12}, \quad \tau_{23} = \tau_{32} = 2\mu d_{23}, \quad \tau_{31} = \tau_{31} = 2\mu d_{31},$$

or in index notation $\tau_{ij} = (-p + \mu' \nabla \cdot \mathbf{v}) \delta_{ij} + 2\mu d_{ij}$.

The acceleration is arrived at by first writing the convective terms in a correct vector form. As can be verified by writing out the components in Cartesian coordinates and comparing, that form is

$$(\mathbf{v} \cdot \nabla)\mathbf{v} = -\mathbf{v} \times (\nabla \times \mathbf{v}) + 0.5\nabla |\mathbf{v}|^2 = -\mathbf{v} \times \text{curl } \mathbf{v} + 0.5\text{grad} |\mathbf{v}|^2. \qquad (A.7.10)$$

The first term on the right is found in an arbitrary orthogonal system by applying equation (A.7.7), the second by use of equation (A.7.4).

The continuity equation is given by

$$\frac{\partial \rho}{\partial t} + \frac{1}{h_1 h_2 h_3} \left[\frac{\partial(h_2 h_3 \rho v_1)}{\partial y_1} + \frac{\partial(h_3 h_1 \rho v_2)}{\partial y_2} + \frac{\partial(h_1 h_2 \rho v_3)}{\partial y_3} \right] = 0. \qquad (A.7.11)$$

Since the Navier-Stokes equations involve taking the second derivative of a vector, necessary for the viscous terms, the procedure is much longer and more tedious than for the continuity equation or the Laplace equation. The acceleration terms can be found from equation (A.7.10), and the pressure gradient from equation (A.7.4). The remaining terms involve the operation $\nabla^2 \mathbf{v}$. Writing this out is a long process and would require several pages of equations. Instead, note that

$$\nabla^2 \mathbf{v} = \nabla(\nabla \cdot \mathbf{v}) - \nabla \times (\nabla \times \mathbf{v}), \qquad (A.7.12)$$

and equations (A.5.7) and (A.5.8) tell us how to compute the divergence and curl. The Navier-Stokes equations then are of the form

$$\rho \left[\frac{\partial \mathbf{v}}{\partial t} - \mathbf{v} \times \boldsymbol{\omega} + \nabla \left(\frac{1}{2} |\mathbf{v}|^2 \right) \right] = -\nabla p + \rho \mathbf{g} + \mu \left[\nabla(\nabla \cdot \mathbf{v}) - \nabla \times \boldsymbol{\omega} \right]. \qquad (A.7.13)$$

To illustrate the previous, consider the next two familiar coordinate systems.

A.7.1   Cylindrical Polar Coordinates

The element of length in this coordinate system is

$$ds^2 = dr^2 + (r d\theta)^2 + dz^2$$

with

$$(y_1, y_2, y_3) - (r, \theta, z)$$

The polar coordinates are expressed in terms of Cartesian coordinates by $r = \sqrt{x_1^2 + x_2^2}$, $\theta = \tan^{-1} x_2/x_1$, $z = x_3$.

Comparing equation (A.7.10) with equation (A.7.3), we see that $h_1 = h_3 = 1$, $h_2 = y_1 = r$. Then

$$\nabla^2 \phi = \frac{1}{r} \frac{\partial}{\partial r} \left(r \frac{\partial \phi}{\partial r} \right) + \frac{1}{r^2} \frac{\partial^2 \phi}{\partial \theta^2} + \frac{\partial^2 \phi}{\partial z^2}, \qquad (A.7.14)$$

$$\nabla \cdot \mathbf{V} = \frac{1}{r} \frac{\partial(r V_r)}{\partial r} + \frac{1}{r} \frac{\partial V_\theta}{\partial \theta} + \frac{\partial V_z}{\partial z}, \qquad (A.8.15)$$

$$\nabla \times \mathbf{V} = \left(\frac{1}{r} \frac{\partial V_z}{\partial \theta} - \frac{\partial V_\theta}{\partial z} \right) \mathbf{e}_r + \left(\frac{\partial V_r}{\partial z} - \frac{\partial V_z}{\partial r} \right) \mathbf{e}_\theta + \frac{1}{r} \left(\frac{\partial(r V_\theta)}{\partial r} - \frac{\partial V_r}{\partial \theta} \right) \mathbf{e}_z, \qquad (A.8.16)$$

$$d_{rr} = \frac{\partial u}{\partial r}, \quad d_{\theta\theta} = \frac{1}{r}\frac{\partial v}{\partial \theta} + \frac{u}{r}, \quad d_{zz} = \frac{\partial w}{\partial z},$$

$$2d_{r\theta} = 2d_{\theta r} = r\frac{\partial}{\partial r}\left(\frac{v}{r}\right) + \frac{1}{r}\frac{\partial u}{\partial \theta}, \quad 2d_{rz} = 2d_{zr} = \frac{\partial u}{\partial z} + \frac{\partial w}{\partial r},$$

$$2d_{z\theta} = 2d_{\theta z} = \frac{1}{r}\frac{\partial w}{\partial \theta} + \frac{\partial v}{\partial z}. \tag{A.8.17}$$

$$r^2 = x^2 + y^2, \quad \theta = \tan^{-1}\frac{y}{x}, \quad z = z.$$

The continuity equation is then

$$\frac{\partial \rho}{\partial t} + \frac{1}{r}\frac{\partial(\rho r u)}{\partial r} + \frac{1}{r}\frac{\partial(\rho v)}{\partial \theta} + \frac{\partial(\rho w)}{\partial r} = 0. \tag{A.7.18}$$

The Navier-Stokes equations for constant density and viscosity are

$$\rho\left(\frac{\partial u}{\partial t} + u\frac{\partial u}{\partial r} + \frac{v}{r}\frac{\partial u}{\partial \theta} + w\frac{\partial u}{\partial z} - \frac{v^2}{r^2}\right) = -\frac{\partial p}{\partial r} + \rho g_r + \mu\left(\nabla^2 u - \frac{u}{r^2} - \frac{2}{r^2}\frac{\partial v}{\partial \theta}\right),$$

$$\rho\left(\frac{\partial v}{\partial t} + u\frac{\partial v}{\partial r} + \frac{v}{r}\frac{\partial v}{\partial \theta} + w\frac{\partial v}{\partial z} + \frac{uv}{r^2}\right) = -\frac{1}{r}\frac{\partial p}{\partial \theta} + \rho g_\theta + \mu\left(\nabla^2 v - \frac{v}{r^2} + \frac{2}{r^2}\frac{\partial u}{\partial \theta}\right),$$

$$\tag{A.7.19}$$

$$\rho\left(\frac{\partial w}{\partial t} + u\frac{\partial w}{\partial r} + \frac{v}{r}\frac{\partial w}{\partial \theta} + w\frac{\partial w}{\partial z}\right) = -\frac{\partial p}{\partial z} + \rho g_\theta + \mu\nabla^2 w,$$

where the Laplacian ∇^2 is as given in equation (A.8.11) and $(u, v, w) = (v_r, v_\theta, v_z)$.

A.7.2   Spherical Polar Coordinates

The element of length in this coordinate system is

$$ds^2 = dR^2 + (R d\beta)^2 + (R\sin\beta \, d\vartheta)^2 \text{ with } (y_1, y_2, y_3) = (R, \beta, \vartheta) \text{ and so}$$

$$h_1 = 1, \quad h_2 = y_1 = R, \quad h_3 = y_1 \sin y_2 = R\sin\beta.$$

In terms of Cartesian coordinates, $R^2 = x_1^2 + x_2^2 + x_3^2$, $\beta = \cos^{-1}\frac{x_3}{R}$, $\theta = \tan^{-1}\frac{x_2}{x}$. Then

$$\nabla\phi = \frac{\partial\phi}{\partial R}\mathbf{e}_R + \frac{1}{R}\frac{\partial\phi}{\partial\beta}\mathbf{e}_\beta + \frac{1}{R\sin\beta}\frac{\partial\phi}{\partial\theta}\mathbf{e}_\theta, \tag{A.7.20}$$

$$\nabla^2\phi = \frac{1}{R^2}\frac{\partial}{\partial R}\left(R^2\frac{\partial\phi}{\partial R}\right) + \frac{1}{R^2\sin\beta}\frac{\partial}{\partial\beta}\left(\sin\beta\frac{\partial\phi}{\partial\beta}\right) + \frac{1}{R^2\sin^2\beta}\frac{\partial^2\phi}{\partial\theta^2}, \tag{A.7.21}$$

$$\nabla\cdot\mathbf{v} = \frac{1}{R^2}\frac{\partial(R^2 u)}{\partial R} + \frac{1}{R\sin\beta}\frac{\partial(v\sin\beta)}{\partial\beta} + \frac{1}{R\sin\beta}\frac{\partial w}{\partial\theta}, \tag{A.7.22}$$

$$\nabla\times\mathbf{v} = \frac{1}{R\sin\beta}\left(\frac{\partial(w\sin\beta)}{\partial\beta} - \frac{\partial v}{\partial\theta}\right)\mathbf{e}_R + \frac{1}{R\sin\beta}\left(\frac{\partial u}{\partial\theta} - \frac{\partial(wR\sin\beta)}{\partial R}\right)\mathbf{e}_\beta$$

$$+ \frac{1}{R}\left(\frac{\partial(Rv)}{\partial R} - \frac{\partial u}{\partial\beta}\right)\mathbf{e}_\theta, \tag{A.7.23}$$

$$d_{RR} = \frac{\partial u}{\partial R}, \quad d_{\beta\beta} = \frac{1}{R}\frac{\partial v}{\partial\beta} + \frac{u}{R}, \quad d_{\theta\theta} = \frac{1}{R\sin\beta}\frac{\partial w}{\partial\theta} + \frac{u}{R} + \frac{v\cot\beta}{R},$$

$$2d_{\beta\theta} = 2d_{\theta\beta} = \frac{\partial}{\partial \beta}\left(\frac{w}{\sin\beta}\right) + \frac{1}{R\sin\beta}\frac{\partial v}{\partial \theta},$$

$$2d_{\theta R} = 2d_{R\theta} = \frac{1}{R\sin\beta}\frac{\partial u}{\partial \theta} + R\frac{\partial}{\partial R}\left(\frac{w}{R}\right), \qquad \text{(A.7.24)}$$

$$2d_{R\beta} = 2d_{\beta R} = R\frac{\partial}{\partial R}\left(\frac{v}{R}\right) + \frac{1}{R}\frac{\partial u}{\partial \beta}.$$

The continuity equation is given by

$$\frac{\partial \rho}{\partial t} + \frac{1}{R^2}\frac{\partial\left(\rho R^2 u\right)}{\partial R} + \frac{1}{R\sin\beta}\frac{\partial\left(\rho v \sin\beta\right)}{\partial \beta} + \frac{1}{R\sin\beta}\frac{\partial\left(\rho w\right)}{\partial \theta} = 0. \qquad \text{(A.7.25)}$$

The Navier-Stokes equations for constant density and viscosity are

$$\rho\left(\frac{\partial u}{\partial t} + u\frac{\partial u}{\partial R} + \frac{v}{R}\frac{\partial u}{\partial \beta} + \frac{w}{R\sin\beta}\frac{\partial u}{\partial \theta} - \frac{v^2 + w^2}{R^2}\right)$$
$$= -\frac{\partial p}{\partial R} + \rho g_R + \mu\left(\nabla^2 u - \frac{2u}{R^2} - \frac{2}{R^2}\frac{\partial v}{\partial \beta} - \frac{2v\cot\beta}{R^2} - \frac{2}{R^2\sin\beta}\frac{\partial w}{\partial \theta}\right),$$

$$\rho\left(\frac{\partial v}{\partial t} + u\frac{\partial v}{\partial R} + \frac{v}{R}\frac{\partial v}{\partial \beta} + \frac{w}{R\sin\beta}\frac{\partial v}{\partial \theta} + \frac{uv - w^2\cot\beta}{R^2}\right)$$
$$= -\frac{1}{R}\frac{\partial p}{\partial R} + \rho g_\beta + \mu\left(\nabla^2 v - \frac{v}{R^2\sin\beta} + \frac{2}{R^2}\frac{\partial u}{\partial \beta} - \frac{2\cot\beta}{R^2\sin\beta}\frac{\partial w}{\partial \theta}\right), \qquad \text{(A.7.26)}$$

$$\rho\left(\frac{\partial w}{\partial t} + u\frac{\partial w}{\partial R} + \frac{v}{R}\frac{\partial w}{\partial \beta} + \frac{w}{R\sin\beta}\frac{\partial w}{\partial \theta} + \frac{uw + vw\cot\beta}{R^2}\right)$$
$$= -\frac{1}{R\sin\beta}\frac{\partial p}{\partial \theta} + \rho g_\theta + \mu\left(\nabla^2 w - \frac{w}{R^2\sin^2\beta} + \frac{2}{R^2\sin\beta}\frac{\partial u}{\partial \theta} + \frac{2\cot\beta}{R^2\sin\beta}\frac{\partial v}{\partial \theta}\right),$$

where the Laplacian ∇^2 is as given in equation (A.8.19) and $(u, v, w) = (v_R, v_\beta, v_\theta)$.

A.8   Tensors in General Coordinates

In Cartesian coordinates, dealing with tensors is particularly simple because the base vectors are constants and thus for a vector (as an example)

$$\mathbf{V} = V_j\,\mathbf{g}_j. \qquad \text{(A.8.1)}$$

In general coordinates, the base vectors vary from point to point, and when differentiating a tensor, the base vectors have to be differentiated as well. This leads to greater complications than might at first be expected, since more than one set of base vectors can be defined and their direction cosines will generally differ. One starts to deal with what are called **contravariant** and **covariant** base vectors and tensor components, depending on the type of base vector that is used. Further, the various tensor components and base vectors will generally differ in their physical dimensions. (For example, in cylindrical polar coordinates two of the covariant components of the velocity vector will have dimensions of length/time, while the third will have dimensions of just reciprocal time.) In our work, we are interested in dealing with the **physical components**—that is, components that all have the same physical dimensions. Unfortunately, physical components do not obey the simple transformation laws of tensors.

The starting point for a study of general tensors is geometry. As in studying rates of deformation, consider two adjacent points separated by a vector distance **dr**. (Boldface will be used to denote both vectors and tensors of higher order.) If we have a coordinate system (x^1, x^2, x^3), we can write

$$d\mathbf{r} = \frac{\partial \mathbf{r}}{\partial x^i} dx^i. \tag{A.8.2}$$

This implies that the vectors $\partial \mathbf{r}/\partial x^i$ are locally tangent to the x coordinate system. So a good choice for base vectors tangent to our coordinate system is

$$\mathbf{g}_i = \frac{\partial \mathbf{r}}{\partial x^i}. \tag{A.8.3}$$

Since the magnitude ds of the distance between the two neighboring points is given by

$$ds^2 = d\mathbf{r} \cdot d\mathbf{r} = \mathbf{g}_i dx^i \cdot \mathbf{g}_j dx^j = \mathbf{g}_i \cdot \mathbf{g}_j dx^i dx^j = g_{ij} dx^i dx^j, \tag{A.8.4}$$

the g_{ij} given by $g_{ij} = \mathbf{g}_i \cdot \mathbf{g}_j$ is suitable as the covariant components of the symmetric *metric tensor*. Depending on the coordinate system, the dimensions of the various g_{ij} components can vary. In spherical polar coordinates, for instance, one has dimensions of length, while two are dimensional.

Returning to equation (A.8.4), we could have written

$$ds^2 = \frac{\partial s}{\partial x^i} \frac{\partial s}{\partial x^j} dx^i dx^j. \tag{A.8.5}$$

Comparison with equation (A.8.4) implies that an alternate definition of the covariant metric tensor components is

$$g_{ij} = \frac{\partial s}{\partial x^i} \frac{\partial s}{\partial x^j}. \tag{A.8.6}$$

If we introduce a second coordinate system **y** at the same point we are considering, notice that using the product rule of calculus, we can write

$$d\mathbf{r} = \mathbf{g}_i(x) dx^i = \frac{\partial \mathbf{r}}{\partial x^i} \frac{\partial x^i}{\partial y^j} dy^j = \mathbf{g}_j(y) dy^j.$$

Thus,

$$\mathbf{g}_j(y) = \frac{\partial x^i}{\partial y^j} \mathbf{g}_i(x). \tag{A.8.7}$$

This is the transformation law for covariant components of a vector. In general, the transformation law for covariant components of a tensor of order n is given by

$$A_{\alpha\beta\gamma\cdots}(y) = \frac{\partial x^i}{\partial y^\alpha} \frac{\partial x^j}{\partial y^\beta} \frac{\partial x^k}{\partial y^\gamma} \cdots A_{ijk\cdots}(x), \tag{A.8.8}$$

where there are n subscripts on the A and n partial derivatives. The partial derivatives act like direction cosines.

We have assumed here that a metric measuring the distance between two neighboring points exists in our space. Such a space is called a ***Riemannian space***. Also, our discussion is limited to three-dimensional spaces. Nonmetric spaces exist, but their applicability to fluid mechanics is doubtful. Generalization to higher dimensions is trivial.

Contravariant base vector components can also be defined using cross-products and the alternating tensor. Letting

$$\mathbf{g}_i \times \mathbf{g}_j = e_{ijk}\mathbf{g}^k, \qquad \text{(A.8.9)}$$

this definition serves the purpose of having the new base vectors orthogonal to the old, but they will not necessarily be tangent to the coordinate system. Equation (A.8.10) can be solved to explicitly give the contravariant base vectors as

$$\mathbf{g}^1 = \mathbf{g}_2 \times \mathbf{g}_3/\sqrt{g}, \quad \mathbf{g}^2 = \mathbf{g}_3 \times \mathbf{g}_1/\sqrt{g}, \quad \mathbf{g}^3 = \mathbf{g}_1 \times \mathbf{g}_2/\sqrt{g},$$

$$\text{where } \sqrt{g} = \mathbf{g}_1 \cdot (\mathbf{g}_2 \times \mathbf{g}_3), \quad g = \det(g_{ij}). \qquad \text{(A.8.10)}$$

In a manner similar to that just used, the components of the symmetric conjugate metric tensor are defined by

$$g^{ij} = \mathbf{g}^i \cdot \mathbf{g}^j. \qquad \text{(A.8.11)}$$

The general transformation law for contravariant components is given by

$$A^{\alpha\beta\gamma\cdots}(y) = \frac{\partial y^\alpha}{\partial x^i}\frac{\partial y^\beta}{\partial x^j}\frac{\partial y^\gamma}{\partial x^j}\cdots A^{ijk\cdots}(x). \qquad \text{(A.8.12)}$$

What is the relationship between the contravariant and covariant tensor components? With g denoting the determinant of g_{ij} as previously, let Δ^{ij} denote the cofactor of g_{ij}. From equation (A.8.10) we have $g^{ij} = \Delta^{ij}/g$, so

$$g_{ij}g^{jk} = g_{ij}\Delta^{jk}/g = \delta_i^k, \qquad \text{(A.8.13)}$$

where the δ are the components of the Kronecker delta. Thus, the contravariant gs play the role of the inverse of the covariant gs. It can be shown in a similar fashion that

$$\mathbf{g}_i = g_{ij}\mathbf{g}^j \quad \text{and} \quad \mathbf{g}^i = g^{ij}\mathbf{g}_j. \qquad \text{(A.8.14)}$$

The metric tensors can also be used to raise and lower indices. Since it can be shown from the transformation laws that $\mathbf{V} = V_i\mathbf{g}^i = V^i\mathbf{g}_i$, it follows that

$$V_i = g_{ij}V^j \quad \text{and} \quad V^i = g^{ij}V_j. \qquad \text{(A.8.15)}$$

Similarly for a second-order tensor,

$$V_{ij} = g_{ik}g_{jl}V^{kl} \quad \text{and} \quad V^{ij} = g^{ik}g^{jl}V_{kl}, \qquad \text{(A.8.16)}$$

and so on for higher-order tensors. ***Mixed components*** could also be introduced by operations such as

$$V_i^k = g_{il}V^{lk} \quad \text{and} \quad V_k^i = g^{jl}V_{kl}, \qquad \text{(A.8.17)}$$

Notice that in all of the preceding, the superscripts are the same on both sides of the equals sign, as are the subscripts. This must always be true for the equation to be valid. For example, the invariants used in the Cayley-Hamilton theorem for second-order symmetric tensors in general coordinates are

$$I = A_i^i, \quad II = \frac{1}{2}\left(A_{ij}A^{ji} - I^2\right),$$

$$III = \frac{1}{6}\left(2A_j^iA_k^jA_i^k - 3A_j^iA_i^jI + I^3\right) = \det\left(A_j^i\right).$$

They can be used to find the inverse of A (the inverse is defined such that $\mathbf{A}^{-1}\mathbf{A} = \boldsymbol{\delta}$) by

$$\left(A^{-1}\right)_i^k = \left(A_{ij}A^{jk} - A_i^k I - \delta_i^k II\right)/III.$$

Components of tensors obey all of the familiar associative and distributive laws.

With this somewhat tedious explanation out of the way, we can proceed to the calculus of tensors. Starting with the vector $\mathbf{A} = A^i\mathbf{g}_i$, it follows that $d\mathbf{A} = (\mathbf{g}_i\frac{\partial A^i}{\partial x^j} + A^i\frac{\partial \mathbf{g}_i}{\partial x^j})dx^j$.

It must be that the derivatives of the base vectors can be written in terms of the base vectors, so let

$$\frac{\partial \mathbf{g}_i}{\partial x^j} = \begin{Bmatrix} \alpha \\ i\ j \end{Bmatrix}\mathbf{g}_\alpha.$$

We then have

$$d\mathbf{A} = \left(\mathbf{g}_i\frac{\partial A^i}{\partial x^j} + A^i\begin{Bmatrix} \alpha \\ i\ j \end{Bmatrix}\mathbf{g}_\alpha\right)dx^j = \mathbf{g}_i\left(\frac{\partial A^i}{\partial x^j} + A^\alpha\begin{Bmatrix} i \\ \alpha\ j \end{Bmatrix}\right)dx^j.$$

We see that the quantities in parentheses are a valid components of a tensor and therefore define the covariant derivative of contravariant components by

$$A^i\big|_j = \frac{\partial A^i}{\partial x^j} + A^\alpha\begin{Bmatrix} i \\ \alpha\ j \end{Bmatrix}. \tag{A.8.18}$$

By a similar calculation, find that define the covariant derivative of covariant components by

$$A_i\big|_j = \frac{\partial A_i}{\partial x^j} - A_\alpha\begin{Bmatrix} \alpha \\ i\ j \end{Bmatrix}. \tag{A.8.19}$$

Contravariant derivatives are defined by $A^i\big|^j = A^i\big|_k g^{kj}$.

The symbols introduced to relate the derivatives of the base vectors to the base vectors themselves are called ***Christoffel symbols*** and are given by

Christoffel symbol of second kind: $\begin{Bmatrix} k \\ i\ j \end{Bmatrix} = \begin{Bmatrix} k \\ j\ i \end{Bmatrix} = [i\ j, \alpha]\,g^{\alpha k},$

Christoffel symbol of first kind: $[i\ j, k] = [j\ i, k] = \dfrac{1}{2}\left(\dfrac{\partial g_{ik}}{\partial x^j} + \dfrac{\partial g_{jk}}{\partial x^i} - \dfrac{\partial g_{ij}}{\partial x^k}\right).$

$$\tag{A.8.20}$$

Christoffel symbols are not components of any tensor, but when combined with the partial derivatives as in equations (A.8.18) and (A.8.19), they produce tensor components. They are responsible for the additional terms we have seen when working in cylindrical and spherical coordinates. Since even with symmetry, there are possibly 21 different components to be calculated, when necessary, the wise person resorts to symbolic processors for their calculation!

With this in hand, representing the Navier-Stokes equations in general coordinates is elementary: Just replace partial derivatives by covariant or contravariant derivatives. The continuity equation then becomes

$$\frac{\partial \rho}{\partial t} + \left(\rho v^i\right)\big|_i = 0, \tag{A.8.21}$$

and the momentum equation is

$$\rho\left(\frac{\partial v^i}{\partial t} + v^j v^i\big|_j\right) = \tau^i_j\big|^j + \rho g^i,$$

$$\tau^i_j = \left(-p + \mu' v^k\big|_k\right)\delta^i_j + \mu\left(v^i\big|_j + v_j\big|^i\right).$$

$\qquad$(A.8.22)

It is easy to write equations in general coordinates: Start with the equations in Cartesian coordinates, replace partial spatial derivatives with covariant derivatives, and pay attention to being consistent in subscripts and superscripts. And it is relatively easy to operate on them and manipulate them, but reducing them to any given coordinate system than the Cartesian one (the Christoffel symbols vanish in Cartesian coordinates) requires a good deal of patience.

One convenient device for computation of the Christoffel symbols is to start with equation (A.8.4) and to write it as

$$T = \left(\frac{ds}{dt}\right)^2 = g_{ij}v^i v^j, \quad \text{where } v^i = \frac{dx^i}{dt}.$$

$\qquad$(A.8.23)

Next, compute

$$f_i = \frac{1}{2}\left[\frac{d}{dt}\left(\frac{\partial T}{\partial v^i}\right) - \frac{\partial T}{\partial x^i}\right].$$

$\qquad$(A.8.24)

It can be shown that

$$f_i = g_{ij}\frac{d^2 x^j}{dt^2} + [j\,k,\,i]\,v^j v^k.$$

$\qquad$(A.8.25)

Comparison of the coefficients of $v^j v^k$ in equations (A.8.24) and (A.8.25) yield the Christoffel symbols more conveniently than by computation directly from the definition.

An account of how to determine *physical components* of tensors can be found in Truesdell (1953). Briefly, physical base vectors are defined by

$$\mathbf{g}(i) = \mathbf{g}_i / \sqrt{g_{(ii)}} = g_{ij}\mathbf{g}^j / \sqrt{g_{(ii)}},$$

$\qquad$(A.8.26)

where the summation convention is suspended on indices enclosed in parentheses. For a vector, then, since

$$\mathbf{V} = V^i \mathbf{g}_i = V^i \sqrt{g_{(ii)}}\,\mathbf{g}(i) = V_j g^{ij}\sqrt{g_{(ii)}}\,\mathbf{g}(i),$$

$\qquad$(A.8.27)

it follows that the physical components of $\mathbf{V}$ are given by

$$V(i) = V^i \sqrt{g_{(ii)}} = V_j g^{ij}\sqrt{g_{(ii)}}.$$

$\qquad$(A.8.28)

Physical components of higher-order tensors can be computed in a similar manner.

References

Abbott, I. H., & von Doenhoff, A. E., **Theory of Wing Sections**, Dover Press, New York, 1959.

Alexandrou, A. N., Numerical simulation of separated flows past bluff bodies, PhD dissertation, U. Michigan, 1986.

Arya, S. P. S., & Plate, E. J., J. Atmos. Sci., vol. **26**, page 656 ff, 1969.

Baldwin, B. S., & Lomax, H., Thin layer approximation and algebraic model for separated turbulent flows, AIAA paper 78–257, 1978.

Batchelor, G. K., Kolmogoroff's theory of locally isotropic turbulence, Proc. Cambridge Phil. Soc., vol. **43**, part 4, 1947.

Batchelor, G. K., & Townsend, A. A., Decay of isotropic turbulence in the initial period, Proc. Roy. Soc. London series A, vol. **193**, pages 539–558, 1948.

Batchelor, G. K., **An Introduction to Fluid Mechanics**, Cambridge University Press, Cambridge, 1967.

Batchelor, G. K., **The Theory of Homogeneous Turbulence**, 2nd edition, Cambridge University Press, Cambridge, 1971.

Bénard, H., Les tourbillons cellulaires dans une nappe liquide, Rev. Gén. Sciences Pure Appl. **II**, pages 1261–1271 and 1309–1328, 1900.

Bénard, H., Les tourbillons cellulaires dans une nappe liquide transportant de la chaleur par convection en régime permanent, Ann. Chim. Phys., vol. **23**, pages 62–144, 1901.

Benjamin, T. B., & Ursell, F., The stability of the plane free surface of a liquid in vertical periodic motion, Proc. Roy. Soc. London series A, vol. **225**, page 505 ff, 1954.

Binnie, A. M., Waves in an open oscillating tank, Engineering, vol. 151, page 224 ff, 1941.

Birkhoff, G., **Hydrodynamics**, Princeton University Press, Princeton New Jersey, (1953).

Blasius, H., Grenzschichtenin Flussigkeiten mit kleiner Reibung, Zeit. Math. Phys., vol. **56**, pages 1–37, 1908. (English translation: NACA TM 125 pages.)

Blasius, H., Functionentheoretische Methoden in der Hydrodynamik, Zeit. Math. Phys., vol. **58**, page 90 ff, 1910.

Bradbury, L. J. S., The structure of a self-preserving turbulent plane jet, J. Fluid Mech., vol. **23**, pages 31–64, 1965.

Bradshaw, P., Ferriss, D. H., & Johmon, R. F., Turbulence in the noise-producing region of a circular jet, J. Fluid Mech., vol. **19**, pages 591–624, 1964.

Bradshaw, P., The understanding and prediction of turbulent flow, J. AIAA, vol. **76**, pages 403–418, 1972.

Bradshaw, P., editor, **Turbulence**, 2nd edition, vol. **12** in **Topics in Applied Physics**, Springer-Verlag, Berlin, 1978.

Bradshaw, P., Cebeci, T., & Whitelaw, J. H., **Engineering Calculation Methods for Turbulent Flow**, Academic Press, New York, 1981.

Brennen, C., A numerical solution of axisymmetric cavity flows, J. Fluid Mech., vol. **37**, pages 671–688, 1969.

Bulirsch, R., & Stoer, J., Numerical treatment of ordinary differential equations by extrapolation methods, Numer. Math., vol. **8**, pages 1–13, 1966.

Carnahan, B., Luther, H. A., & Wilkes, J. O., **Applied Numerical Methods**, J. Wiley & Sons, New York, 1969.

Cebeci, T., & Smith, A. M. O., Numerical solution of the turbulent boundary layer equations, Douglas Aircraft division report DAC 33735, 1967.

Cebeci, T., & Smith, A. M. O., Analysis of turbulent boundary layers, in **Appl. Math. & Mech.**, vol. **15**, Academic Press, New York, 1974a.

Cebeci, T., Calculation of three-dimensional boundary layers 1: swept infinite cylinder and small crossflow, J. AIAA, vol. **12**, pages 779–786, 1974b.

Cebeci, T., & Abbott, D. E., Boundary layers on a rotating disk, J. AIAA, vol. **13**, pages 829–832, 1975.

Cebeci, T., et al., Calculation of three-dimensional compressible boundary layers on arbitrary wings, Proc. NASA-Langley Conf. Aerodyn. Anal. Requiring Adv. Computers, 1975.

Champagne, F., Frieke, C. A., LaRue, J. C., & Wyngard, J., Flux measurements, flux-estimate techniques, and fine-scale turbulence measurements in the unstable surface layer over land, J. Atmos. Sci., vol. **34**, pages 515–530, 1977.

Chandrasekhar, S., **Hydrodynamic and Hydromagnetic Stability**, Clarendon Press, Oxford, 1961.

Chevray, R., & Kovasznay, L. S. G., Turbulence measurements in the wake of a thin flat plate, J. AIAA, vol. **7**, page 1641 ff, 1969.

Chorin, A. J., Vortex sheet approximation of boundary layers, J. Comp. Physics, vol. **27**, pages 428–442, 1978.

Chow, C-Y., **An Introduction to Computational Fluid Mechanics**, J. Wiley, New York, 1979.

Christiansen, J. P., Numerical simulations of hydrodynamics by the method of point vortices, J. Comp. Physics, vol. **13**, pages 363–379, 1973.

Cole, J., **Perturbation Methods in Applied Mathematics**, Ginn-Blaisdell, New York, 1968.

Coleman, B. D., Markovitz, H., & Noll, W., **Viscometric Flows of Non-Newtonian Fluids, Theory and Experiment**, Springer-Verlag, New York, 1966.

Coles, D., A note on Taylor instability in circular Couette flow, J. Appl. Mech., vol. **34**, pages 529–534, 1967.

Coutte, M., Études sur le frottement des liquides, Ann. Chim. Phys., vol. **6**, pages 433–510, 1880.

Craya, A., Contribution a l'analyse de la turbulence associée a des vitesses moyennes, Publications Scientifiques du Ministere de l'Air no. 345, 1958.

Crocco, L., Atti di Guidonia XVII, vol. **7**, page 118 ff, 1939. Translated in Aeron. Res. Council (Britain) Rept. 4,482, 1939.

Cummins, W. E., The forces and moments on a body moving in an arbitrary potential stream, U.S. Navy David Taylor Model Basin Report 708, 1953.

Curl, M., & Graebel, W. P., Application of invariant imbedding techniques to flow instability problems, SIAM J. Appl. Math., vol. **23**, pages 380–394, 1972.

Davey, A., On the numerical solution of difficult eigenvalue problems, J. Computational Physics, vol. **24**, pages 331–338, 1977.

Davey, A., On the removal of the singularities from the Riccati method, J. Computational Physics, vol. **30**, pages 137–144, 1979.

Dean, W. R., Note on the motion of fluids in a curved pipe, Phil. Mag., vol. **4**, pages 208–223, 1927.

Dean, W. R., Fluid motion in a curved channel, Proc. Roy. Soc. London series A, vol. **121**, pages 402–420, 1928.

Edwards, S. F., The theoretical dynamics of homogeneous turbulence, J. Fluid Mech., vol. **18**, pages 239–273, 1964.

Edwards, S. F., Turbulence in hydrodynamics and plasma physics, Int. Conf. Plasma Physics, Trieste, Int. Atomic Energy, Vienna, page 595 ff, 1965.

Einstein, A., A new determination of molecular dimensions, Ann. der Physik IV, band **19**, page 289 ff, 1906.

Einstein, A., Correction to my paper "A new determination of molecular dimensions," Ann. der Physik IV, band **24**, 1911.

Ekman, V. W., On the influence of the earth's rotation on ocean currents, Arkiv Mat. Astr. Fys., vol. **2**(11), pages 1–52, 1905.

Falkner, V. M., & Skan, S. W., Aero. Res. Council Rept. and Memo. no. 1314, 1930.

Favre, A., Kovasznay, L. S. G., Dumas, R., Gaviglio, J., & Coantic, M., La turbulence en mecanique des fluides, Gauther-Villars, Paris, 1976.

Ferziger, J. H., **Numerical Methods for Engineering Applications**, J. Wiley, New York, 1981.

Forbes, L. K., An algorithm for 3-dimensional free-surface problems in hydrodynamics, J. Comp. Physics, vol. **82**, pages 330–347, 1989.

Fujii, T., & Imura, H., Natural convection from a plate with arbitrary inclination, Int. J. Heat Mass Transfer, vol. **15**, pages 755–767, 1972.

Goldstein, S., The steady flow of viscous fluid past a fixed spherical obstacle at small Reynolds numbers, Proc. Roy. Soc. London series A, vol. **123**, pages 225–235, 1929. (A numerical error was corrected by D. Shanks, J. Math. & Physics, vol. **34**, page 36 ff, 1955.)

Goldstein, S., Concerning some solutions of the boundary layer equations in hydrodynamics, Proc. Camb. Phil. Soc., vol. **26**, part 1, pages 1–30, 1930.

Goldstein, S., editor, **Modern Developments in Fluid Dynamics**, vol. **I & II**, Oxford University Press, London, 1938. Reprinted by Dover Publications, Inc., New York, 1965.

Goldstein, S., **Lectures on Fluid Mechanics**, John Wiley Interscience, New York, 1960.

Graebel, W. P., The stability of a Stokesian fluid in Couette flow, Physics of Fluids, vol. **4**, pages 362–368, 1961.

Graebel, W. P., The hydrodynamic stability of a Bingham fluid in Couette flow, IUTAM Symposium, **Second-Order Effects in Elasticity, Plasticity, and Fluid Dynamics**, Pergamon Press, New York, 1964.

Graebel, W. P., On determination of the characteristic equations for the stability of parallel flows, J. Fluid Mech., vol. **24**, pages 497–508, 1966.

Grant, H. L., Stewart, R. W., & Moillet, A., Turbulence spectra from a tidal channel, J. Fluid Mech., vol. **12**, pages 241–268, 1962.

Granville, P. S., Baldwin-Lomax factors for turbulent boundary layers in pressure gradients, J. AIAA, vol. **25**, pages 1624–1627, 1987.

Hamel, G., Jahresbericht der Deutschen Mathematiker-Vereinigung, vol. **25**, page 34 ff, 1917.

Hanjalic, K., & Launder, B. E., Fully developed asymmetric flow in a plane channel, J. Fluid Mech., vol. **51**, pages 301–336, 1972a.

Hanjalic, K., and Launder, B. E., A Reynolds stress model of turbulence and its application to thin shear flows, J. Fluid Mech., vol. **52**, pages 609–628, 1972b.

Hartree, D. R., On an equation occurring in Falkner and Skan's approximate treatment of the equations of the boundary layer, Proc. Camb. Phil. Soc., vol. **33**, pages 223–239, 1937.

Heisenberg, W., Über Stabilität und Turbulenz von Flüssigkeitsströmen, Annal. Physik Leipzig, vol. **74**, pages 557–627, 1924.

Herring, J. R., & Kraichnan, R. H., Comparison of some approximations for isotropic turbulence, Proc. Symp. Statistical Models and Turbulence, San Diego, Springer-Verlag, Berlin, 1972.

Hess, J. L., & Smith, A. M. O., Calculation of non-lifting potential flow about arbitrary three-dimensional bodies, Douglas Aircraft Report no. ES 40622, 1962.

Hess, J. L., & Smith, A. M. O., Calculation of potential flow about arbitrary bodies, Progress in Aeronautical Sciences, vol. **8**, Pergamon Press, 1967.

Hiemenz, K., Die Grenzschicht an einem in den gleichförmigen Flüssigkeitsstrom eingetauchten geraden Kreiszylinder, thesis, U. Göttingen, 1911.

Hinze, J. O., **Turbulence, an Introduction to its Mechanism and Theory**, 2nd edition, McGraw-Hill, New York, 1975.

Hocking, L. M., The behavior of clusters of spheres falling in a viscous fluid—part 2. Slow motion theory, J. Fluid Mech., vol. **20**, pages 129–139, 1964.

Homann, F., Der Einfluss grösser Zähigkeit bei der Strömung um den Zylinder und um die Kugel, Z. Angew. Math. Mech., vol. **16**, pages 153–164, 1936.

Hong, S-K., Unsteady separated flow around a two-dimensional bluff body near a free surface, PhD dissertation, U. Michigan, 1988.

Howard, L. N., Note on a paper of John W. Miles, J. Fluid Mech., vol. **10**, pages 509–512, 1961.

IMSL Math/Library, FORTRAN Subroutines for Mathematical Applications., IMSL, Houston, 1987.

Inui, T., Study on wave-making resistance of ships, Soc. Naval Arch. of Japan, 60th anniversary series, vol. **2**, pages 173–355, Tokyo, 1957.

Jammer, M., **Concepts of Mass in Contemporary Physics and Philosophy**, Princeton University Press, Princeton, 1999.

Jeffrey, G. B., Phil. Mag. (6), vol. **29**, page 455 ff, 1915.

Johnson, D. S., Velocity and temperature fluctuation measurements in a turbulent boundary layer downstream of a stepwise discontinuity in wall temperature, J. Applied Mech., vol. **26**, pages 325–336, 1959.

Joseph, D. D., **Stability of Fluid Motions**, Springer Tracts in Natural Philosophy, vols. **27 & 28**, Springer, Berlin, 1976.

Kaplun, S., The role of coordinate systems in boundary layer theory, Z. Angew. Math. Phys., vol. **5**, pages 111–135, 1954.

Kaplun, S., Low Reynolds number flow past a circular cylinder, J. Math. & Mech., vol. **6**, pages 595–603, 1957.

Kaplun, S., **Fluid Mechanics and Singular Perurbations**, editors. Lagerstrom, Howard, & Liu, Academic Press, New York, 1967.

Kaplun, S., & Lagerstrom, P. A., Asymptotic expansions of Navier-Stokes solutions for small Reynolds numbers, J. Math. Mech., vol. **6**, pages 585–593, 1957.

Kármán, T. von, Über Laminaire und Turbulente Reibung, Z. Angew. Math. Mech., vol. **1**, pages 262–268, 1921.

Kármán, T. von, Berechnung der Druckverteilung an Luftschiffkörpern, Abhandl. Aero-dynamischen Inst. Tech. Hoch. Aachen, vol. **6**, pages 3–17, 1927.

Kármán, T. von, On the theory of turbulence, Proc. Natl. Acad. Sci. U.S., vol. **23**, page 98 ff, 1937a.

Kármán, T. von, The fundamentals of the statistical theory of turbulence, J. Aeron. Sci., vol. **4**, page 131 ff, 1937b.

Kármán, T. von, Some remarks on the statistical theory of turbulence, Proc. 5th Intl. Congr. Appl. Mech., Cambridge, MA, page 347 ff, 1938.

Kelvin, Lord (Thomson, W.), **Popular Lectures and Addresses**, vol. **3**, pages 481–488, Macmillan, London, 1891.

Klebanoff, P. S., Characteristics of turbulence in a boundary layer with zero pressure gradient, NACA. Rept. no. 1247, 1955.

Kline, S. J., Cantwell, B. J., & Lilly, G. M., **Proceedings of the 1980–1981 AFOSR-HTTM Stanford Conference on Complex Turbulent Flows**, Stanford Univ., Stanford, 1981.

Kolmogoroff, A. N., On degeneration of isotropic turbulence in an incompressible viscous liquid, Compt. Rend. Acad. Sci. U. S. S. R., vol. **31**, page 538 ff, 1941.

Korteweg, D. J., & de Vries, G., On the change in form of long waves advancing in a rectangular canal and on a new type of long stationary waves, Philosophical Magazine, 5th Series, vol. **39**, pages 433–443, 1895.

Koschmeider, E. L., **Bénard Cells and Taylor Vortices**, Cambridge University Press, Cambridge, 1993.

Kraichnan, R. H., Irreversible statistical mechanics of incompressible hydromagnetic turbulence, Phys. Rev., vol. **109**, pages 1407–1422, 1958.

Kraichnan, R. H., The structure of isotropic turbulence at very high Reynolds numbers, J. Fluid Mech., vol. **5**, pages 497–543, 1959.

Kraichnan, R. H., Dynamics of non-linear systems, J. Math. Physics, vol. **2**, pages 124–148, 1961.

Kraichnan, R. H., Kolmogorov's hypotheses and Eulerian turbulence theory, Physics Fluids, vol. **7**, pages 1723–1734, 1964.

Kraichnan, R. H., Lagrangian-history closure approximation for turbulence, Physics Fluids, vol. **8**, pages 575–598, 1965.

Kraichnan, R. H., Isotropic turbulence and inertial range structure in the abridged LHDI approximation, Physics Fluids, vol. **9**, pages 1728–1752, 1966.

Kraichnan, R. H., Convergents to turbulent functions, J. Fluid Mech., vol. **41**, pages 189–218, 1970.

Kraichnan, R. H., An almost-Markovian Galilean-invariant turbulence model, J. Fluid Mech., vol. **47**, pages 513–524, 1971.

Kwak, D., Reynolds, W. C., Ferziger, J. H., Three dimensional time-dependent computation of turbulent flow, Stanford U. Dept. Mech. Eng. Report TF-5, 1975.

Lagally, M., Berechnung der Krafte und Momente, die strömende Flüssigkeiten auf ihre Begrenzung ausüben Zeit, Z. Angew. Math. Mech., vol. **2**, page 409, 1922.

Lagerstrom, P. A., & Cole, J. D., Examples illustrating expansion procedures for the Navier-Stokes equations, J. Rat. Mech. Anal., vol. **4**, pages 817–882, 1955.

Lamb, H., **Hydrodynamics,** 6th edition, MacMillan, New York, 1932, reprinted by Dover Publications Inc., New York, 1945.

Landau, L. D., A new exact solution of the Navier-Stokes equation, C. R. Acad. Sci. U. S. S. R., vol. **43**, page 286 ff, 1944.

Landau, L. D., Doklady Acad. Sci. U. S. S. R., vol. **43**, page 286 ff, 1944.

Landau, L. D., & Lifshitz, E. M., **Fluid Mechanics, Course of Theoretical Physics**, vol. **6**, Pergamon Press, London, 1959.

Landweber, L., On a generalization of Taylor's virtual mass relation for Rankine bodies, Quart. Appl. Math., vol. **14**, page 51 ff, 1956.

Landweber, L., & Yih, C.-S., Forces, moments, and added masses for Rankine bodies, J. Fluid Mech., vol. **1**, pages 319–336, 1956.

Launder, B. E., Reece, G. J., & Rodi, W., Progress in the development of a Reynolds-stress turbulence closure, J. Fluid Mech., vol. **68**, pages 537–566, 1975.

Lax, P. D., Weak solutions of non-linear hyperbolic equations and their numerical computations, Comm. Pure Appl. Math., vol. **7**, pages 157–193, 1954.

Leith, C. E., Atmospheric predictability, J. Atmos. Sci., vol. **28**, page 145 ff, 1971.

Lesieur, M., Turbulence homogene et isotrope, aspects phenomenologiques et theories analytiques, Institute National Polytechnique de Grenoble, 1978.

Leslie, D. C., **Developments in the Theory of Turbulence**, Clarendon Press, Oxford, 1973.

Levich, V. G., **Physico-chemical Hydrodynamics**, Prentice-Hall, New York, 1962.

Lieberman, L. N., The second coefficient of viscosity, Phys. Rev., vol. **75**, pages 1415–1422, 1955.

Lin, C. C., On the stability of two-dimensional parallel flows, Proc. Nat. Acad. Sci. U.S., vol. **30**, pages 316–323, 1944.

Lin, C. C., On the stability of two-dimensional parallel flows, parts I-III, Quart. Appl. Math., vol. **3**, pages 117–142, 218–234, 277–301, 1945–1946.

Lin, C. C., **The Theory of Hydrodynamic Stability**, Cambridge University Press, Cambridge, 1955.

Lin, S. C., & Lin, S. C., Study of strong temperature mixing in subsonic grid turbulence, Physics Fluids, vol. **16**, pages 1587–1598, 1973.

MacCormack, R. W., The effect of viscosity in hypervelocity impact cratering, AIAA paper 69-354, 1969.

MacCormack, R. W., A new method for solving the equations of compressible viscous flow, AIAA paper 81-011, 1981.

Macosko, C. W., **Rheology: Principles, Measurements, and Applications**, Wiley-VCH, New York, 1994.

Mader, C. L., **Numerical Modeling of Water Waves**, University of California Press, Berkeley, 1988.

Mangler, W., Zusammenhang zwischen ebenen und rotationssymmetrischen Gren-zschichten in kompressiblen Flüssigkeiten, Z. Angew. Math. Mech., vol. **28**, pages 97–103.

McConnell, A. J., **Applications of Tensor Analysis**, reprinted by Dover Press, New York, 1957.

McLachlan, N. W., **Theory and Application of Mathieu Functions**, Oxford Press, Oxford, 1947.

McMillan, E. L., Chem. Eng. Prog., vol. **44**, page 537 ff, 1948.

Mellor, G. L., & Herring, H. J., A survey of the mean turbulent field closure methods, J. AIAA, vol. **11**, pages 590–599, 1973.

Meyer, R. E., Theory of water-wave refraction, in **Advances in Applied Mechanics**, vol. **19**, pages 53–141, Academic Press, New York, 1979.

Milinazzo, F., & Saffman, P. G., The calculation of large Reynolds number two-dimensional flow using discrete vortices with random walk, J. Comp. Physics, vol. **23**, pages 380–392, 1977.

Millionshchtikov, M., On the theory of homogeneous isotropic turbulence, C. R. Acad. Sci. U. S. S. R., vol. **32**, page 615 ff, 1941.

Millsaps, K., & Pohlhausen, K., J. Aero. Sci., vol. **20**, page 187 ff, 1953.

Milne-Thomson, L. M., **Theoretical Hydrodynamics**, 3rd edition, The MacMillan Co., New York, 1955.

Mises, R. von, Bernerkungen zur Hydrodynamik, Z. Angew. Math. Mech., vol. **7**, pages 425–431, 1927.

Mollo-Christensen, E., Physics of turbulent flow, J. AIAA, vol. **9**, page 1217 ff, 1971.

Moran, J., **An Introduction to Theoretical and Computational Aerodynamics**, John Wiley & Sons, New York, 1984.

Norris, L. H., & Reynolds, W. C., Turbulent channel flow with a moving wavy boundary, Stanford Univ. Dept. Mech. Eng. Rept. FM-10, 1975.

Orr, W. McF., The stability or instability of the steady motions of a liquid, Proc. Roy. Irish Acad., vol. **A 27**, pages 9–26, 69–138, 1906–1907.

Orszag, S. A., Analytical theories of turbulence, J. Fluid Mech., vol. **41**, pages 363–386, 1970.

Orszag, S. A., Accurate solution of the Orr-Sommerfeld stability equation, J. Fluid Mech., vol. **50**, pages 689–703, 1971.

Orszag, S. A., & Patterson, G. S., Numerical simulation of three-dimensional homogeneous isotropic turbulence, Phys. Rev. Lett., vol. **28**, pages 76–79, 1972.

Oseen, C. W., Über die Stokessche Formel und über die verwandte Aufgabe in der Hydrodynamik, Ark Mat., Astron., Fysik, vol. **6**(29), 1910.

Oseen, C. W., **Hydrodynamik**, Akademische Verlagsgesellschaft, Leipzig, 1927.

Ostrach, S., An analysis of laminar free-convection flow and heat transfer about a flat plate parallel to the direction of the generating body force, NACA Report 1111, 1953.

Pern, L., & Gebhart, B., Natural convection boundary layer flow over horizontal and slightly inclined surfaces, Int. J. Heat Mass Transfer, vol. **16**, pages 1131–1146, 1972.

Peters, A. S., A new treatment of the ship wave problem, Comm. Pure & Appl. Math., vol. **2**, pages 123–148, 1949.

Pohlhausen, E., Der Warmeaustausch Zwischen Festen Korpen und Flüssigkeit mit kleiner Reiburg und kleiner Zeit, Z. Angew. Math. Mech., vol. **1**, page 115 ff, 1921a.

Pohlhausen, E., Zur näherungsweisen Integration der Differentialgleichen der laminaren Reibungsschicht, Z. Angew. Math. Mech., vol. **1**, pages 252–268, 1921b.

Prandtl, L., Applications of modern hydrodynamics to aeronautics, translated into English in NACA Report No. 116, 1921.

Prandtl, L., Über Flüssigkeitsbewegung bei sehr kleiner Reibung, Proc. 3rd Intern. Math. Congress, Heidelberg, 1904. Reprinted and translated into English in Tech. Memo N.A.C.A. No. 452, 1928.

Prandtl, L., Über ein neues Formelsystem für die ausgebidate Turbulenz, Nachrichten von der Akad. der Wissenschaft in Göttingen, 1945.

Prandtl, L., **Essentials of Fluid Mechanics**, Hafner Publishing, New York, 1952.

Press, W. H., Flannery, B. P., Teukolsky, S. A., & Vetterling, W. T., **Numerical Recipes—The Art of Scientific Computing**, Cambridge U. Press, Cambridge, 1986.

Proudman, I., & Pearson, J. R. A., Expansions at small Reynolds numbers for the flow past a sphere and a circular cylinder, J. Fluid Mech., vol. **2**, pages 237–262, 1957.

Proudman, I., & Reid, W. H., On the decay of a normally distributed and homogeneous turbulent field, Phil. Trans. R. Soc. A, vol. **247**, page 163 ff, 1954.

Rayleigh, Baron (John William Strutt), **The Theory of Sound**, vol. **1 & 2**, MacMillan, London, 1878. Reprinted by Dover Publications, New York, 1945.

Rayleigh, Baron (John William Strutt), On convection currents in a horizontal layer of fluid when the higher temperature is on the under side, Phil. Mag., vol. **32**, pages 529–546, 1916a.

Rayleigh, Baron (John William Strutt), On the dynamics of revolving fluids, Proc. Roy. Soc. London series A, vol. **293**, pages 148–154, 1916b.

Reynolds, O., On the dynamical theory of incompressible viscous fluids and the determination of the criterion, Phil. Trans. Royal Soc. London series A, vol. **186**, pages 123–164, 1895.

Reynolds, W. C., Recent advances in the computation of turbulent flow, Adv. Chem. Engrg., vol. **9**, pages 193–246, 1974.

Reynolds, W. C., Computation of turbulent flows, Ann. Rev. Fluid Mech., vol. 8, pages 183–208, 1976.

Rich, B. R., An investigation of heat transfer from an inclined plate in free convection, Trans. ASME, vol. **75**, pages 489–499, 1953.

Roache, P. J., **Computational Fluid Dynamics**, Hermosa Publishing, Albuquerque, 1976.

Roberts, K. V., & Christiansen, J. P., Topics in Computational Fluid Mechanics, **Computational Physics Communications**, vol. **3**, pages 14–32, North Holland Publishing, 1972.

Rodi, W., **Turbulence Models and their Application in Hydraulics**, International Association for Hydraulic Research, Delft, Netherlands, 1980.

Rosenhead, L., editor, **Laminar Boundary Layers**, Oxford University Press, London, 1963.

Rotta, J., Statistical theory of non-homogeneous turbulence, translation by W. Rodi in Imp. Coll. Rep. TWF/TN/38, TWF/TN/39, 1951.

Rotta, J., Turbulent boundary layers in incompressible flow, **Progress in Aeronautical Sciences**, vol. **2** (editors A. Ferri, D. Kuchemann, and L. H. G. Steme), pages 1–221, McMillan, New York, 1962.

Rouse, H., editor, **Advanced Mechanics of Fluids**, John Wiley & Sons, New York, 1959.

Rubbert, P. E., & Saaris, G. R., Review and evaluation of a three-dimensional lifting potential flow computational method for arbitrary configurations, AIAA paper 72-188, 10th Aerospace Science Meeting, 16 pages, 1972.

Russell, J. Scott, Report of the Committee on Waves. Report of the 7th meeting of the Brit. Assoc. for the Adv. of Science, Liverpool, pages 417–496, John Murray, London, 1838.

Russell, J. Scott, Report on Waves. Report of the 14th meeting of the Brit. Assoc. for the Adv. of Science, York, pages 311–390, John Murray, London, 1844.

Saffman, P. G., & Wilcox, D. C., Turbulence-model predictions for turbulent boundary layers, J. AIAA, vol. **12**, pages 541–546, 1974.

Saffman, P. G., **Vortex Dynamics**, Cambridge University Press, Cambridge, 1992.

Schlichting, H., **Boundary-Layer Theory**, 7th edition, McGraw-Hill, New York, 1979.

Schmidt, E., & Beckmann, W., Das Temperatur-und Geschwindigkeitsfeld von einer Warme Abgebenden Senkrechten Platte bei naturlicher Konvection, Forsch-Ing.-Wes., vol. **1**, page 391 ff, 1930.

Schmitt, E., & Wenner, K., Warmeabgabe überden Umfangeinesangeblasen geheizten Zylinders. Forsch. Gebiete Ingenieurwesens, VDI, vol. **12**, pages 65–73, 1941.

Schubauer, G. B., & Skramstad, H. K., Laminar boundary-layer oscillations and transition on a flat plate, J. Aeron. Sci., vol. **14**, pages 69–78, 1947; also Nat. Advisory Comm. Aeron. Rrpt. 909, 1948, originally issued as Natl. Advisory Comm. Aeron. Advance Confidential Rept., 1943.

Schwartz, L. W., Non-linear solution for an applied overpressure on a moving stream, J. Engineering Maths., vol. **15**, pages 147–156, 1981.

Scott, M. R., J. Comp. Physics, vol. **12**, page 334 ff, 1973.

Scott, M. R., **Invariant Imbedding and its Application to Ordinary Differential Equations**, Addison-Wesley, Reading, MA, 1973.

Shen, S. F., Calculated amplified oscillations in plane Poiseuille and Blasius flows, J. Aeronaut. Sci., vol. **21**, pages 62–64, 1954.

Sloan, D. M., Eigenfunctions of systems of linear ordinary differential equations with separated boundary conditions using Riccati transformations, J. Computational Physics, vol. **24**, pages 320–330, 1977.

Smith, G. D., **Numerical Solutions of Partial Differential Equations: Finite Difference Methods**, Oxford University Press, Oxford, 1978.

Sommerfeld, A., Ein Beitrag zur hydrodynamischen Erklärung der turbulenten Flüssigkeitsbewegung, Proc. 4th Intr. Congress Math. Rome, pages 116–124, 1908.

Sparrow, E. M., & Gregg, J. L., Laminar free convection from a vertical flat plat with uniform surface heat flux, Trans. ASME, vol. **78**, pages 435–440, 1956.

Squire, H. B., On the stability fro three-dimensional disturbances of viscous fluid flow between parallel wall, Proc. Roy. Soc. London series A, vol. **142**, pages 621–628, 1933.

Squire, H. B., The round laminar jet, Quart. J. Mech. Appl. Math., vol. **4**, pages 321–329, 1951.

Stewartson, K., Further solutions of the Falkner-Skan equations, Proc. Camb. Phil. Soc., vol. **503**, pages 454–465, 1954.

Stewartson, K., On a flow near the trailing edge of a flat plate, Proc. Roy. Soc. London series A, vol. **306**, pages 275–290, 1968.

Stewartson, K., On a flow near the trailing edge of a flat plate II, Mathematika, vol. **16**, pages 106–121, 1969.

Stoker, J. J., **Water Waves**, Interscience Publishers, New York, 1957.

Stokes, G. G., On the theory of internal friction of fluids in motion, Camb. Trans., vol. **8**, page 287 ff, 1845.

Stokes, G. G., On the effect of the internal friction of fluids on the motion of pendulums, Cambridge Phil. Trans. IX, vol. **8**, 1851.

Stratford, B. S., Flow in the boundary layer near separation, Aeronautical Research Council London, RM-3002, 1954.

Sulen, P. L., Sesieur, M., & Frisch, V., Le test field model interprete comme methods de fermenture des equations de la turbulence, Ann. Geophys., vol. **31**, pages 487–496, 1975.

Tailland, A., & Mathield, J., Jet parietal, J. Mechanique, vol. **6**, pages 103–131, 1967. .

Taylor, G. I., Stability of a viscous liquid contained between two rotating cylinders, Phil. Trans. Roy. Soc. London series A, vol. **223**, pages 289–343, 1923.

Taylor, G. I., The energy of a body moving in an infinite fluid, with an application to airships, Proc. Roy. Soc. London series A, vol. **120**, page 13 ff, 1928.

Taylor, G. I., Statistical theory of turbulence parts I-V, Proc. Roy. Soc. London series A, vol. **151**, page 421 ff, 1935a.

Taylor, G. I., Statistical theory of turbulence, Proc. Roy. Soc. London series A, vol. **156**, page 307 ff, 1935b.

Taylor, G. I., Statistical theory of isotropic turbulence, J. Aeronaut. Sci., vol. **4**, page 311 ff, 1937.

Taylor, G. I., The spectrum of turbulence, Proc. Roy. Soc. London series A, vol. **164**, page 476 ff, 1938a.

Taylor, G. I., Some recent developments in the study of turbulence, Proc. 5th Internat. Congr. Appl. Mech., Cambridge, Mass., page 294 ff, 1938b.

Tennekes, H., & Lumley, J. L., **A First Course in Turbulence**, MIT Press, Cambridge, 1972.

Thomson, J. E., Warsi, Z. U. A., & Mastin, C. W., **Numerical Grid Generation: Foundations and Applications**, Elsevier Science Co., New York, 1985.

Thwaites, B., Approximate calculation of the laminar boundary layer, Aero. Quart., vol. **1**, pages 245–280, 1949.

Tomotika, S., & Aoi, T., The steady flow of viscous fluid past a sphere and circular cylinder at small Reynolds numbers, Quart. J. Mech. Appl. Math., vol. **3**, pages 140–161, 1950.

Townsend, A. A., The structure of the turbulent boundary layer, Proc. Cambridge Phil. Soc., vol. **47**, page 375 ff, 1951.

Townsend, A. A., **The Structure of Turbulent Shear Flow**, Cambridge U. Press, London, 1956.

Truesdell, C., The mechanical foundations of elasticity and fluid dynamics, J. Rational Mech. Anal., vol. **1**, pages 125–300, 1952.

Truesdell, C., Z. Angew. Math. Mech., vol. **33**, page 345 ff, 1953.

Truesdell, C., **Six Lectures on Modern Natural Philosophy**, Springer-Verlag, Berlin, 1966.

Tucker, H. J., & Reynolds, A. J., The distortion of turbulence by irrotational plane strain, J. Fluid Mech., vol. **32**, pages 657–673, 1968.

Uberoi, M. S., Equipartition of energy and local isotropy in turbulent flows, J. Appl. Physics, vol. **28**, pages 1165–1170, 1957.

Ursell, F., On Kelvin's ship-wave pattern, J. Fluid Mech., vol. **8**, pages 418–431, 1960.

Van Dyke, M., **Perturbation Methods in Fluid Mechanics**, Parabolic Press, Stanford, 1964.

Van Dyke, M., **An Album of Fluid Motion**, Parabolic Press, Stanford, 1982.

Vliet, G. C., Natural convection local heat transfer on constant heat flux surfaces, J. Heat Transfer, vol. **9**, pages 511–516, 1969.

Webster, C. A. G., An experimental study of turbulence in a density-stratified shear flow, J. Fluid Mech., vol. **19**, pages 221–245, 1964.

Wehausen, J. V., & Laitone, E. V., Surface waves, **Encyclopedia of Physics (Handbuch der Physik)**, vol. **9**, part 3. Springer-Verlag, Berlin, 1960.

Wilcox, D. C., **Turbulence Modeling for CFD**, ISBN 1-928729-10-X, 2nd edition, DCW Industries, 1998.

Wilks, G., & Sloan, D. M., Invariant imbedding, Riccati transformations, and eigenvalue problems, J. Inst. Math. Applied, vol. **18**, pages 99–116, 1976.

Wilks, G., & Bramley, J. S., On the computation of eigenvalues arising out of perturbations of the Blasius profile, J. Computational Physics, vol. **24**, pages 303–319, 1977.

Willmarth, W. W., Structure of turbulence in boundary layers, in **Advances in Applied Mechanics**, vol. **15**, pages 159–254, 1979.

Wolfshtein, M., The velocity and temperature distribution in one-dimensional flow with turbulence augmentation and pressure gradient, Int. J. Heat & Mass Transfer, vol. **12**, page 301 ff, 1969.

Wygnanski, I., & Fiedler, H. E., Two-dimensional mixing region, J. Fluid Mech., vol. **41**, pages 327–363, 1970.

Yih, C.-S., On the flow of a stratified fluid, Proc. 3rd U.S. Nat. Cong. Appl. Mech., pages 857–861, 1958.

Yih, C.-S., **Fluid Mechanics**, West River Press, Ann Arbor, 1977.

Zhou, Q-N., & Graebel, W. P., Numerical simulation of interface-drain interactions, J. Fluid Mech., vol. **221**, pages 511–532, 1990.

Index

Printed and bound by CPI Group (UK) Ltd, Croydon, CR0 4YY

03/10/2024

01040317-0017